Transdisciplinary Advancements in Cognitive Mechanisms and Human Information Processing

Yingxu Wang
University of Calgary, Canada

Information Science
REFERENCE

Senior Editorial Director: Kristin Klinger
Director of Book Publications: Julia Mosemann
Editorial Director: Lindsay Johnston
Acquisitions Editor: Erika Carter
Development Editor: Michael Killian
Production Editor: Sean Woznicki
Typesetters: Mike Brehm, Keith Glazewski, Natalie Pronio, Jennifer Romanchak, Milan Vracarich Jr.
Print Coordinator: Jamie Snavely
Cover Design: Nick Newcomer

Published in the United States of America by
Information Science Reference (an imprint of IGI Global)
701 E. Chocolate Avenue
Hershey PA 17033
Tel: 717-533-8845
Fax: 717-533-8661
E-mail: cust@igi-global.com
Web site: http://www.igi-global.com/reference

Library of Congress Cataloging-in-Publication Data

Transdisciplinary Advancements in Cognitive Mechanisms and Human Information Processing / Yingxu Wang, editor.
p. cm.
Includes bibliographical references and index.
Summary: "This book examines innovative research in the emerging, multidisciplinary field of cognitive informatics, portraying the connections between natural science and informatics that are investigated in this fundamental collection of cognitive informatics research"--Provided by publisher.
ISBN 978-1-60960-553-7 (hardcover) -- ISBN 978-1-60960-554-4 (ebook) 1. Neural computers. 2. Cognitive science. I. Wang, Yingxu.
QA76.87.T73 2011
006.3'2--dc22

2011013281

British Cataloguing in Publication Data
A Cataloguing in Publication record for this book is available from the British Library.

All work contributed to this book is new, previously-unpublished material. The views expressed in this book are those of the authors, but not necessarily of the publisher.

Table of Contents

Detailed Table of Contents

physiological state given artistic patterns authored through specifically developed tools; and, lastly, a layer for presenting the music and lighting changes in the physical dance environment.

We propose that Information and Communication Technology should deliver a new experience to the user. We call this experience Kansei Experience. To deliver it we advocate the reliance on a new computing paradigm called cultural computing. Within this paradigm, we develop the concept of Kansei Media and how it could be implemented via Kansei Mediated Interaction. Kansei Media is about sharing implicit knowledge such as feelings, emotions and moods. We aim for a Kansei Experience, rendered by Kansei Media, which relates to reality, and enhances it to enlighten the user. To do so we investigate the aesthetics of the experience we wish to produce. Finally, to develop the concept of Kansei Mediated Experience we refer two cultures (Western and Eastern) and use famous stories from both as means of delivering guidelines for the implementation of Kansei Experience.

This chapter addresses issues of distributing multimedia presentations in an ambient intelligent environment, examines the existing technologies and proposes IPML, a markup language that extends SMIL for distributed settings. It uses a metaphor of play, with which the timing and mapping issues in distributed presentations are covered in a natural way. A generic architecture for playback systems is also presented, which covers the timing and mapping issues of presenting an IPML script in heterogeneous ambient intelligent environments.

One of the goals of ubiquitous computing technologies is to provide an adaptable and personal content at any time and in any context. As a consequence a user-centered design is required. The goal of this research is to develop new game plays and new narration principles for Multiplayer Ubiquitous Game. We aim to formalize a narrative mechanism to generate events which can stimulate the user's physical actions with the real world, and social communications with other players. Based on the analysis of the relationship between the real world and the virtual world, a narration adaptive to the user's profile is proposed. A prototype using these principles has been developed using off the shell services available on location-based mobile phones.

Recent researches in both cognitive informatics and computational intelligence are interested in the human perceptual senses of spatiality, time, and motion, which are fundamental cognitive life functions according to the Layered Reference Model of the Brain (LRMB). This chapter presents the cognitive process of human perceptual senses on spatiality, time, and motion. The sense of spatiality is investigated into the coordinate system, orientations, and cognitive maps, followed by the development of the mathematical model and the cognitive process of human spatial senses. The sense of time with the biological clocks, cognitive clocks, and their mathematical models are analyzed in order to explain the cognitive process of human time sense. On the basis of the formal models of senses of spatiality and time, the sense of motion is modeled as a complex sense incorporating both of spatiality and time. Then, the cognitive, mathematical, and process models of the sense of motion are rigorously established. This work provides a theoretical framework for the rigorous implementation of the intelligent behaviors of cognitive computers, autonomous agent systems, and robots in cognitive informatics and computational intelligence.

It is recognized that the internal mechanisms for visual information processing are based on semantic inferences where visual information is represented and processed as visual semantic objects rather than direct images or episode pictures in the long-term memory. This chapter presents a cognitive informatics theory of visual information and knowledge processing in the brain. A set of cognitive principles of visual perception is reviewed particularly the classic gestalt principles, the cognitive informatics principles, and the hypercolumn theory. A visual frame theory is developed to explain the visual information processing mechanisms of human vision, where the size of a unit visual frame is tested and calibrated based on vision experiments. The framework of human visual information processing is established in order to elaborate mechanisms of visual information processing and the compatibility of internal representations between visual and abstract information and knowledge in the brain.

In this chapter we perform some experiments to study how an automatic system learns a set of rules from its interaction with an artificial environment. In particular, we are interested in comparing these capabilities to the skills shown by humans to learn the same rules in similar conditions. We perform this analysis by conducting two experiments. On the one hand, we observe the evolution of the automatic learning system in terms of its performance along time. At the beginning, the system does not know the rules, but it can observe the positive/negative results of its decisions. As its knowledge about the environment becomes more precise, its performance improves. On the other hand, seventy students faced the same artificial environment in the same conditions, though this time the experiment was presented as a game. The objective of the game consists in gaining points, but the rules of the game are not known a priori. So, there is a clear incentive for finding them out. We use these experiments to compare the learning curves of both humans and automatic systems, and we use this information to analyze the similarities/differences between both learning processes. In particular, we are interested in assessing how close the automatic system is from passing the Turing test.

We present pattern classification methods based upon nonlinear and combinational optimization techniques, specifically, radial basis functions (RBF) and simulated annealing (SA), to classify masses in mammograms as malignant or benign. Combinational optimization is used to pre-estimate RBF parameters, namely, the centers and spread matrix. The classifier was trained and tested, using the leave-one-out procedure, with shape, texture, and edge-sharpness measures extracted from 57 regions of interest (20 related to malignant tumors and 37 related to benign masses) manually delineated on mammograms by a radiologist. The classifier's performance, with preestimation of the parameters, was evaluated in terms of the area Az under the receiver operating characteristics curve. Values up to Az = 0.9997 were obtained with RBF-SA with pre-estimation of the centers and spread matrix, which are better than the results obtained with pre-estimation of only the RBF centers, which were up to 0.9470. Overall, the results with the RBF-SA method were better than those provided by standard multilayer perceptron neural networks.

Quotient space theory (QST), a new granule computing tool dealing with imprecise, incomplete and uncertain knowledge, uses a triplet, including the universe, its structure and attributes, to describe a problem space or simply a space. As one of important theories of granular computing (GrC), QST is very helpful to the study of cognitive informatics (CI). This chapter summarizes the quotient space's

Section 4

is developed by Lagrange Formalism. In terms of the proposed model, an adaptive integrated tracking controller, based on the computed torque control (CTC) method and the radial basis function neural-network (RBFNN), is presented subsequently. Although CTC is an effective motion control strategy for mobile manipulators, it requires precise models. To handle the unmodeled dynamics and the external disturbance, a RBFNN, serving as a compensator, is adopted. This proposed controller combines the advantages of CTC and RBFNN. Simulation results show the correctness of the proposed model and the effectiveness of the control approach.

In this chapter, we modulate an interchange commerce system based on the economic concept of utility function. A cognitive agent controls the interchanges of the clients in her market. When interchanges are not possible any more, the agent becomes a client of a higher market, giving place to a hierarchical market system. Now, she behaves according to what she has learned from her clients. Apart from physical resources, intangible goods such as knowledge are also interchanged. This cooperative and competitive structure is formalized via process algebra.

The electrocardiographic (ECG) signal is a transthoracic manifestation of the electrical activity of the heart and is widely used in clinical applications. This chapter describes an unbiased linear adaptive filter (ULAF) to attenuate high-frequency random noise present in ECG signals. The ULAF does not contain a bias in its summation unit and the filter coefficients are normalized. During the adaptation process, the normalized coefficients are updated with the steepest-descent algorithm to achieve efficient filtering of noisy ECG signals. A total of 16 ECG signals were tested in the adaptive filtering experiments with the ULAF, the least-mean-square (LMS), and the recursive-least-squares (RLS) adaptive filters. The filtering performance was quantified in terms of the root-mean-squared error (RMSE), normalized correlation coefficient (NCC), and filtered noise entropy (FNE). A template derived from each ECG signal was used as the reference to compute the measures of filtering performance. The results indicated that the ULAF was able to provide noise-free ECG signals with an average RMSE of 0.0287, which was lower than the second-best RMSE obtained with the LMS filter. With respect to waveform fidelity, the ULAF provided the highest average NCC (0.9964) among the three filters studied. In addition, the ULAF effectively removed more noise, measured by FNE, in comparison with the LMS and RLS filters in most of the ECG signals tested. The issues of adaptive filter setting for noise reduction in ECG signals are discussed at the end of this chapter.

Preface

1. INTRODUCTION

Cognitive Informatics (CI) is a transdisciplinary enquiry of computer science, information science, cognitive science, and intelligence science that investigates into the internal information processing mechanisms and processes of the brain and natural intelligence, as well as their engineering applications in cognitive computing ((Wang, 2002a, 2003, 2006, 2007b, 2009a, 2009b; Wang and Kinsner, 2006; Wang and Wang, 2006; Wang and Chiew, 2010; Wang, Kinsner, and Zhang, 2009b; Wang et al., 2002, 2006, 2008, 2009b, 2009c; Baciu et al., 2009; Chan et al., 2004; Kinsner et al., 2005; Patel et al., 2003; Yao et al., 2006; Zhang et al., 2007; Sun et al., 2010). CI is a cutting-edge and multidisciplinary research area that tackles the fundamental problems shared by computational intelligence, modern informatics, computer science, AI, cybernetics, cognitive science, neuropsychology, medical science, philosophy, formal linguistics, and life science (Wang, 2002a, 2003, 2007b, 2009a, 2010a, 2010c). The development and the cross fertilization among the aforementioned science and engineering disciplines have led to a whole range of extremely interesting new research areas known as CI, which investigates the internal information processing mechanisms and processes of the natural intelligence – human brains and minds – and their engineering applications in computational intelligence. CI is a new discipline that studies the natural intelligence and internal information processing mechanisms of the brain, as well as processes involved in perception and cognition. CI forges links between a number of natural science and life science disciplines with informatics and computing science.

Definition 1. *Cognitive informatics* (CI) is a transdisciplinary enquiry of computer science, information science, cognitive science, and intelligence science that investigates into the internal information processing mechanisms and processes of the brain and natural intelligence, as well as their engineering applications in cognitive computing.

The IEEE series of International Conferences on Cognitive Informatics (ICCI) has been established since 2002 (Wang, 2002). The inaugural ICCI event in 2002 was held at University of Calgary, Canada (ICCI'02) (Wang et al., 2002), followed by the events in London, UK (ICCI'03) (Patel et al., 2003); Victoria, Canada (ICCI'04) (Chan et al., 2004); Irvine, USA (ICCI'05) (Kinsner et al., 2005); Beijing, China (ICCI'06) (Yao et al., 2006); Lake Tahoe, USA (ICCI'07) (Zhang et al., 2007); Stanford University, USA (ICCI'08) (Wang et al., 2008); Hong Kong (ICCI'09) (Baciu et al., 2009); and Tsinghua University, Beijing (ICCI'10) (Sun et al., 2010). Since its inception, ICCI has been growing steadily in its size, scope, and depth. It attracts worldwide researchers from academia, government agencies, and industry practitioners. The conference series provides a main forum for the exchange and cross-fertilization of ideas in the new research field of CI toward revealing the cognitive mechanisms and processes of human information processing and the approaches to mimic them in cognitive computing.

This chapter explores the cutting-edge field of CI and its applications in cognitive computing. The theoretical framework of CI is described in Section 2 on the architecture of CI, the abstract intelligence theory of CI, and denotational mathematics for CI. Inspirations of CI to theories for cognitive computing and technologies for cognitive computers are presented in Sections 3. The relationship among abstract, natural, and artificial intelligence is formally elaborated in Section 4 where abstract intelligence provides a theoretical foundation for understanding other forms of natural and artificial intelligence. Applications of CI and cognitive computers are described in Section 5.

2. THE THEORETICAL FRAMEWORK OF COGNITIVE INFORMATICS

It is recognized that *information* is any property or attribute of the natural world that can be distinctly elicited, generally abstracted, quantitatively represented, and mentally processed. Information is the third essence of the natural world supplementing matter and energy. *Informatics* is the science of information that studies the nature of information, its processing, and ways of transformation between information, matter and energy.

The theoretical framework of CI (Wang, 2007b) encompasses: a) fundamental theories of cognitive informatics; b) abstract intelligence; and c) denotational mathematics. An intensive review on *The Theoretical Framework of Cognitive Informatics* was presented in (Wang, 2007b), which provides a coherent summary of the latest advances in the transdisciplinary field of CI and an insightful perspective on its future development.

Fundamental Theories of CI: The theories of informatics and their perceptions on the object of information have evolved from the classic information theory, modern informatics, to cognitive informatics in the last six decades. *Conventional information theories* (Shannon and Weaver, 1949; Bell, 1953; Goldman, 1953), particularly Shannon's information theory (Shannon, 1948) known as the first-generation informatics, study signals and channel behaviors based on statistics and probability theory. *Modern informatics* studies information as properties or attributes of the natural world that can be generally abstracted, quantitatively represented, and mentally processed. The first- and second-generation informatics put emphases on external information processing, which overlook the fundamental fact that human brains are the original sources and final destinations of information, and any information must be cognized by human beings before it is understood, comprehended, and consumed. This observation leads to the establishment of the third-generation informatics, a term coined by Wang in 2002 as cognitive informatics (CI) in the keynote in (Wang, 2002a), which is defined as the science of cognitive information that investigates into the internal information processing mechanisms and processes of the brain and natural intelligence, and their engineering applications via an interdisciplinary approach.

Fundamental theories developed in CI covers the Information-Matter-Energy-Intelligence (IME-I) model (Wang, 2007a), the Layered Reference Model of the Brain (LRMB) (Wang et al., 2006), the Object-Attribute-Relation (OAR) model of internal information representation in the brain (Wang, 2007c), the cognitive informatics model of the brain (Wang and Wang, 2006), natural intelligence (Wang, 2007b), and neuroinformatics (Wang, 2007b). Recent studies on LRMB in cognitive informatics reveal an entire set of cognitive functions of the brain and their cognitive process models, which explain the functional mechanisms and cognitive processes of the natural intelligence with 43 cognitive processes at seven layers known as the sensation, memory, perception, action, meta-cognitive, meta-inference, and higher cognitive layers (Wang et al., 2006).

Abstract Intelligence (αI): The studies on αI form a human enquiry of both natural and artificial intelligence at reductive levels of the neural, cognitive, functional, and logical layers from the bottom up (Wang, 2009a). αI is the general mathematical form of intelligence as a natural mechanism that transfers information into behaviors and knowledge.

The *Information-Matter-Energy-Intelligence* (IME-I) model as shown in Fig. 1 states that the natural world (*NW*) which forms the context of human and machine intelligence is a dual: one aspect of it is the *physical* world (*PW*), and the other is the *abstract* world (*AW*), where *intelligence* (αI) plays a central role in the transformation between information (*I*), *matter* (*M*), and *energy* (*E*). In the IME-I model as shown in Fig. 1, αI plays an irreplaceable role in the transformation between information, matter, and energy, as well as different forms of internal information and knowledge. Typical paradigms of αI are natural intelligence, artificial intelligence, machinable intelligence, and computational intelligence, as well as their hybrid forms. The studies in CI and αI lay a theoretical foundation toward revealing the basic mechanisms of different forms of intelligence. As a result, cognitive computers may be developed, which are characterized as knowledge processors beyond those of data processors in conventional computing.

Denotational Mathematics (DM): The needs for complex and long-series of causal inferences in cognitive computing, αI, computational intelligence, software engineering, and knowledge engineering have led to new forms of mathematics collectively known as denotational mathematics (Wang, 2002b, 2007a, 2008a, 2008e, 2009d, 2010e; Wang, Zadeh and Yao, 2009).

Definition 2.*Denotational Mathematics (DM)* is a category of expressive mathematical structures that deals with high-level mathematical entities beyond numbers and sets, such as abstract objects, complex relations, perceptual information, abstract concepts, knowledge, intelligent behaviors, behavioral processes, and systems.

It is recognized that the maturity of a scientific discipline is characterized by the maturity of its mathematical (meta-methodological) means because the nature of mathematics is a generic meta-methodological science (Wang, 2008a). In recognizing mathematics as the *metamethodology* of all sciences and engineering disciplines, a set of DMs have been created and applied in CI, αI, AI, soft computing, computational intelligence, and computational linguistics. Typical paradigms of DM are such as *concept algebra* (Wang, 2008c), *system algebra* (Wang, 2008d; Wang, Zadeh and Yao, 2009), *real-time process algebra* (Wang, 2002b, 2007a, 2008b), *granular algebra* (Wang, 2009e), *visual semantic algebra* (Wang, 2009c), *inference algebra* (Wang, 2010a, 2010b), and fuzzy inferences (Zadeh, 1965, 1975, 2008). DM

Figure 1. The IME-I model and roles of abstract intelligence in CI

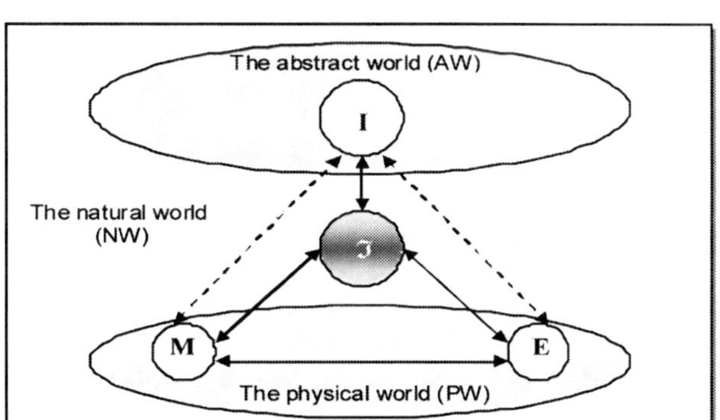

provides a coherent set of contemporary mathematical means and explicit expressive power for CI, αI, AI, and computational intelligence.

3. COGNITIVE COMPUTING AND COGNITIVE COMPUTERS

The term *computing* in a narrow sense is an application of computers to solve a given problem by imperative instructions; while in a broad sense, it is a process to implement the instructive intelligence by a system that transfers a set of given information or instructions into expected intelligent behaviors. The latest advances and engineering applications of CI have led to the emergence of cognitive computing and the development of cognitive computers that perceive, reason, and learn. *Cognitive Computing* is an emerging paradigm of intelligent computing methodologies and systems based on cognitive informatics that implements computational intelligence by autonomous inferences and perceptions mimicking the mechanisms of the brain (Wang, 2002a, 2007d, 2009b; Wang et al., 2010).

Definition 3.*Cognitive Computing* (CC) is a novel paradigm of intelligent computing methodologies and systems that implements computational intelligence by autonomous inferences and perceptions mimicking the mechanisms of the brain.

Computing systems and technologies can be classified into the categories of *imperative*, *autonomic,* and *cognitive* computing from the bottom up. The imperative computers are a traditional and passive system based on stored-program controlled behaviors for data processing (Wang, 2009b). The autonomic computers are goal-driven and self-decision-driven machines that do not rely on instructive and procedural information (Pescovitz, 2002; Wang, 2007d). Cognitive computers are more intelligent computers beyond the imperative and autonomic computers, which embody major natural intelligence behaviors of the brain such as thinking, inference, and learning.

Definition 4. A *cognitive computer* (cC) is an intelligent computer for knowledge processing that perceive, learn, and reason.

Recent studies in cognitive computing reveal that the computing power in computational intelligence can be classified at four levels: *data, information, knowledge,* and *intelligence* from the bottom up. Traditional von Neumann computers are designed to implement imperative data and information processing by stored-program-controlled mechanisms. However, the increasing demand for advanced computing technologies for knowledge and intelligence processing in the high-tech industry and everyday lives require novel cognitive computers for providing autonomous computing power mimicking the natural intelligence of the brain.

A cC is a type of intelligent computers that are capable of autonomous inference and learn. cCs are an emerging technology towards novel computer architectures and advanced intelligent computing behaviors for cognitive knowledge processing and autonomous learning based on contemporary denotational mathematics. cCs provide a general computing platform that extends computational intelligence from data/information processing to knowledge/intelligence processing. In seeking the contemporary mathematical means for cCs as well as for internal knowledge representation and manipulations, a set of denotational mathematics has been developed as described in Section 2. Denotational mathematics creates a coherent set of contemporary mathematical means and explicitly expressive power for rigorously modeling and implementing machine inference and learning processes for cCs.

The *essences* of computing are both its *data objects* and their predefined computational *operations*. From these facets, different computing paradigms may be comparatively analyzed as follows:

a. Conventional computers
  - *Data objects*: abstract bits and structured data
  - *Operations*: logic, arithmetic, and functions

b. Cognitive computers (cC)
  - *Data objects*: words, concepts, syntax, and semantics
  - *Basic operations*: syntactic analyses and semantic analyses
  - *Advanced operations*: concept formulation, knowledge representation, comprehension, learning, inferences, and causal analyses

The above analyses indicate that cC is an important extension of conventional computing in both data objects modeling capabilities and their advanced operations at the abstract level of concept beyond bits. Therefore, cC is an intelligent knowledge processor that is much closer to the capability of human brains thinking at the level of concepts rather than bits. It is recognized that the basic unit of human knowledge in natural language representation is a concept rather than a word (Wang, 2008c, 2010d, 2010e), because the former conveys the structured semantics of a word with its intention (attributes), extension (objects), and relations to other concepts in the context of a knowledge network.

It is noteworthy that, although the semantics of words may be ambiguity, the semantics of concepts is always unique and precise in cC. For example, the word, "bank", is ambiguity because it may be a notion of a financial institution, a geographic location of raised ground of a river/lake, and/or a storage of something. However, the three individual concepts derived from bank, i.e., b_o = bank(organization), b_r = bank(river), and b_s = bank(storage), are precisely unique, which can be formally described in concept algebra (Wang, 2008c, 2010e) for cC as shown in Fig. 2. In the examples of concepts, a generic framework of a concept is represented by the following model known as an abstract concept c, i.e.:

$$c \triangleq (O, A, R^c, R^i, R^o) \tag{1}$$

where

- O is a nonempty set of objects of the concept, $O = \{o_1, o_2, ..., o_m\} \subseteq ÞO$, where $ÞO$ denotes a power set of abstract objects in the universal discourse U., $U = (O, A, R)$.
- A is a nonempty set of attributes, $A = \{a_1, a_2, ..., a_n\} \subseteq ÞA$, where $ÞA$ denotes a power set of attributes in U.
- $R^c = O \times A$ is a set of internal relations.
- $R^i \subseteq C' \times c$ is a set of input relations, where C′ is a set of external concepts in U.
- $R^o \subseteq c \times C'$ is a set of output relations.

A set of denotational mathematics for cC and CC (Wang, 2002b, 2008b, 2008c, 2008d, 2009c, 2009d, 2009e; Wang et al., 2009a), particularly *concept algebra* (Wang, 2008b), has been developed by Wang during 2000 to in 2009. CA provides a set of 8 relational and 9 compositional operations for abstract concepts. A Cognitive Learning Engine (CLE) that serves as the "CPU" of cCs is under developing on the basis of concept algebra, which implements the basic and advanced cognitive computational operations of concepts and knowledge for cCs. The work in this area may also lead to a fundamental solution to computational linguistics, Computing with Natural Language (CNL), and Computing with Words (CWW) as Zadeh proposed (Zadeh, 1975, 2008).

Figure 2. Formal concepts modeled by concept algebra

$b_o\mathbf{ST} \triangleq (A, O, R^c, R^i, R^o)$
$= (b_o\mathbf{ST}.A = \{\text{organization, company, financial business, money, deposit, withdraw, invest, exchange}\},$
$b_o\mathbf{ST}.O = \{\text{international\_bank, national\_bank, local\_bank, investment\_bank, ATM}\}$
$b_o\mathbf{ST}.R^c = O \times A,$
$b_o\mathbf{ST}.R^i = \mathfrak{K} \times b_o\mathbf{ST},$
$b_o\mathbf{ST}.R^o = b_o\mathbf{ST} \times \mathfrak{K}$
)

$b_r\mathbf{ST} \triangleq (A, O, R^c, R^i, R^o)$
$= (b_r\mathbf{ST}.A = \{\text{sides of a river, raised ground, a pile of earth, location}\},$
$b_r\mathbf{ST}.O = \{\text{river\_bank, lake\_bank, canal\_bank}\}$
$b_r\mathbf{ST}.R^c = O \times A,$
$b_r\mathbf{ST}.R^i = \mathfrak{K} \times b_r\mathbf{ST},$
$b_r\mathbf{ST}.R^o = b_r\mathbf{ST} \times \mathfrak{K}$
)

$b_s\mathbf{ST} \triangleq (A, O, R^c, R^i, R^o)$
$= (b_s\mathbf{ST}.A = \{\text{storage, container, place, organization}\},$
$b_s\mathbf{ST}.O = \{\text{information\_bank, resource\_bank, blood\_bank}\}$
$b_s\mathbf{ST}.R^c = O \times A,$
$b_s\mathbf{ST}.R^i = \mathfrak{K} \times b_s\mathbf{ST},$
$b_s\mathbf{ST}.R^o = b_s\mathbf{ST} \times \mathfrak{K}$
)

4. ABSTRACT INTELLIGENCE VS. ARTIFICIAL INTELLIGENCE

It is conventionally deemed that only mankind and advanced species possess intelligence. However, the development of computers, robots, software agents, and autonomous systems indicates that intelligence may also be created or embodied by machines and man-made systems. Therefore, it is one of the key objectives in cognitive informatics and intelligence science to seek a coherent theory for explaining the nature and mechanisms of both natural and artificial intelligence.

Definition 5.*Intelligence* is an ability to acquire and use knowledge and skills, or to inference in problem solving.

It is a profound human wonder on how conscious intelligence is generated as a highly complex cognitive state in human mind on the basis of biological and physiological structures. How natural intelligence functions logically and physiologically? How natural and artificial intelligence are converged on the basis of brain, software, and intelligence science?

Definition 6. *Abstract intelligence* is the general form of intelligence as an abstract mathematical model that transfers information into behaviors and knowledge.

Abstract intelligence is also a discipline of human enquiries.

Definition 7. The *discipline of abstract intelligence* (αI) studies the foundations of intelligence science focusing the core properties of intelligence as a natural mechanism that transfers information into behaviors and knowledge.

In the *narrow sense*, αI is a human or a system ability that transforms information into behaviors. While, in the *broad sense*, αI is any human or system ability that autonomously transfers the forms of abstract information between *data, information, knowledge,* and *behaviors* in the brain or systems.

The studies on αI form a field of enquiry for both natural and artificial intelligence at the reductive levels of neural, cognitive, functional, and logical from the bottom up (Wang, 2009a). The paradigms of αI are such as natural, artificial, machinable, and computational intelligence. With the clarification of the intension and extension of the concept of αI, its paradigms or concrete forms in the real-world can be derived as summarized in Table 1.

Definition 8. The *behavioral model* of αI, §αIST, is an abstract logical model denoted by a set of parallel processes that encompasses the *imperative* intelligence I_I, *autonomic* intelligence I_A, and *cognitive* intelligence I_C from the bottom-up, i.e.:

$$
\begin{aligned}
\S\alpha\mathbf{IST} \triangleq\ & (\Im_I, \Im_A, \Im_C) \\
= \{\ & (B_e, B_t, B_{int}) && //\ \mathrm{I_I}\ \text{- Imperative intelligence} \\
& \|\ (B_e, B_t, B_{int}, B_g, B_d) && //\ \mathrm{I_A}\ \text{- Autonomic intelligence} \\
& \|\ (B_e, B_t, B_{int}, B_g, B_d, B_p, B_{\inf}) && //\ \mathrm{I_C}\ \text{- Cognitive intelligence} \\
\}\ &
\end{aligned}
\tag{2}
$$

According to Definition 8, the relationship among the three forms of intelligence is as follows:

$$
\Im_I \subseteq \Im_A \subseteq \Im_C \tag{3}
$$

Both Eqs. 2 and 3 indicate that any lower layer intelligence and behavior is a subset of those of a higher layer. In other words, any higher layer intelligence and behavior is a natural extension of those of lower layers.

Table 1. Taxonomy of Abstract Intelligence and Its Embodying Forms

No.	Form of intelligence	Embodying means	Paradigms
1	Natural intelligence (NI)	Naturally grown biological and physiological organisms	Human brains and brains of other well developed species
2	Artificial intelligence (AI)	Cognitively-inspired artificial models and man-made systems	Intelligent systems, knowledge systems, decision-making systems, and distributed agent systems
3	Machinable intelligence (MI)	Complex machine and wired systems	Computers, robots, autonomic circuits, neural networks, and autonomic mechanical machines
4	Computational intelligence (CoI)	Computational methodologies and software systems	Expert systems, fuzzy systems, autonomous computing, intelligent agent systems, genetic/evolutionary systems, and autonomous learning systems

It is noteworthy that all paradigms of αI share the same cognitive informatics foundation as described in the following theorems, because they are an artificial or machine implementation or embodiment of αI.

Theorem 1. The *compatible intelligent capability* state that *natural intelligence* (NI), *artificial intelligence* (AI), *machinable intelligence* (MI), and *computational intelligence* (CoI), are compatible by sharing the same mechanisms of αI, i.e.:

$$CoI \cong MI \cong AI \cong NI \cong \alpha I \quad (4)$$

On the basis of Theorem 1, the differences between NI, AI, MI, and CoI are only distinguishable by: (a) The means of their implementation; and (b) The extent of their intelligent capability.

Corollary 1. The *inclusive intelligent capability* states that all real-world paradigms of intelligence are a subset of αI, i.e.:

$$CoI \subseteq MI \subseteq AI \subseteq NI \subseteq \alpha I \quad (5)$$

Corollary 1 indicates that AI, CoI, and MI are dominated by NI and αI. Therefore, one should not expect a computer or a software system to solve a problem where human cannot. In other words, no AI or computer systems may be designed and/or implemented for a given problem where there is no solution being known collectively by human beings as a whole. Further, Theorem 1 and Corollary 1 explain that the development and implementation of AI rely on the understanding of the mechanisms and laws of NI.

5. APPLICATIONS OF CI AND CC

The studies in CI and αI lay a theoretical foundation toward revealing the basic mechanisms of different forms of intelligence (Wang, 2010c). As a result, cognitive computers may be developed, which are characterized as a knowledge processor beyond those of data processors in conventional computing. Key applications in the above cutting-edge fields of CI and CC can be divided into two categories. The first category of applications uses informatics and computing techniques to investigate problems of intelligence science, cognitive science, and brain science, such as abstract intelligence, memory, learning, and reasoning. The second category of applications includes the areas that use cognitive informatics theories to investigate problems in informatics, computing, software engineering, knowledge engineering, and computational intelligence. CI focuses on the nature of information processing in the brain, such as information acquisition, representation, memory, retrieval, creation, and communication. Through the interdisciplinary approach and with the support of modern information and neuroscience technologies, mechanisms of the brain and the mind may be systematically explored based on the theories and cognitive models of CI.

Because CI and CCs provide a common and general platform for the next generation of cognitive computing, a wide range of applications of CI, αI, CC, and DM are expected toward the implementation of highly intelligent machinable thought such as formal inference, symbolic reasoning, problem solving, decision making, cognitive knowledge representation, semantic searching, and autonomous learning. Some expected innovations that will be enabled by CCs are as follows, *inter alia*: a) An *inference machine* for complex and long-series of reasoning, problem solving, and decision making beyond traditional logic and if-then-rule based technologies; b) An *autonomous learning system* for cognitive knowledge acquisition and processing; c) A novel *search engine* for providing comprehendable and formulated knowledge via the Internet; d) A *cognitive medical diagnosis system* supporting evidence-based medical care and clinical practices; e) A *cognitive computing node* for the next generation of the intelligent Internet; and f) A *cognitive processor* for implementing cognitive robots and cognitive agents.

6. CONCLUSION

This chapter has summarized the latest development in cognitive informatics, abstract intelligence, denotational mathematics, cognitive computing, and cognitive computers. The theoretical framework of cognitive informatics and cognitive computing has been reviewed. The context and relations among the aforementioned fields have been elaborated. A set of applications in the cutting-edge areas has been reported.

REFERENCES

Baciu, G., Y. Yao, Y. Wang, L.A. Zadeh, K. Chan, and W. Kinsner eds. (2009): *Proceedings of the 8th IEEE International Conference on Cognitive Informatics* (ICCI'09), Hong Kong, IEEE Computer Society Press, Los Alamitos, CA., June.

Bell, D. A. (1953). *Information Theory*. London: Pitman.

Chan, C., W. Kinsner, Y. Wang, and D.M. Miller eds. (2004), *Cognitive Informatics: Proc. 3rd IEEE International Conference* (ICCI'04), IEEE CS Press, Victoria, Canada, August.

Goldman, S. (1953). *Information Theory*. Englewood Cliffs, NJ: Prentice-Hall.

Kinsner, W., D. Zhang, Y. Wang, and J. Tsai eds. (2005), *Cognitive Informatics: Proc. 4th IEEE International Conference* (ICCI'05), IEEE CS Press, Irvine, California, USA, August.

Patel, D., S. Patel, and Y. Wang eds. (2003), *Cognitive Informatics: Proc. 2nd IEEE International Conference* (ICCI'03), IEEE CS Press, London, UK, August.

Pescovitz, D. (2002). Autonomic computing: Helping computers help themselves. *IEEE Spectrum*, *39*(9), 49–53. doi:10.1109/MSPEC.2002.1030968

Shannon, C. E. (1948). A Mathematical Theory of Communication. *The Bell System Technical Journal*, *27*, 379–423, 623–656.

Shannon, C. E., & Weaver, W. (1949). *The Mathematical Theoryof Communication*. Urbana: Illinois University Press.

Sun, F., Y. Wang, J. Lu, B. Zhang, W. Kinsner, and L.A. Zadeh eds. (2010), *Proceedings of the 9th IEEE International Conference on Cognitive Informatics* (ICCI'10), Tsinghua University, Beijing, IEEE Computer Society Press, Los Alamitos, CA., July.

Wang, Y. (2002a): Keynote: On Cognitive Informatics, *Proc. 1st IEEE International Conference on Cognitive Informatics* (ICCI'02), Calgary, Canada, IEEE CS Press, August, 34-42.

Wang, Y. (2002b). The Real-Time Process Algebra (RTPA). *Annals of Software Engineering*, *14*, 235–274. doi:10.1023/A:1020561826073

Wang, Y.: (2003), On Cognitive Informatics, *Brain and Mind: A Transdisciplinary Journal of Neuroscience and Neurophilosophy*, 4(2), pp.151-167.

Wang, Y. (2006), Keynote: Cognitive Informatics - Towards the Future Generation Computers that Think and Feel, *Proc. 5th IEEE International Conference on Cognitive Informatics* (ICCI'06), Beijing, China, IEEE CS Press, July, pp. 3-7.

Wang, Y.: (2007a), *Software Engineering Foundations: A Software Science Perspective*, CRC Series in Software Engineering, Vol. II, Auerbach Publications, NY, USA, July.

Wang, Y. (2007b). The Theoretical Framework of Cognitive Informatics. *International Journal of Cognitive Informatics and Natural Intelligence*, *1*(1), 1–27. doi:10.4018/jcini.2007010101

Wang, Y. (2007c). The OAR Model of Neural Informatics for Internal Knowledge Representation in the Brain. *International Journal of Cognitive Informatics and Natural Intelligence*, *1*(3), 64–75. doi:10.4018/jcini.2007070105

Wang, Y. (2007d). Towards Theoretical Foundations of Autonomic Computing. *International Journal of Cognitive Informatics and Natural Intelligence*, *1*(3), 1–16. doi:10.4018/jcini.2007070101

Wang, Y. (2008a), On Contemporary Denotational Mathematics for Computational Intelligence, *Transactions of Computational Science*, 2, Springer, June, 6-29.

Wang, Y. (2008b). RTPA: A Denotational Mathematics for Manipulating Intelligent and Computational Behaviors. *International Journal of Cognitive Informatics and Natural Intelligence*, *2*(2), 44–62. doi:10.4018/jcini.2008040103

Wang, Y. (2008c). On Concept Algebra: A Denotational Mathematical Structure for Knowledge and Software Modeling. *International Journal of Cognitive Informatics and Natural Intelligence*, *2*(2), 1–19. doi:10.4018/jcini.2008040101

Wang, Y. (2008d). On System Algebra: A Denotational Mathematical Structure for Abstract System Modeling. *International Journal of Cognitive Informatics and Natural Intelligence*, *2*(2), 20–42. doi:10.4018/jcini.2008040102

Wang, Y. (2009a). On Abstract Intelligence: Toward a Unified Theory of Natural, Artificial, Machinable, and Computational Intelligence. *International Journal of Software Science and Computational Intelligence*, *1*(1), 1–18. doi:10.4018/jssci.2009010101

Wang, Y. (2009b). On Cognitive Computing. *International Journal of Software Science and Computational Intelligence*, *1*(3), 1–15. doi:10.4018/jssci.2009070101

Wang, Y. (2009c). On Visual Semantic Algebra (VSA): A Denotational Mathematical Structure for Modeling and Manipulating Visual Objects and Patterns. *International Journal of Software Science and Computational Intelligence*, *1*(4), 1–15. doi:10.4018/jssci.2009062501

Wang, Y. (2009d). Paradigms of Denotational Mathematics for Cognitive Informatics and Cognitive Computing. *Fundamenta Informaticae*, *90*(3), 282–303.

Wang, Y. (2009e), Granular Algebra for Modeling Granular Systems and Granular Computing, *Proc. 8th IEEE International Conference on Cognitive Informatics* (ICCI'09), Hong Kong, IEEE CS Press, June, pp. 145-154.

Wang, Y. (2010a), Keynote: Cognitive Computing and World Wide Wisdom (WWW+), *Proc. 9th IEEE Int'l Conf. Cognitive Informatics* (ICCI'10), Tsinghua Univ., Beijing, IEEE CS Press, July, pp. 4-5.

Wang, Y. (2010b), Keynote: Cognitive Informatics and Denotational Mathematics Means for Brain Informatics, *1st Int'l Conference on Brain Informatics* (ICBI'10), Toronto, Aug., pp. 2-13.

Wang, Y. (2010c). Cognitive Robots: A Reference Model towards Intelligent Authentication. *IEEE Robotics and Automation*, *17*(4), 54–62. doi:10.1109/MRA.2010.938842

Wang, Y. (2010d). On Formal and Cognitive Semantics for Semantic Computing. *International Journal of Semantic Computing*, *4*(2), 83–118. doi:10.1142/S1793351X10000833

Wang, Y. (2010e). On Concept Algebra for Computing with Words (CWW). *International Journal of Semantic Computing*, *4*(3), 331–356. doi:10.1142/S1793351X10001061

Wang, Y., Baciu, G., Yao, Y., Kinsner, W., Chan, K., & Zhang, B. (2010). Perspectives on Cognitive Informatics and Cognitive Computing. *International Journal of Cognitive Informatics and Natural Intelligence*, *4*(1), 1–29. doi:10.4018/jcini.2010010101

Wang, Y., & Chiew, V. (2010). On the Cognitive Process of Human Problem Solving. *Cognitive Systems Research: An International Journal, Elsevier*, *11*(1), 81–92. doi:10.1016/j.cogsys.2008.08.003

Wang, Y., R. Johnston, and M. Smith eds. (2002), *Cognitive Informatics: Proc. 1st IEEE International Conference* (ICCI'02), IEEE CS Press, Calgary, AB, Canada, August.

Wang, Y., & Kinsner, W. (2006). Recent Advances in Cognitive Informatics. *IEEE Transactions on Systems, Man and Cybernetics. Part C, Applications and Reviews*, *36*(2), 121–123. doi:10.1109/TSMCC.2006.871120

Wang, Y., Kinsner, W., Anderson, J. A., Zhang, D., Yao, Y., & Sheu, P. (2009c). A Doctrine of Cognitive Informatics. *Fundamenta Informaticae*, *90*(3), 203–228.

Wang, Y., Kinsner, W., & Zhang, D. (2009b). Contemporary Cybernetics and its Faces of Cognitive Informatics and Computational Intelligence, *IEEE Trans. on System, Man, and Cybernetics. Part B*, *39*(4), 823–833.

Wang, Y., & Wang, Y. (2006). Cognitive Informatics Models of the Brain. *IEEE Transactions on Systems, Man and Cybernetics. Part C, Applications and Reviews*, *36*(2), 203–207. doi:10.1109/TSMCC.2006.871151

Wang, Y., Wang, Y., Patel, S., & Patel, D. (2006). A Layered Reference Model of the Brain (LRMB), *IEEE Trans. on Systems, Man, and Cybernetics. Part C*, *36*(2), 124–133.

Wang, Y., Zadeh, L. A., & Yao, Y. (2009a). On the System Algebra Foundations for Granular Computing. *International Journal of Software Science and Computational Intelligence*, *1*(1), 64–86. doi:10.4018/jssci.2009010105

Wang, Y., Zhang, D., Latombe, J.-C., & Kinsner, W. (Eds.). (2008), Proceedings of the 7th IEEE International Conference on Cognitive Informatics (ICCI'08), Stanford University, IEEE Computer Society Press, Los Alamitos, CA., August.

Yao, Y. Y., Shi, Z., Wang, Y., & Kinsner, W. (Eds.). (2006), *Cognitive Informatics: Proc. 5th IEEE International Conference* (ICCI'06), IEEE CS Press, Beijing, China, July.

Zadeh, L. A. (1965), Fuzzy Sets and Systems, in J. Fox ed., *Systems Theory*, Polytechnic Press, Brooklyn NY, 29-37.

Zadeh, L. A. (1975). Fuzzy Logic and Approximate Reasoning. *Syntheses*, *30*, 407–428. doi:10.1007/BF00485052

Zadeh, L. A. (2008), Toward Human Level Machine Intelligence – Is It Achievable? Proceedings of the 7th IEEE International Conference on Cognitive Informatics (ICCI'08), Stanford University, IEEE Computer Society Press, Los Alamitos, CA., August, pp. 1.

Zhang, D., Wang, Y., & Kinsner, W. (Eds.). (2007), Proceedings of the 6th IEEE International Conference on Cognitive Informatics (ICCI'07), Lake Tahoe, IEEE Computer Society Press, Los Alamitos, CA., August. Yingxu Wang *University of Calgary*

Acknowledgment

The author would like to thank all contributors for this book based on selected papers published in *International Journal of Cognitive Informatics and Natural Intelligence* (IJCINI) during 2009. I acknowledge the support of IEEE Computer Society, The IEEE ICCI Steering Committee, and editors of IJCINI at IGI.

Section 1

Chapter 1
A Cognitive Informatics Reference Model of Autonomous Agent Systems (AAS)

Yingxu Wang
University of Calgary, Canada

ABSTRACT

Despite the fact that the origin of software agent systems has been rooted in autonomous artificial intelligence and cognitive psychology, their implementations are still based on conventional imperative computing techniques rather than autonomous computational intelligence. This paper presents a cognitive informatics perspective on autonomous agent systems (AAS's). A hierarchical reference model of AAS's is developed, which reveals that an autonomous agent possesses intelligent behaviors at three layers known as those of imperative, autonomic, and autonomous from the bottom up. The theoretical framework of AAS's is described from the facets of cognitive informatics, computational intelligence, and denotational mathematics. According to Wang's abstract intelligence theory, an autonomous software agent is supposed to be called as an intelligent-ware, shortly, an intelware, parallel to hardware and software in computing, information science, and artificial intelligence.

INTRODUCTION

A *software agent* is an intelligent software system that autonomously carries out robotic and interactive applications based on goal-driven cognitive mechanisms. The studies on software agent are rooted in the essences of computing science and cognitive science such as automata theory (von Neumann, 1946, 1958, 1963, 1966; Shannon, 1956), Turing machines (Turing, 1950), cognitive psychology (Newell, 1990; Sternberg, 1997; Anderson and Rosenfeld, 1998; Matlin, 1998), artificial intelligence (*McCarthy,* 1955, 1963; McCulloch, 1943, 1965; Barr and Feigenbaum, 1981), computational intelligence (Poole et al., 1997; Wang, 2008a), and decision theories (Wald, 1950; Newell and Simon, 1972; Berger et al., 1990; Bronson and Naadimuthu, 1997; Wang and Ruhe, 2007; Wang, 2008b).

DOI: 10.4018/978-1-60960-553-7.ch001

The history towards software agents may be traced back to the work as early as in the 1940s. J. *McCarthy,* W. McCulloch, *M.L. Minsky, N. Rochester, and C.E. Shannon* proposed the term *Artificial Intelligence* (AI) (*McCarthy,* 1955, 1963; McCulloch, 1943, 1965). S.C. Kleene analyzed the relations of *automata* and nerve nets (Kleene, 1956). Then, Bernard Widrow developed the technology of *artificial neural networks* in the 1950s (Widrow and Lehr, 1990). The concepts of *robotics* (Brooks, 1970) and *expert systems* (Giarrantans and Riley, 1989) were developed in the 1970s and 1980s, respectively. In 1992, the notion of *genetic algorithms* was proposed by J.H. Holland (Holland, 1992). Then, *distributed artificial intelligence* and *intelligent system* technologies emerged since late 1980s (Bond and Gasser, 1988; Kurzweil, 1990; Chaib-Draa et al., 1992; Meystel and Albus, 2002, Meystel and Albus, 2002).

The origin of the term *autonomous agent* is based on Carl Hewitt and his colleagues' *artificial intelligence actor models* proposed in 1973 (Hewitt et al., 1973, 1991). Then, as a novel approach of artificial intelligence, agent technologies have been proliferated since the early 1990s (Foner, 1993; Genesereth and Ketchpel, 1994; Hayes-Roth, 1995; Axelrod, 1997; Huhns and Singh, 1997; Wooldridge and Jenings, 1995; Wooldridge, 2002, Wang, 2003b). Pattie Maes perceived that a software agent is a process that lives in the world of computers and networks and that can operate autonomously to fulfill a set of tasks (Maes, 1991). Dimitris N. Chorafas described a software agent as a new software paradigm of things that think (Chorafas, 1998). Software agents are characterized by knowledge, learning, reasoning, and adaptation, which are rational to the extent that their behaviors are predictable by given goals and the solution environment (Russell and Norvig 1995; Poole, Mackworth, and Goebel 1997; Nilsson 1998).

Multi-agent systems are proposed in (Wittig, 1992; Wellman, 1999) as distributed intelligent systems (Bond and Gasser, 1988) in which each node is an autonomous software agent. The key technology of autonomous agent systems is how a variety of heterogeneous agents allocate their roles, coordinate their behaviors, share their resources, and communicate their information, beliefs, and needs (Maes, 1991). The interaction mechanisms of multi-agent systems, such as cooperation, negotiation, belief reconciliation, information sharing, and distributed decision making, are identified as important issues in the design and implementation of multi-agent systems.

Autonomic computing is one of the fundamental technologies of software agents, which is a mimicry and simulation of the natural intelligence possessed by the brain using general computers. Autonomic computing was first proposed by IBM in 2001, where it is perceived that "Autonomic computing is an approach to self-managed computing systems with a minimum of human interference. The term derives from the body's autonomous nervous system, which controls key functions without conscious awareness or involvement (IBM, 2006)." Various studies on autonomic computing have been reported following the IBM initiative (Kephart and Chess, 2003; Murch, 2004; Wang, 2004).

According to Wang's *abstract intelligence* theory (Wang, 2008a, 2009), software agents are a paradigm of abstract and computational intelligence, which is a subset of or an application-specific virtual brain. Behaviors of a software agent are mirrored human behaviors. Therefore, a software agent may be more accurately named as an *intelligent-ware*, shortly, an *intelware*, parallel to hardware and software in computing, information science, and artificial intelligence. In this notion, intelware will be treated as a synonym of an autonomous agent system.

This paper presents a coherent theoretical framework of autonomous agent systems (AAS's) or *intelware* from the facets of cognitive informatics, computational intelligence, and denotational mathematics. The nature of software agents and intelware is elaborated. A reference model of AAS with intelligent behaviors at three layers known as those of imperative, autonomic, and autonomous is developed from the bottom up. The theoretical framework of AAS's/intelware is presented on the basis of cognitive informatics and computational intelligence theories. A set of denotational mathematics is introduced in order to provide a fundamental mathematical means for formally and rigorously dealing with the highly complicated architectures and intricate behaviors of AAS's and intelware.

THE NATURE OF SOFTWARE AGENTS AND INTELWARE

Definition 1. *A software agent, or more actually an intelware, is an intelligent software system that autonomously carries out robotistic and interactive applications based on goal-driven cognitive mechanisms.*

On the basis of Definition 1, an autonomous agent is a software agent that possesses high-level autonomous ability and behaviors beyond conventional imperative computing technologies.

Definition 2. *An Autonomous Agent System (AAS) is a composition of distributed agents that possesses autonomous computing and decision making abilities as well as interactive communication capability to peers and the environment.*

The classification of agent/intelware technologies can be described in Table 1, where *I* and *O* denote the inputs/outputs of a given AAS. When both input event (I) and output behavior (O) are constant, it denotes a *routine* intelware; while when both *I/O* are variable, it represents the most complicated *autonomous* intelware. Otherwise, the combinations of variable event/constant behavior and constant event/variable behaviors indicate an *algorithmic* or *autonomic* intelware, respectively.

Table 1. Classification of intelware / AAS's

		Behavior (O)	
		Constant	**Variable**
Event (I)	Constant	*Routine*	*Autonomic*
	Variable	*Algorithmic*	*Autonomous*

In Table 1, the routine and algorithmic AAS's may be implemented by computational imperative behaviors. However, the autonomic AAS's should be implemented by autonomic computing, as that of the autonomous AAS's by autonomous mechanisms and behaviors.

THE REFERENCE MODEL OF INTELWARE/AUTONOMOUS AGENT SYSTEMS

The reference model of intelware/AAS's is a hierarchical model with three layers known as those of imperative, autonomic, and autonomous behaviors. This section elaborates the mathematical models of the imperative and intelligent behaviors of intelware/AAS's in the layered reference model of agent intelligence.

The Hierarchical Behavioral Model of Intelware/AAS's

Behaviorism is a doctrine of psychology and cognitive informatics that describes the association between a given stimulus and an observed response of human brains and AAS's. Cognitive informatics reveals that human and AAS behaviors may be classified into four categories known as the *perceptive, cognitive, instructive,* and *reflective* behaviors (Wang, 2007b).

The reference model of AAS's (RMAAS) is a hierarchical behavioral model of agent intelligence as illustrated in Figure 1. In the RMAAS model, the hierarchy of agent behaviors can be divided into the imperative, autonomic, and

Figure 1. The hierarchical reference model of autonomous agent systems (RMAAS)

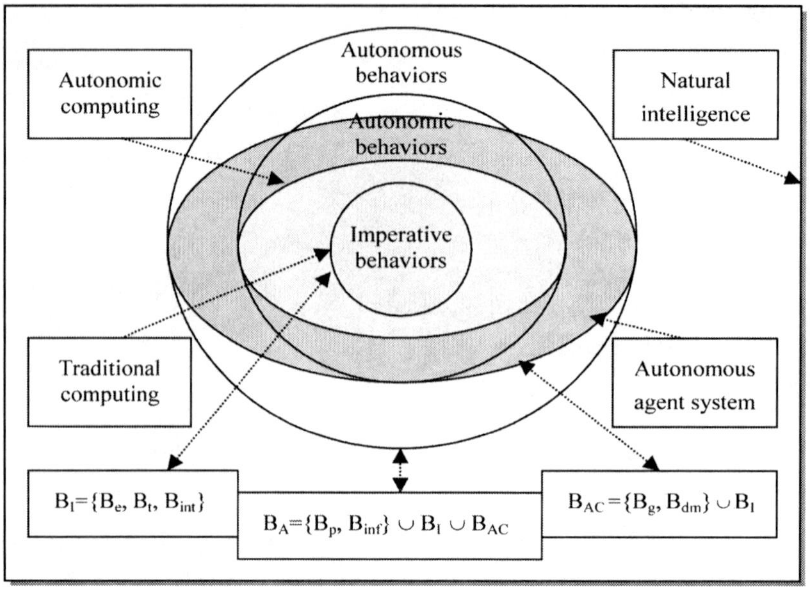

autonomous layers. Conventional computing machines are implemented only by imperative behaviors. However, the autonomic computing systems and AAS's are implemented by advanced cognitive behaviors. *Imperative computing* is an enclosure of instructive and passive behaviors. The *autonomic computing* is an enclosure of internally motivated behaviors beyond those of the imperative space. The *autonomous computing* is an enclosure of perceptive- and inference-driven behaviors beyond those of both imperative and autonomic computing. More formal descriptions of the three types of behaviors of AAS's will be presented in the following subsections.

The Imperative Behavioral Layer of Intelware/AAS's

According to the RMAAS model as illustrated in Figure 1, the imperative behavioral intelligence of intelware and AAS's can be formally modeled and elaborated in this subsection.

Definition 3. *The imperative behavioral layer of AAS's, B_I, is a set of instruction-based behaviors such as the event-driven behaviors (B_e), time-driven behaviors (B_t), and interrupt-driven behaviors (B_{int}), i.e.:*

$$B_I \triangleq \{B_e, B_t, B_{int}\} \tag{1}$$

An imperative system implemented with B_I may do nothing unless a specific program is loaded, in which the stored program transfers a general-purpose computer to a specific intelligent application. The imperative system is a passive system that implements deterministic, context-free, and stored-program controlled behaviors.

Definition 4. *An event is an abstract variable that represents an external stimulus to a system or the occurring of an internal change of status, such as an action of users, an updating of the environment, and a change of the value of a control variable.*

The types of events that may trigger a behavior can be classified into operational (@*e*S), time (@*t*TM), and interrupt (@*int*◉) events, where @ is the *event prefix*, and S, TM, and ⊙ the type suffixes, respectively. The *interrupt event* is a kind of special event that models the interruption of an executing process, the temporal handover of controls to an Interrupt Service Routine (ISR), and the return of control after its completion.

Definition 5. *An interrupt, denoted by* $\lightning$*, is a parallel process relation in which a running process P is temporarily held by another higher priority process Q via an interrupt event @int*◉ *at the interrupt point* ◉*, and the interrupted process will be resumed when the high priority process has been completed, i.e.:*

$$P \lightning Q \triangleq P \,||\, (@int\circledbullet \nearrow Q \searrow \circledbullet) \quad (2)$$

where $\nearrow$ and $\searrow$ denote an interrupt service and an interrupt return, respectively.

In general, all types of events, including the operational events, timing events, and interrupt events, are captured by the system in order to dispatch a designated behavior.

Definition 6. *An event-driven behavior* B_e*, denoted by* $\hookrightarrow_e$*, is an imperative process in which the ith behavior in term of a designated process* P_i *is triggered by a predefined event* $@e_iS$*, i.e.:*

$$B_e \triangleq \mathop{\mathrm{R}}_{i=1}^{n} (@e_i\mathbf{S} \hookrightarrow_e P_i) \quad (3)$$

where the big-R notation is a mathematical calculus that denotes a sequence of repetitive/iterative behaviors or a set of recurring structures (Wang, 2007a).

Definition 7. *A time-driven behavior* B_t*, denoted by* $\hookrightarrow_t$*, is an imperative process in which the ith behavior in term of process* P_i *is triggered by a predefined point of time* $@t_iTM$*, i.e.:*

$$B_t \triangleq \mathop{\mathrm{R}}_{i=1}^{n} (@t_i\mathbf{TM} \hookrightarrow_t P_i) \quad (4)$$

where $@t_1TM$ may be a system timing or external timing event.

Definition 8. *An interrupt-driven behavior* B_{int}*, denoted by* $\hookrightarrow_{int}$*, is an imperative process in which the ith behavior in term of process* P_i *is triggered by a predefined system interrupt* $@int_i$◉*, i.e.:*

$$B_{int} \triangleq \mathop{\mathrm{R}}_{i=1}^{n} (@int_i \odot \hookrightarrow_{int} P_i) \quad (5)$$

As a summary, an imperative computing system can be described as follows.

Definition 9. *An Imperative Computing (IC) system is a passive system that implements deterministic, context-free, and stored-program controlled behaviors.*

The Autonomic Behavioral Layer of Intelware/AAS's

According to the RMAAS model as illustrated in Figure 1, the autonomic behavioral intelligence of intelware and AAS's can be formally modeled and elaborated in this subsection.

Definition 10. *The autonomic behavioral layer of AAS's,* B_C*, is a set of internally motivated and self- generated behaviors such as the goal-driven behaviors (*B_g*) and decision-driven behaviors (*B_d*) on the basis of the imperative layer* B_I*, i.e.:*

$$\begin{aligned} B_C &\triangleq \{B_g, B_d\} \cup B_I \\ &= \{B_e, B_r, B_{int}, B_g, B_d\} \end{aligned} \quad (6)$$

Definition 11. *A goal-driven behavior* B_g*, denoted by* $\hookrightarrow_g$*, is an autonomic process in which the ith behavior in term of process* P_i *is generated by the system itself, rather than be given, corresponding to the goal* $@g_iST$*, i.e.:*

$$B_g \triangleq \mathop{R}_{i=1}^{n} (@g_i\mathbf{ST} \hookrightarrow_g P_i) \tag{7}$$

where the goal, denoted by gST, is a triple, i.e.:

$$gST = (P, \Omega, \Theta) \tag{8}$$

in which that P = {$p_1, p_2, \ldots, p_n$} is a finite nonempty set of purposes or motivations, Ω is a finite set of constraints for the goal, and Θ is the environment of the goal.

Definition 12. *A decision-driven behavior B_d, denoted by $\hookrightarrow_d$, is an autonomic process in which the ith behavior in term of process P_i is generated by a given decision $@d_iST$, i.e.:*

$$B_d \triangleq \mathop{R}_{i=1}^{n} (@d_i\mathbf{ST} \hookrightarrow_d P_i) \tag{9}$$

where the decision, denoted by dST, is a selected alternative a Ω from a nonempty set of alternatives $\mathcal{A}$, based on a given set of criteria C, i.e.:

$$d = f(\mathcal{A}, C)$$

$$= f\!: \mathcal{A} \times C \rightarrow \mathcal{A},\ \mathcal{A} \neq \varnothing \tag{10}$$

Definition 13. *An Autonomic Computing (AC) system is an intelligent system that implements nondeterministic, context-dependent, and adaptive behaviors based on goal- and decision-driven mechanisms.*

The autonomic systems do not rely on instructive and procedural information, but are dependent on internal status and willingness that formed by long-term historical events and current rational or emotional goals (Wang, 2007d).

The Autonomous Behavioral Layer of Intelware/AAS's

According to the RMAAS model as illustrated in Figure 1, the autonomous behavioral intelligence of intelware and AAS's can be formally modeled and elaborated in this subsection.

Definition 14. *The autonomous behavioral layer of AAS's, B_A, is a set of autonomously generated behaviors by internal cognitive processes such as the perception-driven behaviors (B_p) and inference-driven behaviors (B_{inf}) on the basis of the imperative space B_I and the autonomic space B_C, i.e.:*

$$\begin{aligned} B_A &\triangleq \{B_p, B_{inf}\} \cup B_I \cup B_C \\ &= \{B_e, B_t, B_{int}, B_g, B_d, B_p, B_{inf}\} \end{aligned} \tag{11}$$

The new forms of behaviors covered in the autonomous layer can be elaborated as follows.

Definition 15. *A perception-driven behavior B_p, denoted by $\hookrightarrow_p$, is a cognitive process in which the ith behavior in term of process P_i is generated by the result of a perceptive process $@p_iPC$, i.e.:*

$$B_p \triangleq \mathop{R}_{i=1}^{n} (@p_i\mathbf{PC} \hookrightarrow_p P_i) \tag{12}$$

where PC stands for a type of process, and the perception result pPC is an outcome of the cognitive process of perception that an AAS may generate.

Inferences are cognitive processes that reason about a possible causality from given premises based on known causal relations between a pair of cause and effect proven true by empirical arguments, theoretical inferences, or statistical regulations.

Definition 16. *An inference-driven behavior B_{inf} denoted by $\hookrightarrow_{inf}$ is a cognitive process in which the ith behavior in term of process P_i is generated by the result of an inference process @inf$_i$PC, i.e.:*

$$B_{inf} \triangleq \underset{i=1}{\overset{n}{R}} (@inf_i\mathbf{PC} \hookrightarrow_{inf} P_i) \quad (13)$$

where formal inferences can be classified into the deductive, inductive, abductive, and analogical categories, as well as modal, probabilistic, and belief theories (Wang, 2007e).

As shown in Definition 16 and Figure 1, an AAS implemented on B_A extends the conventional behaviors B_I and B_C to more powerful and intelligent behaviors, which are generated by internal and autonomous processes such as the perception and inference processes. With the possession of all the seven forms of intelligent behaviors in B_A, the AAS may advance closer to the intelligent power of human brains.

Relationships between the Agent Behaviors of Intelware/AAS's at the Three Layers of RMAAS

Contrasting Definitions 3, 10, and 14, the following relationships among the three-layer agent intelligent behaviors can be established on the basis of the RMAAS model as illustrated in Figure 1.

Theorem 1. *The relationships of the imperative behaviors B_I, autonomic behaviors B_C, and cognitive behaviors B_A of intelware or AAS's are hierarchical and inclusive, i.e.:*

$$B_I \subseteq B_C \subseteq B_A \quad (14)$$

Theorem 1 and Definition 14 indicate that any lower layer behavior of an intelware or AAS is a subset of those of a higher layer. In other words, any higher layer behavior is a natural extension of those of lower layers as shown in Figure 1. Therefore, the *necessary and sufficient conditions of AAS's, C_{AAS}*, are the possession of all behaviors at the three layers.

Corollary 1. *The behavioral model of intelware or AAS, §AASST, can be logically modeled by a set of parallel processes that encompasses the imperative behaviors B_I, autonomic behaviors B_C, and autonomous behaviors B_A from the bottom-up, i.e.:*

$$\begin{aligned}
\S AAS\mathbf{ST} &\triangleq (B_I, B_C, B_A) \\
&= \{ \ (B_e, B_t, B_{int}) \qquad // \ B_I \\
&\quad \ || \ (B_e, B_t, B_{int}, B_g, B_d) \qquad // \ B_C \\
&\quad \ || \ (B_e, B_t, B_{int}, B_g, B_d, B_p, B_{\inf}) \quad // \ B_A \\
&\quad \}
\end{aligned} \quad (15)$$

where || denotes a parallel relation in RTPA.

THEORETICAL FOUNDATIONS OF INTELWARE/AAS'S

Recent research reveals that the foundations of agent technologies root in cognitive informatics, denotational mathematics, and computational intelligence (Wang, 2002a, 2003b, 2008a). Along with the latest advances in cognitive informatics, non-imperative autonomous agent systems known as *intelware* and *cognitive computers* are emerging. This section explores the theoretical foundations of AAS's and intelware. The latest development of fundamental theories and technologies underpinning AAS's and intelware are highlighted.

Denotational Mathematics for AAS's

Applied mathematics can be classified into two categories known as *analytic* and *denotational* mathematics (Wang, 2002b, 2007a, 2008a,

Table 2. Paradigms of denotational mathematics

No.	Paradigm	Structure	Mathematical entities	Algebraic operations	Usage
1	Concept algebra	$CA \triangleq (C, OP, \Theta) = (\{O, A, R^c, R^i, R^o\}, \{\bullet_r, \bullet_c\}, \Theta_C)$	$c \triangleq (O, A, R^c, R^i, R^o)$	$\bullet_r \triangleq \{\leftrightarrow, \nleftrightarrow, \prec, \succ, =, \cong, \sim, \triangleq\}$ $\bullet_c \triangleq \{\Rightarrow, \overset{-}{\Rightarrow}, \overset{+}{\Rightarrow}, \overset{\sim}{\Rightarrow}, \uplus, \pitchfork, \Lleftarrow, \vdash, \mapsto\}$	Algebraic manipulations on abstract concepts
2	System algebra	$SA \triangleq (S, OP, \Theta) = (\{C, R^c, R^i, R^o, B, \Omega\}, \{\bullet_r, \bullet_c\}, \Theta)$	$S \triangleq (C, R^c, R^i, R^o, B, \Omega, \Theta)$	$\bullet_r \triangleq \{\nleftrightarrow, \leftrightarrow, \prod, =, \sqsubseteq, \sqsupseteq\}$ $\bullet_c \triangleq \{\Rightarrow, \overset{-}{\Rightarrow}, \overset{+}{\Rightarrow}, \overset{\sim}{\Rightarrow}, \boxminus, \uplus, \pitchfork, \Lleftarrow, \vdash\}$	Algebraic manipulations on abstract systems
3	Real-time process algebra (RTPA)	$RTPA \triangleq (\mathfrak{T}, \mathfrak{P}, \mathfrak{N})$	$\mathfrak{P} \triangleq \{:=, \blacklozenge, \Rightarrow, \Leftarrow, \nLeftarrow, >, <, \mid>, \mid<, @, \triangleq, \uparrow, \downarrow, !\otimes, \boxtimes, \S\}$ $\mathfrak{T} \triangleq \{\mathbf{N}, \mathbf{Z}, \mathbf{R}, \mathbf{S}, \mathbf{BL}, \mathbf{B}, \mathbf{H}, \mathbf{P}, \mathbf{TI}, \mathbf{D}, \mathbf{DT}, \mathbf{RT}, \mathbf{ST}, @e\mathbf{S}, @t\mathbf{TM}, @int\odot, ⓢs\mathbf{BL}\}$	$\mathfrak{R} \triangleq \{\to, \curvearrowright, \mid, \mid\ldots\mid\ldots, R^*, R^+, R^i, \circlearrowleft, \rightarrowtail, \parallel, \oiint, \mid\mid\mid, \gg, ↯, \hookrightarrow_t, \hookrightarrow_e, \hookrightarrow_i\}$	Algebraic manipulations on abstract processes

2008c). The former are mathematical structures that deal with functions of variables as well as their operations and behaviors; while the latter are mathematical structures that formalize rigorous expressions and inferences of system architectures and behaviors with abstract concepts, complex relations, and dynamic processes. The denotational and expressive needs in cognitive informatics, computational intelligence, software engineering, and knowledge engineering have led to new forms of mathematics collectively known as denotational mathematics.

Definition 17. *Denotational mathematics is a category of expressive mathematical structures that deals with high-level mathematical entities beyond numbers and simple sets, such as abstract objects, complex relations, behavioral information, concepts, knowledge, processes, intelligence, and systems.*

The term denotational mathematics is first introduced by Yingxu Wang in the emerging discipline of cognitive informatics (Wang, 2002a, 2007a, 2008c). Typical paradigms of denotational mathematics are comparatively presented in Table 1, where their structures, mathematical entities, algebraic operations, and usages are contrasted. The paradigms of denotational mathematics as shown in Table 1 are *concept algebra* (Wang, 2008d), *system algebra* (Wang, 2008e), and *Real-Time Process Algebra* (RTPA) (Wang, 2002b, 2008f).

The emergence of denotational mathematics is driven by the practical needs in cognitive informatics, computational intelligence, computing science, software science, and knowledge engineering, because all these modern disciplines study complex human and machine behaviors and their rigorous treatments. Among the new forms of denotational mathematics, *concept algebra* is designed to deal with the abstract mathematical structure of concepts and their representation and manipulation in knowledge engineering. *System algebra* is created to the rigorous treatment of abstract systems and their algebraic relations and operations. RTPA is developed to deal with series of behavioral processes and architectures of human and systems.

Denotational mathematics provides a powerful mathematical means for modeling and formalizing AAS's. Not only the architectures of AAS's, but also their dynamic behaviors can be rigorously and systematically manipulated by denotational mathematics. Applications of denotational mathematics in cognitive informatics and computational intelligence have been elaborated with a wide range of real-world case studies (Wang, 2008a, 2008c), which demonstrate that denotational mathematics is an ideal mathematical means for dealing with concepts, knowledge, behavioral processes, and human/machine intelligence in ASS's and intelware.

Cognitive Informatics Theories of AAS's

Cognitive informatics is the transdisciplinary enquiry of cognitive and information sciences that investigates into the internal information processing mechanisms and processes of the brain and natural intelligence, and their engineering applications via an interdisciplinary approach (Wang, 2002a, 2003a, 2003b, 2006, 2007a, 2007b, 2007c, 2007d). According to the abstract intelligence theory (Wang, 2008a, 2009), because cognitive informatics investigates the internal information processing mechanisms and processes of the brain and natural intelligence, its research results underlie the engineering applications of AAS's. Cognitive informatics reveals that artificial intelligence (AI) is a subset of natural intelligence (NI) (Wang, 2007a, 2007b). Therefore, AAS's may be referred to the natural intelligence and behavioral mechanisms of human beings.

A Layered Reference Model of the Brain (LRMB) is developed (Wang, et al., 2006) that reveals the logical model of NI and a coherent set of cognitive mechanisms. LRMB presents a systematical view toward the formal description and modeling of architectures and behaviors of AAS's, which are created to extend human capability, reachability, and/or memory capacity. The LRMB model explains the functional mechanisms and cognitive processes of the natural intelligence with 39 cognitive processes at seven layers known as the *sensation, memory, perception, action, meta-cognitive, meta-inference,* and *higher cognitive layers* from the bottom up. LRMB elicits the core and highly repetitive recurrent cognitive processes from a huge variety of life functions, which may shed light on the study of the fundamental mechanisms and interactions of complicated mental processes as well as AAS's, particularly the relationships and interactions between the inherited and the acquired life functions as well as those of the subconscious and conscious cognitive processes. The cognitive model of the brain can be used as a reference model for goal- and inference-driven technologies in AAS's.

Definition 18. *The cognitive model of the kernel of an AAS or intelware, AAS_k can be described as a real-time intelligent system with an inherited Agent Operating System AOS and a set of Agent Intelligent Behaviors AIB in parallel, i.e.:*

$$AAS_k \triangleq \text{AOS} \parallel \text{AIB} \qquad (16)$$

Definition 19. *The Cognitive Models of Memory (CMM) states that the architecture of human memory is parallel configured by the Sensory Buffer Memory (SBM), Short-Term Memory (STM), Long-Term Memory (LTM), Conscious Status Memory (CSM), and Action-Buffer Memory (ABM), i.e.:*

Figure 2. The generic abstract intelligence model (GAIM)

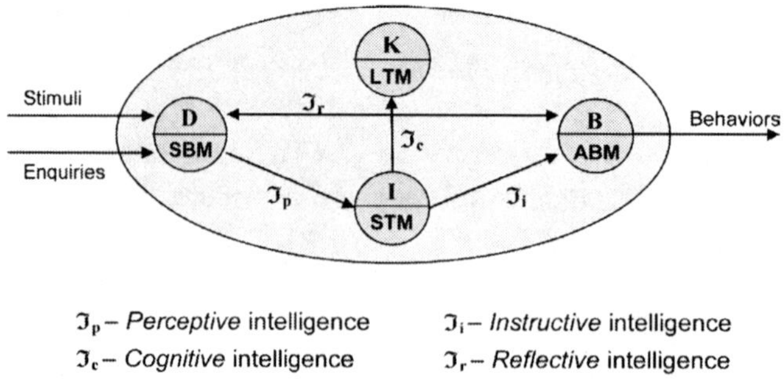

$$\begin{array}{l} CMM \triangleq (\ LTM \\ \quad\quad || \ STM \\ \quad\quad || \ CSM \\ \quad\quad || \ SBM \\ \quad\quad || \ ABM \\ \quad\quad) \end{array} \qquad (17)$$

The CMM model provides a neural informatics foundation of natural intelligence. With the CMM model, the broad sense of an AAS, AAS', can be described by mimicking the abstract architecture and mechanisms of the brain.

Definition 20. *The cognitive model of AAS's, AAS, is represented by a real-time intelligent system that encompasses the intelware and the CMM as well as their interactions, i.e.:*

$$\begin{array}{l} AAS \triangleq \ Intelware \\ \quad\quad || \ CMM \\ \quad\quad = (\ AOS \\ \quad\quad\quad || \ AIB \\ \quad\quad\quad) \\ \quad\quad || \ (\ LTM \\ \quad\quad\quad || \ STM \\ \quad\quad\quad || \ CSM \\ \quad\quad\quad || \ SBM \\ \quad\quad\quad || \ ABM \\ \quad\quad\quad) \end{array} \qquad (18)$$

Eq. 18 indicates that although *intelware* is considered the center of AAS's, the memories are essential to enable it to properly function, and to keep temporary and permanent results physiologically retained and retrievable.

Computational Intelligence Theories of AAS's

According to the abstract intelligence theory (Wang, 2008a, 2009), intelligence is perceived as the driving force or the ability to acquire and use knowledge and skills, or to reason in problem solving. It was conventionally perceived that only human beings possess higher-level intelligence. However, the development of computers, robots, intelligent systems, and AAS's indicates that intelligence may also be created or implemented by machines and man-made systems.

Definition 21. *Intelligence, in the narrow sense, is a human or a system ability that transforms information into behaviors; and in a broad sense, it is any human or system ability that autonomously transfers the forms of abstract information between data, information, knowledge, and behaviors in the brain.*

Definition 22. *The Generic Abstract Intelligence Model (GAIM), as shown in Figure 2, represents abstract intelligence in four forms known as the perceptive, cognitive, instructive, and reflective* *intelligence, corresponding to the specific forms of cognitive information and their memories.*

The GAIM indicates that different forms of intelligence are the driving force that transfers between a pair of abstract objects in the brain such as *data* (*D*), *information* (*I*), *knowledge* (*K*), and *behavior* (*B*). It is noteworthy that each abstract object is physiologically retained in a particular type of memories as given in the CMM model. This is the neural informatics foundation of natural intelligence, and the physiological evidences of why natural intelligence can be classified into four forms as shown in Figure 2.

According to Definitions 21 and 22, *computational intelligence* is a paradigm of abstract intelligence. Computational intelligence models human intelligence by computational methodologies and cognitively inspired models.

Definition 23. *The computational intelligence model of AAS's and intelware, §AASST, is a parallel structure represented by the Agent Operating System (AOSST) and a set of agent intelligence represented by the Agent Intelligent Behaviors (AIBST), as shown in Figure 3.*

The GAIM and §AASST model reveal that NI and AI share the same cognitive informatics foundations on the basis of abstract intelligence. The *compatible intelligent capability* states that NI, AI, AAS's, and intelware are compatible by sharing the same mechanisms of intelligent capability and behaviors. In other words, at the logical level, NI of the brain shares the same mechanisms as those of AI and computational intelligence. The differences between NI and AI are only distinguishable by the means of implementation and the extent of intelligent ability. Therefore, the studies on NI and AI in general, and intelware and AAS's in particular, may be unified into a coherent framework based on cognitive informatics and computational intelligence, which are formalized by denotational mathematics.

CONCLUSION

This paper has presented a coherent theoretical framework of Autonomous Agent Systems (AAS), known as *intelware,* from the facets of cognitive informatics, computational intelligence, and denotational mathematics. A reference model of AAS has been developed with three-layer intelligent behaviors known as the imperative, autonomic, and autonomous agent intelligence from the bottom up. It has been recognized that the characteristics of an AAS is its perception-driven and inference-driven behaviors beyond the imperative and autonomic ones as provided by conventional imperative and autonomic computing.

In order to formally and rigorously deal with the highly complicated architectures and intricate behaviors of intelware and AAS's, a new mathematical means known as denotational mathematics has been developed. Typical paradigms of denotational mathematics have been introduced such as concept algebra, system algebra, and RTPA. The findings of this work, particularly the necessary and sufficient conditions of imperative and autonomous computing, and the abstract intelligence model of natural and artificial intelligence, have formed a solid foundation for explaining and developing advanced autonomous computing systems and their engineering applications.

ACKNOWLEDGMENT

The author would like to acknowledge the support of the Natural Sciences and Engineering Research Council of Canada (NSERC) to this work. The author would like to thank the valuable comments and suggestions of the reviewers and colleagues.

Figure 3. The computational intelligence model of AAS

The Computational Intelligent Model of AAS

$$\S\text{AAS}\mathbf{ST} \triangleq \text{AOS}\mathbf{ST} \qquad \text{// Agent operating system}$$

$$\parallel \text{AIB}\mathbf{ST} \qquad \text{// Agent intelligent behaviors}$$

$$= \{ \quad \text{// AOS}\mathbf{ST}$$

$$<\text{SE}\mathbf{ST}\text{: } \mathop{R}\limits_{ptr\mathbf{P}=0}^{n-1} \text{SENSORS[ptr}\mathbf{P}]\mathbf{ST}> \qquad \text{// Layer 1: Sensation engine}$$

$$\parallel <\text{ME}\mathbf{ST}\text{: } \mathop{R}\limits_{addr\mathbf{P}=0}^{5} \text{MEM[addr}\mathbf{P}]\mathbf{ST}> \qquad \text{// Layer 2: Memory engine}$$

$$= <\text{SBM}\mathbf{ST} \parallel \text{STM}\mathbf{ST} \parallel \text{CSM}\mathbf{ST} \parallel \text{LTM}\mathbf{ST} \parallel \text{ABM}\mathbf{ST}>$$

$$\parallel <\text{PE}\mathbf{ST}\text{: } \mathop{R}\limits_{i\mathbf{P}=0}^{7} \text{PROC[i}\mathbf{N}]\mathbf{ST}> \qquad \text{// Layer 3: Perception engine}$$

$$= <\text{Attention}\mathbf{ST} \parallel \text{Motivation}\mathbf{ST} \parallel \text{Emotion}\mathbf{ST} \parallel \text{Attitude}\mathbf{ST}$$
$$\parallel \text{SensOfSpatiality}\mathbf{ST} \parallel \text{SensOfTime}\mathbf{ST} \parallel \text{SensOfMotion}\mathbf{ST}>$$

$$\parallel <\text{AE}\mathbf{ST}\text{: } \mathop{R}\limits_{ptr\mathbf{P}=0}^{n_{SERVO}\mathbf{H}-1} \text{SERVOS[ptr}\mathbf{P}]\mathbf{ST}> \qquad \text{// Layer 4: Action engine}$$

$$\parallel <\text{CE}\mathbf{ST}\text{: } \mathop{R}\limits_{i\mathbf{P}=0}^{10} \text{PROC[i}\mathbf{N}]\mathbf{ST}> \qquad \text{// Layer 5: Meta-cognition engine}$$

$$= <\text{ObjectIdentification}\mathbf{ST} \parallel \text{Abstraction}\mathbf{ST} \parallel \text{ConceptEstablishment}\mathbf{ST}$$
$$\parallel \text{Search}\mathbf{ST} \parallel \text{Categorization}\mathbf{ST} \parallel \text{Comparison}\mathbf{ST} \parallel \text{Memorization}\mathbf{ST}$$
$$\parallel \text{Qualification}\mathbf{ST} \parallel \text{Quantification}\mathbf{ST} \parallel \text{Selection}\mathbf{ST}>$$

$$\parallel <\text{IE}\mathbf{ST}\text{: } \mathop{R}\limits_{i\mathbf{P}=0}^{6} \text{PROC[i}\mathbf{N}]\mathbf{ST}> \qquad \text{// Layer 6: Meta-inference engine}$$

$$= <\text{Deduction}\mathbf{ST} \parallel \text{Induction}\mathbf{ST} \parallel \text{Abduction}\mathbf{ST} \parallel \text{Analogy}\mathbf{ST}$$
$$\parallel \text{Analysis}\mathbf{ST} \parallel \text{Synthesis}\mathbf{ST}>$$

$$\parallel <\text{HCE}\mathbf{ST}\text{: } \mathop{R}\limits_{i\mathbf{P}=0}^{7} \text{PROC[i}\mathbf{N}]\mathbf{ST}> \qquad \text{// Layer 7: Higher cognition engine}$$

$$= <\text{Comprehension}\mathbf{ST} \parallel \text{Learning}\mathbf{ST} \parallel \text{Planning}\mathbf{ST} \parallel \text{ProblemSolving}\mathbf{ST}$$
$$\parallel \text{DecisionMaking}\mathbf{ST} \parallel \text{Creation}\mathbf{ST} \parallel \text{PattenRecognition}\mathbf{ST}>$$

$$\parallel <\S\text{t}\mathbf{TM}> \qquad \text{// Relative clock}$$

$$\}$$

$$\parallel \{ \quad \text{// AIB}\mathbf{ST}$$

$$\parallel < \mathop{R}\limits_{k\mathbf{N}=0}^{n_e\mathbf{N}-1} @e_k\mathbf{S} \hookrightarrow P_k\mathbf{ST}> \qquad \text{// } \textit{Event}\text{-driven behaviors } (B_e)$$

$$\parallel < \mathop{R}\limits_{k\mathbf{N}=0}^{n_t\mathbf{N}-1} @t_k\mathbf{TM} \hookrightarrow P_k\mathbf{ST}> \qquad \text{// } \textit{Time}\text{-driven behaviors } (B_t)$$

$$\parallel < \mathop{R}\limits_{k\mathbf{N}=0}^{n_{int}\mathbf{N}-1} @int_k\odot \hookrightarrow P_k\mathbf{ST}> \qquad \text{// } \textit{Interrupt}\text{-driven behaviors } (B_{int})$$

$$\parallel < \mathop{R}\limits_{k\mathbf{N}=0}^{n_t\mathbf{N}-1} @g_k\mathbf{ST} \hookrightarrow P_k\mathbf{ST}> \qquad \text{// } \textit{Goal}\text{-driven behaviors } (B_g)$$

$$\parallel < \mathop{R}\limits_{k\mathbf{N}=0}^{n_t\mathbf{N}-1} @d_k\mathbf{ST} \hookrightarrow P_k\mathbf{ST}> \qquad \text{// } \textit{Decision}\text{-driven behaviors } (B_d)$$

$$\parallel < \mathop{R}\limits_{k\mathbf{N}=0}^{n_t\mathbf{N}-1} @p_k\mathbf{ST} \hookrightarrow P_k\mathbf{ST}> \qquad \text{// } \textit{Perception}\text{-driven behaviors } (B_p)$$

$$\parallel < \mathop{R}\limits_{k\mathbf{N}=0}^{n_{int}\mathbf{N}-1} @inf_k\mathbf{ST} \hookrightarrow P_k\mathbf{ST}> \qquad \text{// } \textit{Inference}\text{-driven behaviors } (B_{inf})$$

$$\}$$

REFERENCES

Anderson, J. A., & Rosenfeld, E. (Eds.). (1988). *Neurocomputing: Foundations of Research*, Cambridge.

Axelrod, R. (1977). *The Complexity of Cooperation: Agent-Based Models of Competition and Collaboration*. Princeton, NJ: Princeton Univ. Press.

Barr, A., & Feigenbaum, E. A. (Eds.). (1981). *The Handbook of Artificial Intelligence* (*Vol. 1*). Stanford and Los Altos, CA: HeurisTech Press and Kaufmann.

Berger, J. (1990). *Statistical Decision Theory – Foundations, Concepts, and Methods*. Springer-Verlag.

Bond, A. H., & Gasser, L. (1988). *Readings in Distributed Artificial Intelligence*. San Mateo, CA: Morgan Kaufmann.

Bronson, R., & Naadimuthu, G. (1997). *Schaum's Outline of Theory and Problems of Operations Research* (2nd ed.). NY: McGraw-Hill.

Brooks, R. A. (1970). New Approaches to Robotics. *American Elsevier, NY*, *5*, 3–23.

Chaib-Draa, B. Moulin, R. Mandiau, and P. Millot. (1992). Trends in Distributed Artificial Intelligence. *Artificial Intelligence Review*, *6*, 35–66. doi:10.1007/BF00155579

Chorafas, D. N. (1998). *Agent Technology Handbook*. NY: McGraw-Hill.

Foner, L. (1993), What ís an Agent, Anyway? A Sociological Case Study, *Agents Memo 93-01*, MIT Media Lab, Cambridge, MA.

Genesereth, M. R., & Ketchpel, S. P. (1994). Software Agents. *Communications of the ACM*, *37*(7), 48–53. doi:10.1145/176789.176794

Giarrantans, J., & Riley, G. (1989). *Expert Systems: Principles and Programming*. Boston: PWS-KENT Pub. Co.

Hayes-Roth, B. (1995). An Architecture for Adaptive Intelligent Systems. *Artificial Intelligence*, *72*(1-2), 329–365. doi:10.1016/0004-3702(94)00004-K

Hewitt, C., Bishop, R., & Steiger, R. (1973), A Universal Modular Actor Formalism for Artificial Intelligence, *Proc. 3rd Int. Joint Conf. on Artificial Intelligence*, Stanford, CA, Aug.

Hewitt, C., & Inman, J. (1991). *DAI Betwixt and Between: From Intelligent Agents to Open Systems Science*. IEEE Trans. on System, Man, and Cybernetics, Nov/Dec.

Holland, J. H. (1992). Genetic Algorithms. *Scientific American*, *267*, 66–72. doi:10.1038/scientificamerican0792-66

Huhns, M., & Singh, M. (Eds.). (1997). *Readings in Agents*. San Francisco: Kaufmann.

IBM. (2006), *Autonomous Computing White Paper: An Architectural Blueprint for Autonomous Computing*, 4th ed., June, 1-37.

Jennings, N. R. (2000). On Agent-Based Software Engineering. *Artificial Intelligence*, *17*(2), 277–296. doi:10.1016/S0004-3702(99)00107-1

Kephart, J., & Chess, D. (2003). The Vision of Autonomic Computing, IEEE. *Computer*, *26*(1), 41–50. doi:10.1109/MC.2003.1160055

Kleene, S.C. (1956), Representation of Events by Nerve Nets, in C.E. Shannon and J. McCarthy eds., *Automata Studies*, Princeton Univ. Press, 3-42.

Kurzweil, R. (1990). *The Age of Intelligent Machines*. Cambridge, MA: MIT Press.

Maes, P. (Ed.). (1991). *Designing Autonomous Agents: Theory and Practice from Biology to Engineering and Back*. London: The MIT press.

Matlin, M. W. (1998). *Cognition* (4th ed.). Orlando, FL: Harcourt Brace College Publishers.

McCarthy, J. (1963). *Situations, Actions, and Causal Laws, Memo 2*. Stanford, CA: Stanford University Artificial Intelligence Project.

McCarthy, J., Minsky, M. L., Rochester, N., & Shannon, C. E. (1955), *Proposal for the 1956 Dartmouth Summer Research Project on Artificial Intelligence*, Dartmouth College, Hanover, NH, USA, http://www.formal.stanford.edu/jmc/history/dartmouth/*dartmouth.html.*

McCulloch, W. S. (1965). *Embodiments of Mind*. Cambridge, MA: MIT Press.

McCulloch, W. S., & Pitts, W. (1943). A Logical Calculus of the Ideas Immanent in Nervous Activity. *The Bulletin of Mathematical Biophysics*, *5*, 115–137. doi:10.1007/BF02478259

Meystel, A. M., & Albus, J. S. (2002). *Intelligent Systems, Architecture, Design, and Control*. John Wiley & Sons.

Murch, R. (2004). *Autonomic Computing*. London: Person Education.

Newell, A. (1990). *Unified Theories of Cognition*. Cambridge, MA: Harvard University Press.

Newell, A., & Simon, H. A. (1972). *Human Problem Solving*. NJ: Prentice-Hall Englewood Cliffs.

Nilsson, N. J. (1998). *Artificial Intelligence: A New Synthesis*. San Mateo, CA: Morgan Kaufmann.

Poole, D., Mackworth, A., & Goebel, R. (1997). *Computational Intelligence: A Logical Approach. Oxford*. Oxford, UK: Oxford University Press.

Russell, S. J., & Norvig, P. (1995). *Artificial Intelligence: A Modern Approach*. Englewood Cliffs, NJ: Prentice-Hall.

Shannon, C. E. (Ed.). (1956). *Automata Studies*. Princeton: Princeton University Press.

Sternberg, R. J. (1997). The Concept of Intelligence and the its Role in Lifelong Learning and Success. *The American Psychologist*, *52*(10), 1030–1037. doi:10.1037/0003-066X.52.10.1030

Turing, A. M. (1950). Computing Machinery and Intelligence. *Mind*, *59*, 433–460. doi:10.1093/mind/LIX.236.433

von Neumann, J. (1946), The Principles of Large-Scale Computing Machines, reprinted in *Annals of History of Computers*, 3(3), 263-273.

von Neumann, J. (1958). *The Computer and the Brain, Yale Univ*. New Haven: Press.

von Neumann, J. (1963), *General and Logical Theory of Automata,* A.H. Taub ed., Collected Works, Vol. 5, Pergamon, 288-328.

von Neumann, J., & Burks, A. W. (1966). Theory of Self-Reproducing Automata, Univ. of Illinois Press. *Urbana (Caracas, Venezuela)*, IL.

Wald, A. (1950). *Statistical Decision Functions*. John Wiley & Sons.

Wang, Y. (2002a), Keynote: On Cognitive Informatics, *Proc. 1st IEEE International Conference on Cognitive Informatics* (ICCI'02), Calgary, Canada, IEEE CS Press, August, 34-42.

Wang, Y. (2002b), The Real-Time Process Algebra (RTPA), *Annals of Software Engineering: An International Journal*, 14, USA, 235-274.

Wang, Y. (2003a), Cognitive Informatics: A New Transdisciplinary Research Field, *Brain and Mind: A Transdisciplinary Journal of Neuroscience and Neurophilosophy,* 4(2), 115-127.

Wang, Y. (2003b), Keynote: Cognitive Informatics Models of Software Agent Systems, Proc. 1st International Conference on Agent-Based Technologies and Systems (ATS'03), Univ. of Calgary Press, Calgary, Canada, August, 25.

Wang, Y. (2004), Keynote: On Autonomic Computing and Cognitive Processes, *Proc. 3rd IEEE International Conference on Cognitive Informatics* (ICCI'04), Victoria, Canada, IEEE CS Press, August, 3-4.

Wang, Y. (2006), Keynote: *Cognitive Informatics - Towards the Future Generation Computers that Think and Feel, Proc. 5th IEEE International Conference on Cognitive Informatics* (ICCI'06), Beijing, China, IEEE CS Press, July, 3-7.

Wang, Y. (2007a). *Software Engineering Foundations: A Software Science Perspective (Vol. II)*. Aurebach Publications, NY, USA: CRC Book Series in Software Engineering.

Wang, Y. (2007b), Keynote: Cognitive Informatics Foundations of Nature and Machine Intelligence, *Proc. 6th International Conference on Cognitive Informatics* (ICCI'07), IEEE CS Press, Lake Tahoe, CA., Aug., 3-12.

Wang, Y. (2007c). The Theoretical Framework of Cognitive Informatics, *International Journal of Cognitive Informatics and Natural Intelligence. IGI, USA*, *1*(1), 1–27.

Wang, Y. (2007d), Exploring Machine Cognition Mechanisms for Autonomic Computing, *International Journal on Cognitive Informatics and Natural Intelligence*, March, 1(2), i - v.

Wang, Y. (2007e), The Cognitive Processes of Formal Inferences, *International Journal of Cognitive Informatics and Natural Intelligence*, IGI, USA, Dec., 1(4), 75-86.

Wang, Y. (2008a), Keynote: On Abstract Intelligence and Its Denotational Mathematics Foundations, *Proc. 7th IEEE International Conference on Cognitive Informatics* (ICCI'08), Stanford University, CA., USA, IEEE CS Press, August, 5-15.

Wang, Y. (2008b), Toward a Generic Mathematical Model of Abstract Game Theories, *Transactions of Computational Science, 2,* Springer, June, 205-223.

Wang, Y. (2008c), On Contemporary Denotational Mathematics for Computational Intelligence, *Transactions of Computational Science, 2,* Springer, June, 6-29.

Wang, Y. (2008d), On Concept Algebra: A Denotational Mathematical Structure for Knowledge and Software Modeling, *International Journal of Cognitive Informatics and Natural Intelligence,* IGI, USA, April, 2(2), 1-19.

Wang, Y. (2008e), On System Algebra: A Denotational Mathematical Structure for Abstract System modeling, *International Journal of Cognitive Informatics and Natural Intelligence,* IGI, USA, April, 2(2), 20-42.

Wang, Y. (2008f), RTPA: A Denotational Mathematics for Manipulating Intelligent and Computational Behaviors, *International Journal of Cognitive Informatics and Natural Intelligence,* IGI, USA, April, 2(2), 44-62.

Wang, Y. (2009). On Abstract Intelligence: Toward a Unified Theory of Natural, Artificial, Machinable, and Computational Intelligence, *International Journal of Software Science and Computational Intelligence,* IGI, USA, Jan., 1(1), 1-18.

Wang, Y. and G. Ruhe (2007), The Cognitive Process of Decision Making, *International Journal of Cognitive Informatics and Natural Intelligence,* IGI, USA, March, 1(2), 73-85.

Wang, Y., Y. Wang, S. Patel, and D. Patel (2006), A Layered Reference Model of the Brain (LRMB), *IEEE Trans. on Systems, Man, and Cybernetics (C),* March, 36(2), 124-133.

Wellman, M. P. (1999), Multiagent Systems, in R.A. Wilson and C.K. Frank eds., *The MIT Encyclopedia of the Cognitive Sciences*, MIT Press, MA.

Widrow, B., & Lehr, M. A. (1990), 30 Years of Adaptive Neural Networks: Perception, Madeline, and Backpropagation, *Proc. of the IEEE*, Sept., 78(9), 1415-1442.

Wittig, T. (Ed.). (1992). *ARCHON: An Architecture for Multi-Agent Systems*. London: Ellis Horwood.

Wooldridge, M. (2002). *An Introduction to Multiagent Systems*. John Wiley & Sons.

Wooldridge, M., & Jennings, N. (1995). Intelligent Agents: Theory and Practice. *The Knowledge Engineering Review*, *10*(2), 115–152. doi:10.1017/S0269888900008122

This work was previously published in International Journal of Cognitive Informatics and Natural Intelligence, Volume 3, Issue 1, edited by Yingxu Wang, pp. 1-16, copyright 2009 by IGI Publishing (an imprint of IGI Global).

Chapter 2
Autonomic Agent Systems: Categorical Models and Behaviors

Phan Cong Vinh
FPT University, Vietnam

ABSTRACT

A new computing paradigm is currently on the spot: interaction based on series of actions. Most of autonomic agent systems (AASs) exploit this type of interaction to self-adjust their autonomous behaviors as a fundamental operational paradigm. At an interaction interface, actions evolve over time, hence series of actions occurs as a royal candidate for modeling, specifying, programming, and verifying AASs. For considering AASs, series of actions and adaptation relations; our formal approach consists, in particular, of categorical models and behaviors such that, firstly , AASs, series of actions and adaptation relations will categorically be modeled to provide algebraic frameworks for development of reasoning on their behaviors and, secondly, categorical behaviors of AASs, series of action and adaptation relations will be investigated and developed taking advantage of their categorical models.

INTRODUCTION

For *autonomic agent systems* (AASs), autonomic computing is a generic property delineating capability to self-adjust their goal-driven computational behaviors without direct human interventions. Autonomic computing has been described as the set of concepts, technologies, and tools that enable AASs to become more self-managing. This potentiality is often related to possessing learning capabilities through analysis of past behaviors and interactions (Vinh, 2007, 2009a, 2009b, 2009c, 2009d, 2010; Vinh & Bowen, 2007, 2008). Autonomic computing has intensely been studied by various areas of engineering including agent systems, computational intelligent systems and human orientated systems (Vinh, 2010; Denko, Yang & Zhang, 2009; Jin & Liu, 2004; Pacheco,

DOI: 10.4018/978-1-60960-553-7.ch002

2004; Witkowski & Stathis, 2004; Parashar & Hariri, 2006; Wang, 2007b; Ko, Gupta, & Jo, 2007; Yang & Liu, 2007; Butera, 2007; Calisti, Meer, & Strassner, 2008). With regard to AASs, autonomic computing (Wang, 2007b) and cognitive informatics (Wang, 2007a) have been set as two major pillars to support such systems. By the latest developments of autonomic computing (Vinh, 2009b; Denko, Yang & Zhang, 2009; Parashar & Hariri, 2006; Calisti, Meer, & Strassner, 2008) and cognitive informatics (Wang & Kinsner, 2006), AASs are now at a crucial point in their evolution, marked by research activities being booming (Vinh, 2009d; Topaloglu & Bayrak, 2008; S.A. DeLoach & Matson, 2008; F.E. Walter & Schweitzer, 2008; K. Zoethout & Molleman, 2008). AASs pose new challenges for the development and application of autonomic computing techniques, due to their special characteristics including: *nondeterminism, context-awareness and goal- and inference-driven adaptability* (Wang, 2007b).

AASs are agent systems, which implement autonomic computing mechanisms such as: nondeterminism, context-awareness and goal- and inference-driven adaptability. For autonomic computing techniques applicable to AASs, a new computing paradigm is currently on the spot: interaction based on series of actions. Most of AASs exploit this type of interaction to self-adjust their autonomous behaviors as a fundamental operational paradigm. At an interaction interface, actions evolve over time, hence series of actions occurs as a royal candidate for modeling, specifying, programming, and verifying AASs.

In this chapter, we focus on modeling AASs, series of action and adaptation relations, and then developing reasoning on their behaviors. Our formal approach consists mainly of *categorical models and behaviors* such that,

- *Firstly*, algebraic frameworks of AASs, series of action and adaptation relations will be constructed for development of reasoning on their behaviors using categorical language and,
- *Secondly*, categorical behaviors of AASs, series of action and adaptation relations will be considered taking advantage of their categorical models where the behavior-oriented notions will be formed for our approach.

OUTLINE

The chapter is a reference material for readers who already have a basic understanding of AASs and are now ready to know the novel approach for formalizing self-* in AASs using categorical language. Formalization is presented in a straightforward fashion by discussing in detail the necessary components and briefly touching on the more advanced components. Several notes explaining how to use the formal aspects, including justifications needed in order to achieve the particular results, are presented.

We attempt to make the presentation as self-contained as possible, although familiarity with the notion of self-* in AASs is assumed. Acquaintance with the algebra and the associated notion of categorical language is useful for recognizing the results, but is almost everywhere not strictly necessary.

The rest of chapter is organized as follows: Section of Autonomic Computing as Self-* presents the notion of autonomic computing (AC). In section of Preliminaries, we recall some concepts from the category theory used in the chapter. Section of Categorical Model and behavior of AASs presents categorical approach to model and behavior of AASs. In section of Categorical Model and Behavior of Series of Actions in AASs, model and behavior of actions are developed by categorical approach. Quantitative behavior of series of actions is presented in next section. Section of Categorical Model and Behavior of Series of Adaptation Relations is a place to abstract model and develop behavior of series of adaptation relations. Some discussions and further work are considered in section of Notes and Remarks. Finally, a short summary is given in section of Conclusions.

AUTONOMIC COMPUTING AS SELF-*

Autonomic computing (AC) imitates and simulates the natural intelligence possessed by the human autonomic nervous system using generic computers. This indicates that the nature of software in AC is the simulation and embodiment of human behaviors, and the extension of human capability, reachability, persistency, memory, and information processing speed (Wang, 2007a). AC was first proposed by IBM in 2001 where it is defined as

"Autonomic computing is an approach to self-managed computing systems with a minimum of human interference. The term derives from the body's autonomic nervous system, which controls key functions without conscious awareness or involvement" (IBM, 2001).

AC is generally described as self-*. Formally, let self-* be the set of self-_'s. Each self-_ to be an element in self-* is called a *self-* facet*. That is,

$$\text{self-}* = \{\text{self-}\_ \mid \text{self-}\_ \text{ is a self-}* \text{ facet}\} \tag{1}$$

We see that self-CHOP is composed of four self-* facets of self-configuration, self-healing, self-optimization and self-protection. Hence, self-CHOP is a subset of self-*. That is, self-CHOP = {self-configuration, self-healing, self-optimization, self-protection} ⊂ self-*. Every self-* facet must satisfy some certain criteria, so-called *self-* properties*. In (Wolf & Holvoet, 2006), T.D. Wolf and T. Holvoet classified the self-* properties in autonomic networks.

In its AC manifesto, IBM proposed eight facets setting forth an AC system (ACS) known as *self-awareness, self-configuration, self-optimization, self-maintenance, self-protection (security and integrity), self-adaptation, self-resource-allocation* and *open-standard-based* (IBM, 2001). Kinsner pointed out that these facets indicate that IBM perceives AC is a mimicry of human nervous systems (Kinsner, 2007). In other words, self-awareness (consciousness) and non-imperative (goal-driven) behaviors are the main features of ACSs (Wang, 2007a).

PRELIMINARIES

In this section, we recall some concepts from the category theory (Asperti & Longo, 1991 ; Bergman, 1998 ; Adamek, Herrlich, & Strecker, 2009 ; Levine, 1998 ; Lawvere & Schanuel, 1997) used in this chapter.

What is a Category?

Category as a Graph

A category C can be viewed as a graph *(Obj(**C**), Arc(**C**),s,t)*, where

- *Obj(**C**)* is the set of nodes we call *objects*,
- *Arc(**C**)* is the set of edges we call *morphisms* and
- *Arc(**C**)→Obj (**C**):* are two maps called *source* (or *domain*) and *target* (or *codomain*), respectively.

We write $f{:}x \rightarrow y$ when f is in *Arc(C)*, $s(f)=X$, and $t(f)=y$.

Explanation on Terminology

An object in the category is an algebraic structure such as a set. We are probably familiar with some notations for finite sets: *{Student A, Atudent B, Student C}* is a name for the set whose three elements are *Student A, Student B and Student C*. Note that the order in which the elements are listed is irrelevant.

A morphism *f* in the category consists of three things: a set *x*, called the source of the morphism; a set *y*, called the target of the morphism and a rule assigning to each element *x* in the source an element *y* in the target. This *y* is denoted by *f(x)*, read "*f* of *x*". Note that the morphism is also called the *map*, *function*, *transformation*, *operator* or *arrow*. For example, let *x*=:*{Student A, Atudent B, Student C}*, *y*= *{Math, Physics, Chemistry, History}* and let *f* assign each student his or her favorite subject. The following internal diagram is an illustration.

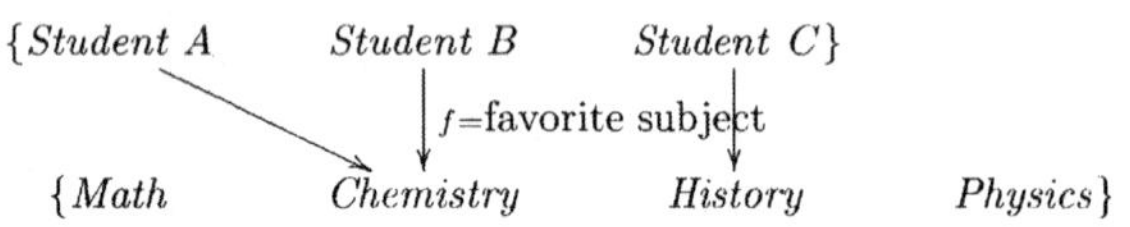

(2)

This states that the favorite subject of the *Student C* is *History*, written by *f(Student C)=History*, while *Student A* and *Student B* prefer *Chemistry*. There are some important properties of any morphism

- From each element in the source:*{Student A, Atudent B, Student C}* there is exactly one arrow leaving.
- To an element in the target *{Math, Physics, Chemistry, History}* there may be zero, one or more arrows arriving.

It is possible that the source and target of the morphism could be the same set. The following internal diagram is an example.

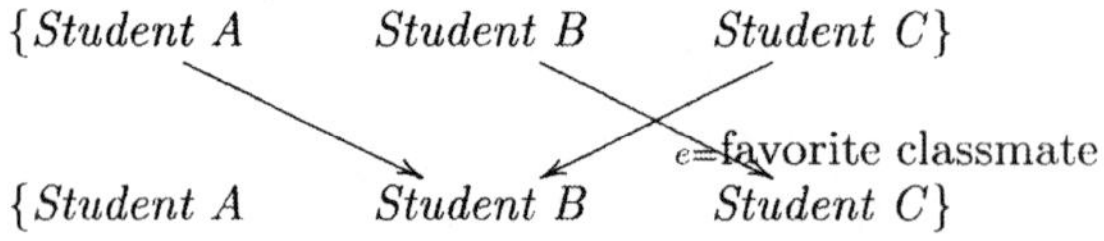

(3)

and, in the case, the morphism is called an *endomorphism* whose representation is available as in

(4)

Identity Morphism and Composition of Morphisms

Associated with each object *X* in *Obj(**C**)*, there is a morphism $1_x = x \rightarrow x$, called the *identity* morphism on *X*, and to each pair of morphisms *f*:*x*→*y* and *g*,y→Z there is an associated morphism *f*;*g*:*x*→Z, called the *composition* of *f* with *g*. The representations in (5) include the external diagrams of identity morphism and composition of morphisms.

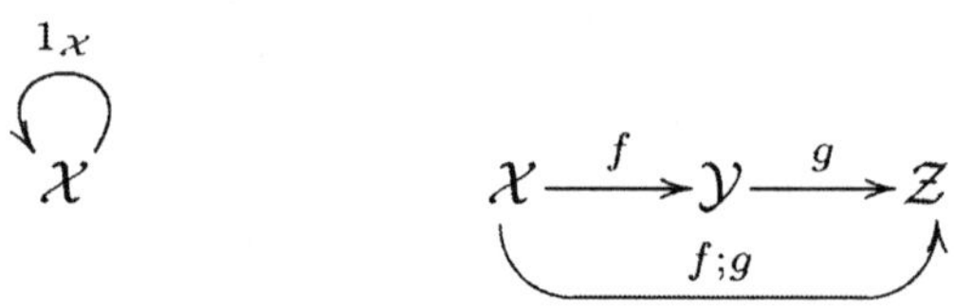

(5)

Explanation on Terminology

Here are the corresponding internal diagrams of the identity morphism.

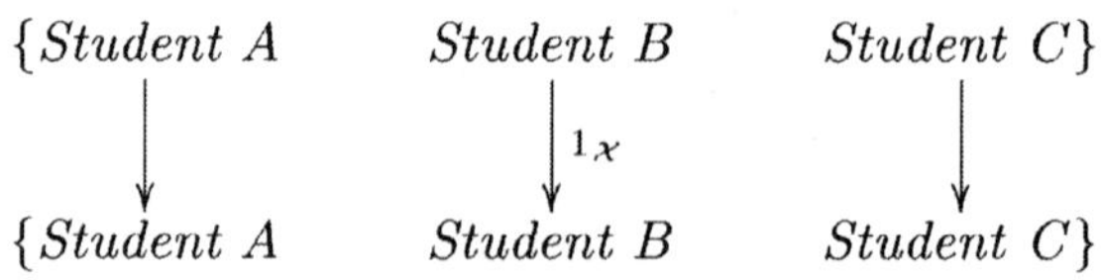

(6)

or

(7)

and here, the composition of morphisms is described in the internal diagram

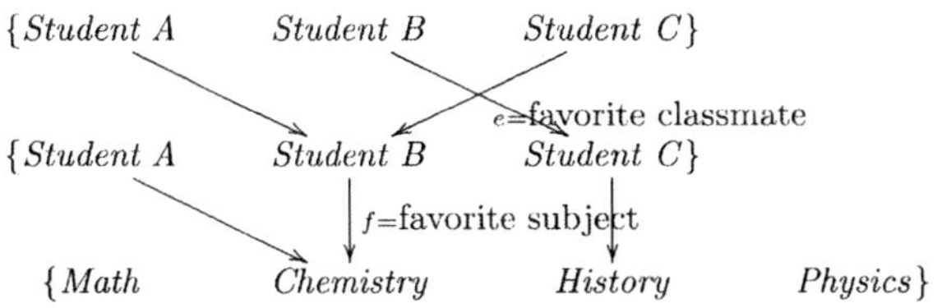

(8)

or, in the external diagram $x \xrightarrow{e} x \xrightarrow{f} y$. From diagram (8), we can obtain answers for the question "What should each student support to his or her favorite classmate for subject?" In fact, the answers are such as "*Student A* likes *Student B*, *Student B* likes *Chemistry*, so *Student A* should support *Chemistry*", "*Student B* likes *Student C*, *Student C* likes *History*, so *Student B* should support *History*" and *Student C* likes *Student B*, *Student B* likes *Chemistry*, so *Student C* should support *Chemistry*."

The composition of two morphisms *e* and *f* means that *e* and *f* are combined to obtain a third morphism $x \xrightarrow{e;f} y$. This is represented in the following internal diagram.

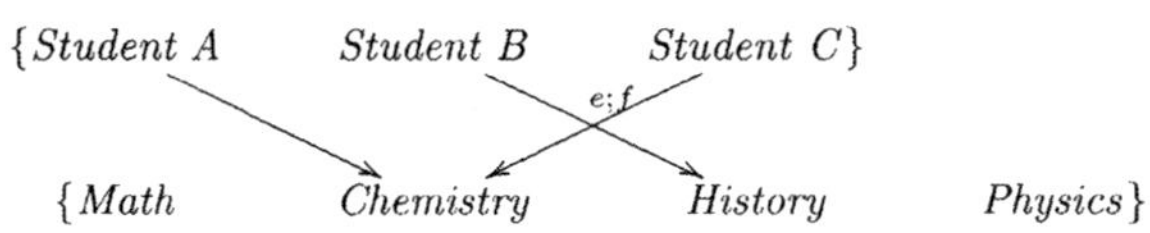

(9)

where, for example, *e;f(Student B)-History* is read as "the favorite subject of the favorite classmate of *Student B* is *History*."

Identity and Associativity for Composition of Morphisms

The following equation must hold for all objects *x,y* in *Obj(**C**)* and morphism *f:x→y* in *Arc(**C**)*:

$$Identity: \quad 1_x; f = f = f; 1_y$$

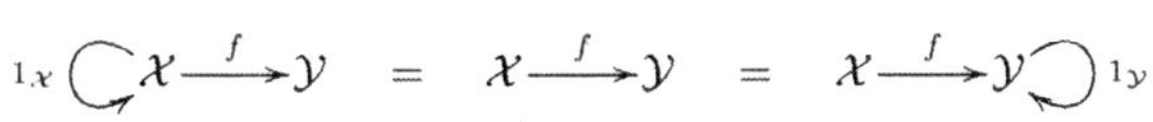

(10)

The following equation must hold for all objects *x,y*, and *Z* in *Obj(**C**)* and morphisms *f:x→y,g:y→Z*, and *h:Z→T* in *Arc(**C**)*:

$$Associativity: \quad (f;g);h = f;(g;h)$$

(11)

Functor

Functor is a special type of mapping between categories. Functor from a category to itself is called an *endofunctor*. Note that, in this chapter, when the notion of endofunctor dominates throughout in use, then we can name them as the functor, for short, without any confusion. The functors are also viewed as morphisms in a category, whose objects are smaller categories.

A *multifunctor* is a generalization of the functor concept to *n* arguments. Specially, a *bifunctor* is a multifunctor with $n = 2$.

There are two kinds of functors distinguished by the way they treat morphisms to be *covariant* and *contravariant*. A functor $\top$ is covariant if for each source morphism $x \xrightarrow{f} y$ the target morphism has the form $\top X \xrightarrow{\top f} \top y$. A functor $\top$ is contravariant if for each source morphism $x \xrightarrow{f} y$ the target morphism has the form $\top X \xleftarrow{\top f} \top y$.

Homomorphism

Let $\top$ be a functor with algebraic objects such as algebras *a:* $\top$*x→x* and *b:* $\top$*y→y*. A *homomorphism of algebras* (also called a *map of algebras*) from *(x,a)* to *(Y,b)* is a function *f:x→y* between the carrier sets *x* and *y* such that the equation $a;f = \top f;b$ holds. That is, the following diagram commutes$\top$

$$\begin{array}{ccc} \mathsf{T}\,\mathcal{X} & \xrightarrow{a} & \mathcal{X} \\ {\scriptstyle \mathsf{T}f}\downarrow & & \downarrow{\scriptstyle f} \\ \mathsf{T}\,\mathcal{Y} & \xrightarrow{b} & \mathcal{Y} \end{array} \tag{12}$$

Isomorphism

A morphism *f:x→y* in the category **C** is an *isomorphism* if there exists a morphism *g:y→x* in that category such that *f:g-1x* and *g;f=1y.*

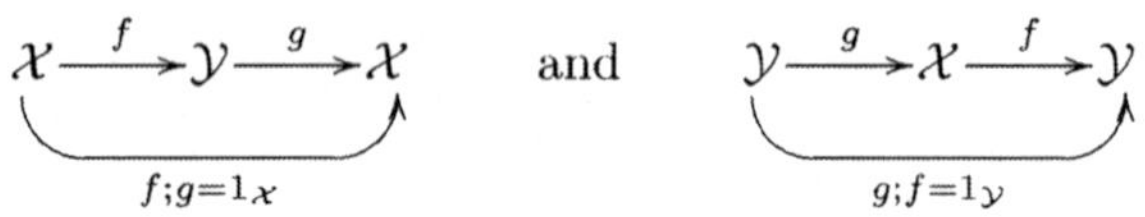

(13)

That is, if the following diagram commutes.

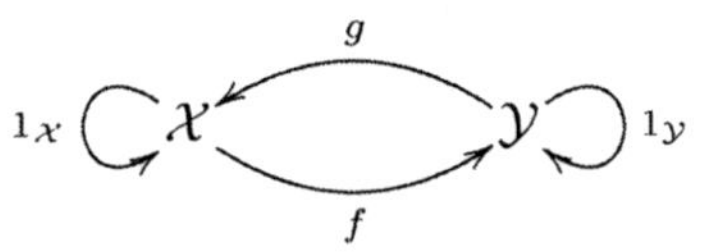

(14)

Natural Isomorphism

Let **C** and **C′** be two categories. Consider a parallel pair

$$\mathbf{C} \underset{\mathsf{T}'}{\overset{\mathsf{T}}{\rightrightarrows}} \mathbf{C}' \tag{15}$$

of functors of the same variance. Two functors of T and T′ are *naturally equivalent* (also called *naturally isomorphic*) if there is an inverse pair of natural isomorphisms between them. In other words, the inverse pair of natural isomorphisms between T and T′ is a pair

$$\mathsf{T} \underset{\zeta_{-}}{\overset{\eta_{-}}{\rightleftarrows}} \mathsf{T}' \tag{16}$$

such that for each object *X* in **C** the morphisms

$$\mathsf{T}\,\mathcal{X} \underset{\zeta_{\mathcal{X}}}{\overset{\eta_{\mathcal{X}}}{\rightleftarrows}} \mathsf{T}'\,\mathcal{X} \tag{17}$$

are an inverse pair of isomorphisms in **C′**.

T-Algebra

Let **Cat** be a category, *A* an object in *Obj*(**Cat**), T:Cat→Cat an endofunctor and *f* a morphism $\mathsf{T}(\mathcal{A}) \xrightarrow{f} \mathcal{A}$; then T-*algebra* is a pair $\langle \mathcal{A}, f \rangle$.

Obj(**Cat**), is called a *carrier* of the algebra and T a *signature* of the algebra.

T-*homomorphism* between two T-algebras $\langle \mathcal{A}, f \rangle$ and $\langle \mathcal{B}, g \rangle$ is a morphism $\mathcal{A} \xrightarrow{\psi} \mathcal{B}$ such that

$$f;\psi = \mathsf{T}(\psi);g \tag{18}$$

It is equivalent to saying that the following diagram commutes

$$\begin{array}{ccc} \mathsf{T}(\mathcal{A}) & \xrightarrow{f} & \mathcal{A} \\ {\scriptstyle \mathsf{T}(\psi)}\downarrow & & \downarrow{\scriptstyle \psi} \\ \mathsf{T}(\mathcal{B}) & \xrightarrow{g} & \mathcal{B} \end{array} \tag{19}$$

Categorical Model and Behavior of Autonomic Agent Systems

Autonomic agent system (AAS) is thought of as a typical autonomic computing system to be "*an intelligent system that implements nondeterministic, context-dependent, and adaptive behaviors based on goal- and inference-driven mechanisms*" (Wang, 2007b). Hence, for abstracting behaviors of AASs, we start with considering *context-dependent adaptive agent systems* (CAASs) where we first examine *deterministic context-dependent adaptive agent systems* (DCAASs) and then extend to *nondeterministic context-dependent adaptive agent systems* (NCAASs) by categorical approach in this section.

Deterministic Context-Dependent Adaptive Agent Systems

DCAASs we want to abstract are intuitionally multiple partial morphism applications, such as

$$s_0 \xrightarrow{\sigma_0} s_1 \xrightarrow{\sigma_1} s_2 \xrightarrow{\sigma_2} s_3 \cdots \tag{20}$$

where

- All indexes are in the set $T(=\mathbb{N})$ of times,
- For all i in T, s_i are configurations of a CAAS in the set, denoted by *Sys*, of configurations and
- For all i in $T,^{\sigma_i}$ are actions/transformations in the set, denoted by *Action*, of actions/transformations which make adaptation of the configuration s_i to become the configuration s_i+1.

The meaning of (20) is understood as

$$\ldots\ s_2(s_1(s_0())) = \ldots\ s_2(s_1(\sigma_0)) = \ldots\ s_2(\sigma_1) = \ldots\ \sigma_2 \tag{21}$$

The adaptation process in (20) can also be descriptively drawn as

$$s_0()\ \sigma_0\ \sigma_1\ \sigma_2\ \cdots \longmapsto s_1(\sigma_0)\ \sigma_1\ \sigma_2\ \cdots \longmapsto \sigma_0\ s_2(\sigma_1)\ \sigma_2\ \cdots \tag{22}$$

or, in another representation

$$\xrightarrow{s_0} \sigma_0\ \sigma_1\ \sigma_2\ \cdots \longmapsto \sigma_0 \xrightarrow{s_1} \sigma_1\ \sigma_2\ \cdots \longmapsto \sigma_0\ \sigma_1 \xrightarrow{s_2} \sigma_2\ \cdots \tag{23}$$

Note that in (22) and (23), we want to represent the above-mentioned adaptation process of DCAAS based on series of actions where each step of the process is an application of unary partial morphism $1 \xrightarrow{s_i} Sys$ on $1 \xrightarrow{\sigma_{i-1}} Action$, for all i in T.

The adaptation process, in (22) and (23), describes the computation of the DCAAS including the adaptation steps to change CAAS *configurations*. We can specify a CAAS configuration at an adaptation step to be a member of set $Sys \rightarrow Action^n$ with $n \geqslant 0 \geqslant$, where $Action^n$ is defined by

$$Action^n = \underbrace{Action \longrightarrow Action \longrightarrow \ldots \longrightarrow Action}_{n \text{ times}} \quad \text{for } n \geqslant 0 \tag{24}$$

All the categorical models we want to present are based on mapping a CAAS configuration in the above-mentioned specification to another. For example, a specific DCAAS can be specified by the following morphism:

$$Adap : (Sys \longrightarrow Action) \longrightarrow Sys \tag{25}$$

(i.e., *Adap:*$(Sys \rightarrow Action^1) \rightarrow (Sys \rightarrow Action^0)$ or denoted by *Adap(Sys→Action, Sys))*another specific DCAAS can be specified by

$$Adap : (Sys \longrightarrow Action) \longrightarrow (Sys \longrightarrow Action) \tag{26}$$

(i.e., this specification of *Adap* is explicitly written as *Adap:*$(Sys \rightarrow Action^1) \rightarrow (Sys \rightarrow Action^1)$ or denoted by *Adap:(Sys→Action, Sys→Action)*) again, we can also specify another specific DCAAS as

$$Adap : (Sys \longrightarrow Action^n) \longrightarrow (Sys \longrightarrow Action) \tag{27}$$

(i.e., this specification of *Adap* is explicitly written as *Adap:*$(Sys \rightarrow Action^n) \rightarrow (Sys \rightarrow Action^1)$ or denoted by *Adap:*$(Sys \rightarrow Action^n, Sys \rightarrow Action)$) and we can, in the completely same way, do for any other specific DCAAS.

The above-mentioned morphisms, namely *Adap*, are called *adaptation relations*.

Generally, a DCAAS based on series of actions is a tuple $\langle Sys, Action, Out, Adap \rangle$ consisting of the following components

- *Sys* is the set of CAAS configurations,
- *Action* is the set of actions/transformations on CAAS configurations,
- *Out* is an output morphism defined by

$$Out : Sys \longrightarrow \mathbb{R} \tag{28}$$

and

- *Adap* is an adaptation morphism defined by, for all $n,m \geqslant 0$,

$$Adap : (Sys \longrightarrow Action^n) \longrightarrow (Sys \longrightarrow Action^m) \tag{29}$$

Sometimes, for the adaptation morphism in (29), if $n = 1$ and $m = 0$ then the notations $s \xrightarrow{\sigma} s'$ and $S \overset{y}{\rightsquigarrow}$ are used to denote $Adap(s \rightarrow \sigma) = s'$ and $Out(s) = y$, respectively.

Note that both *Sys* and *Action* may be infinite. If both *Sys* and *Action* are finite, then we have a finite DCAAS, otherwise we have an infinite DCAAS.

Nondeterministic Context-Dependent Adaptive Agent Systems

NCAASs we want to model are intuitionally multiple partial morphism applications, such as

$$s_0 \xrightarrow{\sigma_0|x_0} s_1 \xrightarrow{\sigma_1|x_1} s_2 \xrightarrow{\sigma_2|x_2} s_3 \cdots \tag{30}$$

where

- All indexes i in T, s_i and σi are similar in meaning to the ones mentioned in (20)
- For all i in T, x_i are real numbers that can be thought of as the *multiplicity* (or *weight*) with which the adaptation from s_i to s_{i+1} occurs.

The first steps of the adaptation process in (30) can also be descriptively drawn as

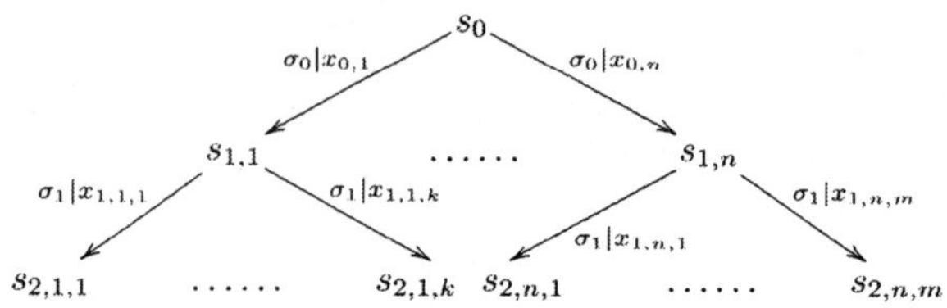

(31)

where

For the first step,

$$1 \xrightarrow{s1} (\{s_{1,1,\ldots,} s_{1,n}\} \xrightarrow{\subset} Sys)$$

and

$$1 \xrightarrow{x_0} (\{x_{0,1,\ldots,} s_{0,n}\} \xrightarrow{\subset} \mathbb{R})$$

For the second step,

$$1 \xrightarrow{s_2} (\{s_{2,1,1,\ldots,} s_{2,1,k}\} \cup \ldots \cup \{s_{2,n,1,\ldots,} s_{2,n,m}\} \xrightarrow{\subset} Sys)$$

and

$$1 \xrightarrow{x_1} (\{x_{1,1,1,\ldots,} x_{1,1,k}\} \cup \ldots \cup \{x_{1,n,1,\ldots,} x_{1,n,m\}} \xrightarrow{\subset} \mathbb{R})$$

and the meaning of (30) is viewed as the following morphism.

$$Adap : (Sys \longrightarrow Action) \longrightarrow (Sys \longrightarrow \mathbb{R}) \tag{32}$$

The adaptation morphism *Adap* is nondeterministic. This can be explained as follows: *Adap* assigns to each CAAS configuration in *Sys*→*Action* a morphism *Sys*→*R* that can be seen as a kind of *nondeterministic* CAAS *configuration* (or so-called *distributed* CAAS *configuration*) and specifies for any CAAS configuration in *Sys* a multiplicity (or weight) *Adap(Sys→Action)(Sys)* in R.

This nondeterminism of NCAASs makes change in all the categorical models mentioned in subsection of Deterministic Context-dependent Adaptive Agent Systems. For example, a specific NCAAS can be specified by the following morphism:

$$Adap : (Sys \longrightarrow Action) \longrightarrow (Sys \longrightarrow \mathbb{R}) \tag{33}$$

(i.e., an explicit specification of *Adap* is *Adap: (Sys→Action*^{1}*)→((Sys →Action*^{0}*)*→R) or denoted by *Adap((Sys→Action), (Sys*→R)))another specific NCAAS can be specified by

$$Adap : (Sys \longrightarrow Action) \longrightarrow ((Sys \longrightarrow Action) \longrightarrow \mathbb{R}) \tag{34}$$

(i.e., an explicit specification of *Adap* is *Adap: (Sys→Action*^{1}*)→((Sys →Action*^{1})→R) or denoted by *Adap((Sys→Action), ((Sys→Action)* →R)))again, we can also specify another specific NCAAS as

$$Adap : (Sys \longrightarrow Action^n) \longrightarrow ((Sys \longrightarrow Action) \longrightarrow \mathbb{R}) \tag{35}$$

(i.e., an explicit specification of *Adap* is *Adap: (Sys→Action*^{n}*)→((Sys →Action*^{1})→R) or denoted by *Adap((Sys→Action*^{n}*), ((Sys→Action*)→R)))and we can, in the completely same way, do for any other specific NCAAS.

Generally, an NCAAS based on series of actions is a tuple $\langle Sys, Action, Out, Adap \rangle$ consisting of the following components

- *Sys* is the set of CAAS configurations,
- *Action* is the set of actions/transformations on CAAS configurations,
- *Out* is an output morphism defined by

$$Out : Sys \longrightarrow \mathbb{R} \tag{36}$$

and

- *Adap* is an adaptation morphism defined by, for all *n,m*≥ 0,

$$Adap : (Sys \longrightarrow Action^n) \longrightarrow ((Sys \longrightarrow Action^m) \longrightarrow \mathbb{R}) \tag{37}$$

Sometimes, for the adaptation morphism in (37) if $n = 1$ and $m = 0$ then the notations $s \xrightarrow{\sigma|x} s'$ and 978-1-60960-553-7.ch002.f60.tif are used to denote *Adap(s→σ)(s´)*=x´ and O*ut(s)* = *y,* respectively.

As a result, we get a significant relationship between DCAASs and NCAASs as presented by the following theorem.

Theorem 1 (Relationship between DCAASs and NCAASs) DCAASs *are just of specific* NCAASs.*In other words, using categorical language,* DCAASs $\xrightarrow{\subset}$ NCAASs.

Proof: In fact, by the adaptation morphism in (37) of NCAASs, let *f* be the morphism *f:Sys*→R and the finite set

$$\mathbb{R}(Sys) = \{1 \xrightarrow{s} Sys \mid f(s) \neq 0\} \xrightarrow{\subseteq} Sys$$

Hence it follows that when

$$\exists!1 \xrightarrow{s} Sys : f(s) = 1$$

but $\forall s' \neq s : f(s') = 0$ (i.e., the set R*(Sys)* is a singleton set of CAAS configuration with weight of 1. Note that the notation ∃! is read as "exist only") then (37) becomes the adaptation morphism of DCAASs as in (29). In other words, in this context, NCAASs will become DCAASs.Q.E.D.

From now on, unless explicitly stated otherwise, we will just consider series of actions on NCAASs where all treatments can also be applied to series of actions on DCAASs because, as we know, DCAASs are just of specific NCAASs as justified in theorem 1.

CATEGORICAL MODEL AND BEHAVIOR OF SERIES OF ACTIONS

Adaptation process of NCAASs that is pictorially drawn in diagram (30) can be separated into two complementary parts as follows:

$$s_0 \xrightarrow{\sigma_0} s_1 \xrightarrow{\sigma_1} s_2 \xrightarrow{\sigma_2} s_3 \ldots \qquad (38)$$

and

$$s_0 \xrightarrow{x_0} s_1 \xrightarrow{x_1} s_2 \xrightarrow{x_2} s_3 \ldots \qquad (39)$$

On the one hand, diagram (38) emphasizes $1 \xrightarrow{\sigma_i} Action$, for all *i* in *T*, in the adaptation process. This allows us to abstract conveniently sequence of σi as series of actions in this section. On the other hand, diagram (39) gives rise to $1 \xrightarrow{x_i} \mathbb{R}$, for all *i* in *T*, as weights of the series of actions in the adaptation process and weight-based quantitative behaviors of the series of actions will be evaluated in section of Quantitative Behavior of Series of Actions.

A number of different notations are in use for denoting series of actions.

$$sa = (\sigma_0, \sigma_1, \sigma_2, \ldots) \qquad (40)$$

is a common notation which specifies a series of actions *sa* which is indexed by the natural numbers. We are also accustomed to

$$sa = (\sigma_i) \qquad \forall i \in T \qquad (41)$$

Informally, series of actions can be understood as a rope on which we hang up a sequence of actions/transformations for display. Hence it follows that

Definition 1. (Series of actions) *For morphisms* $1 \xrightarrow{i} T$ *and* $1 \xrightarrow{\sigma_i} Action$, *there exists a unique morphism* $T \xrightarrow{sa} Action$ such that the equation i;sa=σi holds. This is described by the following commutative diagram

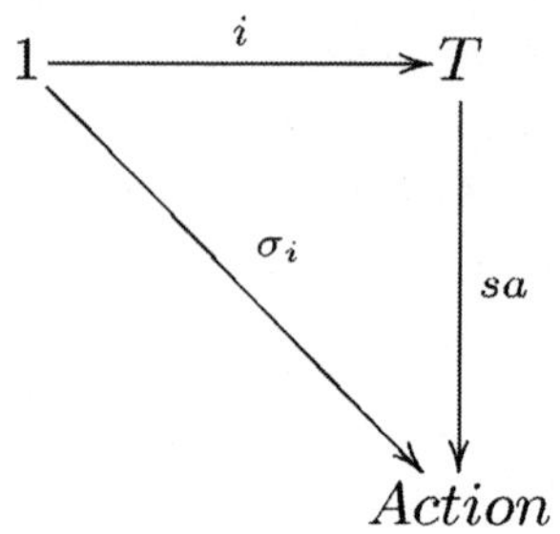

(42)

Morphism $T \xrightarrow{sa} Action$ defines a series of actions.

Note that morphism $T \xrightarrow{sa} Action$ is understood as

$$\forall i[i \in T \Rightarrow \exists! \sigma_i[\sigma_i \in Action \;\&\; sa(i) = \sigma_i]]$$

In other words, $T \xrightarrow{sa} Action$ generates series of actions as an infinite sequence of $sa(0) = \sigma_0, sa(1) = \sigma_1, \ldots, s(i) = \sigma_i, \ldots$ which is written as $(sa(0), sa(1), \ldots, sa(i), \ldots)$ or $(\sigma_0, \sigma_1, \ldots \sigma_i, \ldots)$

Definition 2. (Set of series of actions) *Given* $T \xrightarrow{sa} Action$ then the set of series of actions, denoted by Action$^\omega$, is defined by

$$Action^\omega = \{sa \mid T \xrightarrow{sa} Action\} \qquad (43)$$

The following result stems immediately from definitions 1 and 2

$$\frac{T \xrightarrow{sa} Action}{1 \xrightarrow{sa} Action^\omega} \qquad (44)$$

Rule (44) means that for each morphism $T \xrightarrow{sa} Action$, there is a morphism $1 \xrightarrow{sa} Action^\omega$ generating member in *Action*$^\omega$. That is morphism $T \xrightarrow{sa} Action$ generates series of actions and $1 \xrightarrow{sa} Action^\omega$ constructs the set of series of actions.

For series of actions, we define a mechanism to generate them. This mechanism consists of an object T equipping with structural morphisms $1 \xrightarrow{0} T \xrightarrow{succ} T$ with the property that for $Action$ and any $1 \xrightarrow{\sigma_0} Action$, $Action \xrightarrow{next\_up} Action$ there exists a unique morphism $T \xrightarrow{sa} Action$ such that the following diagram commutes

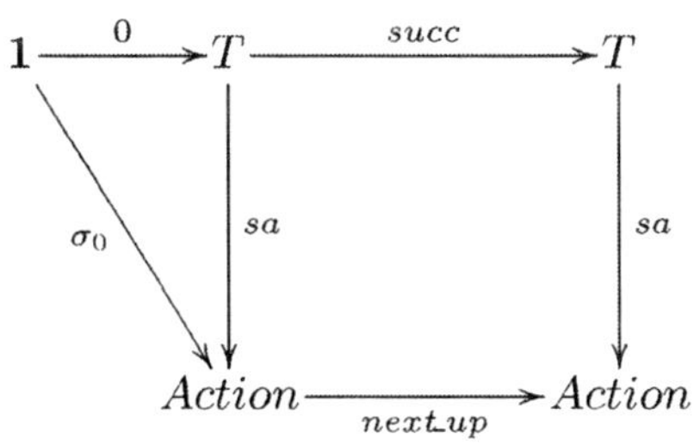

(45)

Definition 3. (Construction of series of actions) *We define a construction morphism of series of actions, denoted by* $\rhd\lhd$, such that

$$Action \times [T \xrightarrow{sa} Action] \xrightarrow{\rhd\lhd} [T \xrightarrow{sa} Action]$$

This means that

$$\rhd\lhd\ (A \times B \xrightarrow{f \times g} C \times D = A \rhd\lhd B \xrightarrow{f \rhd\lhd g} C \rhd\lhd D$$

It follows that any series of actions $T \xrightarrow{sa} Action$ can be represented in a format including two parts of *head* and *tail* that are connected by "$\rhd\lhd$" such that

$$T \xrightarrow{sa} Action \equiv \overbrace{1 \xrightarrow{0} T \xrightarrow{sa} Action}^{\sigma_0} \bowtie \overbrace{1 \xrightarrow{i>0} T \xrightarrow{sa} Action}^{\sigma_{i>0}}$$

where

$$\overbrace{1 \xrightarrow{0} T \xrightarrow{sa} Action}^{\sigma_0} = sa(0)$$

and

$$\overbrace{1 \xrightarrow{i>0} T \xrightarrow{sa} Action}^{\sigma_{i>0}} = (sa(1), sa(2), \ldots)$$

are called head and tail, respectively.

Definition 4. (Head of series of actions) *We define a head construction morphism, denoted by* $1 \overset{0}{\Rightarrow} (-)$, *such that*

$$1 \overset{0}{\Rightarrow} (-) : [T \xrightarrow{sa} Action] \to Action$$

This states that

$$\forall (a \rhd\lhd s)[(a \rhd\lhd s) \in [T \xrightarrow{sa} Action] \Rightarrow$$

$$\exists! \sigma_0 [\sigma_0 \in Action \ \&\ 1 \overset{0}{\Rightarrow} (a \rhd\lhd s) = a = \sigma_0]]$$

It follows that

$$1 \overset{0}{\Rightarrow} (T \xrightarrow{sa} Action) = 1 \xrightarrow{0} T \xrightarrow{sa} Action$$

Definition 5. (Tail of series of actions) *We define a tail construction morphism, denoted by* $(-)'$, *such that*

$$(-)' : [T \xrightarrow{sa} Action] \to [T \xrightarrow{sa} Action]$$

This means that

$$\forall (a \rhd\lhd s)[(a \rhd\lhd s) \in [T \xrightarrow{sa} Action \Rightarrow$$

$$\exists! \langle \sigma_1, \sigma_2, \ldots \rangle [\langle \sigma_1, \sigma_2, \ldots \rangle \in$$

$$[T \xrightarrow{sa} Action] \ \&\ (a \rhd\lhd s)' = s = \langle \sigma_1, \sigma_2, \ldots \rangle]]$$

As a convention, $(-)^{\langle n \rangle}$ denotes applying recursively the $(-)'\, n$ times. Thus, specifically, $(-)^{\langle 2 \rangle}, (-)^{\langle 1 \rangle} and (-)^{\langle 0 \rangle}$ stand for $((-)')', (-)'$ and $(-)$, respectively.

It follows that the first member of series of actions $T \xrightarrow{sa} Action$ is given by

$$1 \overset{0}{\Rightarrow} ((T \xrightarrow{sa} Action)') \equiv 1 \xrightarrow{1} T \xrightarrow{sa} Action$$

and, in general, for every $k \in T$ the *k*-th member of series of actions $T \xrightarrow{sa} Action$ is provided by

$$1 \overset{0}{\Longrightarrow} ((T \xrightarrow{sa} Action)^{\langle k \rangle}) \equiv 1 \xrightarrow{k} T \xrightarrow{sa} Action \quad (46)$$

Series of actions to be an infinite sequence of all σ_i is viewed and treated as single mathematical entity, so the derivative of series of actions

$$T \xrightarrow{sa} Action$$

is given by

$$(T \xrightarrow{sa} Action).$$

Now using this notation for derivative of series of actions, we can specify series of actions $T \xrightarrow{sa} Action$ as

- Initial value: $1 \xrightarrow{0} T \xrightarrow{sa} Action$ and
- Differential equation:

$$((T \xrightarrow{sa} Action)^{\langle n \rangle})' = (T \xrightarrow{sa} Action)^{\langle n+1 \rangle}$$

The initial value of $T \xrightarrow{sa} Action$ is defined as its first element $1 \xrightarrow{0} T \xrightarrow{sa} Action$, and the derivative of series of actions, denoted by

$$(T \xrightarrow{sa} Action)',$$

is defined by

$$((T \xrightarrow{sa} Action)^{\langle n \rangle})' = (T \xrightarrow{sa} Action)^{\langle n+1 \rangle}$$

for any integer $n \in T$. In other words, the initial value and derivative equal the head and tail of $T \xrightarrow{sa} Action$, respectively. The behavior of a series of actions $T \xrightarrow{sa} Action$ consists of two aspects: it allows for the observation of its initial value $1 \xrightarrow{0} T \xrightarrow{sa} Action$; and it can make an evolution to the new series of actions $(T \xrightarrow{sa} Action)'$, consisting of the original series of actions from which the first element has been removed. The initial value of $(T \xrightarrow{sa} Action)'$, which is

$$1 \overset{0}{\Rightarrow} ((T \xrightarrow{sa} Action)') \equiv 1 \xrightarrow{1} T \xrightarrow{sa} Action$$

can in its turn be observed, but note that we have to move from $T \xrightarrow{sa} Action$ to $(T \xrightarrow{sa} Action)'$ first in order to do so. Now a behavioral differential equation defines a series of actions by specifying its initial value together with a description of its derivative, which tells us how to continue.

Every member $\sigma_i \in Action$ can be considered as a series of actions in the following manner. For every $\sigma_i \in Action$, a unique series of actions is defined by morphism *f*:

$$1 \xrightarrow{\sigma_i} Action \xrightarrow{f} Action^{\omega}, \quad \text{with composite } 1 \xrightarrow{(\sigma_i, o, o, \ldots)} Action^{\omega}$$

Such that the following equation holds

$$\sigma_i ; f = (\sigma_i, o, o, \ldots)$$

with o denoting empty member (or null member) in *Action* and $(\sigma_i, o, o, \ldots)$ in $Action^{\omega}$.

Note that, for any $T \xrightarrow{sa1} Action$ and $T \xrightarrow{sa2} Action$, we have

$sa1 = sa2 \text{ iff } 1 \xrightarrow{n} T \xrightarrow{sa1} Action = 1 \xrightarrow{n} T \xrightarrow{sa2} Action$
(47)

with every $n \in T$.

Definition 6 (Bisimulation) *Bisimulation on $Action^{\omega}$ is a relation, denoted by ~, between series of actions $T \xrightarrow{sa1} Action$ and $T \xrightarrow{sa2} Action$ such that*

$$\frac{sa1 \sim sa2}{1 \overset{0}{\Longrightarrow} (sa1) = 1 \overset{0}{\Longrightarrow} (sa2) \text{ and } (sa1)' \sim (sa2)'}$$
(48)

For validating whether *sa1 = sa2*, a powerful method is so-called *proof principle of coinduction* (Rutten, 2001) that states as follows:

Theorem 2 (Coinduction)

$$\frac{sa1 \sim sa2}{sa1 = sa2} \quad (49)$$

Proof: Hence in order to prove the equality of two series of actions *sa1* and *sa2*, it is sufficient to establish the existence of a bisimulation relation *sa1~sa2*. In fact, for two series of actions *sa1* and *sa2* and a bisimulation *sa1~sa2*. We have

By induction on $n \geqslant 0$ and bisimulation ~

$$\vdash \ sa1^{\langle n \rangle} = sa2^{\langle n \rangle}$$

By bisimulation ~

$$\vdash 1 \overset{0}{\Rightarrow} (sa1^{\langle n \rangle}) = 1 \overset{0}{\Rightarrow} (sa2^{\langle n \rangle})$$

By identity (46)

$$\vdash 1 \xrightarrow{n} sa1 = 1 \xrightarrow{n} sa2$$

with all $n \geqslant 0$.

It follows that, by (47), sa1=sa2. Note that the notation ├ denotes an inference. By this way, when we write a ├ b, for example, which means that b is true because of a that we have. Moreover, when we write a ├ b ├ c ├ that is understood as a chain inference by which a justification happens. Q.E.D.

As a consequence, using coinduction we can establish the validity of the equivalence between series of actions $T \xrightarrow{sa11} Action$ and $T \xrightarrow{sa12} Action$ in $Action^{\omega}$.

Theorem 3 (Generating series of actions) *For all sa in $Action^{\omega}$.*,

$$sa = 1 \overset{0}{\Longrightarrow} (sa) \bowtie (sa)' \quad (50)$$

Proof: This stems from the coinductive proof principle (49). In fact, it is easy to check the following bisimulation $sa \sim 1 \overset{0}{\Rightarrow} (sa) \rhd\lhd (sa)'$. It follows that $sa = 1 \overset{0}{\Rightarrow} (sa) \rhd\lhd (sa)'$ Q.E.D.

QUANTITATIVE BEHAVIOR OF SERIES OF ACTIONS

For quantitative behaviors of series of actions, we consider again the diagram (39) in section of Categorical Model and Behavior of Series of Actions. Let $B(-)$ be the quantitative behavior of series of actions in $Action^{\omega}$ in NCAAS.

Definition 7 (Quantitative behavior) *For morphisms $1 \xrightarrow{i} T$ and $1 \xrightarrow{x_i} \mathbb{R}$ there exists a unique morphism $T \xrightarrow{B(-)} \mathbb{R}$ such* that the equation $i; B(-) = x_i$ holds. This is described by the following commutative diagram

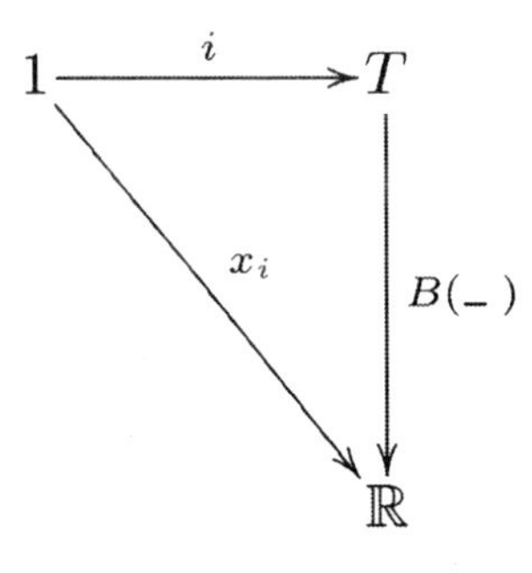

(51)

Morphism $T \xrightarrow{B(-)} \mathbb{R}$ defines a quantitative behavior of series of actions in $Action^{\omega}$.

Let $\mathbb{R}^{\omega} = \{B(-) \mid B(-) : T \to \mathbb{R}$. In this context, any x in R can also be represented as *constant* $(x,0,0,0,\ldots)$ in R^{ω}, denoted by $[x]$, where $x, 0$ in R and its behavioral differential equation is $[x]'=[0]$. For every x in R, a unique morphism f is defined by

$$1 \xrightarrow{\;x\;} \mathbb{R} \xrightarrow{\;f\;} \mathbb{R}^{\omega}, \qquad [x]=(x,0,0,\ldots)\colon 1 \to \mathbb{R}^{\omega}$$

Such that the following equation holds $x;f=(x,0,0,\ldots)$.

Sum (denoted by ⊕) between $B1(-)=(x_0,x_1,\ldots)$ and $B2(-)=(y_0,y_1,\ldots)$ in R^{ω} is defined by

$$\oplus : [T \xrightarrow{B(-)} \mathbb{R}] \times [T \xrightarrow{B(-)} \mathbb{R}] \to [T \xrightarrow{B(-)} \mathbb{R}]$$

$((B1(-) \oplus B2(-))(0), \ldots, (B1(-) \oplus B2(-)0(n), \ldots) = (x_0+y_0, \ldots x_n+y_n, \ldots)$

Convolution product (denoted by ⊕) between $B1(-)=(x_0,x_1,\ldots)$ and $B2(-)=(y_0,y_1,\ldots)$ in R^{ω} is defined by

$$\oplus : [T \xrightarrow{B(-)} \mathbb{R}] \times [T \xrightarrow{B(-)} \mathbb{R}] \to [T \xrightarrow{B(-)} \mathbb{R}]$$

$((B1(-) \oplus B2(-))(0), \ldots, (B1(-) \oplus B2(-)0(n), \ldots) =$

$$(x_0 \times y_0, \ldots, \sum_{k=0}^{n} x_{n-k} \times y_k, \ldots)$$

Let $X=(0,1,0,0,\ldots)$ in R^{ω} be *formal variable*, then it is easy to verify $X \oplus B(-)=(B(-))', X \oplus B(-))'=B(-)$ and

$X \oplus B(-)=B(-) \oplus X$.

Let $B(s_0)$ in R^{ω} be the quantitative behavior of series of actions at CAAS configuration s_0 in *Sys* in NCAAS. We can define $B(s_0)$ in the following way.

Let s_1 in $\{s_{1,1,\ldots,} s_{1,n}\}$ and s_0 be the CAAS configurations for which

$$Adap(s_0 \to \sigma_0)(s_1 \to \sigma_1) \neq 0$$

Then the quantitative behavior $B(s_0)$ of series of actions at CAAS configuration s_0 can be completely defined by the following system of behavioral differential equations, one for each CAAS configuration $s_{0,} s_1$ in $\{s_{1,1,\ldots,} s_{1,n}\}$ and so on.

$$\begin{cases} \text{Differential equation:} & (B(s_0))' = Adap(s_0 \longrightarrow \sigma_0)(s_{1,1} \longrightarrow \sigma_1) \otimes B(s_{1,1}) \\ & \oplus \ldots \oplus Adap(s_0 \longrightarrow \sigma_0)(s_{1,n} \longrightarrow \sigma_1) \otimes B(s_{1,n}) \\ \text{Initial value:} & 1 \overset{0}{\Longrightarrow} (B(s_0)) = Out(s_0) \end{cases} \tag{52}$$

Theorem 4 (Existing a unique solution) *System of behavioral differential equations* (52) *exists a unique solution. That is, for each $B(s_0)$ there exists a unique morphism denoted by* (*the same notation*) $B(s_0) : (\mathbb{R}^{\omega})^n \to \mathbb{R}^{\omega}$ satisfying (52).

Proof: Let *TermSet* be the set of all instances of right side in the equation (52). The uniqueness of solution stems, in fact, from there exists a unique homomorphism $h: Termset \to R^{\omega}$, which assigns to each member in *TermSet* a sequence of data in R^{ω}. Then, this sequence of data is what we define to be the effect of the morphism $B(s_0) : (\mathbb{R}^{\omega})^n \to \mathbb{R}^{\omega}$ Q.E.D.

It follows from theorem 3 in section of Categorical Model and Behavior of Series of Actions that, for the quantitative behavior, representation (50) of series of actions becomes $B(s_0) = 1 \overset{0}{\Rightarrow} (B(s_0)) \oplus (X \oplus (B(s_0))')$. Thus, system of behavioral differential equations (52) is equivalent to the following system of equations.

$$\begin{aligned} B(s_0) \;=\; & Out(s_0) \oplus Adap(s_0 \longrightarrow \sigma_0)(s_{1,1} \longrightarrow \sigma_1) \otimes X \otimes B(s_{1,1}) \oplus \ldots \oplus \\ & Adap(s_0 \longrightarrow \sigma_0)(s_{1,n} \longrightarrow \sigma_1) \otimes X \otimes B(s_{1,n}) \end{aligned} \tag{53}$$

CATEGORICAL MODEL AND BEHAVIOR OF SERIES OF ADAPTATION RELATIONS

In this section, we construct monoids of adaptation relations and then series of adaptation relations.

Adaptation Monoid and Series of Adaptation Relations

Adap$_n$*(Sys,Sys)*, with $n \geqslant 0$, is the set of adaptation relations $\geqslant Adap: (Sys \to Action^n \to) \to (Sys \to Action^n \to) \to \to$. In other words, given $n \geqslant 0$ then

$$\begin{aligned}&\mathrm{Adap}_n(Sys, Sys)\\&= \{Adap : (Sys \to Action^n) \to (Sys \to Action^n)\}\\&= \{Adap_i(Sys \to Action^n, Sys \to Action^n),\\&\text{for all } i \text{ in } T\end{aligned}$$

Note that, in the current context, we write $Adap_i^{(n)}$ as a notation of $Adap_i(Sys \to Action^n, Sys \to Action^n)$. Thus, we have

$$\mathrm{Adap}_n(Sys, Sys) = \{Adap_i^{(n)}, \text{ for all } i \text{ in } T\}$$

This set with the composition operation "; " satisfies two following properties:

Relation composition: Let f and g be members of $\mathrm{Adap}_n(Sys, Sys)$, then the relation composition $(f;g)$: $(Sys \to Action^n) \to (Sys \to Action^n)$ is as

$$g:(f: (Sys \to Action^n) \to (Sys \to Action^n))(Sys \to Action^n)$$

In other words, let $f = Adap_i^{(n)}$ and $g = Adap_j^{(n)}$ then

$$(Adap_i^{(n)}; Adap_j^{(n)}) = Adap_j(Adap_i^{(n)}, Sys \to Action^n)$$

Identity relation: There exists an identity relation $1_{Sys \to Action^n}$ (or 1_n, for short) $\mathrm{Adap}_n(Sys, Sys)$ for $1_n;f = f;1_n = f$ to be held. In other words, this can be specified by

$$\begin{aligned}&Adap_i(1_n, Sys \to Action^n)\\&= Adap_i(Sys \to Action^n, 1_n)\\&= Adap_i(Sys \to Action^n, Sys \to Action^n)\\&= Adap_i^{(n)}\end{aligned}$$

Thus, $\mathrm{Adap}_n(Sys, Sys)$ with the composition operation ";" is called *adaptation monoid*. Moreover, the monoid $\mathrm{Adap}_n(Sys, Sys)$ is also a monoid category including only one object to be the set

$$\{Adap_i : (Sys \to Action^n) \to (Sys \to Action^n)\}$$

every member of the set is a morphism, and by the composition operation the associativity and identity on the morphisms are completely satisfied.

Definition 8 (*Series of adaptation relations*) *For morphisms* $1 \xrightarrow{i} T$ *and*, $1 \xrightarrow{Adap_i^{(n)}} \mathrm{Adap}_n(Sys, Sys)$ there exists a unique morphism such $T \xrightarrow{f} \mathrm{Adap}_n(Sys, Sys)$ that the equation $i; f = Adap_i^{(n)}$ holds. This is described by the following commutative diagram

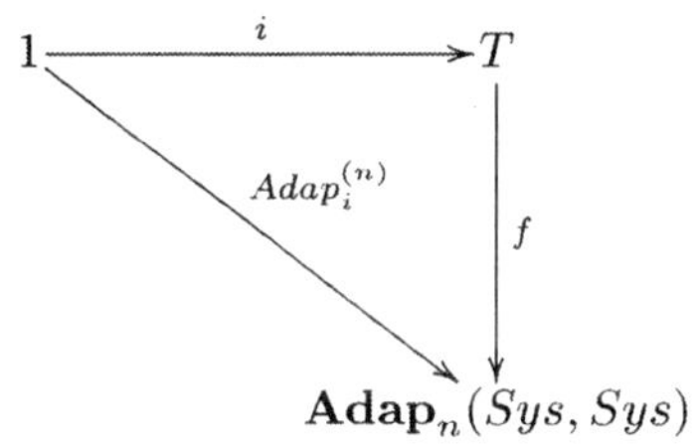

(54)

Morphism $T \xrightarrow{f} \mathrm{Adap}_n(Sys, Sys)$ defines a series of adaptation relations.

This is understood that series of adaptation relations also satisfies the behavioral properties of series of actions in section of Categorical Model and Behavior of Series of Actions. However, behavior of series of adaptation relations has its specific properties which we consider from now on.

A Category of Monoids $\mathrm{Adap}_i(Sys, Sys)$

By the adaptation monoids $\mathrm{Adap}_i(Sys, Sys)$, for all $i \geqslant 0$, we can construct **Series(Adap)**(*Sys,Sys*) to be a monoids category. In fact, **Series(Adap)** (*Sys,Sys*) is constructed as follows:

*Obj(***Series(Adap)**(*Sys,Sys*)) is the set of adaptation monoids $\mathrm{Adap}_i(Sys, Sys)$ for all $i \geqslant 0$. In other words,

*Obj(***Series(Adap)**(*Sys,Sys*)) $=\{\mathbf{Adap}_i(Sys,Sys), \forall i \geqslant 0\}$

Associated with each object $\mathrm{Adap}_i(Sys, Sys)$ in *Obj(***Series(Adap)**(*Sys,Sys*)), we define a morphism

$$\mathrm{Adap}_i(Sys, Sys) \xrightarrow{1_{\mathrm{Adap}_i(Sys,Sys)}} \mathrm{Adap}_i(Sys, Sys)$$

the identity morphism on $\mathrm{Adap}_i(Sys, Sys)$ such that

$$\mathrm{Adap}_i(Sys, Sys) \xrightarrow{1_{\mathrm{Adap}_i(Sys,Sys)}=1i} \mathrm{Adap}_i(Sys, Sys)$$

or

$$\{Adap_k^{(i)}, \text{for all } k \text{ in } T\} \xrightarrow{1_{\mathrm{Adap}_i(Sys,Sys)}=1i} \{Adap_k^{(i)}, \text{for all } k \text{ in } T\}$$

and to each pair of morphisms $\mathrm{Adap}_i(Sys, Sys) \xrightarrow{f} \mathrm{Adap}_j(Sys, Sys)$ and $\mathrm{Adap}_j(Sys, Sys) \xrightarrow{g} \mathrm{Adap}_k(Sys, Sys)$ such that

$$Adap_i(Sys, Sys) \xrightarrow{f=1_i \times Action^{j-i}} Adap_j(Sys, Sys)$$

and

$$Adap_j(Sys, Sys) \xrightarrow{g=1_j \times Action^{k-j}} Adap_k(Sys, Sys)$$

there is an associated morphism $\mathrm{Adap}_i(Sys, Sys) \xrightarrow{f;g} \mathrm{Adap}_k(Sys, Sys)$, the composition of *f* with *g* such that

$$Adap_i(Sys, Sys) \xrightarrow{f;g=1_i \times Action^{k-i}} Adap_k(Sys, Sys)$$

For all objects in *Obj(***Series(Adap)**(*Sys,Sys*)) and the morphisms

$$Adap_i(Sys, Sys) \xrightarrow{f=1_i \times Action^{j-i}} Adap_j(Sys, Sys)$$

$$Adap_j(Sys, Sys) \xrightarrow{g=1_j \times Action^{k-j}} Adap_k(Sys, Sys)$$

and

$$Adap_k(Sys, Sys) \xrightarrow{h=1_k \times Action^{m-k}} Adap_m(Sys, Sys)$$

in *Arc***Series(Adap)**(*Sys,Sys*), the following equations hold:

Associativity: $(f;g); h=f;(g;h)=1_i \times Action^{m-i}$

Identity: $1_{\mathbf{Adap}i(Sys,Sys)};f=f=f;1_{\mathbf{Adap}j(Sys,Sys}$

(i.e., $1_i, 1_i \times Action^{j-i}=1_i \times Action^{j-i} 1_i \times Action^{j-l};1_j$)

As a result, the above-mentioned monoid morphisms can be diagrammatically drawn such as

$$\mathbf{Adap}_i(Sys, Sys) \xrightarrow{1_i \times Action^{\pm k}} \mathbf{Adap}_{i \pm k}(Sys, Sys) \tag{55}$$

or

$$\{Adap_l^{(i)}, \textit{ for all } l \textit{ in } T\} \xrightarrow{1_i \times Action^{\pm k}} \{Adap_l^{(i+k)}, \textit{ for all } l \textit{ in } T\}$$

Some consequences coming from constructing **Series(Adap)**(*Sys,Sys*) are stated by the following corollaries.

Corollary 1. *All monoid morphisms of***Series(Adap)**(*Sys,Sys*) *is monoid isomorphisms*

Proof: In fact, this result immediately stems from (55). For every pair of monoid morphisms $\mathrm{Adap}_i(Sys, Sys)$ and $\mathrm{Adap}_j(Sys, Sys)$ in *Arc***Series(Adap)**(*Sys,Sys*), we have

$$\mathbf{Adap}_i(Sys,Sys) \underset{1_j \times Action^{i-j}}{\overset{1_i \times Action^{j-i}}{\rightleftarrows}} \mathbf{Adap}_j(Sys,Sys), \quad \text{with loops } 1_i \text{ on } \mathbf{Adap}_i(Sys,Sys) \text{ and } 1_i \text{ on } \mathbf{Adap}_j(Sys,Sys) \tag{56}$$

These morphisms are exactly isomorphic. Q.E.D.

Corollary 2. *Isomorphisms between any pair of monoids in* **Series(Adap)**(*Sys,Sys*) *are isomorphisms between the pair of* AASs.

Proof: This comes from the fact that each object of category **Series(Adap)**(*Sys,Sys*) is just an AAS.Q.E.D.

From the above-mentioned justification of **Series(Adap)**(*Sys,Sys*), it is possible to derive $\mathrm{Adap}_i(Sys, Sys)$ for every $i \geqslant 0$. Derivation of every $\mathrm{Adap}_i(Sys, Sys)$ is simplified by the following inference rules of categorical programming:

$$\frac{}{1 \xrightarrow{\mathbf{Adap}(Sys,Sys)} Obj(\mathbf{Series}(\mathbf{Adap})(Sys,Sys))} \tag{57}$$

This means that it is always able to select an object $\mathrm{Adap}(Sys, Sys)$ from *Obj***Series(Adap)**(*Sys,Sys*)) which we have constructed. Note that,

$$\begin{aligned} &\mathrm{Adap}(Sys, Sys) \\ &= \{\mathrm{Adap}\colon (Sys \to Action^0) \to (Sys \to Action^0)\} \\ &= \{Adap : Sys \to Sys\} \end{aligned}$$

$$\frac{\left(\begin{array}{c} 1 \xrightarrow{\mathbf{Adap}(Sys,Sys)} Obj(\mathbf{Series}(\mathbf{Adap})(Sys,Sys)) \\ \text{and} \\ \mathbf{Adap}(Sys,Sys) \xrightarrow{1_0 \times Action^i} \mathbf{Adap}_i(Sys,Sys) \end{array}\right)}{1 \xrightarrow{\mathbf{Adap}_i(Sys,Sys)} Obj(\mathbf{Series}(\mathbf{Adap})(Sys,Sys))} \tag{58}$$

It means that given $\mathrm{Adap}(Sys, Sys)$ we can compute $\mathrm{Adap}_i(Sys, Sys)$ for every *i*. Note that $\mathrm{Adap}(Sys, Sys) \xrightarrow{1_0} \mathrm{Adap}(Sys, Sys)$.

$$\frac{\left(\begin{array}{c} 1 \xrightarrow{\mathbf{Adap}_i(Sys,Sys)} Obj(\mathbf{Series}(\mathbf{Adap})(Sys,Sys)) \\ \text{and} \\ \mathbf{Adap}_i(Sys,Sys) \xrightarrow{1_i \times Action^{j-i}} \mathbf{Adap}_j(Sys,Sys) \end{array}\right)}{1 \xrightarrow{\mathbf{Adap}_j(Sys,Sys)} Obj(\mathbf{Series}(\mathbf{Adap})(Sys,Sys))} \tag{59}$$

This rule states that given $\mathrm{Adap}_i(Sys, Sys)$ we can compute $\mathrm{Adap}_j(Sys, Sys)$ for every $j \neq i$.

From the construction of **Series(Adap)** (*Sys,Sys*), we see that every $\mathrm{Adap}_i(Sys, Sys)$ can be formed in the unifying way and, moreover, it is also easy to see that we can write code in virtually any implementation language for performing the construction procedure based on these three rules. The major point is that we have gained the substantial aspects of the construction procedure without any excessive inclination towards a specific implementation detail; it is at a high abstract level. This is quite helpful when we want to prove properties of the construction. In fact, we can prove

Theorem 5. *Every object* $\mathrm{Adap}_i(Sys, Sys)$ *can be constructed* by any other object in **Series(Adap)** (*Sys,Sys*)

Proof: Applying rule in (58) and/or rule in (59) to construct every object $\mathrm{Adap}_i(Sys, Sys)$ from another object in **Series(Adap)**(*Sys,Sys*) Q.E.D.

This is certainly a property we expect of any construction procedure.

Theorem 6. Series(Adap)(*Sys,Sys*) *is a complete graph*

Proof: In fact, this is a consequence stemming from theorem 5.Q.E.D.

This is indeed a property of our abstract construction mechanism.

NOTES AND REMARKS

By the results obtained from this chapter, we discuss some further work being able to be extended in the future. In fact, we can reach the following:

- The sets $Sys \rightarrow Action^{i \geq 0}$ of CAAS configurations specify a category. Actually, let **Cat(DCAAS)** be such a category of the sets of CAAS configurations, whose structure is constructed as follows:
- Each set $Sys \rightarrow Action^{i \geq 0}$ of CAAS configurations defines an object. That is,
   - *Obj***Cat(DCAAS))** $= \{Sys \rightarrow Action^{i \geq 0}\}$
- Each *Adap* defines a morphism. That is,
- *Arc***Cat(DCAAS))** $= \{Adap: (Sys \rightarrow Action^{i \geq 0}) \rightarrow (Sys \rightarrow Action^{j \geq 0})\}$
- It is easy to check that associativity in (11) and identity in (10) on all *Adaps* are satisfied.
- Further about the category **Cat(DCAAS)**, which we have constructed. The category **Cat(DCAAS)** equipped with structure $(Sys \rightarrow Action^{i \geq 0}) \rightarrow (Sys \rightarrow Action^{j \geq 0}) \rightarrow R$ defines a category **Cat(NCAAS)** of the sets of CAAS configurations.
- Structure $(Sys \rightarrow Action^{i \geq 0}) \rightarrow (Sys \rightarrow Action^{j \geq 0}) \rightarrow R$ in category **Cat(NCAAS)** defines an algebra, so-called *Adap*-algebra(NCAAS). This originates from definition on T-algebra in section of Preliminaries, where functor T is defined such that T = *Adap* with
- $Adap: (Sys \rightarrow Action^{i \geq 0}) \rightarrow (Sys \rightarrow Action^{j \geq 0}) \rightarrow R$.
- With the above-mentioned result, we obtain a compact formal definition of NCAASs such that each *Adap*-algebra(NCAAS) specifies an NCAAS.
- Moreover, by our approach, each computational behavior defined by Y. Wang in (Wang, 2007b) is just an algebraic object of category. In addition, imperative computing systems, adaptive computing systems and autonomic computing systems are instances of a functor on such the category.
- In the considered context, we apply the category theory, which deals in an abstract way with algebraic objects and relationships between them for specifying interaction behaviors in ASSs. For modeling, analyzing and verifying the interaction behaviors, category theory is much better-approaching than other ones such as process algebras, FSM or UML. In fact, the categorical approach becomes more powerful because process algebras and FSM are just of algebraic objects of category and UML is really a semi-formal approach. Categories were first described by Samuel Eilenberg and Saunders Mac Lane in 1945 (Lawvere & Schanuel, 1997), but have since grown substantially to become a branch of modern mathematics. Category theory spreads its influence over the development of both mathematics and theoretical computer science. The categorical structures themselves are still the subject of active research, including work to increase their range of practical applicability.

CONCLUSION

In this chapter, using categorical language, models and behaviors of the autonomic agent systems (AASs), series of actions and series of adaptation relations have been interpreted. In other words, the behavior-oriented notions of the AASs, series of actions and series of adaptation relations have denotationally been formalized.

We have started with investigating deterministic and nondeterministic context-dependent adaptive agent systems (DCAASs and NCAASs) where all the categorical models and behaviors are based on mapping a CAAS configuration to another. In the specification, we have considered a CAAS configuration of the systems at every adaptation step to be a member of $Sys \rightarrow Action^{i \geq 0}$.

Series of actions has been modeled to provide an algebraic framework for examining behaviors of series of actions in the DCAASs and NCAASs. Specifically, series of actions has formally been viewed as the morphism $T \xrightarrow{sa} Action$ in our approach. Behaviors of series of actions have quantitatively been developed taking advantage of their algebraic models where quantitative behaviors of series of actions have formally been developed based on the morphism $T \xrightarrow{B(-)} \mathbb{R}$, and then the system of behavioral differential equations has been constructed for evaluating solution.

For adaptation relations, every $\mathrm{Adap}_i(Sys, Sys)$ has been constructed as an adaptation monoid to shape series of adaptation relations. By the adaptation monoids $\mathrm{Adap}_i(Sys, Sys)$, we have formed **Series(Adap)**(*Sys,Sys*) to be a category of adaptation monoids for behavioral reasoning on series of adaptation relations.

ACKNOWLEDGMENT

Thank you to the colleagues and anonymous reviewers for their helpful comments and valuable suggestions which have contributed to the final preparation of the chapter.

REFERENCES

Adamek, J., Herrlich, H., & Strecker, G. (2009). *Abstract and Concrete Categories*. Dover Publications.

Asperti, A., & Longo, G. (1991). *Categories, Types and Structures*. M.I.T. Press.

Bergman, G. M. (1998). *An Invitation to General Algebra and Universal Constructions. 15 the Crescent, Berkeley CA 94708*. US: Henry Helson.

Butera, W. (2007, 9-11 July). Text Display and Graphics Control on a Paintable Computer. In G. Serugendo, J. Flatin, & M. Jelasity (Eds.), *Proceedings of 1st international conference on self-adaptive and self-organizing systems (saso'07)* (pp.45–54). Boston, Massachusetts, USA: IEEE Computer Society Press.

Calisti, M., Meer, S., & Strassner, J. (Eds.). (2008). *Advanced Autonomic Networking and Communication*. Springer-Verlag. doi:10.1007/978-3-7643-8569-9

De, S. A., Loach, W. O., & Matson, E. (2008, February). A Capabilities-based Model for Adaptive Organizations. *Autonomous Agents and Multi-Agent Systems, 16*(1), 13–56. doi:10.1007/s10458-007-9019-4

Denko, M. K., Yang, L. T., & Zhang, Y. (Eds.). (2009). *Autonomic Computing and Networking* (1st ed.). Springer USA. (452 pages)

IBM. (2001). Autonomic Computing Manifesto. Retrieved from http://www.research. ibm.com/autonomic/.

Jin, X., & Liu, J. (2004, April). From Individual Based Modeling to Autonomy Oriented Computation. In Nickles, M., Rovatsos, M., & Weiss, G. (Eds.), *Agents and computational autonomy: Potential, risks, and solutions* (*Vol. 2969*, pp. 151–169). Springer Berlin. doi:10.1007/978-3-540-25928-2_13

Kinsner, W. (2007, January). Towards Cognitive Machines: Multiscale Measures and Analysis. [IJCINI]. *International Journal of Cognitive Informatics and Natural Intelligence, 1*(1), 28–38. doi:10.4018/jcini.2007010102

Ko, S., Gupta, I., & Jo, Y. (2007, 9-11 July). Novel Mathematics-Inspired Algorithms for Self-Adaptive Peer-to- Peer Computing. In G. Serugendo, J. Flatin, & M. Jelasity (Eds.), *Proceedings of 1st international conference on self-adaptive and self-organizing systems (saso'07)* (pp. 3–12). Boston, Massachusetts, USA: IEEE Computer Society Press.

Lawvere, F., & Schanuel, S. (1997). *Conceptual Mathematics: A First Introduction to Categories* (1st ed.). Cambridge University Press.

Levine, M. (1998). Categorical Algebra. In Benkart, G., Ratiu, T., Masur, H., & Renardy, M. (Eds.), *Mixed motives* (*Vol. 57*, pp. 373–499). USA: American Mathematical Society.

Pacheco, O. (2004, April). Autonomy in an Organizational Context. In Nickles, M., Rovatsos, M., & Weiss, G. (Eds.), *Agents and computational autonomy: Potential, risks, and solutions* (*Vol. 2969*, pp. 195–208). Springer Berlin. doi:10.1007/978-3-540-25928-2_16

Parashar, M., & Hariri, S. (Eds.). (2006). *Autonomic Computing: Concepts, Infrastructure and Applications* (1st ed.). CRC Press.

Rutten, J. (2001). Elements of Stream Calculus (An Extensive Exercise in Coinduction). [Elsevier Science Publishers Ltd.]. *Electronic Notes in Theoretical Computer Science*, 45.

Topaloglu, U., & Bayrak, C. (2008, February). Secure Mobile Agent Execution in Virtual Environment. *Autonomous Agents and Multi-Agent Systems*, *16*(1), 1–12. doi:10.1007/s10458-007-9018-5

Vinh, P. (2007). Homomorphism between AOMRC and Hoare Model of Deterministic Reconfiguration Processes in Reconfigurable Computing Systems. *Scientific Annals of Computer Science* (XVII), 113-145.

Vinh, P. (2009a, May). Formal Aspects of Self-* in Autonomic Networked Computing Systems. In Denko, M. K., Yang, L. T., & Zhang, Y. (Eds.), *Autonomic Computing and Networking* (1st ed., pp. 381–410). Springer, USA. doi:10.1007/978-0-387-89828-5_16

Vinh, P. (2009b). *Dynamic Reconfigurability in Reconfigurable Computing Systems: Formal Aspects of Computing* (1st ed.). Saarbrucken, Germany: VDM Verlag Dr. Muller.

Vinh, P. (2009c, May). Formalizing Parallel Programming in Large Scale Distributed Networks: From Tasks Parallel and Data Parallel to Applied Categorical Structures. In F. Xhafa (Ed.), *Parallel Programming, Models and Applications in Grid and P2P Systems* (1st ed., Vol. 17, pp. 24–53). IOS Press.

Vinh, P. (2009d, January). Categorical Approaches to Models and Behaviors of Autonomic Agent Systems. [IJCiNi]. *International Journal of Cognitive Informatics and Natural Intelligence*, *3*(1), 17–33. doi:10.4018/jcini.2009010102

Vinh, P. (2010, May). Aspect-Oriented Self-configuring P2P Networking in Mobile Environments: A Formal Specification and Verification. In P. Alencar and D. Cowan (Ed.), *Handbook of Research on Mobile Software Engineering: Design, Implementation and Emergent Applications* (1st ed.). IGI Global.

Vinh, P., & Bowen, J. (2007, 6–8 June). A Formal Approach to Aspect-Oriented Modular Reconfigurable Computing. In *Proceedings of 1st ieee & ifip international symposium on theoretical aspects of software engineering (tase)* (pp. 369–378). Shanghai, China: IEEE Computer Society Press.

Vinh, P., & Bowen, J. (2008, June). Formalization of Data Flow Computing and a Coinductive Approach to Verifying Flowware Synthesis. *LNCS Transactions on Computational Science*, *4750*(1), 1–36. doi:10.1007/978-3-540-79299-4_1

Walter, F.E., S. B., & Schweitzer, F. (2008, February). A Model of a Trust-based Recommendation System on a Social Network. *Autonomous Agents and Multi-Agent Systems*, *16*(1), 57–74. doi:10.1007/s10458-007-9021-x

Wang, Y. (2007a, January). The Theoretical Framework of Cognitive Informatics. [IJCiNi]. *International Journal of Cognitive Informatics and Natural Intelligence*, *1*(1), 1–27. doi:10.4018/jcini.2007010101

Wang, Y. (2007b, July–September). Toward Theoretical Foundations of Autonomic Computing. [IJCiNi]. *International Journal of Cognitive Informatics and Natural Intelligence*, *1*(3), 1–16. doi:10.4018/jcini.2007070101

Wang, Y., & Kinsner, W. (2006, March). Recent Advances in Cognitive Informatics. *IEEE Transactions on Systems, Man and Cybernetics. Part C, Applications and Reviews*, *36*(2), 121–123. doi:10.1109/TSMCC.2006.871120

Witkowski, M., & Stathis, K. (2004, April). A Dialectic Architecture for Computational Autonomy. In Nickles, M., Rovatsos, M., & Weiss, G. (Eds.), *Agents and computational autonomy: Potential, risks, and solutions* (*Vol. 2969*, pp. 261–273). Springer Berlin. doi:10.1007/978-3-540-25928-2_21

Wolf, T. D., & Holvoet, T. (2006) *Autonomic Computing: Concepts, Infrastructure and Applications* (1st ed.). A Taxonomy for Self-* Properties in Decentralized Autonomic Computing (pp. 101–120). CRC Press.

Yang, B., & Liu, J. (2007, 9-11 July). An Autonomy Oriented Computing (AOC) Approach to Distributed Network Community Mining. In G. Serugendo, J. Flatin, & M. Jelasity (Eds.), *Proceedings of 1st international conference on self-adaptive and self-organizing systems (saso'07)* (pp. 151–160). Boston, Massachusetts, USA: IEEE Computer Society Press.

Zoethout, K., W. J., & Molleman, E. (2008, February). Task Dynamics in Self-organising Task Groups: Expertise, Motivational, and Performance Di erences of Specialists and Generalists. *Autonomous Agents and Multi-Agent Systems*, *16*(1), 75–94. doi:10.1007/s10458-007-9022-9

Chapter 3
Concept of Symbiotic Computing and its Agent-Based Application to a Ubiquitous Care-Support Service

Takuo Suganuma
Tohoku University, Japan

Kenji Sugawara
Chiba Institute of Technology, Japan

Tetsuo Kinoshita
Tohoku University, Japan

Fumio Hattori
Ritsumeikan University, Japan

Norio Shiratori
Tohoku University, Japan

ABSTRACT

In this paper, a concept of "symbiotic computing" is formalized to bridge gaps between Real Space (RS) and Digital Space (DS). Symbiotic computing is a post-ubiquitous computing model based on an agent-oriented computing model that introduces social heuristics and cognitive functions into DS to bridge the gaps. The symbiotic functions and agent-based architecture of symbiotic applications are also discussed. Based on the concept, functions, and architecture of symbiotic applications, we develop an agent-based care-support service to enable supervision of persons by their families and friends easily while protecting privacy. In this application system, a hierarchical structure of multi-agents is organized dynamically using heuristics in agents based on the situation of a watched person and watching persons. The system appropriately alters the contents and quality of the live video. The flexible system construction scheme using a multiagent framework facilitates the symbiosis of RS and DS by bridging the gaps in the care-support service domain.

INTRODUCTION

The growth of Web technology enables many people to get various kinds of information from the internet using high-performance search engines and recently advanced web services. People have come to live in virtual communities with related social networks as they do in real communities. Furthermore, the recent rapid growth of the ubiquitous technology provides more convenient services to people who can make full use of them [Lyytinen, 2002]. The technologies have led from a traditional society to a modern information network society in which people can exchange information easily and efficiently via the internet.

On the other hand, emerging problems have come with the internet society: the digital divide, security, network-based crimes, and so on. These problems have been caused by social and human difficulties rather than by computer and network technology per se. Nevertheless, technology should tackle these difficult problems by inclusion of sociality and humanity into computing models.

Symbiotic computing, which we propose, provides a framework to bridge gaps between Real Space (RS) and Digital Space (DS). We consider that a gap, which causes some problems in our internet society, exists because of the lack of mutual cognition between RS and DS. That is, people cannot receive its advanced services without IT skills, and DS cannot provide a service that is suitable to a person depending on that person's situation and preference. Furthermore, DS cannot provide a safe and secure service without heuristics to describe a person's activities in a society, such as customs, laws, and expertise. We define symbiotic computing based on ubiquitous computing, Web computing, and models of AI and Cognitive Informatics [Wang1, 2007]. A model of cognitive properties to include human factors and social relations into information processing was discussed by Wang [Wang, 2005; Wang2, 2007]. The concept of symbiotic computing was also defined based on so-called calm computing proposed by Mark Weiser [Weiser, 1996]. The word "symbiotic" is adopted to describe the mutually interrelated character of RS and DS, which enables mutual cognition to provide safe and suitable services for RS.

A model of the symbiotic computing in this paper is formalized based on an agent-oriented model to include social heuristics and cognitive functions into DS to bridge the gaps. To realize the concept of symbiotic computing in a real situation, advanced abilities of systems such as flexibility, adaptability, and sociality are required to construct applications that bridge the gaps. For that reason, we use a multiagent-based framework for symbiotic applications. In our framework, agents construct and reconstruct organizations dynamically according to the situations of RS—locations, preferences, requirements, and conditions of users—and the status of devices and networks around the users. The agent organization, which is customized to diverse situations in RS, can integrate necessary and sufficient functional elements and information in DS; it can also provide situation-aware services to users in RS. This flexible system construction scheme is a basic function of symbiotic computing [Suganuma 2003]; it plays an important role in facilitating the symbiosis of RS and DS by bridging the gaps.

In this paper, we describe the application of symbiotic computing to care-support systems for supervision of elderly people or children. This is a typical example of the gap that prevents a symbiotic relationship between persons and communities and between persons and DS. For this reason, we decided to develop a care-support system for supervision as an experimental system of symbiotic computing. Users of the system, a watching persons and a watched person, are supported in their supervising activity using live streaming video. According to the users' locations, requirements for supervision will differ such as privacy, and the quality and status of video display and capture devices. To serve those differing requirements, system elements

Figure 1. Gaps and symbiosis between "real space" and "digital space"

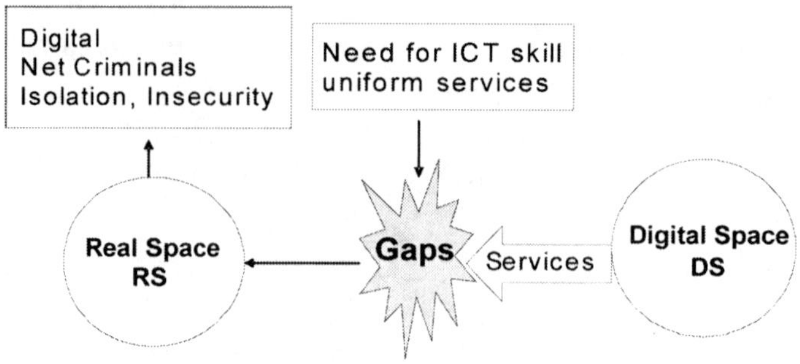

in DS such as software, devices, and networks are organized dynamically to provide safe and convenient supervision services to users. By using this application example, we present the power of the proposed concept, model, and architecture, and illustrate how symbiotic computing works to realize symbiosis between DS and RS.

CONCEPTS OF SYMBIOTIC COMPUTING

Figure 1 shows gaps which isolate Real Space (RS) from a Digital Space (DS) with various installed IT services and resources [Shiratori, 2007]. Here, RS is the space of real world where people are living actually. While DS is the cyber space which is comprised of digital equipment such as PCs, electric appliances, small communicators, networks, information resources, services, etc. To date, the RS has invested a huge amount of social capital and human resources to develop better DS services, which are expected to help people to enjoy satisfactory services in their daily lives and work places. However, because of the gaps, people feel disappointed because they cannot receive satisfactory and suitable services that they require. Therefore, we define the gaps by the different expectations and disappointments of the RS.

One reason for the gaps is a lack of functionality to find relationships between human requests and information resources and services for which RS developed DS, rather than a lack of technologies to develop advanced functions in DS. For example, to find the suitable web site from huge amount of web pages in DS, by using web search engine, user has to provide some appropriate keywords. This keyword selection will greatly effect to the quality of the search result depending on the user's experiences in the search engine. To support the users, some search engine provides the advanced search functions that can specify optional search conditions; however, it sometimes confuses the users because of its complex operations. As a result the user cannot obtain the benefit of the advanced function. The salient implication is that DS technologies are sufficient to fill the gaps now, but their functions are not.

To bridge the gaps, we propose symbiosis between RS and DS so that users in RS feel trust, security, and a sense of ease when they interact with the system. Symbiosis between RS and DS enables people to enjoy rich services of DS with only a slight cost of time and work because DS recognizes their requests without their operations. It then automatically provides suitable services according to their request, personality, and situation.

Figure 2 shows a conceptual schema of the Sense of Symbiosis, which is introduced by a feeling of trust, security, and a sense of ease which are brought about by the feeling of a user that a process of information services in the Digital Space is always related correctly to an intention and an action of the user in Real Space.

Figure 2. User's feeling of symbiosis between RS and DS based on mutual cognition

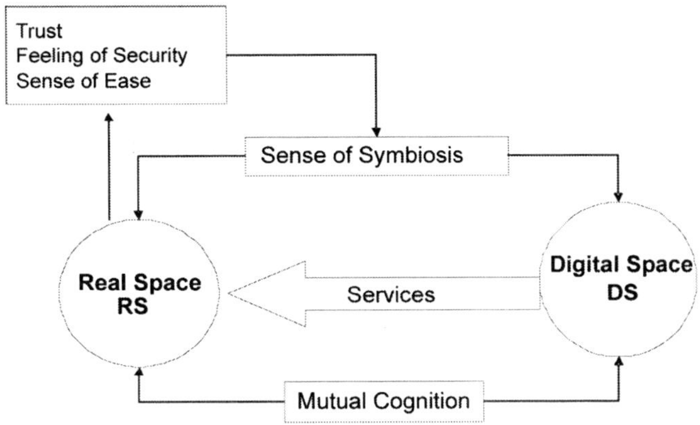

The sense of symbiosis is realized through Mutual Cognition Functions, Perceptual Functions, and Social Functions, which are depicted in Figure 3. Perceptual Functions are used by systems to recognize intentions and actions of users from the data acquired by ubiquitous devices. Social Functions are also used by systems to infer risks and intention of the user using heuristics and social knowledge.

A concept of mutual cognition is also defined by users' feelings of trust that the user is always cared for by a system which the user knows well. A user intensifies the feeling of trust through experiments of the following process of the mutual cognition. In the schema portrayed in Figure 3, mutual cognition is defined as a relation between a personal feeling that "I know that you are taking care of me while not invading my privacy." and a system behavior that assures "I know that you understand I watch over you to help you do something you want to do without risk."

Figure 3. Concept of mutual cognition

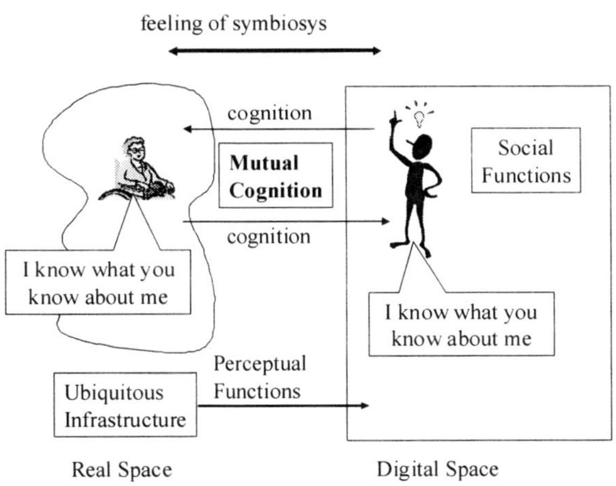

A concept of symbiotic computing has been proposed to solve the gap problems from a technical point of view. The concept of the symbiosis implies biological terms of "the relationship between two different living creatures that live close together and depend on each other in particular ways, each getting particular benefits from the other" [Oxford, 2000]. We apply this concept to a relation between RS and DS to bridge the gaps so that users and society easily obtain maximum benefits from DS.

SYMBIOTIC FUNCTIONS AND SYMBIOTIC APPLICATIONS

Definition of Symbiotic Computing

Symbiotic computing is a model to compute social heuristics and users' activities using data from ubiquitous devices and WWW to include feelings of safety, security, and trust into an advanced information-processing model. Figure 4 shows that symbiotic computing is defined by a

Figure 4. Functional space of symbiotic computing spanned by symbiotic functions

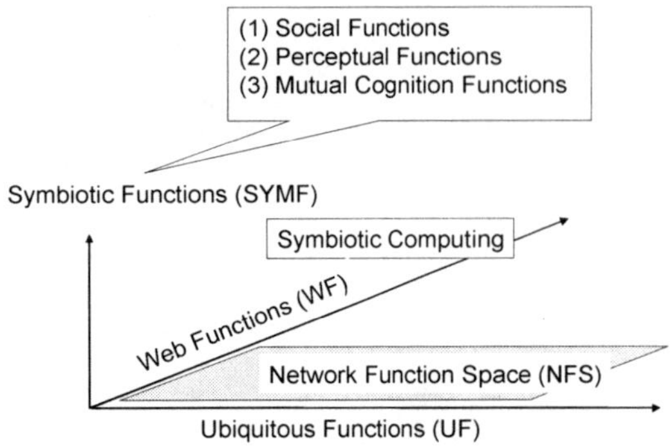

functional space which is spanned by Network Function Space (NFS) and Symbiotic Functions (SYMF) [Shiratori, 2007]. The NFS is a function space which is spanned by the Ubiquitous Functions (UF), Web Functions (WF), and other functions to establish advanced network functions such as a next generation network, an adaptive and flexible network, and so on. The SYMF is also a function space, proposed newly in this paper, which is spanned by Perceptual Functions, Social Functions, and Mutual Cognition Functions.

Perceptual Functions

Perceptual Functions (PFs) can be categorized into the following two classes of functions:

1. To acquire data from ubiquitous devices embedded in RS through NFS, and to manage the data securely for licensed users or software without violation of individual privacy. For example, a collection of signals acquired by RFID system; consists of tags that transmit unique id by using radio frequency wave, and sensors that receive the wave, related to a user A is managed as a data file which is accessible by licensed users and licensed software.
2. To send awareness mined from the above data to Mutual Cognition Function and applications. The awareness is transformed based on a cognitive model, which is formalized as an perceptual function related to an application domain. For example, physical location data of a user collected by the RFID system is transformed to a name of a place which a user occupies: a house or a region.

Social Functions

Social Functions (SF) are categorized into the following two classes of functions:

1. To acquire social information and heuristics related to communities from RS through the PFs and from DS through Web mining functions in SFs. For example, a name of a social event and its schedule can be mined from Web data.
2. To reuse and distribute social information stored in DB/KB in the SF to applications, PFs and Mutual Cognition Functions. The KBs are also modeled by traditional semantic network models, a production model, and so on. Multi-agent systems in the SFs are used to cooperate with DBs and KBs to solve problems depending on each application domain [Hattori, 1999].

Mutual Cognition Functions

A mutual cognition function between a user and a system is realized by a Mutual Cognition Function Agent (MCFA), which is designed as a multi-agent system to watch and support the user. Figure 5 shows that an MCFA deals with the mutual cognition process with a user model based on the process flow.

Figure 5 shows that a process of mutual cognition comprises four processes. The first process of mutual cognition is to recognize a user's behavior from the time series of the awareness from PFs. The next process is to infer the user's request or the user's intention from the behavior using information related to the user's customs or schedule

Figure 5. Model of a mutual cognition function

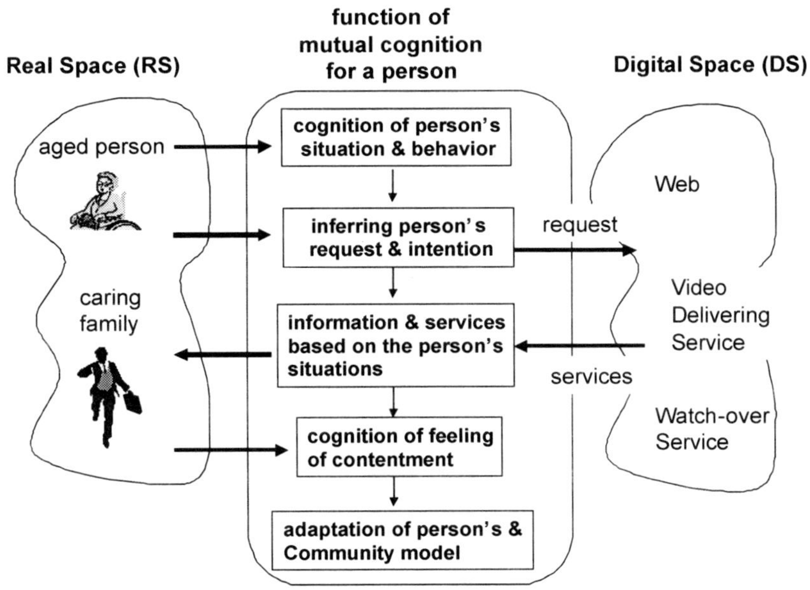

of the day. According to the request, the MCFA sends a request to the DS. In the third process, the MCFA provides a service which is adapted to the user's situation, as derived from awareness from the PF. In the fourth process, the MCFA recognizes the user's feeling of contentment for the service from awareness derived from the PF. Finally, models of the user and the communities related to the user are adapted to the result of the service, thereby improving the mutual cognition process. A user model is stored in the MCFA and is adapted based on the interaction.

Agent-Based Architecture of Symbiotic Applications

Figure 6 depicts an architecture of a Symbiotic Application (SYMA). An SYMA comprises an Application Function (AF), SF, PF, MCF, and Network Functions (NFs) [Shiratori, 1996]. An NF is a collection of collaborative agents to realize functions which are represented as a basic plane surface in Figure 4, which includes the ubiquitous functions (UF), the Web functions (WF), the broadband network functions, multimedia functions and other network functions used by PF, MCF, and SF. The NF also includes "flexible network functions", which we also proposed and developed [Shiratori, 1996] to provides adaptive functions of network services to changes of its environment.

Agents in UF acquire data of signals from sensor networks, which consist of several various ubiquitous devices. Agents send those data to agents in the PF. Agents in the UF also control ubiquitous devices to communicate with users in RS based on requests from PF.

An AF is a multiagent system which supports users working and living based on an application logic [Sugawara, 2005]. An MCF, which is designated to a user, serves as a user interface to acquire a user's request to the AF depending on the application logic. The AF provides suitable service for the user, adapting to the user's situation and personality.

Users' requests change depending on the situation; the quality of the service of the SYMA also changes depending on the quality of the information and communication devices surrounding the user. Therefore, the SYMA must change

Figure 6. An architecture of a symbiotic application

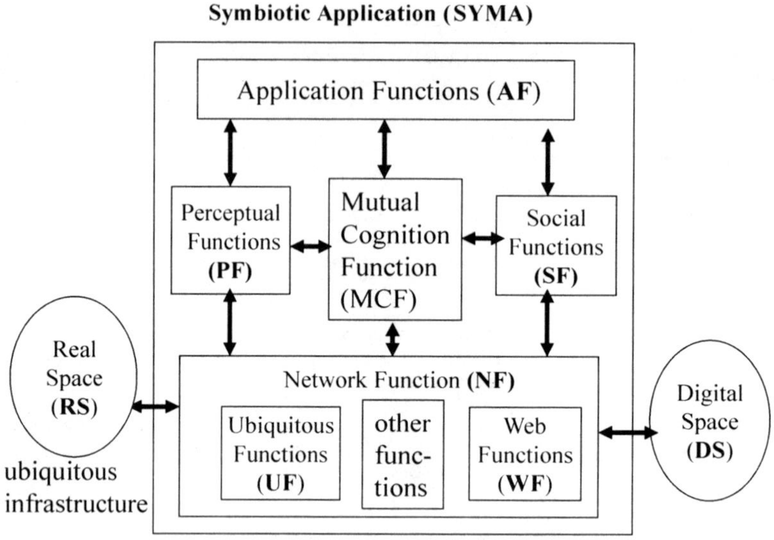

the functions and quality of service to meet the request dynamically. As represented in Figure 6, an AF controls changes of its components and the organization of an SYMA dynamically to do so. An agent-based architecture is introduced to adapt its functions and quality of service dynamically to a user's situation and environment. An example of adaption of the organization of a multiagent system is described in section 6 in this paper.

AGENT-BASED FRAMEWORK FOR SYMBIOTIC APPLICATIONS

To develop an SYMA based on the agent-based architecture, as mentioned in section 3, a framework named ADIPS was proposed to develop agent-based systems which have the following properties [Kinoshita, 1998].

1. Wrapping-based Model of an Agent

Figure 7 shows that an ADIPS agent is an autonomous module which is generated based on the wrapping approach. An ADIPS agent consists of a wrapper module and a module called a base

Figure 7. Model of an ADIPS agent

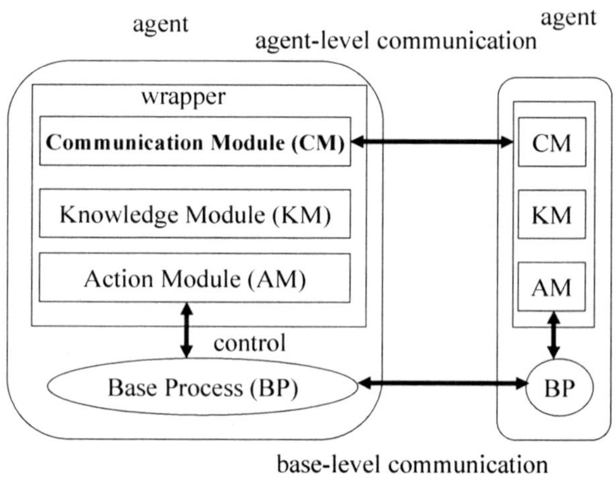

process, which is a set of resources such as programs, devices, web pages, databases, knowledge bases, and actuators. An operation to integrate a base process into an agent with a wrapper module is called "agentification". An advantage of the agentification is that distributed base processes in different computers can mutually communicate using an Agent Communication Language (ACL). The ADIPS framework provides an organization-oriented ACL and protocol for base processes to communicate and cooperate through the wrappers.

Figure 8. Repository-based development of multi-agent systems

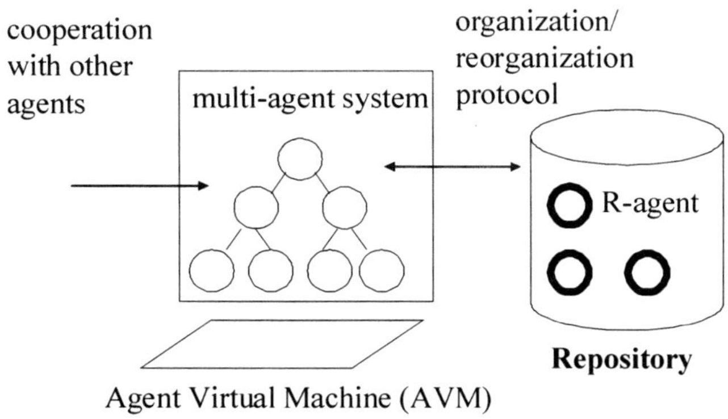

2. Repository-based Development of Multi-agents

An ADIPS agent is an autonomous module because it is unique, persistent, and pro-active in an ADIPS environment constructed of distributed Agent Virtual Machines (AVM). An ADIPS agent can produce its replica autonomously and at the request of another agent. The replica is registered, with a unique identifier, as an autonomous agent.

The ADIPS framework provides a special type of AVM for programmers. Called a repository, it stores reusable and sharable agents, named R-agents, which are used to develop a multi-agent system dynamically according to a request of a user or an agent. An R-agent is designed and programmed by a programmer; it is then reused by users. The ADIPS agents have protocols named organization and re-organization protocols.

When an agent working in an AVM in Figure 8 sends the organization or re-organization protocol to a Repository to make a new organization of agents in the AVM, the R-agents in the Repository start to design the ordered organization cooperatively. The protocol was designed based on the contract net protocol [Smith, 1980]. For example, an agent working in AVM sends a message of request of "generate-function" to a repository. The R-agents start making an organization using the organization protocol based on the Contract Net Protocol to generate a multi-agent system achieving a function in a directed AVM.

3. Repository-based Adaptation

A multi-agent system working on an AVM or on distributed AVMs has a property to adapt its functions to change of the social or system requirements easily and rapidly [Fujita, 1998]. This property gives the DS a cost-effective evolutional mechanism to serve the RS progressively. Figure 8 also shows a framework of adaptation of a multi-agent system using the reorganization protocol defined in the ADIPS framework.

The ADIPS framework was implemented as a programming language named DASH consisting of distributed agent virtual machines; DASH-AVM was developed using the Java language. In addition, IDEA is an environment for developing multi-agent systems efficiently over distributed DASH-AVMs [Uchiya, 2007]. Experimental SYMAs such as a ubiquitous care-support service in this paper have been implemented using DASH.

APPLICATION OF SYMBIOTIC COMPUTING TO UBIQUITOUS CARE-SUPPORT SERVICES

In this section, we discuss a gap between requirements for a supervised person and supervising people living in RS and supervisory services implemented in DS; we design the experimentally symbiotic application shown in Figure 6.

Some research groups have attempted to apply a real-time multimedia watching system to a pervasive computing environment. In one such study, streaming video is delivered from the selected camera closest to the target people using physical position information of the target [Takemoto, 2005; Silva, 2005]. Another research direction is on flexible displays, which seamlessly play streaming video from the nearest display to an observer based on location information [Cui, 2004; Lohse, 2005]. This kind of technique is sometimes called "service migration".

These studies have the function of selecting existing cameras and displays based on a user's location information. However, when the watching person has an advanced requirement related to video quality or privacy, it is not guaranteed to fulfill the requirement in these existing systems. For example, in the following cases, it is expected to cope with the requirement. These are examples of the gaps in this application domain:

1. The watcher wants to view the watched person's facial color or expression at high resolution.
2. The watcher wants to view the watched person comprehensively, as smoothly as possible.
3. The watched person does not want to be monitored during private times, etc.

Existing studies have merely switched the cameras and the displays because they only consider a user's location information. Therefore, even if the video quality does not meet the requirement, the camera/display that is the nearest to the user is selected. It is necessary to understand that multimedia communication in a ubiquitous computing environment, which includes resource-sensitive devices and unstable wireless links, can be a big challenge.

To achieve the goal, a dynamic system construction mechanism is promising because it can consider contexts of diverse system elements such as device status, network congestion, and software availability, as well as the locations of the users of both sides. Furthermore, the users' requirements and social relations are expected to be taken into account in systematic manner. In addition, robust and resilient software development infrastructure is required to handle reconfiguration and extension of system components in a pervasive environment where elements are radically changeable.

Taking these points into consideration, we have been developing a ubiquitous care-support system, named uEyes, which provides gentle supervisory services to both sides of the watched person and watching people in a ubiquitous computing environment. We describe the details of design and implementation of uEyes based on Symbiotic Computing and show results of initial experiments to verify the effect of the computing model, in the following sections.

AGENT-BASED DESIGN OF UEYES

Model of uEyes Based on Symbiotic Computing

Figure 9 shows a design of uEyes based on the SYMA Architecture. In the figure, functional requirements of uEyes are mapped onto the functions in SYMA. Here, PF includes management of users' locations of watched and watching persons and their requirements. Furthermore, SF involves basic knowledge of human relationships and activities of daily life for the users. In this application domain, Network Function (NF) is richly functional because the supervisory service is fundamentally constructed as a highly distributed and resource-consuming multimedia communication system. In Ubiquitous Function (UF), a location sensor to capture the physical location of a tag is controlled. Web Function WF includes abilities to control real-time multimedia streaming in terms of Quality of Service (QoS), delivery destination, and privacy level. Device Function (DF) is newly

Figure 9. Design of uEyes based on the architecture of symbiotic applications

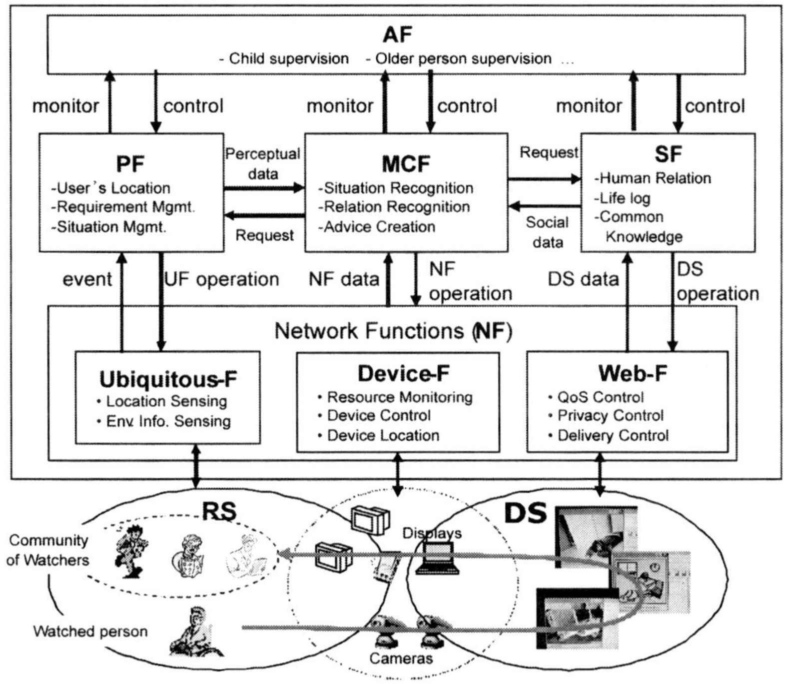

introduced to control display and capture devices, which play an important role as an interaction point of RS and DS. Mutual Cognition Function (MCF) is the core component of this system. It obtains information from other functions and performs recognition related to human relationships and situations of users related to the supervisory task. The results of the recognition are used to operate NF, which actually effects the multimedia streaming operation. Application Function (AF) is for management of overall system behavior, depending on the specific application such as supervision for elderly persons, target objects, children, and pets.

Dynamic Organization of Agents in uEyes

In uEyes, arbitrary combinations of a watched person and watching people should be connected with a live streaming video channel in various kinds of ubiquitous information environments. To realize this, functional elements in PF and NF should be constructed dynamically according to the physical location of users, namely watched and watching persons, available ubiquitous devices around the users, network status, situation of the users, human relationships between the watched and watching persons, etc. For example, suppose that a watched person Z walks through her home. During Z's time in the dining room, where an ultrasonic location sensing system is installed, the sensor should be used to identify Z's location. When Z moves to the bedroom where no location sensor is available, other types of location sensing functions such as light switches, are expected to be included in the system. In addition, when Z is watched by her son, high-resolution capture and display devices and broadband network can be incorporated because we do need not consider the privacy issues in this case.

In Symbiotic Applications, the organization of functional elements, the agent organization, is constructed dynamically using the organiza-

Figure 10. Initial agent organization of uEyes based on the architecture of symbiotic applications

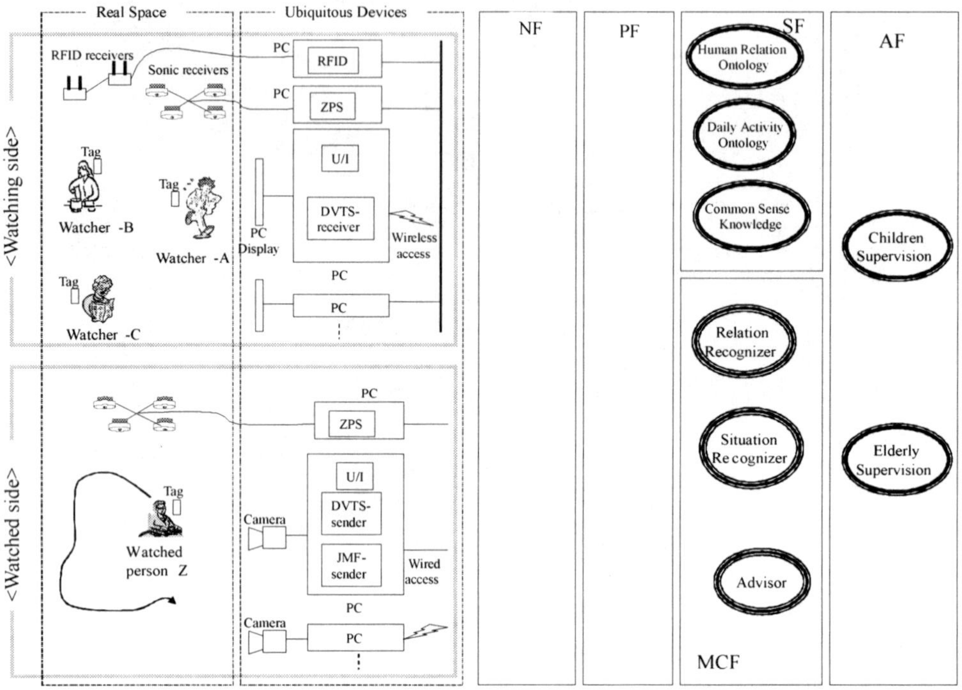

tion and reorganization mechanism proposed in section 5. This dynamic configuration ability is a remarkable feature of our agent-based architecture for SYMA.

Figure 10 represents the initial state of the agent organization of uEyes. Several agents reside in SF, MCF, and AF. The MCF has three agents: Relation Recognizer, Situation Recognizer, and Advisor. The Relation Recognizer agent recognizes the human relationship between the watched person and the watcher in the real world. The Situation Recognizer agent recognizes situations of users in the real world in detail, based on the data of his location and movement acquired by NF agents. The Advisor agent creates QoS advice based on information from the Relation Recognizer and the Situation Recognizer. Then it sends advice to the NF agents to control the video quality directly. The agents in MCF have intelligent ability to infer the situation of RS by wrapping a knowledge-based processing mechanism such as a rule-based system.

The SF agents provide social knowledge related to supervision. The social knowledge includes basic information about human relationships between the watched person and a watching person, the concept of daily activities such as "sleeping"" and "taking meal", and common knowledge related to supervision such as "people sleep in a bedroom at night". The SF agents hold these kinds of knowledge by wrapping an ontology framework and providing useful information with other agents. Application Functions agents, AF agents, are introduced for management of overall system behavior depending on the specific applications. Under management of this agent, agent organization for a specific purpose, for instance, supervision of children or elderly people, is constructed dynamically.

Figure 11 is an example of an agent organization that has been dynamically constructed in NF and PF according to the situation of RS. In this situation, suppose that a watched person Z is supervised by his son A. In the room in which

Figure 11. Agent organization of ueyes dynamically constructed according to situation of RS

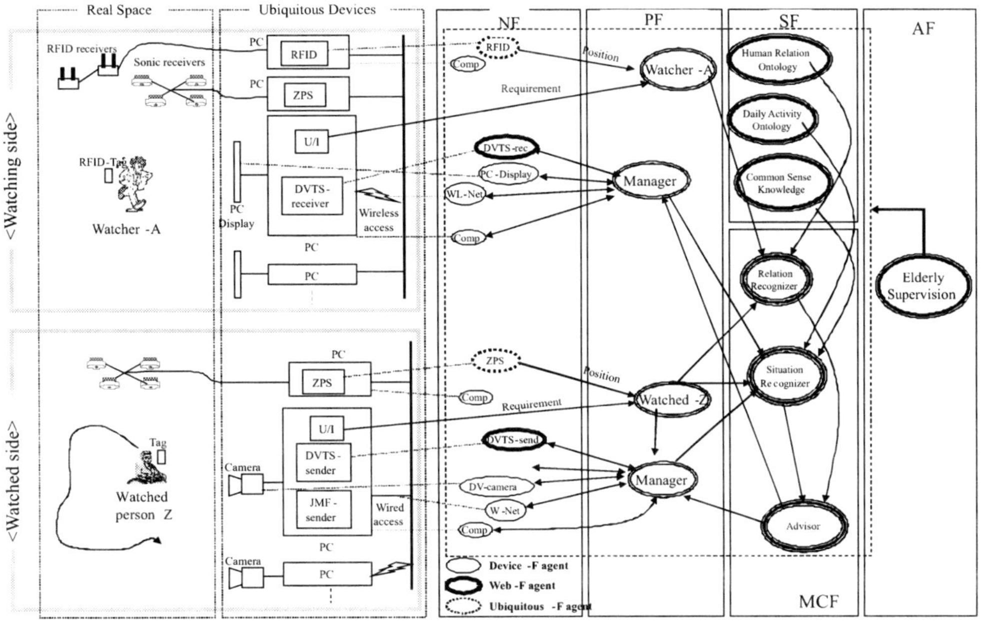

Z is staying, an ultrasonic location sensor system can be available. In the room where A is staying, an RFID system is useful. Therefore, first, these ubiquitous devices are incorporated, and the agent controlling the devices, the ZPS agent and RFID agent, are instantiated in NF. When A and Z enter the rooms, the User agent, i.e., Watcher-A agent and Watched-Z agent are instantiated to PF; they start cooperation among location sensor agents to trace the physical location of the users. From the user's location, Situation Recognizer agent judges the situation of the users. For example, Z sleeps in bedroom in the afternoon, the Situation Recognizer agent recognizes that Z's health condition is bad and sends a warning to the Advisor agent. The Relation Recognizer agent also considers the human relationship between A and Z. In this case, they share a family relationship; for that reason, privacy concerns are not considered. The service organization process starts at this moment. In this situation, high-resolution video is necessary to show the detailed situation of Z to A. Consequently, DVTS-send and DVTS-rec agents, which deliver high-resolution video using DV over IP technology [Ogawa, 2000], are incorporated into the organization as well as a DV-camera agent. However, because a personal computer (PC) near A is connected by wireless network access, the DVTS-send agent specifies the parameter of frame-rate to a low level to maintain high-resolution in low bandwidth. As shown in this example, agents in NF and PF are dynamically organized to match the situation of RS.

Agent Organization Process of uEyes

The NF agents and PF agents described in the previous section work cooperatively to provide a QoS that meets the user's requirements for the observation task and device situation. Here the ADIPS organization and re-organization protocol is used to construct the agent organization. This cooperative behavior is performed with the following steps:

- [*Step-1: IAR update*] Agents with a physical location closely exchange information on each context, and update the Inter-agent Relationship (IAR). The details of IAR are discussed in [Takahashi, 2005] and are omitted in this paper. The IAR is used to organize agents effectively during organization protocol execution based on the Contract Net Protocol. Agents in the same PC potentially have a tight mutual relationship. For example, the DVTS-send agent will have limitation of its video transmission performance through the status of the Comp agent's CPU resource context and WL-Net agent's bandwidth context. The coordination of this tradeoff is prepared to be performed by IAR update process.
- [*Step-2: User information acquisition*] The user agent in the closest PC or handheld device collects a user's information such as requirements and profiles. This information is kept by the User agent in PF; it is used to check whether provided QoS of services are satisfied or not.
- [*Step-3: Agent organization decision in initiate site*] When a user moves into a sensor area, the RFID agent or ZPS agent captures the movement and informs related agents based on the user's location. Then, only the related agents begin negotiation to determine the agent organization based on the IAR.
- [*Step-4: Agent organization decision in opposite site*] Agent organizations of both the sender site and receiver site are presumed to work properly to accomplish the QoS. In this step, coordination between the sender and the receiver site is performed. The SF agents would be activated in this step if privacy and emergency concerns were important.
- [*Step-5: Service provisioning starts*] After the organization is constructed, live video streaming starts by agents in NF, controlling each target such as the camera device, software, access network, display device, etc.
- [*Step-6: User's feedback*] After the end of the service, the user's feedback is collected by the User agent; then IAR is updated based on the user's evaluation.

Here, we describe the process of QoS control related to privacy and emergencies. This process is based on recognition of human relationships and environmental situation by SF and MCF agents.

In the initial phase in Step-4, the Relation Recognizer agent infers the human relationship between the observer and the watched person using their profiles that each User agent maintains. Figure 12 shows that the Relation Recognizer agent utilizes ontology. It represents background knowledge to infer human relationships. In particular, the agent calculates the human network and measures the distance between instances of the observer and the watched person by the ontology. Then the strength of the human relationship is inferred by this distance.

Next, the Situation Recognizer agent recognizes the situation in detail by acquiring information on the user's surrounding environment. This information is obtained from UF agents. For instance, the recognition processing is performed by firing rules that the Situation Recognizer agent has, as shown in Figure 13. In this example, from FACT01 and FACT02, it is recognized that a watched person is in the dining room now by Rule01, and that Tag is located at a position close to the floor. Furthermore, from FACT03 and FACT04, it is understood that the dining room is usually used to take a meal, and that a meal is not usually eaten close to the floor. Therefore, an abnormal situation is recognized by Rule02. Finally, the Advisor agent judges the providing service's quality and privacy level by considering human relationship information from the Relation Recognizer agent and user situation information from the Situation Recognizer agent. Then the Advisor agent directs the Manager agent. The Manager agent chooses the actual provision of QoS to the user by cooperating with NF agents.

Figure 12. Human relationship ontology provided by the human relationship ontology agent

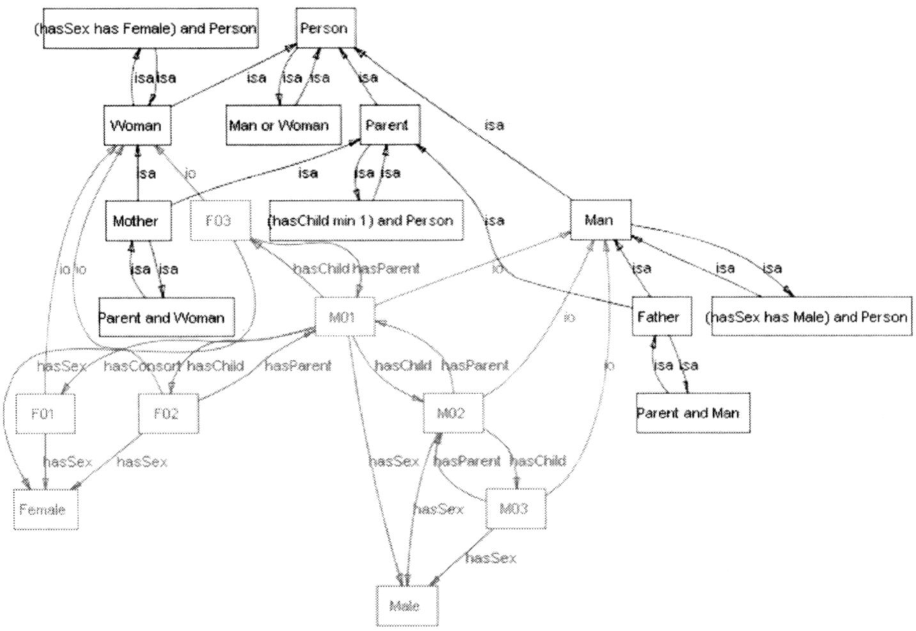

Figure 13. Example of situation recognition knowledge in a situation recognition agent

```
//Tag location
(FACT01 :tag-id 0003 :floor 1 :x 350 :y 1200 :z 50...)
```

```
//Room configuration
(FACT02 :desc room-prop :room-id 002 :floor 1
   :name myDinningRoom
   :role dinning :x1 50 :x2 500 :y1 1000 :y2 1700 ...)
```

```
//Rule whereHeIs
(rule Rule01
      (FACT01 :tag-id 0003 :floor ?floor :x ?x
              :y ?y)
      (FACT02 :floor ?f :name ?name :x1 ?x1 :x2 ?x2
              :y1 ?y1 :y2 ?y2) = ?selected
      (< ?x1 ?x)
      (> ?x2 ?x)
      (< ?y1 ?y)
      (> ?y2 ?y)
      (= ?f ?floor)
      -->
      (modify currentLoc:floor ?f)
      (modify currentLoc:room ?name)
)
```

```
//Room Purpose
(FACT03 :desc room-purpose :role dinning
   :purpose take_meal ...)
```

```
//Position
(FACT04 :desc position :purpose take_meal
   :min_high 1000 ...)
```

```
//Rule User Situation
(rule Rule02
      (currentLoc :room ?name)
      (FACT02 :name ?name) = ?room_c
      (?room_c :role ?role) = ?room_p
      (?room_p :purpose ?purpose)
      (FACT04 :purpose ?purpose)
              = ?selected_p
      (FACT01 :tag-id 0003 :z ?z)
      ~ (< ?selected_p:min_high ?z )
      -->
      (modify situation:status abnormal)
)
```

IMPLEMENTATION AND EXPERIMENTS

Implementation of uEyes

We are developing an application of uEyes to supervise elderly people in the home. Figure 14(a) and (c) portrays a snapshot of user terminal for uEyes. This is a special device that is always brought with an observer. This provides the basic receiving capability for video streaming. Consequently, in cases where other displays cannot be used, this terminal will be selected for receiving the video of watched people because it is the nearest device for displaying to the watcher. In addition, the User agent resides in this terminal to monitor the user's requirements and presence. In this example, we use a handheld PC

Figure 14. Hardware configuration

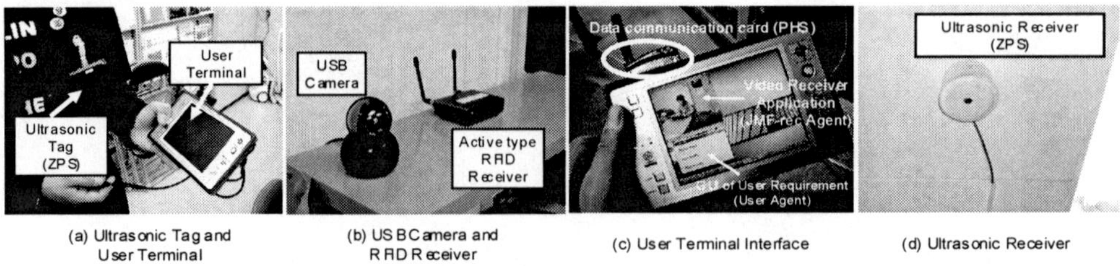

Figure 15. Experimental environment settings

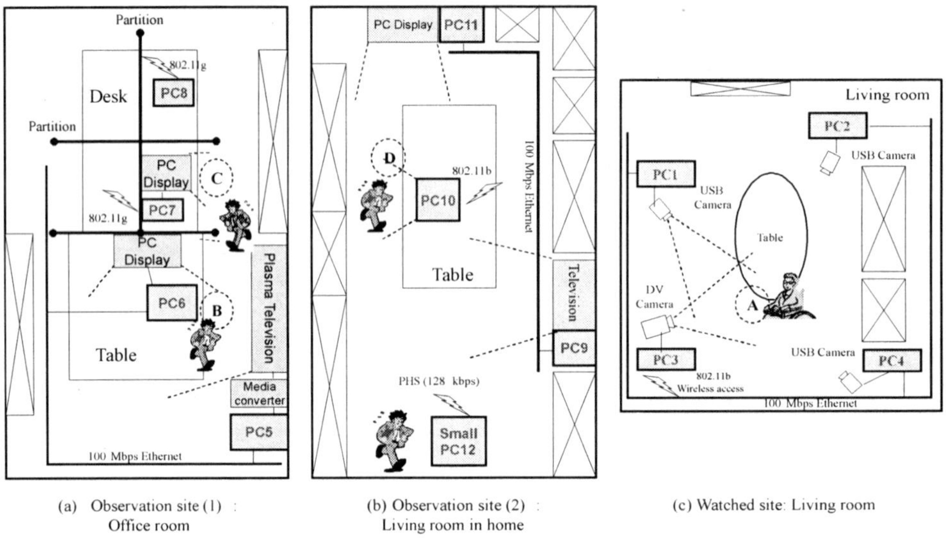

(Microsoft Windows XP, Celeron M 900 MHz, 256 MB memory, VAIO type-U; Sony Corp.) for this device. It is connected to the network with a CF type data communication card (PHS), which provides a link of 128 kbps bandwidth. We use sensors of two kinds to obtain location information of uses in the room. We use Furukawa Sanki's Zone Positioning System (ZPS) [ZPS, 2008] for ultra sonic sensors as shown in Figure 14(a) and (d). Figure 14(a) shows a ZPS tag in the style of a name plate. We also use an RFID system (Fujitsu Software Technologies Ltd.) [LPS, 2007], as shown in Figure 14(b). This is an active-type RFID system using 315 MHz radio frequency. It can recognize the tag location within 2 m from receiver, in a minimum setting.

Experimental Environment and Scenarios

Figure 15 shows the room settings of the watching site. Figure 15(a) depicts the room installation of the office room and Figure 15(b) as that of the living room of his son. Several display devices are used, including PC displays, built-in displays on laptop PCs, and TV sets. They are connected to PCs that are linked by wireless or wired networks. Here, "Small PC12" represents the user terminal described in section 7.1. In terms of the location sensor, a ZPS ultrasonic sensor is used in both (a) and (b). Here, the elderly father is in his living room. His son moves around his living room and office room. In this situation, our system will select the most appropriate camera, a PC with reasonable network connection, and display devices, considering the RS situation. Then suitable quality of live video is displayed on one display according to the son's requirements for the supervision and status of devices.

Experiment

Exp.(1): Experiments with NF Agents and PF Agents

Method: In the experiment, the watching person first specifies a user requirement from "best resolution" and "best smoothness" options, using a user interface on the user's terminal provided by the User agent. This selection is based on the background of the following:

a. The son wants to observe the father's facial color or expression in high resolution of the video because he is worried about status of his father's illness (Exp.(1)-1).
b. The son wants to see his father's full posture as smooth as possible because he is concerned about his father's lower back pain condition (Exp.(1)-2).

Subsequently, the son moves through the rooms. In these experiments, the father's location is fixed for simplification. He is at point "A" in his living room shown in Figure 15(c). Based on the requirements and location of the son, agents work cooperatively to select the most adequate sets of entities based on multiple context information. We observe the behavior of the agents.

Result of Exp.(1)-1: The son specifies the best resolution of the video to observe his father's facial color. Then he moves to the location at point "B" in Figure 15(a). Point "B" is the service area of PC display of PC6 and a television set connected to PC5.

Here, in the case of the location-based scheme, the video service moved to PC display of PC6 from the user terminal because it is judged to be nearest to the PC display. However, the video quality is too low to observe the father's facial color vividly because it is moved with the same video quality parameters as it is in the user terminal. The user terminal was connected to the network with 128 kbps. For that reason, the quality was reduced to save bandwidth resource consumption. This is caused by the lack of device awareness.

In contrast, in the proposed system, an agent organization was constructed. It selected the TV display and DV camera with high resolution to fulfill the user's requirement, as portrayed in Figure 16. In terms of software, DVTS is selected because it can provide high-quality video. Moreover, agents recognized that, at the sender site, PC3 with a DV camera is connected by 11 Mbps wireless link. Therefore, the full frame rate of DVTS transmitting will not be available. A DVTS-send agent tunes its frame rate to meet the network bandwidth. In this case, multiple situations of entities, involving the network, user, software, and hardware are effectively coordinated and user requirements are satisfied. This flexible configuration will reduce the gaps attributable to the inconvenience and quality of services.

Result of Exp.(1)-2: The son specifies the high smoothness of movement of the video to observe his father's health condition in this case. Then he moves to the location at point "D" in Figure 15(b). Point "D" is the service area of portable PC PC10 and PC display of PC11.

Here, in the case of a location-based scheme, the video service moved to display of PC10 from the user terminal because it is judged to be the nearest PC display. However, the frame rate of the video is too low to view the movement of his father's body smoothly. On the other hand, the proposed system selected the PC display of PC11 and the USB camera connected to PC1, with high frame rate to fulfill the user's requirement. In terms of the network context, PC11 is the best because it is connected by a wired link with 100 Mbps. Moreover, agents recognize that PC11 cannot play DVTS video because DVTS software was not installed in PC11. For this reason, the

Figure 16. Result of Exp.(1)-1: Effect of NF and PF agents in case of high quality requirement

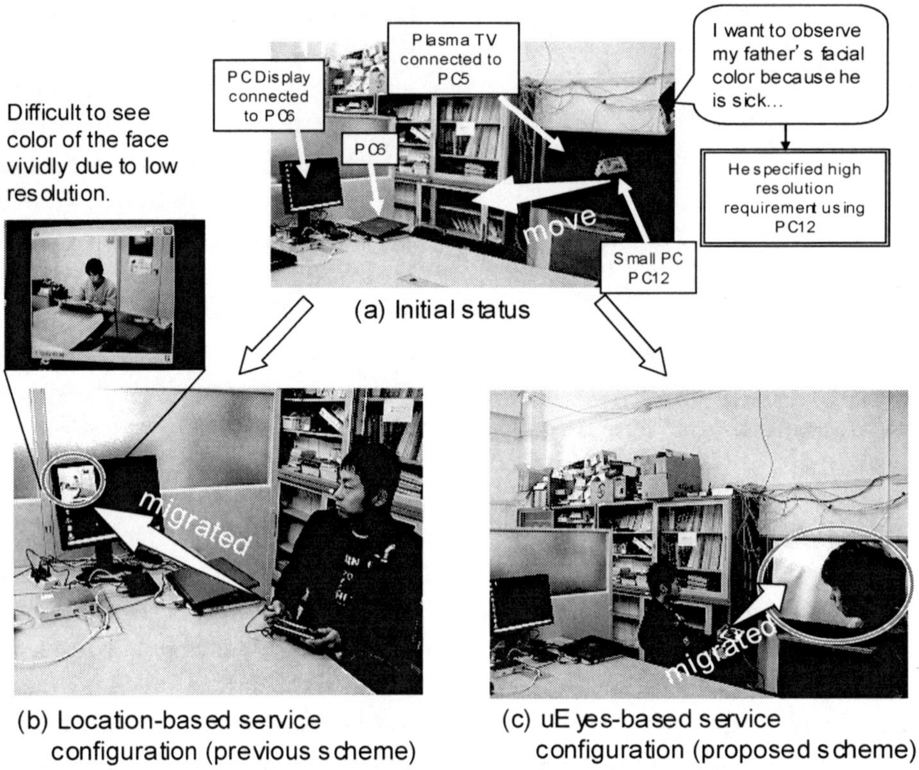

USB camera with JMF-send agent is selected. The JMF-send agent also controls frame rates as high as possible. In this case, multiple situations were deeply considered; user requirements for high smoothness of the video were satisfied by the constructed agent organization.

Exp.(2): Experiments with SF Agents

Next, we performed an experiment to verify the effect when SF agents are introduced. It is expected that it achieves the goals of observation by a community of two or more people because understanding of the interpersonal relationships and recognition of the situation can be done by the SF and MCF agents.

Here, we show an example of observation by several persons: (a) watched person's son, (b) his relative, (c) his neighbor with a good relation, and (d) a person of the same town as him. In a situation where the watched person's privacy is considered, such as when he is in his bed room, agents are organized to deliver appropriate video images. In this case, the JMF-send agent, which can operate image filtering, is incorporated into the organization. Then the JMF-send agent controls the image to provide each display to (a)–(d), as shown in Figure 17. A raw image with high quality is delivered to the son. A lower quality image that portrays only movement is delivered to the relative. The others do not receive an image.

Another experiment is performed to show an example of the display in the emergency situation when a watched person falls, as with a seizure. The Situation Recognizer judges it to be an emergency of a fall, etc., using the elderly person's location information provided by an ultrasonic wave tag, and the background knowledge related to the structure of the house. The display is changed so that many people can know the situation by lowering the privacy level. Concretely, an unretouched image is delivered to the son and the

Figure 17. Result of Exp.(2): Effect of SF Agents in case that it needs to keep watched person's privacy

(a) His son's view (b) His relative's view (c) His neighbor's view (d) View of his friend living nearby

relative, a neighbor with a good relation to the watched person receives a low-quality image that shows only the appearance; a person of the same town is notified by an emergency message without any video image.

Discussion

From experiments described in previous sections, we confirmed the feasibility of proposed system based on Symbiotic Computing. We evaluated that our system can effectively construct a service configuration that matches the requirement, coping not only with the location information, but also with device and network status around users in a ubiquitous environment. In this application, heterogeneous entities like display devices, capture devices, PCs, networks, different kinds of sensors, software components, etc., are wrapped by the wrapping-based model of the ADIPS framework; they are integrated efficiently. This architecture would accelerate the reuse of the components used in this care-support system in the form of agent, by other types of applications. Basically, the modularity, the autonomy and the loose coupling characteristics of the agents would conform to the construction of symbiotic applications. It can adapt to diverse types of entities and scalability of system size. Using this architecture, development and extension of the system can be accomplished easily.

On the other hand, we confirmed the effect of SF agents and MCF agents. Using these agents, uEyes can recognize relationships between people in the real world and detailed situations of the watched person. Based on that information, uEyes can handle QoS parameters and set privacy levels to an adequate level. These social functions contribute to enhancement of security and safety in service provision for non-expert users to bridge the gaps.

In terms of the privacy concerns, this system provides a very simple function to switch between different views of the watched person, based on the pre-defined human relationships and system/network resource status. This application example is shown mainly to describe the overall system behavior by simplifying each individual function. The privacy is one of the most important issues in this kind of system. Also we have to consider more gentle human interface provided to the watched person, including privacy level instruction and confirmation of the list of watching persons. We are investigating these additional functions; however, further discussion is out side the scope of this paper.

This work is assuming the situation where the ubicomp environment matures at a certain level. Thus, as for the deployment of this system in practical situation, we have some limitations in current ubiquitous computing environment in terms of cost and usability. To address these problems, some kinds of automatic configuration mechanism to reduce users' burden is required in installation and operation. The autonomous property of agent can help to realize this mechanism; however we will consider this in our future work.

RELATED WORKS

The proposed model and architecture based on the concept of Symbiotic Computing aims at realizing the post-ubiquitous computing environment. Therefore, our work is related to many other studies in the field of ubiquitous computing. For instance, in EU, a Framework Program for Information Society Technology (IST) has been progressing, showing the direction of the road ahead to realizing the "Ambient Intelligence" towards 2010. Many projects are on going to achieve the goal in the Framework Program FP7. Our work provides a generic model to integrate the existing technologies for ubiquitous computing and to implement them in a systematic manner, according to the basic model of co-existence between real space and digital space. Moreover, the Social Function in our model is a remarkable function that would be a key component to overcome the traditional ubiquitous computing environment. This function can add another important aspect of human and community, such as human relationships, social rules, common sense, social behavior, etc, in order to give the users a pleasant feeling of safety when they have to directly deal with the IT environment.

In addition to the concept, model and architecture of Symbiotic Computing, we give a software infrastructure to implement symbiotic applications based on multiagent technologies. There has been considerable research on frameworks to provide ubiquitous services [Itao, 2004; Minar, 1999; Minar, 2000; Roman, 2002; Minami, 2003; Gribble, 2001]. Ja-Net [Itao, 2004] aims to construct emergent services based on user preferences. Hive [Minar, 1999; Minar, 2000] is a multi-agent platform for dynamically creating services through interaction among agents. Roman et al. proposed a middleware for active spaces [Roman, 2002]. STONE [Minami, 2003] and Ninja [Gribble, 2001] can provide the services based on the service template that is requested by the user. These superior frameworks and service construction schemes are investigated for dynamic cooperation among many kinds of system components to provide user-oriented services. These previous works are based only on user contexts and functional components while concentrating on providing guarantees of coordination and operation, or standardization of the specifications. Therefore, these works are expected to satisfy a particular requirement and limitation including the network and computer resources. We believe that it is much more important to consider QoS for provisioning services, particularly in the ubicomp environment. Compare to the existing works, our system gives the dynamic system construction mechanism, considering not only a user's location, but also multiple contexts of diverse system elements such as device status, network congestion, software availability, etc., as well as users' requirements in a systematic manner. Moreover, a robust and resilient software development infrastructure can be provided to handle multiple reconfiguration and extension of system components in ubicomp environments in which elements are eminently changeable.

In terms of the supervisory application systems, some research groups have investigated the application for health care in ubicomp environments. GerAmi [Corchado, 2008] is an advanced support system with intelligent agent that dynamically schedules nurses' tasks, reports on their activities, and monitors geriatric patient care. In a similar study, researchers have studied the contextual information to recognize user activities to provide opportunistic services to hospital staff, by using a hidden Markov model [Sanchez, 2008].

In one of these attempts, the live video supervisory systems have been studied. They offer detailed information of various target persons with the sense of safety on the watching side, which cannot be obtained by traditional monitoring systems [Rowan, 2005; Boudy, 2006; Kang, 2006], which transmit only the location information and vital information of the watched person. Some research groups are working on how to apply this kind of system to ubicomp environments [Takemoto, 2005; Silva, 2005; Cui, 2004; Lohse, 2005] as described in section 5. Existing systems have merely switched the cameras and the displays because they consider only a user's location information. Therefore, even if the video quality is inadequate, the camera or display that is nearest to the user is selected. Our system can consider a balanced relationship between QoS and privacy. It can handle the contexts of diverse system elements such as device status, network congestion, software availability, etc., as well as all users' locations, to realize suitable QoS. The users' requirements and social relations are also considered systematically to ensure privacy. Both aspects could be handled sufficiently and simultaneously. This successful implementation of the prototype system was lead by the concept of Symbiotic Computing and its system construction based on multiagent technologies.

CONCLUSION

We defined symbiotic computing based on ubiquitous computing, Web computing, and relevant models of AI and Cognitive Informatics. A model of the symbiotic computing in this paper is newly formalized based on an agent-oriented model to incorporate social heuristics and cognitive functions into DS to bridge the gaps. To realize the model, we used an agent-based framework in which agents construct and reconstruct organizations dynamically according to situations of RS. The agent organization, which is customized to diverse situations in RS, can integrate necessary and sufficient functional elements and information in DS. It can provide situation-aware services to users in RS. This flexibility is a core function of symbiotic computing: it plays an important role in facilitating the symbiosis of RS and DS by bridging the gaps.

We applied this concept and model to a support system for supervision of elderly people and children. The experimental results show that the gaps in that domain can be bridged successfully because its dynamically constructed agent organization for supervision considers the situation of users and devices in RS. The system can provide services that specifically address quality and privacy issues.

Future studies will apply this model to application domains other than multimedia supervision such as health-care support, child commuting support, cooperative intellectual work, and so on.

ACKNOWLEDGMENT

This work was partially supported by Sendai Intelligent Knowledge Cluster and the Japan Society for the Promotion of Science, Grants-in-Aid for Scientific Research, 19200005.

REFERENCES

Boudy J., Baldinger J.-L., Delavault F., Muller M., Farin I., Andreao R. V., Torres-Muller S., Serra A., Gaiti D., Rocaries F., Dietrich C., Lacombe A., Steenkeste F., Schaff M., Baer M., Ozguler A., & Vaysse S. (2006) Telemedecine for elderly patient at home: the TelePat project. Proc. of the 4th International Conference on Smart Homes and Health Telematics (ICOST2006)}, 19, pp. 74-81.

Corchado J. M., Bajo J. & Abraham A. (2008) GerAmi: Improving Healthcare Delivery in Geriatric Residences. IEEE Intelligent Systems, 23(2), pp.19-25.

Cui, Y., Nahrstedt, K., & Xu, D. (2004). Seamless User-level Handoff in Ubiquitous Multimedia Service Delivery. Multimedia Tools and Applications Journal, Special Issue on Mobile Multimedia and Communications and m-Commerce, 22, pp. 137-170.

Fujita, S., Hara, H., Sugawara, K., Kinoshita, T., & Shiratori, N. (1998). Agent-based design model of adaptive distributed systems. Applied Intelligence, 9, pp. 57-70.

Gribble S., Welsh M., Behren R., Brewer E., Culler D., Borisov N., Czerwinski S., Gummadi R., Hill J., Joseph A., Katz R., Mao Z., Ross S., & Zhao B. (2001) The Ninja architecture for robust Internet-scale systems and services. Special Issue of Computer Networks on Pervasive Computing, 35(4), pp. 473-497.

Hattori, H., Ohguro, T., Yokoo, M., Matsubara, S., & Yoshida, S. (1999). Socialware: multiagent systems for supporting network communities. CACM, 42(3), pp. 55-61.

Itao T., Tanaka S., Suda T., & Aoyama T. (2004). A Framework for Adaptive UbiComp Applications Based on the Jack-in-the-Net Architecture. Wireless Networks, 10(3), pp. 287-299.

Kang J. M., Yoo T., & Kim H. C. (2006) A Wrist-Worn Integrated Health Monitoring Instrument with a Tele-Reporting Device for Telemedicine and Telecare. IEEE Trans. Instrumentation and Measurement, 55(5), pp. 1655-1661.

Kinoshita, T., & Sugawara, K. (1998). ADIPS Framework for Flexible Distributed Systems. Lecture Notes in Artificial Intelligence 1599, Multiagent Platform, pp. 18-32.

Lohse, M., Repplinger, M., & Slusallek, P. (2005). Dynamic Media Routing in Multi-User Home Entertainment Systems. The Eleventh International Conference on Distributed Multimedia Systems (DMS'2005).

LPS: Local Positioning System. (2007) http://jp.fujitsu.com/group/fst/services/ubiquitous/rfid/index.html (in Japanese)

Lyytinen, K., & Yoo, Y. (2002). Issue and Challenges in Ubiquitous Computing. CACM, 45(12), pp. 63-65.

Minami M., Morikawa H., & Aoyama T. (2003) The Design and Evaluation of an Interface-based Naming System for Supporting Servic Synthesis in Ubiquitous Computing Environment. Journal of IEICEJ, J86-B(5), pp. 777-789.

Minar N., Gray M., Roup O., Krikorian R., & Maes P. (1999) Hive: Distributed Agents for Networking Things. Proc. of the 1st International Symposium on Agent Systems and Applications / 3rd International Symposium on Mobile Agents (ASA/MA '99), pp. 118-129.

Minar N., Gray M., Roup O., Krikorian R., & Maes P. (2000) Hive: Distributed Agents for Networking Things. IEEE Concurrency Magazine, 8(2), pp. 24-33.

Ogawa, A., Kobayashi, K., Sugiura, K., Nakamura, O., & Murai, J. (2000). Design and Implementation of DV based video over RTP. Packet Video Workshop 2000.

Oxford University Press. (2000). Oxford Advanced Learners dictionary.

Roman M., Hess C., Cerqueira R., Ranganathan A., Campbell R. H., & Nahrstedt K. (2002) A Middleware Infrastructure for Active Spaces. IEEE Pervasive Computing, pp. 74-83.

Rowan J. & Mynatt E. D. (2005) Digital Family Portrait Field Trial: Support for Aging in Place," {\it Proc. of ACM SIGCHI Conference on Human Factors in Computing Systems 2005 (CHI2005), pp. 521-530.

Sanchez D, Tentori M. & Favela J. (2008) Activity Recognition for the Smart Hospital. IEEE Intelligent Systems, 23(2), pp.50-57.

Shiratori, N., Suganuma, T., Sugiura, S., Chakraborty, G., Sugawara, K., Kinoshita, T., & Lee, E. S. (1996). Framework of a flexible computer communication network. Computer Communications, 19, pp. 1268-1275.

Shiratori, N. Symbiotic Computing Project. (2007) http://symbiotic.agent-town.com/

Silva, G.C. D., Oh, B., Yamasaki, T., & Aizawa, K. (2005). Experience Retrieval in a Ubiquitous Home. Proc. of the 2nd ACM Workshop on Capture, Archival and Retrieval of Personal Experiences 2005 (CARPE2005), pp. 35-44.

Smith, R. (1980). The Contract Net Protocol, High Level Communication and Control in a Distributed Problem Solver. IEEE Trans. Comp., 29(12), pp. 1104-1113.

Suganuma, T., Imai, S., Kinoshita, T., Sugawara, K., & Shiratori, N. (2003). A Flexible Videoconference System based on Multi-agent Framework. IEEE Trans. on Systems, Man, and Cybernetics part A, 33(5), pp. 633-641.

Sugawara, K. (2005). Agent-based Support System for Project Teaming for Teleworkers. Lecture Notes in Artificial Intelligence 3371, Intelligent Agents and Multi-Agent Systems, pp. 279-290.

Takahashi H., Tokairin, Y., Suganuma, T., & Shiratori, N. (2005). Design and Implementation of An Agent-based middleware for Context-aware Ubiquitous Services. Frontiers in Artificial Intelligence and Applications, New Trends in Software Methodologies, Tools and Techniques, 129, pp. 330-350.

Takemoto, M., Oh-Ishi, T., Iwata, T., Yamato, Y., Tanaka, Y., Tokumoto, S., Shimamoto, N., Kurokawa, A., Sunaga, H., & Koyanagi, K. (2005). Service-composition Method and Its Implementation in Service-provision Architecture for Ubiquitous Computing Environments. IPSJ Journal, 46(2), pp. 418-433.

Uchiya, T., Maemura, T., Hara, H., & Kinoshita, T. (2007). Interactive Design Model of Agent System for Symbiotic Computing. Proc. of 7th IEEE International Conference on Cognitive Informatics.

Wang, Y. (2005). On Cognitive Properties of Human Factors in Engineering. Proc. of ICCI2005, pp. 174-182.

Wang, Y. (2007a) The Theoretical Framework of Cognitive Informatics. The International Journal of Cognitive Informatics and Natural Intelligence (IJCINI), 1(1), pp.1-27.

Wang, Y. (2007b) On Laws of Work Organization in Human Cooperation, The International Journal of Cognitive Informatics and Natural Intelligence (IJCINI), 1(2), pp. 1-15.

Weiser, M. (1996) Ubiquitous Computing. http://www.ubiq.com/hypertext/weiser/UbiHome.html

ZPS: Zone Positioning System. (2008) http://www.furukawakk.jp/products/ (in Japanese)

This work was previously published in International Journal of Cognitive Informantics and Natural Intelligence, Volume 3, Issue 1, edited by Yingxu Wang, pp. 34-56, copyright 2009 by IGI Publishing (an imprint of IGI Global)

Chapter 4
Repository-Based Multiagent Framework for Developing Agent Systems

Takahiro Uchiya
Nagoya Institute of Technology, Japan

Hideki Hara
Chiba Institute of Technology, Japan

Kenji Sugawara
Chiba Institute of Technology, Japan

Tetsuo Kinoshita
Tohoku University, Japan

ABSTRACT

Agent systems have been designed and developed using recent agent technologies. However, design and debugging of these systems remain as difficult tasks of designers because agents have situational and nondeterministic characteristics and because any useful design support facilities have not been provided for designers. To raise the efficiency of the agent system design process, we propose an interactive design method of agent systems founded on an agent-repository-based multiagent framework that emphasizes an important feature of agent design: the use and reuse of existing agents from an agent repository. We propose an interactive design environment of agent system (IDEA) and demonstrate its effectiveness.

INTRODUCTION

Using recent Information and Communication Technology (ICT), people can get much information from Digital Spaces (DSs) such as the internet. Useful workspaces or communities can be constructed in a DS as well in a Real Space (RS) in the physical world. ICT changes the traditional society to modern networked societies where people can exchange information and knowledge freely and easily. However, emerging problems are apparent in internet society, such as the digital divide/e-Gap, security, and network-based crimes. Modern ICT should confront these

DOI: 10.4018/978-1-60960-553-7.ch004

difficult problems and provide solutions by bringing sociality and humanity into computing models. In fact, based on Cognitive Informatics (Wang, 2002), a model of cognitive properties to bring human factors and social relations into information processing was proposed and discussed by Wang and Kinsner (Wang, 2005; Wang, 2006; Wang, 2007).

To overcome these problems, *Symbiotic computing*, which we have proposed, provides a conceptual scheme to bridge an e-Gap between RS and DS (Shiratori, 2005; Suganuma, 2006). We considered that the e-Gap, from which problems arise, results from a lack of mutual cognition between RS and DS: people cannot receive advanced services without IT skills. Consequently, DS cannot provide a service suitable for users depending on their respective situations and preferences. Moreover, DS cannot provide a safe and secure service without heuristics related to a person's activities in a society, such as customs, norms, and expertise.

In our scheme, the agent-based design models of both the symbiotic function (SF) and the symbiotic application system (SAS), which consists of many SFs that support various activities of people, are adopted based on agent-based computing technologies. An SAS operates in an open distributed environment in which RS and DS fluctuate from time to time and provide stable services for people. Therefore, the SAS must deal with such fluctuations autonomously by tuning and changing its structure and functions. The necessary properties of an SAS such as intelligent, flexible, and adaptive properties can be realized easily by composing the SAS as an "Agent System" (Sugawara, 2007). In this paper, we specifically examine agent and multiagent technologies for building an SAS over an open distributed environment.

Generally, software with new characteristics such as autonomy and sociality is called an agent; an information system that uses agents as its components is called an agent system. Agent systems of many kinds have been designed and developed using recent agent technologies. As described above, in our symbiotic computing project, various SF are designed and implemented as software agents; an SAS is also realized as an agent system by selecting and organizing these agents dynamically.

However, the design and debugging of agent systems persists as a difficult problem not only because of the situational and nondeterministic behavioral properties of the agents, but also because of the lack of an effective design method and design-support technologies. To date, we have studied an agent-repository-based multiagent framework called ADIPS, which accumulates the developed agents and agent systems in an agent repository and which enables the dynamic composition and re-composition of agent systems based on this repository (Kinoshita, 1998; Fujita, 1998). Applying the ADIPS/DASH framework, a recent implementation of ADIPS framework with an effective agent repository (Hara, 2002; Uchiya, 2002; Uchiya, 2003), we developed various agent systems in our previous work (Suganuma, 2003; Imai, 2004; Konno, 2004; Kitagata, 2005; Takahashi, 2006; Abar, 2008; Takahashi, 2009; Kim, 2010).

In this paper, we propose a design support facility of agent systems based on the repository-based multiagent framework, thereby providing an efficient and systematic design environment for SAS designers. In section two, we discuss the state of the art of technologies and essential problems of agent system design. In section three, we present an overview of ADIPS/DASH framework as a basis for developing agent systems, and in section four, we propose an interactive design method of agent systems based on ADIPS/DASH framework. In this method, designers can interact with the repository in a design process of target agent system and use/reuse contents of the repository. The method emphasizes the essential features of agent system design based on the repository: (i) systematic use and reuse of existing agents and agent systems accumulated in the repository, and

(ii) cooperative interactions between designers and agents in the repository that support the bottom-up and top-down design process of the target agent system. Next, in section five, we design and implement a design support environment for ADIPS/DASH framework, called the Interactive Design Environment of Agent system (IDEA) (Uchiya, 2007). We evaluate essential functions of IDEA in section six, and conclude the paper in the last section.

PROBLEMS IN AGENT SYSTEM DESIGN

Design Task of Agent System

From a methodological perspective, in general, design schema of two types can be adopted in design of agent systems: top-down design and bottom-up design. The basic features of each design scheme are investigated together with existing design methods and tools.

(1) Top-Down Design

Identically to a conventional software system, a top-down design process comprises the following stages: problem definition, requirements definition, design, implementation, testing, and verification. Often, agents and their organization must be designed sequentially or concurrently to realize the target agent system. Design methods and tools such as ZEUS, JACK, MaSE (DeLoach, 2001), Tropos (Mylopoulos, 2002), and GAIA (Zambonelli, 2003) have been proposed to facilitate such top-down design.

For example, in ZEUS methodologies, the ZEUS agent is developed top-down by executing the following procedure: investigating a problem domain, defining an agent, defining a task, defining an agent organization, defining agent cooperation, generating agent codes, and implementing these tasks. In MaSE methodologies, a MaSE agent is developed top-down by executing the following procedure: capturing goals, applying use cases, refining roles, creating agent classes, constructing a conversation, assembling agent classes, designing the system, and generating the agent codes. Moreover, in GAIA methodologies, the GAIA agent is developed top-down by executing the following procedures: determining the requirements statement, analyzing the role model and interaction model, designing an agent model, a services model, and an acquaintance model, and designing the overall system and implementing the system.

Using these methods, the designers can promote the consistent design of an agent system and its component agents systematically. Therefore, inconvenient results such as competitive situations among agents might be detected and removed easily at the design phase. However, it becomes difficult to address changes of the structure and functions of the whole system after development has finished.

(2) Bottom-Up Design

Bottom-up design subsumes that the required agents can be selected from among a set of existing agents and assigned as components of the target agent system without the design and implementation of new agents. Although finding all agents of the target system is difficult, designers can concentrate on the design of new agents that realize the required new functions if the developed agents are useful and reused as a part of the target agent system with respect to the given requirements. Moreover, in bottom-up design, a function of autonomous and dynamic construction of an agent organization based on the developed agents might be expected at the runtime of the system. In such a case, advanced agent communication and cooperation functions must be provided for agents to be designed. This bottom-up design is an interesting method of agent system development, but useful design support tools based on the method have not been provided for designers.

Although the design and implementation of an agent system can be done in a top-down or bottom-up manner, testing and debugging of the target agent system are difficult because of the non-deterministic and situation-dependent behavior of the multiagent organization. The generate-and-test cycle used in an explorative design process of the agent system should be supported.

Next, we discuss the design support methods based on the following two approaches from a viewpoint of implementation of the agent system: (i) a Programming Approach and (ii) a Framework Approach.

(i) Programming Approach:

In this approach, an agent is designed and implemented in a top-down way using existing programming languages such as Java. Design flexibility can be retained in the design process. Therefore, agents with dedicated architecture like basic type and the reactive type agents can be implemented easily. However, for deliberative type and composite type agents, the burden of designers is expected to be increased because of the design and implementation of the internal mechanisms embedded in all agents.

(ii) Framework Approach:

By providing an agent design support environment based on the specific agent architecture for the designers, the design and implementation of the agents can be done systematically and efficiently. Such an environment, a "framework", provides facilities such as a knowledge representation scheme, a problem-solving function, and the agent communication function for the designers. In recent years, many frameworks such as ADIPS (Kinoshita, 1998; Fujita, 1998), JADE (Bellifemine, 1999), SAGE (Ghafoor, 2004), AgentBuilder (Reticular Systems), JATLite (Jeon, 2000), ZEUS (Nwana, 1999), and JACK (Renquist, 2001) have been proposed and used. These frameworks might provide many functions to design and implement various agents for designers. These frameworks typically provide user-oriented support facilities for design, implementation, debugging, and testing.

Problems of Agent System Design

From the perspective of a state-of-the art of agent system design, we emphasize two problems.

(P1) Under the top-down design method, the designer's burden might increase in direct relation to the size of the target agent system to be designed. For that reason, systematic bottom-up design processing, by which the developed agents are useful or reused, should be supported.

(P2) The costs of both testing and debugging of the agent system become larger than that of the non-agent system. Design support functions by which designers can interact with the target agents flexibly should be required.

To overcome these problems, we propose a design method that supports the bottom-up design process as well as the top-down design process. A prototype of the design support environment is also presented to demonstrate the effectiveness of the proposed method.

REPOSITORY-BASED MULTIAGENT FRAMEWORK

Basic Concept

The ADIPS/DASH framework is a latest repository-based multiagent framework developed by our research group (Figure 1). The essential functions of this framework can be summarized as follows.

- F1. Repository-based multiagent framework for distributed problem solving

Figure 1. ADIPS/DASH: Repository-based multiagent framework

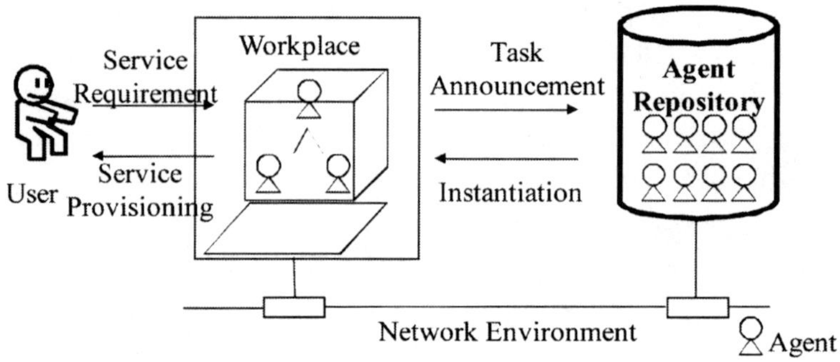

ADIPS/DASH framework provides a multiagent platform over the distributed environment, which consists of the distributed Agent Workplaces (abbreviated as the workplaces) and the Agent Repository (abbreviated as the repository).

The repository manages various DASH agents (abbreviated as the agents) and is responsible to design and realize the multiagent systems based on the users' requests.

A workplace is an agent execution environment on a distributed computer platform and is responsible to monitor and control the behavior of agents realized by the repository.

The user can design various multiagent systems by sending his/her request to the repository and realize the system on the workplace to execute the problem solving tasks.

Hence, this framework is called the Repository-based multiagent framework.

- F2. Building a multiagent system by Agent Repository

A multiagent system is constructed in the repository and delivered to the workplace. The organization procedure in the agent repository (Figure 2) is basically as same as the Contract Net protocol (Smith, 1980), however, the following functions such as the instantiation of organization onto the workplace and the reconstruction of organization at the run time of agent systems, are unique and important features of ADIPS/DASH framework as the developing and runtime environment of agent systems.

A user and/or an agent run on the workplace can send a request to the repository to start a design process of a multiagent system, which provides a required service for the user/agent. Receiving a request, the repository tries to find out suitable agents and construct an agent organization (a multiagent system) using the Extended Contract Net Protocol (ECNP) of ADIPS/DASH framework.

The ECNP is an agent cooperation protocol to deal with the design processes of construct, reconstruct and instantiate the multiagent systems.

When a required agent organization is successfully constructed in the repository, the repository instantiates the agent organization as a multiagent system run on the specified workplaces. Hence, the multiagent system can be realized with respect to the given request in a dynamic way.

- F3. Using multiagent systems on Agent Workplace

When a multiagent system is instantiated on a workplace by the repository, the workplace activates this system to start a problem solving task. The workplace monitors the behavior of these agents and collects runtime information such as execution log.

When the problem solving task of the multiagent system is finished, the agent organization

Figure 2. Organization procedure in the agent repository on the ADIPS/DASH framework

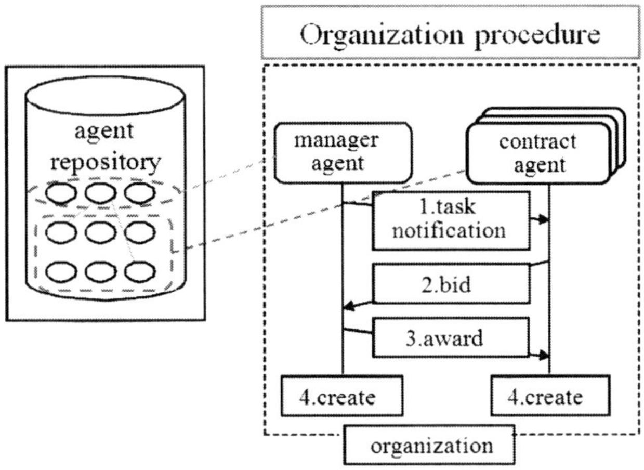

is dissolved by the workplace. All agents of the multiagent system are stopped and removed from the workplace.

On the other hand, it is possible to construct a multiagent system which consists of the active agents run on the distributed workplaces by using ECNP. In this case, a construction process based on ECNP is similar to the original CNP. Moreover, in order to use/reuse the realized multiagent system in the future, the workplace can save the multiagent system in the repository by a request for preservation.

- F4. Reconstruction of multiagent system at the runtime

It is possible to reconstruct the structure and functions of a multiagent system at the runtime of the system based on a request of a user or a problematic agent in the multiagent system. The ECNP provides the functions and protocols for reorganization operation.

- F5. Rule-based agent programming

The design of an agent is to describe the behavior knowledge for cooperative problem solving together with the meta knowledge for managing the agent in the repository.

The behavior knowledge is described as a set of rules using the rule-type knowledge representation format, on the other hand, the meta knowledge is described using frame-type knowledge representation format. The description of the designed agent is called the agent program, which are interpreted and executed using an inference engine (production system type engine) of DASH agent.

- F6. "Rule Set" for reusable knowledge

The general-purpose and/or useful behavior knowledge can be defined as a pre-defined set of rules, called "Rule Set" and stored in the repository, in order to support the agent design task based on the use/reuse of the rule sets.

The agent designer can select and specify the suitable rule sets in an agent program. In the repository, the specified rule sets are included in a knowledge base of the agent. A set of rules for handling a task oriented cooperation protocol is an example of the rule set.

- F7. Wrapping external program

The ADIPS/DASH framework provides a wrapping mechanism for the agent designers to utilize the external programs such as Java programs as the procedural knowledge of the agent.

- F8. Interoperation with other agents

The interoperation mechanism can be included in the ADIPS/DASH framework. Using this mechanism, the DASH agents can communicate with the different type of agents such as FIPA-compliant agents using the ACL messages of DASH agent (Li, 2008).

- F9. Test, debug and validation of multiagent system

As mentioned in previous section, the Interactive Design Environment of Agent system (IDEA) provides a design environment of agent/multiagent system (Uchiya, 2007). The details of IDEA are appeared in next section.

CONCEPT OF INTERACTIVE DESIGN METHOD OF AGENT SYSTEM

Repository-Based Design Support of Agent System

First, we explain the purpose and concept of repository-based design.

Systematic Use and Reuse of Existing Agent Systems Stored in the Repository

Agents and agent systems that have already been designed and used as applications are stored and managed to enable their use and reuse in the design process of a new agent system to support a bottom-up design process. For instance, the repository of the repository-based agent framework is useful as a necessary mechanism (Uchiya, 2003).

Cooperation of Designers and Agents in a Design Process of a Target Agent System to be Designed

The support functions by which designers can interact with stored agents and agents under development are provided for designers to support the generate-and-test cycle in the bottom-up design process. For instance, an interactive simulation of agents' behavior over a virtual distributed environment might be useful for agent testing and debugging.

Interactive Design Support of Agent System

We explain an interactive design method of the agent system. In the following design stages, we specifically address the design and implementation stage from the viewpoint of bottom-up design (Uchiya, 2007).

A. Problem Determination
A problem to be solved is defined in this stage, for instance, using a design specification language and natural language.

B. Requirement Definition
Requests from users are defined and described as requirement specifications of the target agent system. For instance, a state transition diagram or an agent knowledge description language is useful to represent the specifications. For example, a requirement definition is represented in an object-attribute-value form such as "(task:name XYZ:attribute1 value1:attribute2 value2)", which represents a required task XYZ specified using a collection of attribute-value pairs.

C. Design and Implementation
An agent system is designed and implemented based on the following sub-processes.

C-1. Attempts to Reuse the Existing Agent System

[Supportive function] Agent retrieval function

[Designer's task] Retrieve the existing agents from the repository

[Outline] A designer sends a requirement specification of the required function to the repository to use or reuse existing agents and the agent system. According to that requirement, each agent in the repository independently examines whether the given specification is fulfilled or not. Consequently, the designer receives responses from agents in the repository and receives a suitable agent for use in the design.

C-2. Programming of Agent Knowledge and Function

[Supportive function] Agent programming function

[Designer's task] Describe agent behavior knowledge

[Outline] Using an agent knowledge representation language, a designer describes the behavioral knowledge as a set of rules (it is called an agent program). Some functions that can be realized easily as procedures can be designed and implemented using conventional programming language. These procedures are combined with the agents using the wrapping mechanism. A knowledge template, which is a design pattern of agent behavioral knowledge and cooperation protocol such as the ECNP, is provided for designers to ease agent programming.

C-3. Interactive Simulation

[Supportive function] Test and debug function

[Designer's task] Simulate and verify the agent system behavior

[Outline] An interactive simulation function of agents' behavior is provided for a designer to support the generate-and-test cycle in the bottom-up design process. Using a virtual distributed environment defined by the designer, the designer can generate agents from the repository and observe their behavior. The behavioral simulation of an organization of agents can also be done as with a single agent. Moreover, a function of exchanging messages between the designer and the agents to be simulated is provided to test the behavior of agents under development. Using this function, for instance, the designer can play the part of an agent that is under development, thereby simulating its cooperative behavior with other agents in the virtual distributed environment. Using the interactive simulation, the designers can detect and modify the deficits of knowledge and functions of agents quickly by returning to stage C-2.

C-4. Registration of the Agent System to the Repository

[Supportive function] Agent registration function

[Designer's task] Register the designed agent system to the repository

[Outline] The results of the target agent system can be stored in the repository to support subsequent step-wise refinement of agent system development.

D. Test and Verification

The functions of the target agent system are expected to be verified in the real environment.

Comparison of the Proposed Bottom-Up Design Method to the Conventional Top-Down Design Method

The interactive design process of an agent system based on the proposed design method is portrayed in Figure 3. In this design process, the fundamental structure of design processes from stage-A to stage-D resembles that of conventional top-down design processes. However, in stage-C, new functions have been introduced to decrease the designer's burden of development of agent. For example, in the C-1 stage, the existing agent and agent system can be used and reused efficiently as components of the target agent system; in the C-3

Figure 3. Interactive design process of the agent system

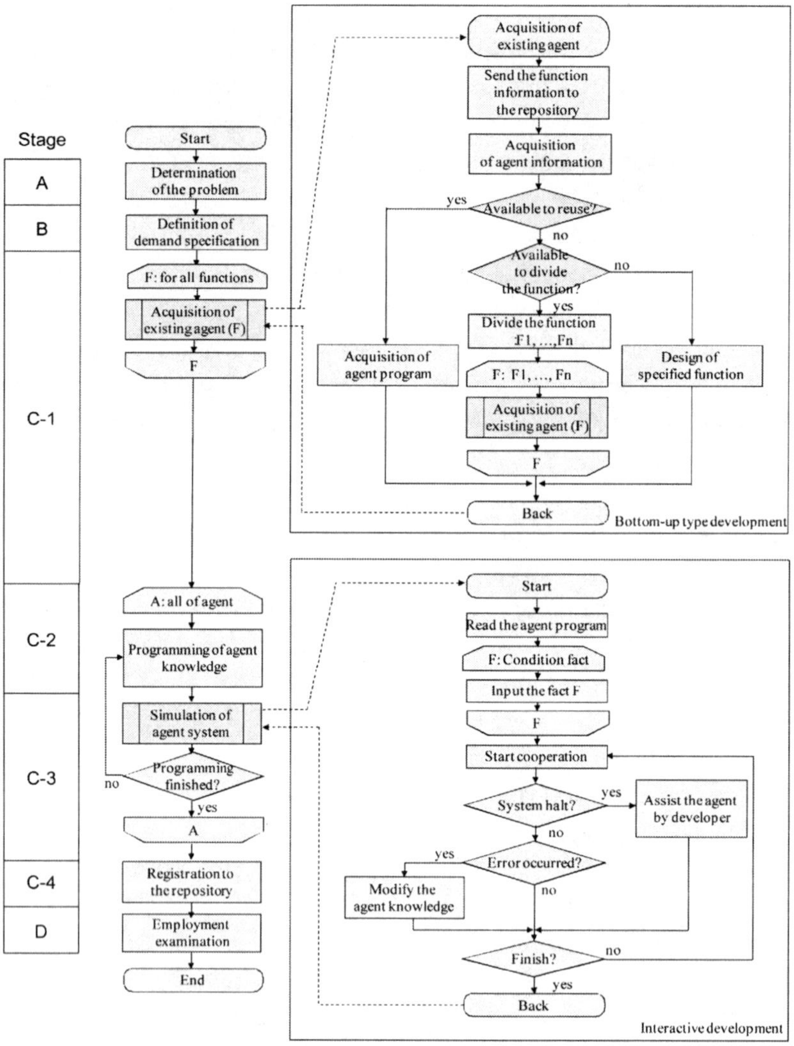

stage, the target agent system and the designer can mutually interact complementarily and execute development both interactively and smoothly.

DESIGN AND IMPLEMENTATION OF "THE INTERACTIVE DESIGN ENVIRONMENT OF AGENT SYSTEM"

According to the design method, we developed a prototype of an interactive design support system called the Interactive Design Environment of Agent system (IDEA) (Uchiya, 2007). We select and use a repository-based multiagent framework, ADIPS/DASH, for implementation of the IDEA prototype.

The functional relations between the ADIPS/DASH framework and IDEA are depicted in Figure 4.

The ADIPS/DASH is used to realize an agent execution environment equipped with the repository mechanism. In addition, IDEA provides

Figure 4. Overview of Interactive Design Environment of Agent system (IDEA)

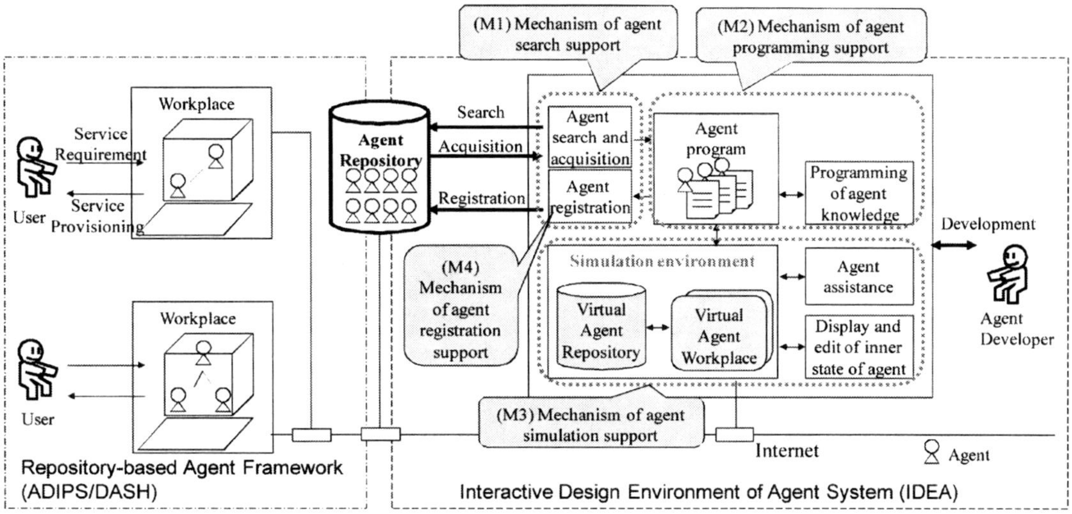

Figure 5. Support of agent search phase

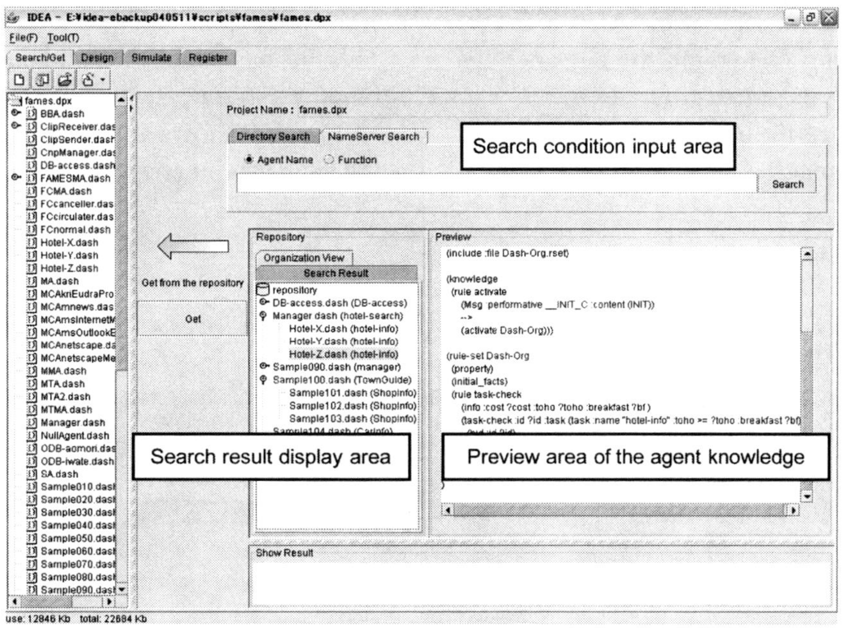

an interactive design environment for designers of agent systems.

The following four mechanisms are introduced into IDEA to support design from C-1 to C-4.

(M1) Mechanism of Agent Search Support

Figure 5 portrays the agent search interface, which supports the C-1 stage. Its three main parts are a search condition input for seeking agents from the repository, a search result display, and a preview of the agent knowledge.

In the search condition input area, a designer inputs the requirement specification of a candidate agent, such as an agent name and a function name. The received message is displayed in the search result display area when a candidate agent, which is detected in the repository, sends a message as its response. The designer can then move an agent into the developer's environment by choosing the agent from a search result window.

(M2) Mechanism of Agent Programming Support

Fig. 6 shows a mechanism for agent programming to support the C-2 stage. This mechanism has an agent-programming editor based on a rule-based knowledge representation of the ADIPS/DASH framework. Using this editor, the designer can describe and test the agent programs.

Figure 7 depicts the basic structure of an agent program: the "property" part holds a set of meta-data descriptions of this agent; the "initial_facts" part holds a set of default facts; and the "knowledge" part holds a set of rules of the agent behavior.

Figure 8 portrays an example of a rule description of behavioral knowledge of a CPU monitoring agent, which is a member of the videoconference agent system. This rule, named "make-report-service-down", specifies that the CPU monitoring agent must send a warning message with the "warning" performative to the videoconference manager agent "VCM.W1.spiral" when the CPU load given by the variable "?data1" exceeds the predefined threshold given by the variable "?thre".

(M3) Mechanism of Agent Simulation Support

This mechanism has some interactive simulation functions to support the C-3 stage.

Figure 9 shows an agent monitor for observing the behavior and organization of the agent in

Figure 6. Support of the agent programming phase

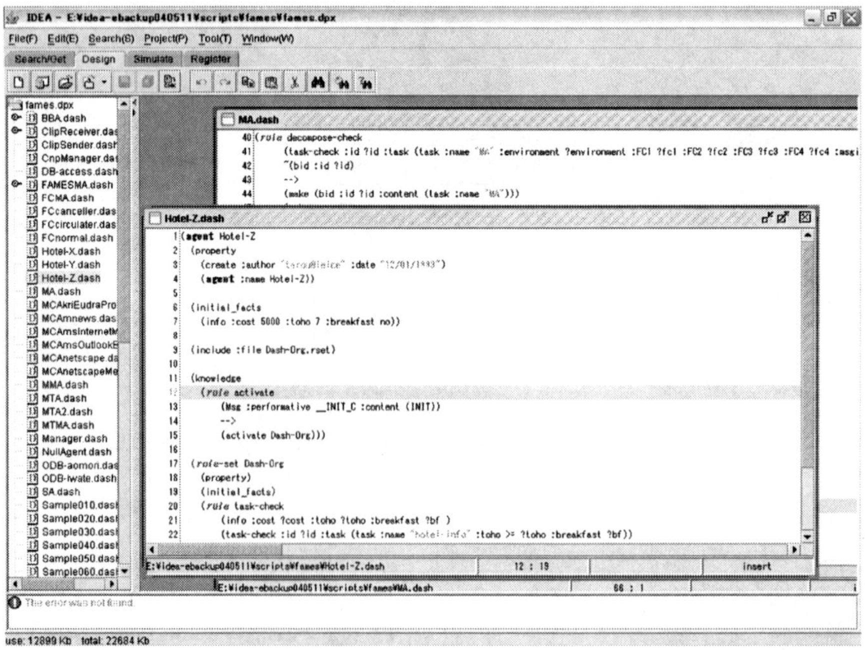

Figure 7. Agent program structure

```
(rule make-report-service-down
      (data :data1 ?data1)
      (threshold :threshold ?thre)
      (> ?data1 ?thre)
      (c-count :count ?cou) = ?cha
      -->
      (send :performative warning :to vCM.w1.spiral
            :content (request-action :action reduce-resource))
      (modify ?cha:count 0)
)
```

the virtual distributed environment. The agent inspector shows the inner states of an agent to monitor and modify the behavioral knowledge of the runtime agent. Furthermore, the ACL editor supports communication between the designer and the agents under development by the ACL messages of the ADIPS/DASH framework.

Using these tools, the designer can monitor and control the behavior of agents in an interactive

Figure 8. Example of agent behavioral knowledge

```
(agent <agentname>
  (property
   (create ...) )
  (initial_facts
   (...) ... (...) )
  (knowledge
   (rule ...) ... (rule ...) ) )
```

Figure 9. Support of the agent simulation phase

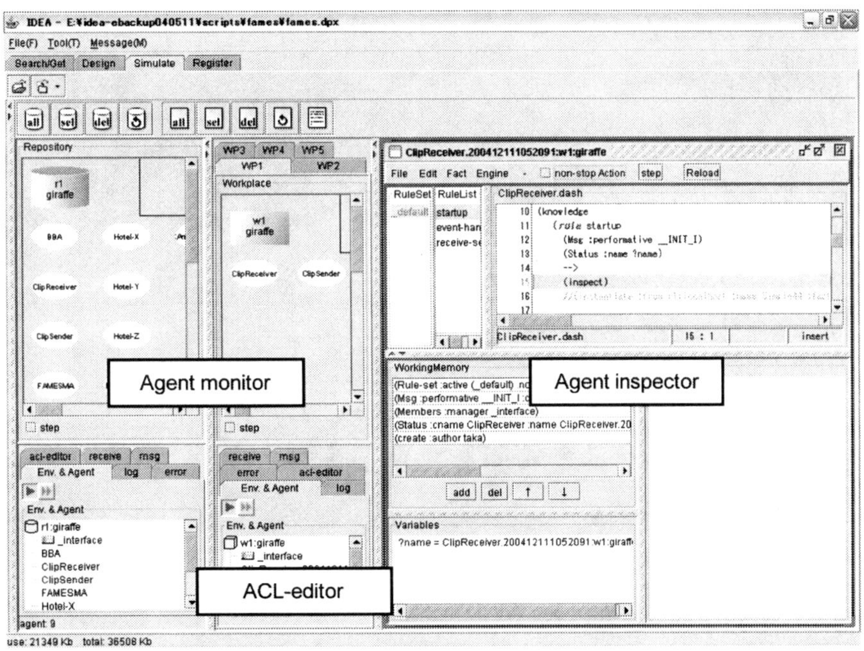

manner during the testing and debugging of agents. Moreover, the designer can access other design stages at any time by selecting the design stage tags shown at the upper part of screen. For instance, by selecting the "Design" tab, the designer goes to the "Design" stage to modify the agent program; then the designer will again view the "Simulation" stage and resume the simulation by reloading the modified agents.

(M4) Mechanism of Agent Registration Support

Figure 10 shows an agent registration support mechanism for the C-4 stage. This mechanism provides an interface to store the completed agent system to the repository.

EXPERIMENT AND EVALUATION

We performed the following evaluations to verify the validity of IDEA based on the interactive design method.

Evaluation of Agent Programming Support

Using an IDEA system prototype, we evaluated the amount of agent knowledge description in bottom-up development. Two cases of agent system development were assessed: (I) without reusing the existing agent system in the repository, and (II) reusing the existing agent system with the IDEA system. The following eight systems were selected for evaluation as deployable applications.

[Application 1] Hotel Selection System

This system addresses the problem of choosing the highest evaluated hotel from among several hotels based on a user request. The agent organization consists of three hotel agents, three evaluation agents, one overall evaluation agent, one secretary agent, and one blackboard agent.

{Agent Organization: Hotel Agent (3), Evaluation Agent (3), Overall Evaluation Agent (1), Secretary Agent (1), Blackboard Agent (1)}

Figure 10. Support of the agent registration phase

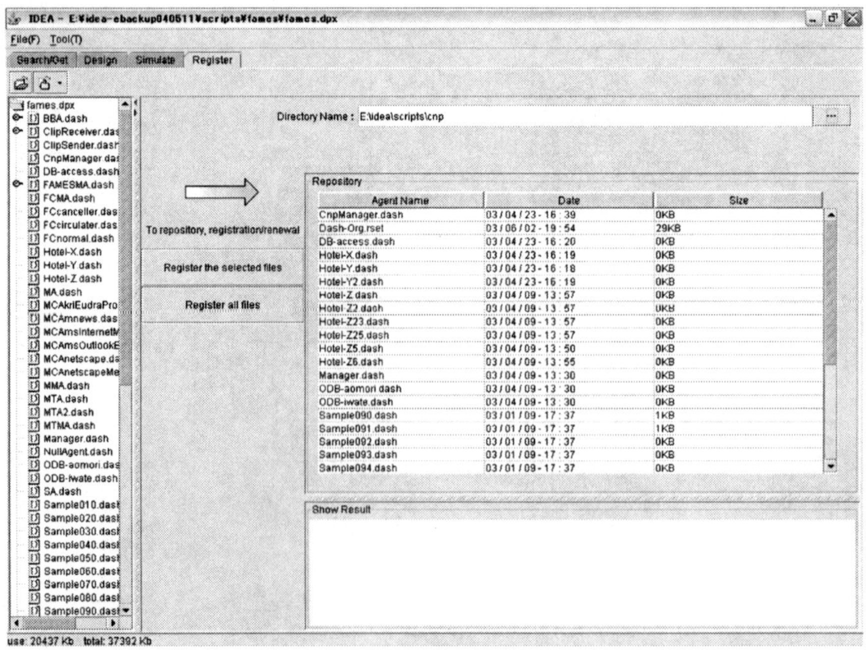

[Application 2] Meeting Schedule Adjustment System

This system adjusts a meeting schedule among several users.

{Agent Organization: Secretary Agent (3), Meeting Room Agent (1)}

[Application 3] Hotel Reservation System

This system reserves a suitable hotel using a Contract Net Protocol (Smith, 1980).

{Agent Organization: Hotel Agent (3), Travel Agency Agent (1), Secretary Agent (1)}

[Application 4] Retrieval System of UNIX Command Manual

This system retrieves the UNIX command manual according to a user request.

{Agent Organization: Interface Agent (1), Command Knowledge Agent (59)}

[Application 5] Retrieval System of Presence Information

This system retrieves the presence information of a user.

{Agent Organization: Retrieval Agent (1), Room Agent (5), Move Management Agent (1), Going out Management Agent (1)}

[Application 6] Ad-Hoc Communication Service (Kitagata, 2005)

This system provides ad-hoc communication services in the ad-hoc environment.

{Agent Organization: Service Agent (4), Service Management Agent (3), Name Resolution Agent (1), Communication Channel Agent (1), Connection Agent (1), Middleware Agent (1)}

[Application 7] Network Management System (Konno, 2004)

This system manages a local area network and detects obstacles autonomously.

{Agent Organization: Dynamic Information Management Agent (72), Static Information Management Agent (8), Organization Agent (8), Interface Agent (1)}

[Application 8] Asynchronous Messaging System (Kitagata, 2000)

This system provides a flexible asynchronous e-mail message service.

{Agent Organization: Manager Agent (6), Task Processing Agent (16), Secretary Agent (1)}

In case (I), the designers created the protocol, the rule set, and the agent knowledge from scratch. Testing and debugging was repeated and the burden on the designers increased.

In contrast, in case (II), the designers can seek reusable agents from the repository and assign the selected agents as some participating agents of the target system. The mean time of retrieving agents was less than 5 s, which can satisfy the needs of agent designers.

Moreover, knowledge templates such as the Contract Net Protocol template were useful. Figure 11 presents the experiment results. We confirmed that the agent knowledge descriptions are reduced to 54%, on average, of their former size. Moreover, the developer simulated and debugged the agent system smoothly because the operational verification of the existing agent had already been performed fully. This result illustrates that the burden imposed by the agent designer's programming work is reduced by support of the functions of agent search and template offered in bottom-up type development.

Figure 11. Evaluation of agent programming support

		Ex.1 Hotel Selection System	Ex.2 Meeting Schedule Adjustment System	Ex.3 Hotel Reservation System	Ex.4 Retrieval System of Manual of UNIX Command	Ex.5 Retrieval System of Presence Information	Ex.6 Ad-hoc Communication Service	Ex.7 Network Management System	Ex.8 Asynchronous Messaging System
No reuse	Total number of description lines	655	385	266	1863	839	12991	15479	35927
Reuse	Total number of developed agents	9	4	5	60	8	11	89	23
	Number of agents that developer reused using the repository	3	1	4	60	7	10	84	23
	Number of agents that are newly developed	6	3	1	0	1	1	5	0
	Total number of description lines [n]	655	385	266	1861	839	13018	15479	35996
	Number of reused lines [n-r]	316	107	107	1331	419	6894	12596	24635
	Number of newly written lines [r]	339	278	159	530	420	6124	2883	11381
	Rate of description curtailment [((n-r)/n)× 100]	48%	28%	40%	71%	50%	53%	81%	68%

Evaluation of Simulation Support

Evaluation of Interactive Simulation Functions

To verify IDEA's effectiveness, we tested the effects of interactive simulation functions: "virtual distributed environment", "exchange messages between designer and agents", "dynamic knowledge modification", and "message log analyzer".

We confirmed that the former three functions are useful via the agent simulation interface portrayed in Figure 9. For instance, the agent monitor of the simulation interface showed dynamic behavior of distributed agents in real time; the designers can observe non-deterministic behavior such as organization, communication, cooperation, and competition of agents.

The ACL-editor tab of this interface enables the designer to send ACL-messages to agents under development. Consequently, the designer can select and run an agent that is a part of the agent system to test and verify the functions of the agent and the agent system smoothly.

The agent inspector shown in Figure 9 enables the designer to modify knowledge of agents dynamically at the runtime of an agent system. Thereby, the designer can test and modify the functions of the agent system both quickly and smoothly.

The latter function, i.e. the message log analyzer, is portrayed in Figure 12. The left panel shows the message flow of the whole agent system; the right panel shows the message sequence of the agent system. They are generated using behavioral logs of agents stored in IDEA. They can be used to detect a bottleneck that restricts message passing of agent systems. They can also determine causes of agents' abnormal behavior.

The experience described above verified the effectiveness of interactive simulation functions.

Evaluation of Simulation Cost

The workload amount was measured as depicted in Figure 13 to evaluate the simulation support function of IDEA. The following measures are introduced.

Figure 12. Message log analyzer of IDEA

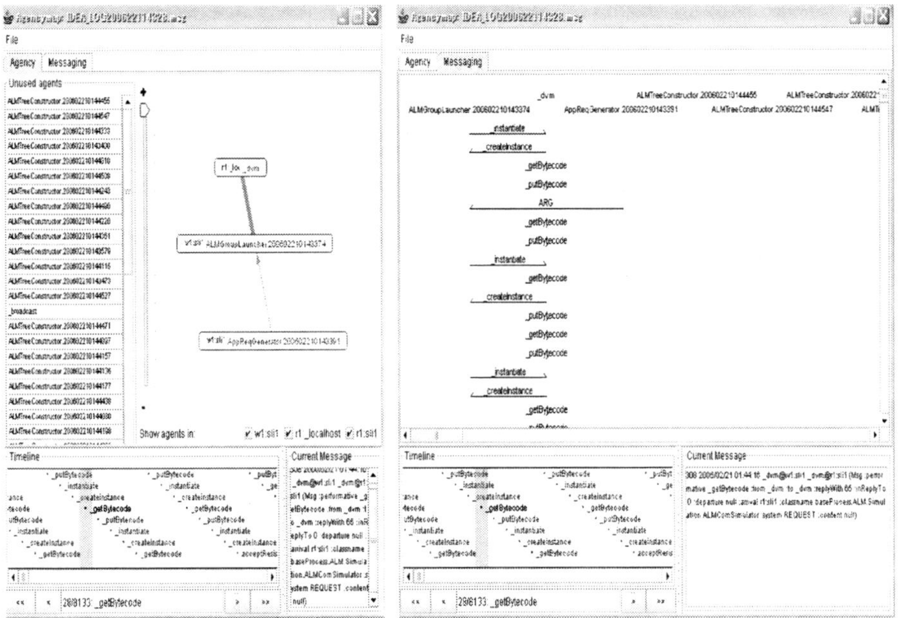

Figure 13. Cooperative action test of agent system

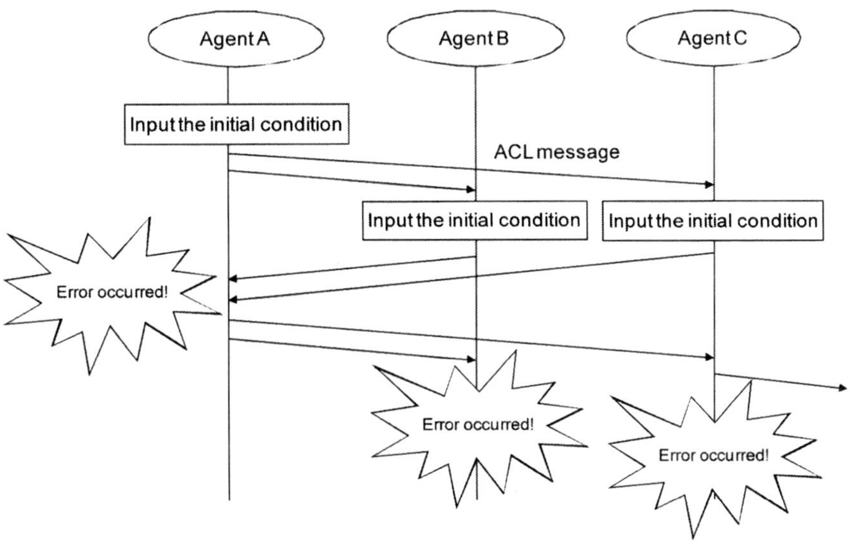

- Number of initial conditions: N
- Number of errors: k
- Time required for completion of correction of the i-th error from a cooperation operation start: T_i $(i \leq k)$
- Required time from the operation start to the end in the case of being errorless: T_{min}
- Required time from the operation start to the end using IDEA: T_a

Figure 14. Evaluation of debugging work reduction

	Number of times of start	Number of times to input the initial condition	Execution time
Normal	$(k + 1)$ 4	$(k + 1) \times N$ 12	$(STi + Tmin)$ 291(s)
Proposed Method	**1**	**3**	(Ta) **137(s)**

The results of this experiment are presented in Figure 14.

Using the agent system presented in [Example 2], we set up the agent system so that three errors might occur: once each at the beginning, middle stage, and the end of operations. Simulation parameters are N=3 and k=3.

First, T_{min} was about 75 s. Next we measured the debugging time without IDEA: the total time was about 291s. Whenever the designer corrects the agent knowledge in the simulation phase, the designer must reboot the agent system and input the initial condition repeatedly. The debugging time therefore increases according to the number of errors.

Finally, we measured the time necessary for debugging with IDEA. The total time of the proposed method is about 137 s because system rebooting is obviated. Moreover, the designer can set the initial conditions using a function of dynamic knowledge modification provided by IDEA. The result shows that the working time can be shorter than that without IDEA.

Through the comparison presented above, we verified that the time of debugging work was shortened, and that work is done flexibly and smoothly using IDEA.

Discussion

An agent system is realized by organizing many agents with respect to the given users' requests, and is often required to reorganize and redesign their structure and functions to address the changes of both the users' requests and the run-time conditions. Consequently, the roles of design methods and tools become important to reduce the burden on designers.

Although top-down, step-wise refinement design processes can be supported by existing methods and tools, the designers must address every aspect of design tasks simultaneously from the problem determination phase to the verification phase. The burden on designers increases concomitantly with the scale of the target agent system. Another important problem is providing suitable design models, design languages, and design functions for designers based on the respective design methodologies. Moreover, a persistent and difficult problem is the support of systematic use and reuse of the existing agents and agent systems in the design processes of the target agent system, considering the functional properties of the agent platforms.

To overcome the problems described above, an interactive design method and IDEA environment were proposed and implemented in this paper. Compared to existing design methods and tools, the proposed method specifically examines the following features: "Systematic use and reuse of existing agents and agent systems stored in the agent repository" and "Cooperation of the designers and the designed agents during the design process of the agent system". The interactive design processes with the design support functions are realized as the IDEA environment to provide a practical reuse-oriented design process for designers. Moving forward and backward in the IDEA design stage corresponds to trial-and-error actions; the designers can proceed with the top-down, bottom-up, or mixed design using the assets stored in the repository.

Then we discuss the scalability of our proposed method. Existing design methods and tools, in general, cannot cope with heterogeneous agents because they implicitly assume single agent architecture. In stark contrast, the proposed method based on the ADIPS/DASH framework enables the use and reuse of heterogeneous agents run on different platforms, such as JADE and SAGE, using the ADIPS/DASH framework agent repository. To do so, we introduce a proxy agent on the agent interoperation mechanism for ADIPS/DASH framework, which is a support facility based on the proposed method. A proxy agent is designed as an ADIPS/DASH agent that corresponds to a heterogeneous agent as a cooperative partner on the different agent platform to retain ACL-level interoperability between an ADIPS/DASH agent and a heterogeneous agent with different agent architecture or a different implementation of the ACL interface. Using proxy agents managed by the agent repository, the designer can find and use heterogeneous agents with useful functions and services as cooperative members of heterogeneous agent systems. Applying the proposed design method, the design and implementation of proxy agents can be implemented similarly to normal ADIPS/DASH agents using the design templates of proxy agents managed by the agent repository.

CONCLUSION

In this paper, we proposed a design support facility of agent systems based on the repository-based multiagent framework to provide an efficient and systematic design environment for agent system designers. A prototype of the Interactive Design Environment of Agent system (IDEA) is implemented based on the proposed method. Experimental results demonstrate the effectiveness of the IDEA in the development of agent systems.

In future works, we will provide an advanced agent search function, a protocol template (design pattern) function, an advanced debugging function, and so on. Moreover, we will expand the proposed design method by accumulating numerous design cases of agent-based symbiotic applications.

REFERENCES

Abar, S., Konno, S., & Kinoshita, T. (2008). Autonomous network monitoring system based on agent-mediated network information. *The International Journal of Computer Science and Network Security*, *8*(2), 326–333.

Bellifemine, F., Poggi, A., & Rimassa, G. (1999). JADE – a FIPA-compliant agent framework. In *Proceedings of Practical Application of Intelligent Agents and Multi Agents* (PAAM '99), (pp.97-108).

DeLoach, S. A., Wood, M., & Sparkman, C. H. (2001). Multiagent systems engineering. *International Journal of Software Engineering and Knowledge Engineering*, *11*(3), 231–258. doi:10.1142/S0218194001000542

Fujita, S., Hara, H., Sugawara, K., Kinoshita, T., & Shiratori, N. (1998). Agent-based design model of adaptive distributed system. *Applied Intelligence*, *9*(1), 57–70. doi:10.1023/A:1008299131268

Ghafoor, A., & Rehman, M. ur, Khan, Z. Abbas, Ali, A., Ahmad, H. Farooq and Suguri, H. (2004). SAGE: next generation multi-agent system. In *Proceedings of Parallel and Distributed Processing Techniques and Applications*, (pp.139-145).

Hara, H., Sugawara, K., Kinoshita, T., & Uchiya, T. (2002). Flexible distributed agent system and its application. In *Proceedings of the Fifth Joint conference of Knowledge-based Software Engineering*, (pp.72-77), IOS Press.

Imai, S., Kitagata, G., Konno, S., Suganuma, T., & Kinoshita, T. (2004). Developing a knowledge-based videoconference system for non-expert users. *Journal of Distance Education Technologies*, *2*(2), 13–26. doi:10.4018/jdet.2004040102

Jeon, H., Petrie, C., & Cutkosky, M. (2000). JATLite: a java agent infrastructure with message routing. *IEEE Internet Computing*, *4*(2), 87–96. doi:10.1109/4236.832951

Kim, H. Kinoshita, T., Lim, Y. and Kim, T. (2010). A bankruptcy problem approach to load-shedding in multiagent-based microgrid operation. Sensors Vol.10, No.10, (pp.8888-8898), MDPI Publishing.

Kinoshita, T., & Sugawara, K. (1998). ADIPS framework for flexible distributed systems. In *Proceedings of Pacific Rim International Workshop on Multi-Agents* (PRIMA'98 in PRICAI'98), (pp.161-175).

Kitagata, G., Matsushima, Y., Hasegawa, D., Kinoshita, T., & Shiratori, N. (2005). An agent-based middleware for communication service on ad hoc network. In *Proceedings of the 19th International Conference on Advanced Information Networking and Application*, (pp.363-367).

Kitagata, G., Sekiba, J., Suganuma, T., Kinoshita, T., & Shiratori, N. (2000). Agent-based flow control mechanism for flexible asynchronous messaging system FAMES. In *Proceedings of the 14th Int. Conf. on Information Networking (ICOIN-14)*, (pp.2B-2.1-8).

Konno, S., Iwaya, Y., Abe, T., & Kinoshita, T. (2004). Design of network management support system based on active information resource. In *Proceedings of the 18th International Conference on Advanced Information Networking and Application*, (pp.102-106).

Li, X., Uchiya, T., Konno, S., & Kinoshita, T. (2008). Proposal for agent platform dynamic interoperation facilitating mechanism. In *Proceedings of the 21th Int. Conf. Industrial, Engineering and Other Applications of Applied Intelligent Systems* (IEA/AIE 2008), LNAI5027, (pp.825-834).

Mylopoulos, J., Kolp, M., & Giorgini, P. (2002). Agent oriented software development. In *Proceedings of the 2nd Hellenic Conference on Artificial Intelligence* (SETN-02).

Nwana, H. S., Ndumu, D. T., Lee, L. C., & Collins, J. C. (1999). ZEUS: a toolkit for building distributed multi-agent systems. *Applied Artificial Intelligence Journal*, *13*(1), 129–186. doi:10.1080/088395199117513

Renquist, N. R., Andrew, H., & Andrew, L. (2001). Jack – summary of an agent infrastructure. In *Proceedings* of *5th International Conference on Autonomous Agents*. Reticular Systems. AgentBuilder – An integrated toolkit for constructing intelligence software agents, available at http://www.agentbuilder.com/

Shiratori, N., et al. (2005). Symbiotic Computing Project, http://symbiotic.agent-town.com/

Smith, R. G. (1980). The contract net protocol: high-level communication and control in a distributed problem solver. *IEEE Transactions on Computers*, *29*(12), 1104–1113. doi:10.1109/TC.1980.1675516

Suganuma, T., Imai, S., Kinoshita, T., Sugawara, K., & Shiratori, N. (2003). A flexible videoconference system based on multiagent framework. *IEEE Trans. on Systems, Man, and Cybernetics – Part A. Systems and Humans*, *33*(5), 633–641.

Suganuma, T., Uchiya, T., Konno, S., Kitagata, G., Hara, H., Fujita, S., et al. (2006). Bridging the E-Gaps: towards post-ubiquitous computing. In *Proceedings of the 20th International Conference on Advanced Information Networking and Applications* (AINA'06), FINA2006 Symposium, Vol.2, (pp.780-784).

Sugawara, K., et al. (2007). A concept of symbiotic computing and its application to telework. In *Proceedings of the IEEE 2007 International Conference on Cognitive Informatics* (ICCI'07), (pp.302-311).

Takahashi, A., Suganuma, T., Abe, T. and Kinoshita, T. (2006). Dynamic construction scheme of multimedia processing system based on multiagent framework. *The International Journal of Wireless and Mobile Computing*, Vol.2, No.1.

Takahashi, H., Izumi, S., Suganuma, T., Kinoshita, T., & Shiratori, N. (2009). Multi-agent system for user-oriented healthcare support. [IJIS]. *The International Journal of Informatic Society*, *1*(3), 32–41.

Uchiya, T., Maemura, T., Li, X., & Kinoshita, T. (2007). Design and Implementation of Interactive design environment of Agent System. In *Proceedings of the 20th Int. Conf. Industrial, Engineering and Other Applications of Applied Intelligent Systems* (IEA/AIE2007), LNAI4570, AAAI/ACM, (pp.1088-1097).

Uchiya, T., Suganuma, T., Kinoshita, T., & Shiratori, N. (2002). An architecture of active agent repository for dynamic networking. In *Proceedings of the First International Joint Conference on Autonomous Agents and Multiagent Systems* (AAMAS2002), (pp.1266-1267).

Uchiya, T., Takeda, A., Suganuma, T., Kinoshita, T., & Shiratori, N. (2003). A method for realizing user-oriented service with repository-based agent framework. In *Proceedings of the 1st Int. Forum on Information and Computer Technology* (IF-ICT2003), IPSJ, (pp.119-124).

Wang, Y. (2002). On cognitive informatics. In *Proceedings of 1st IEEE International Conference on Cognitive Informatics* (ICCI'02), (pp.34-42), Calgary, Canada. IEEE CS Press.

Wang, Y. (2005). On cognitive properties of human factors in engineering. In *Proceedings of the IEEE 2005 International Conference on Cognitive Informatics* (ICCI'05), (pp.174-182). IEEE CS Press.

Wang, Y. (2007). The theoretical framework of cognitive informatics. [IJCiNi]. *International Journal of Cognitive Informatics and Natural Intelligence*, *1*(1), 1–27. doi:10.4018/jcini.2007010101

Wang, Y., & Kinsner, W. (2006). Recent advances in cognitive informatics. *IEEE Transactions on Systems, Man and Cybernetics. Part C, Applications and Reviews*, *36*(2), 121–123. doi:10.1109/TSMCC.2006.871120

Zambonelli, F., Jennings, N., & Wooldridge, M. (2003). Developing multiagent systems: the gaia methodology. *ACM Transactions on Software Engineering and Methodology*, *12*(3), 317–370. doi:10.1145/958961.958963

Chapter 5
An Agent System to Manage Knowledge in CoPs

Juan Pablo Soto
University of Castilla - La Mancha, Spain

Aurora Vizcaíno
University of Castilla - La Mancha, Spain

Javier Portillo-Rodríguez
University of Castilla - La Mancha, Spain

Mario Piattini
University of Castilla - La Mancha, Spain

ABSTRACT

This paper proposes a multi-agent architecture and a trust model with which to foster the reuse of information in organizations which use knowledge bases or knowledge management systems. The architecture and the model have been designed with the goal of giving support to communities of practices which are a means of sharing knowledge. However, members of these communities are currently often geographically distributed, and less trust therefore exists among members than in traditional co-localizated communities of practice. This situation has led us to propose our trust model, which can be used to calculate what piece of knowledge is more trustworthy. The architecture's artificial agents will use this model to recommend the most appropriate knowledge to the community's members.

INTRODUCTION

The need to support knowledge processes in organizations has always existed. However, its importance has definitely increased in the last few years. Recently, the concept of knowledge management suggests a paradox since compared with traditional production factors knowledge is so complex, scattered and hidden that it is rather complicated to manage it.

On the other hand, traditional Knowledge Management Systems (KMS) have received certain criticism as they are often implanted in companies overloading employees with extra work; for instance, employees have to introduce information into the KMS and worry about updating this information. As a result of this, these systems are sometimes not greatly used by the employees since the knowledge that these systems have is often not valuable or on other occasions the knowledge sources do not provide the confidence necessary for employees to reuse the information. Reusing information and not reinventing the wheel are frequently heard arguments. For this purpose, companies create both social and technical networks in order to stimulate knowledge exchange. An essential ingredient of knowledge sharing information in organizations is that of "community of practice", by which we mean groups of people with a common interest where each member contributes knowledge about a common domain (Wenger, 1998). The ability of a community of practice to create a friendly environment for individuals with similar interests and problems in which they can discuss a common subject matter encourages the transfer and creation of new knowledge. Many companies report that such communities help reduce problems caused by lack of communication, and save time by "working smarter" (Wenger et al, 2002). In addition, communities of practice provide their members with the confidence to share information with each other. Moreover, individuals are frequently more likely to use knowledge built by their community team members than that created by members outside their group (Desouza et al, 2006). For these reasons, we consider the modelling of communities of practice into KMS as an adequate method by which to provide these systems with a certain degree of control to measure the confidence and quality of information provided by each member of the community.

In order to carry this out, we have designed a multi-agent architecture in which agents try to emulate human behaviour in communities of practice with the goal of fostering the use and exchange of information where intelligent agents suggest "trustworthy knowledge" to the employees and foster the knowledge flow between them.

The remainder of this work is organized as follows. The next section focuses on community of practice then in section 3 two important concepts related to our work are described: agents and trust. In Section 4 the trust model is presented. Later in section 5 the multi-agent architecture proposed to manage trustworthy KMS is described. In Section 6 a prototype developed to evaluate our architecture is explained in order to illustrate how it could be used. Section 7 describes a preliminary experiment carry out to test this prototype. Section 8 outlined related work and finally, conclusions are presented in Section 9.

COMMUNITIES OF PRACTICE

Intellectual capital and knowledge management are currently growing since knowledge is a critical factor for an organization's competitive advantage (Kautz, 2004). This growth determines organizations' performance by studying how well they manage their most critical knowledge. However, to manage this critical knowledge it has to be known what knowledge is, and although there is no consensus about a knowledge concept (Kakabadse, et al, 2001), there are several definitions of knowledge as in (Ackoff, 1989) and (Davenport et al, 1998). In our case, knowledge is going to be understood as in (Ackoff, 1989), that is, as an appropriate collection of information, such that its intent is to be useful. In order to manage knowledge an important instrument are communities (Gebert et al., 2004; Malhotra, 2000). A community can be defined as a group of socially interacting persons who are mutually tied to one another and regularly meet at a common place (Hillery, 1955). The development of Internet and groupware technologies led to a new kind of community "virtual communities" where members can or not meet one another face to face and they may exchange words and ideas through the mediation of computers networks (Geib et al., 2004).

This type of communities can be divided regarding their objectives and scope into socially-oriented, commercially-oriented and professionally-oriented. We focus our research on the last one which consists of company employees who communicate and share information to support their professional tasks. An special case of professionally-oriented communities are the "Communities of Practice" (CoPs) defined by Wenger et al. (2002) as groups of people who share a concern, a set of problems, or a passion about a topic, and who deepen their knowledge and expertise in this area by interacting on an ongoing basis.

Regarding a knowledge point of view CoPs share values, beliefs, languages, and ways of doing things many companies report that CoPs help reduce problems due to lack of communication, and save time by "working smarter" (Wenger, 2002). Millen et al. (2002) discuss the costs and benefits of CoPs in a study of seven large, geographically dispersed organizations. The study indicates benefits on individual, community and organizational level, like increased access to experts and information resources, increased idea generation and better problem solving, to indications of more successfully executed projects and product innovations. Moreover, individuals are frequently more likely to use knowledge built by their community team members than that created by members outside their group (Desouza et al, 2006), that is, they are more likely to share knowledge with people they trust. However, since in current CoPs people are usually geographically dispersed they do not have a face to face communication and this situation could be a problem since the level of trust between members can decrease and in consequence there could be a lower level of sharing knowledge.

On the other hand, even though CoPs are a focus of knowledge sharing hardly ever there is any quality control of the knowledge generated in the community. Our proposal is to use agent's technology to foster knowledge exchange at communities of practice and to evaluate the suitability of knowledge shared.

AGENTS AND TRUST

Because of the importance of knowledge management, tools to support some of the tasks related to knowledge management have been developed. Different techniques are used to implement these tools. One of them, which is providing to be quite useful, is that of intelligent agents (van-Elst et al, 2003). Software agent technology can monitor and coordinate events, meetings and disseminate information (Balasubramanian et al, 2001). Furthermore, agents are proactive; this means they act automatically when it is necessary. The autonomous behavior of the agents is critical to the goal of this research since agents help to reduce the amount of work that employees have to perform, for instance searching information in a knowledge base. On the other hand one of the main advantages of the agent paradigm is that it constitutes a natural metaphor for systems with purposeful interacting agents, and this abstraction is close to the human way of thinking about their own activities (Wooldridge & Ciancarini, 2001). This foundation has led to an increasing interest in social aspects such as motivation, leadership, culture or trust (Fuentes et al, 2004). Our research is related to this last concept of "trust" since artificial agents can be made more robust, resilient and effective by providing them with trust reasoning capabilities.

For agents to function effectively in a community, they must ensure that their interactions with the other agents are trustworthy. For this reason it is important that each agent is able to identify trustworthy partners with which they should interact and untrustworthy correspondents with which they should avoid interaction. The stability of a community depends on the right balance of trust and distrust.

In literature we found several trust and reputation mechanisms that have been proposed to be used in different domains such as e-commerce (Zacharia et al, 1999), peer-to-peer computing (Wang & Vassileva, 2003), recommender systems (Schafer et al, 1999), etc. In next section we describe the trust model that we propose to be used in our multi-agent architecture.

TRUST MODEL

One of our aims is to provide a trust model based on real world social properties of trust in Communities of Practice (CoPs). An interesting fact is that members of a community are frequently more likely to use knowledge built by their community team members than those created by members outside their group (Desouza et al, 2006). This factor occurs because people trust more in the information offered by a member of their community than in that supplied by a person who does not belong to that community. Of course, the fact of belonging to the same community of practice already implies that these people have similar interests and perhaps the same level of knowledge about a topic. Consequently, the level of trust within a community is often higher than that which exists outside the community. As a result of this, as is claimed in (Desouza et al, 2006), knowledge reuse tends to be restricted within groups. Therefore, people, in real life in general and in companies in particular, prefer to exchange knowledge with "trustworthy people" by which we mean people they trust. For these reasons we consider the implementation of a mechanism in charge of measuring and controlling the confidence level in a community in which the members share information to be of great importance.

Most previous trust models calculate trust by using the users' previous experience with other users but when there is no previous experience, for instance, when a new user arrives to a community, these models cannot calculate a reliable trust value. We propose calculating trust by using four factors that can be stressed depending on the circumstances. These factors are:

- *Position*: employees often consider information that comes from a boss as being more reliable than that which comes from another employee in the same (or a lower) position as him/her (Wasserman & Glaskiewics, 1994). However, this is not a universal truth and depends on the situation. For instance in a collaborative learning setting collaboration is more likely to occur between people of a similar status than between a boss and his/her employee or between a teacher and pupils (Dillenbourg, 1999). In an enterprise this position can be established in different ways by, for instance, using an organizational diagram or classifying the employees according to the knowledge that a person has, as can be seen in Allen's proposal in (Allen, 1984), which distinguishes between:
    - Technological gatekeepers, defined as those actors who have a high level of knowledge interconnectedness with other local firms and also with extra-community sources of knowledge. These basically act by channeling new knowledge into the community and diffusing it locally.
    - External stars, which are highly interconnected with external sources of knowledge but have hardly any interaction with other local firms.

  Such different positions inevitably influence the way in which knowledge is acquired, diffused and eventually transformed within the local area. Because of this, as will later be explained, this factor will be calculated in our research by taking into account a weight that can strengthen this factor to a greater or to a lesser degree.

- *Expertise*: This term can be briefly defined as the skill or knowledge that a person who knows a great deal about a specific thing has. This is an important factor since people often trust experts more than novice employees. In addition, "individual" level knowledge is embedded in the skills and competencies of the researchers, experts, and professionals working in the organization (Nonaka & Takeuchi, 1995). The level of expertise that a person has in a company or in a CoP could be calculated from his/her CV or by considering the amount of time that a person has been working on a topic. This is data that most companies are presumed to have.
- *Previous experience*: This is a critical factor in rating a trust value since, as was mentioned in the definitions of trust and reputation, previous experience is the key value through which to obtain a precise trust value. However, when previous experience is scarce or it does not exist humans use other factors to decide whether or not to trust in a person or a knowledge source. One of these factors is intuition.
- *Intuition*: This is a subjective factor which, according to our study of the state of the art, has not been considered in previous trust models. However, this concept is very important because when people do not have any previous experience they often use their "intuition" to decide whether or not they are going to trust something. Other authors have called this issue "indirect reputation or prior-derived reputation" (Mui et al, 2002). In human societies, each of us probably has different prior beliefs about the trustworthiness of strangers we meet. Sexual or racial discrimination might be a consequence of such prior belief (Mui et al, 2002). We have tried to model intuition according to the similarity between personal profiles: the greater the similarity between one person and another, the greater the level of trust in this person as a result of intuition.

By taking all these factors into account, we have defined our own model with which to rate trust in CoPs, and this is summarized in Figure 1.

The main goal of this model is to rate the level of confidence in an information source or in a provider of knowledge in a CoP.

As the model will be used in virtual communities where people are usually distributed in different locations we have implemented a multi-agent architecture in which each software agent acts on behalf of a person and each agent uses this trust model to analyze which person or piece of knowledge is more trustworthy.

As the number of interactions that an agent will have with other agents in the community will be low in comparison with other scenarios such as auctions we cannot use trust models which need a lot of interactions to obtain a reliable trust value; it is more important to obtain a reliable initial trust value and it is for this reason that we use position, expertise and intuition.

As observed in Figure 1, we use four factors to obtain a trust value, but how do we use these factors? We have classified these four factors

Figure 1. Trust Model

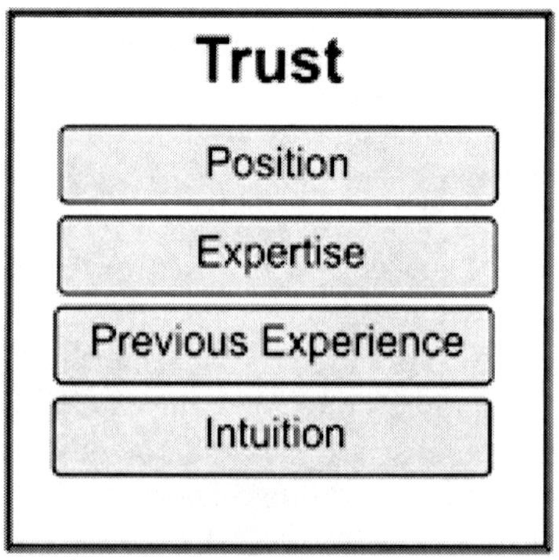

Figure 2. Using the trust model

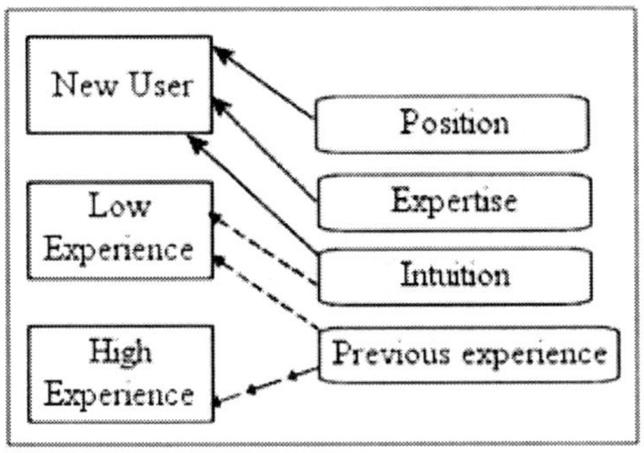

into two groups: objective factors (position and expertise) and subjective factors (intuition and previous experience). The former are given by the company or community and the latter depend on the agent itself and the agent's experience in time. There are four different ways of using these factors, which depend upon the agent's situation (see Figure 2).

- If the agent has no previous experience, for instance because it is a new user in the community, then the agent will use its intuition and the position and expertise of other agents to discover which other agents it can trust.
- When the agent has previous experience obtained through interactions with other agents but this previous experience is low (low number of interactions), the agent calculates the trust value by considering the intuition value and the previous experience value. For instance, if an agent A has a high experience value for agent B because it interacted with B successfully several times but agent A has a low intuition value for agent B (profiles are not very similar), then agent A reduces the value obtained through experience. In this case the agent does not use position and expertise factors (objective factors) because the agent has its own experience and this experience is adjusted with its intuition which is subjective and more personalized.
- When the agent has enough previous experience to consider that the trust value obtained is reliable, then the agent only considers this value.

MULTI-AGENT ARCHITECTURE

In order to give support to CoPs, we have designed a multi-agent architecture that uses the trust model explained in previous section with the goal of recommending trustworthy knowledge in CoPs and therefore fostering the reuse of information generated in these communities. Therefore, we can say that the goals of this architecture are:

- Assists members in identifying trustworthy entities.
- Gives artificial agents the ability to reason about the trustworthiness of other agents or about a knowledge source.
- Encourages knowledge exchange between the community members.
- Provides the confidence necessary to foster the usage of information and knowledge of the community.

Taking these facts into account, we propose a multi-agent architecture which is composed of two levels (see Figure 3): a reactive level and a Deliberative-Social level. The reactive level is considered by other authors as a typical level that a multi-agent system must have (Ushida et al, 1998)(Ushida, 1998). A deliberative level is often also considered as a typical level but a social level is not frequently considered in an explicit way, despite the fact that these systems (multi-agent systems) are composed of several individuals, interactions between them and plans constructed by them. The social level is only considered in those systems that try to simulate social behaviour or those that represent a more generic architecture which has been prepared to represent this or other behaviour. Since we wish to

Figure 3. General architecture

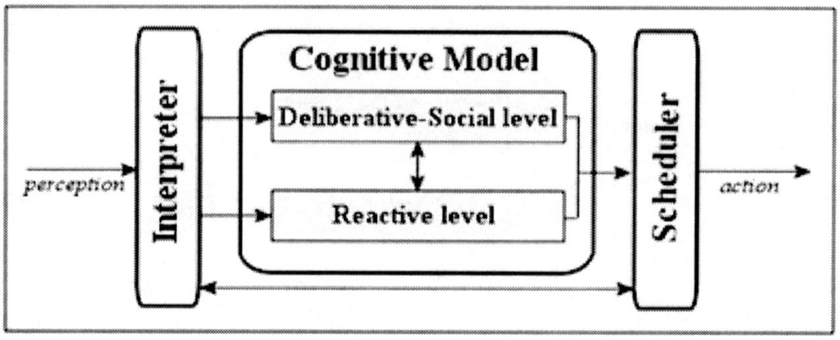

emulate human feelings such as trust, reputation and even intuition we have added a social part that considers the social aspects of a community which takes into account the opinions and behaviour of each of the members of that community. Other previous works have also added a social level. For instance, in (Imbert & de Antonio, 2005) (Imbert, 2005) the authors emulate human emotions such as fear, thirst or bravery but they use an architecture which is made up of three levels: reactive, deliberative and social. In our case the deliberative and the social level are not separate levels because we realised that plans created in the deliberative part involve social interactions so we considered that in our case it would be more efficient to define a level composed of two parts (Deliberative-Social level) instead of considering two separated levels.

- **Reactive level:** This is the agent's capacity to perceive changes in its environment and to respond to these changes at the precise moment at which they happen. It is in this level when an agent will execute the request of another agent without any type of reasoning.
- **Deliberative-Social level:** The agent has a type of behaviour which is orientated towards objectives, that is, it takes the initiative in order to plan its performance with the purpose of attaining its goals. In this level the agent would use the information that it receives from the environment, and from its beliefs and intuitions, to decide which is the best plan of action to follow in order to fulfil its objectives. In this level we have individual goals which refer to the deliberative part and social goals or cooperative goals which refer to the social part.

Two further important components of our architecture are the *Interpreter* and the *Scheduler*. The former is used to perceive the changes that take place and to decide which level must take the initiative depending on the event that the agent perceives. The scheduler indicates how the actions should be scheduled and executed.

Each of the levels of our architecture is described in the following subsections.

Reactive Level

This level must respond at the precise moment at which an event has been perceived (see Figure 4).

For instance when an agent is consulted about its position within the organization or when a user wishes to send the system simple answers. This level is formed of the following modules:

- **Internal model**: As an agent represents a person in a CoP this model stores the user's features, these features will be consulted by other agents in order to calculate trust values. Therefore, this module stores the following parts in the user profile:

Figure 4. Reactive level

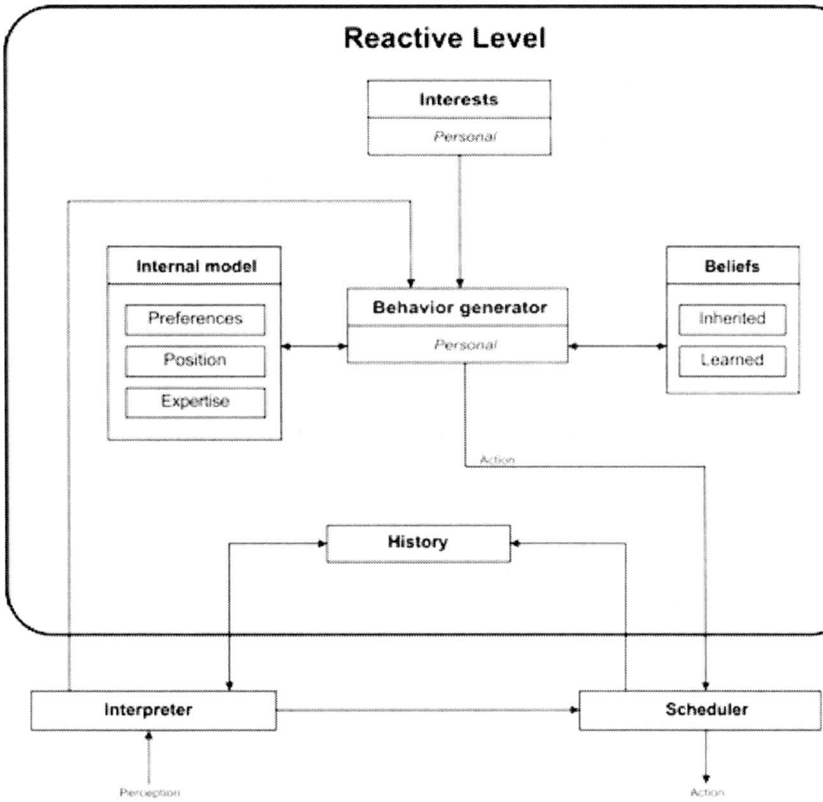

- *Expertise.* This term has been explained in the Trust Model in section 4.
- *Preferences.* In this part we try to represent user preferences by using, for example, the Felder-Silverman test which tell us whether the agent is representing a visual user (one who prefers visual representations of presented information-pictures, diagrams, flow charts,...), a verbal user (who prefers written and spoken explanations) or another kind of user that the Felder-Silverman model supports (Felder & Silverman, 1988; Felder, 1996).
- *Position.* Explained in the Trust Model section.

- **Behaviour generator**: This component is necessary for the development of this architecture since it has to select the agent's behaviour. This behaviour is defined on the basis of the agent's beliefs.
- **Interests:** These are individual interests which represent the user's needs.
- **History:** This component stores the interactions of the agents with the environment.
- **Beliefs**: The beliefs module is composed of inherited beliefs and lessons learned from the agent itself. Inherited beliefs are the organization's beliefs that the agent receives. Examples of this might be an organizational diagram of the enterprise or the philosophy of the company or community. Lessons learned are the lessons that the agent obtains while it interacts with the environment. This interaction can be used to establish parameters

in order to know what the agent can trust (agents or knowledge sources).

Deliberative-Social Level

In this level the agent's behaviour is based on goals, that is, the agent has several defined goals and it tries to achieve these goals by scheduling plans. Due to the fact that we are trying to represent human behaviour in CoPs, it is necessary to bear in mind that this human behaviour must benefit the whole community. Therefore, the agent has to deliberate about its individual goals but it must also act by taking community goals and the community's profit into account. That is why we have considered a social and a deliberative part. The former tries to achieve social goals (community goals) and the latter is more focused upon achieving individual goals.

In this level the agent obtains information about the environment and, by taking into account its interests and intuitions, it decides which plan is the best to achieve its goals (see Figure 5).

The components of the Deliberative-Social architecture are:

- **Interests:** This component represents community interests. These interests are created when the community comes into being. There are some interests that all communities may share such as:
    - Maintaining a constant collaboration of community members.
    - Identifying and maintaining experts in the community
    - Keeping community knowledge updated
    - Maintaining a trustworthy environment in which community members share trustworthy knowledge.

  There are also Personal Interests which influence the whole community such as sharing suitable knowledge.
- **Beliefs:** This module represents a view that the agent has of the environment. In our case these beliefs are composed of the idea that the agent has of the communities and their members. For instance, in this module there is information about the community's topics, in which areas other members are working, etc.
- **Goals Generator:** Depending on the state of the agent this module must decide what the most important goal to be achieved is.
- **Plans Generator:** This module is in charge of evaluating how a goal can be attained and which plans are most convenient if this goal is to be achieved. We should recall that plans are a specification of the actions that an agent may carry out in order to attain its goals.
- **Intuitions:** Intuitions are beliefs that have not been verified but which an agent thinks may be true. According to Mui et al. (2002) (Mui, 2002) intuition has not yet been modelled by agent systems. In this work we have tried to adapt this concept by comparing the agents' profiles (as we mentioned in Section 4) to obtain an initial value of intuition that can be used to form a belief about an agent when the intuition is proved be true. This is another important feature taken into account to calculate a trust value, since when an agent has little o null interaction with another; the agent will use this value to have a value of trust as it was previously explained.
- **History:** This component stores the interactions of the agents with the environment.

In the following section, we will describe a prototype developed to validate our architecture.

PROTOTYPE

In order to test our architecture we have developed a prototype system into which people can

Figure 5. Deliberative-social level

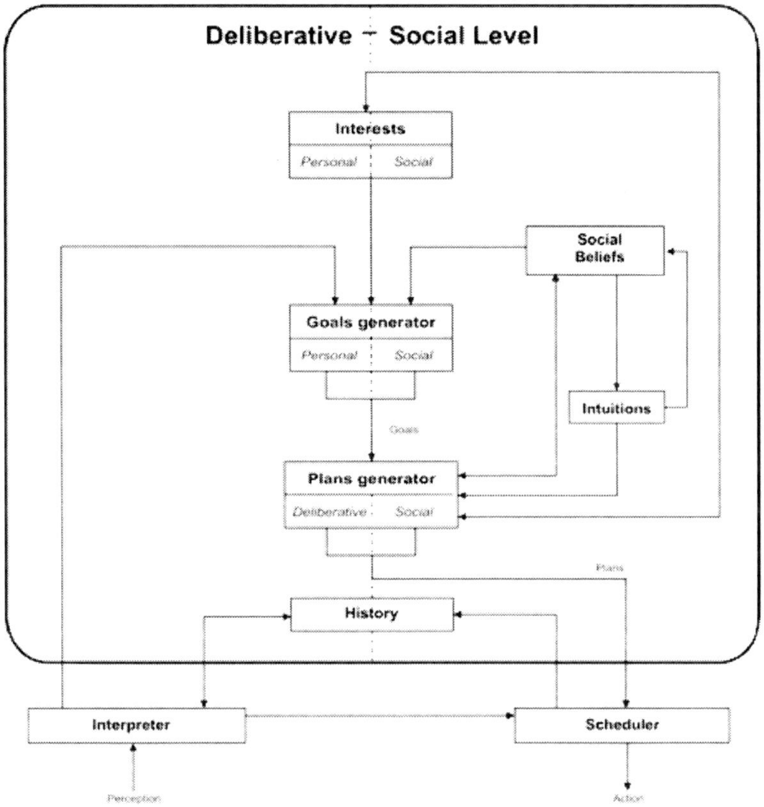

introduce documents and where these documents can also be consulted by other people. The goal of this prototype is to allow software agents to help employees to discover the information that may be useful to them thus decreasing the overload of information that employees often have and strengthening the use of knowledge bases in enterprises. In addition, we try to avoid the situation of employees storing valueless information in the knowledge base.

One feature of this system is that when a person searches for knowledge in a community, and after having used the knowledge obtained, that person then has to evaluate the knowledge in order to indicate whether:

- The knowledge was useful.
- How it was related to the topic of the search (for instance a lot, not too much, not at all).

To design this prototype we have designed a *User Agent* and a *Manager Agent*. The former is used to represent each person that may consult or introduce knowledge in a knowledge base. Therefore, the *User Agent* can assume three types of behavior or roles similar to the tasks that a person may carry out in a knowledge base. The User Agent plays one role or another depending upon whether the person that it represents carries out one of the following actions:

- The person contributes new knowledge to the communities in which s/he is registered. In this case the User Agent plays the role of **Provider**.
- The person uses knowledge previously stored in the community. Then, the User Agent will be considered as a **Consumer**.

Figure 6. Communities of agents

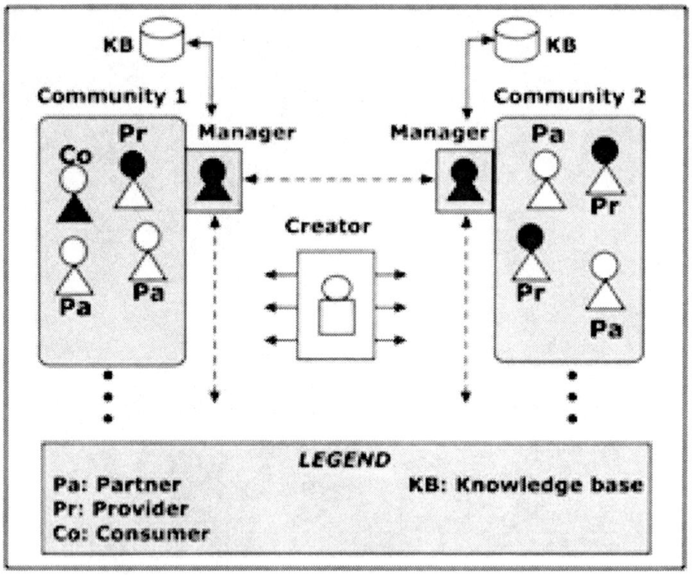

- The person helps other users to achieve their goals, for instance by giving an evaluation of certain knowledge. In this case the role is of a **Partner**. So, Figure 6 shows that in community 1 there are two User Agents playing the role of Partner, one User Agent playing the role of Consumer and another being a Provider.

The second type of agent within a community is called the *Manager Agent* (represented in black in Figure 6) which must manage and control its community.

The prototype provides the options of using community documents and when the documents are used, reputation values can be modified. An user can also propose new topics in the community, etc.

In order to make it easier to search for documents in a community, users can choose one topic from those which are available in the community and the user agent will try to find documents about this topic.

The general idea is to consider those documents which came from trustworthy knowledge sources according to the user's opinion or needs. In order to discover which knowledge sources are trustworthy the user agents will use the trust model. Depending on the context, this trust model can be used in different ways. We are going to consider how the trust model is applied in different situations. First, when agents have previous experience this means that user agents have previously interacted with a knowledge source and they have some feedback (trust values in our case) about it. The second scenario is a more complicated situation in which the agents have no previous experience and therefore do not have trust values for other user agents.

The way in which we apply these factors in the different contexts is as follows:

1. If the agent has no previous experience, for instance, because it is representing a new user in the community, and its user wants to search for documents relating to a topic T, the user agent follows these steps:
   1.1. The user agent makes a request to the other members of the community in order to discover which user agents have documents about topic T.

1.2. The user agent stores the id (identification) in a list of those agents which have documents about T.

1.3. For each agent grouping in the list, the user agent calculates a trust value by using the position, expertise and intuition factors. For instance, the user agent might obtain a list with 10 agents that match the request and for each of these agents, the user agent will obtain information about their positions (to discover, for instance, if the agent represents a boss or a newcomer), their levels of expertise in the community area, in our case there are five possible levels (from novice to expert), and their intuition values in relation to the agent that has made the request (with five values from "totally different" to "totally equals"). In this case the intuition level is calculated by comparing user profiles, that is, if the user agent compares two profiles with very similar characteristics, this means that users, represented by user agents, work in the same area, have similar expertise level,…, etc and consequently the trust value will increase because the user agent "senses" that working with this user will be a successful interaction. So, the user agent's list might contain an agent that represents a newcomer user with a high level of expertise and with similar preferences, or a boss with different preferences and a medium level of expertise in the area concerned. Once the agent has obtained all these values, it calculates a general trust value per agent by combining the different factors, obtaining the lowest value when the agent, for instance, represents a rookie newcomer with a profile which is totally different to that of the requester, and obtaining the highest value when the agent represents a boss with a high level of experience and who has a very similar profile to the agent which is making the request.

1.4. The user agent shows the results which are sorted by trust values, that is, the first documents on the list come from the most trustworthy knowledge sources (in this case the most trustworthy agent with the highest trust values). There are other possibilities, depending on user preferences. The user can choose to sort the list by using level of experience, position or level of intuition. At each request the user will receive a list and from each list the user will obtain information about each factor by the use of star icons and shield icons. For instance, as we can see in Figure 7, the results of the request (sorted by reputation) show a large amount of results, and the first one on the list has five stars in the reputation level and four shields in the position level.

2. If there is a small amount of previous experience and this previous experience is not sufficient for the agent to discover whether the other user agent is trustworthy or not then we combine previous experience with the other three factors. So in this context the user agent follows the following steps when looking for documents about a topic T:

2.1. The user agent makes a request to the members of the community in order to discover which user agents have documents about topic T.

2.2. For each agent (in the requested group) that our agent has previously interacted with, it uses the four factors (position, intuition, expertise and previous experience) to calculate a trust value by using (1).

Figure 7. Showing and sorting results

$$\mathbf{T}_{ij} = w_e * E_j + w_p * P_j + w_i * I_{ij} + (\sum_{j=1}^{n} QC_{ij})/n \quad (1)$$

where E_j is the value of expertise which is calculated according to the degree of experience that the person upon whose behalf the agent acts has in a domain. In this case the domain of the community which the agent wishes to join.

P_j is the value assigned to a person's position. This position is defined in the agent's internal model of the reactive architecture described in Section 4.1.

I_{ij} denotes the intuition value that agent i has in agent j which is calculated by comparing each user's profile.

In addition, previous experience should also be calculated. When an agent i consults information from another agent j, the agent i should evaluate how useful this information was. This value is called QC_{ij} (Quality of j's Contribution in the opinion of i). To attain the average value of an agent's contribution, we calculate the sum of all the values assigned to these contributions and we divide it between their total. In the expression n represents the total number of evaluated contributions. Finally, w_e, w_p and w_i are weights with which the trust value can be adjusted according to the degree of knowledge that one agent has about another.

2.3. For each agent in the group (the results group) that the agent has no previous experience it calculate a trust value as we mentioned in 1.3.

2.4. The user agent shows the results, which are sorted by trust or quality values as in the previous situation.

3. If the user agent has enough previous experience (this is considered when an agent has interacted many times with another. This number of interactions depends on a threshold that can be adjusted to each domain) then the user agent calculates the trust value by only using the previous experience factor. In this case we only consider this factor (experience) because this is the principal factor that humans usually consider when they have to trust somebody/something. That's why this concept is the base of all trust models described in literature as it will be explained in section 7. In this context the user agent follows the following steps when looking for documents about a topic T:

3.1. The user agent follows step 2.1

3.2. For each agent in the group (the results group) the user agent calculates a trust value by using the previous experience factor that is, by using (2) which is the last part of formula (1),

$$(\sum_{j=1}^{n} QC_{ij})/n \quad (2)$$

3.3. The user agent follows step 2.3.
3.4. The user agent shows the results which are sorted by trust or quality values.

These are three possible scenarios that illustrate how the trust model is used. When a person inserts a document in the community, s/he inserts the document and a quality value for that document. If another person uses that document, after using that document, the person who requested it must evaluate its quality. The User Agent compares the value given by the owner with the value given by the consumer to discover whether the two users have the same opinion about the document. If this is so then the previous experience value for the other user increases and if the opposite is true then the previous experience value is decreased. That is, if a user A thinks that a document D has a quality value of 8 and another user B, after using D, thinks that the document has a quality value of 2, the trust value that user B has for user A is decreased. This manner of rating trust helps to detect a problem which is increasing in companies or communities in which employees introduce not valuable information because they are rewarded if they contribute with knowledge in the community. Thus, if a person introduces documents that are not related to the community with the aim of obtaining rewards, the situation can be detected, because when the other person evaluate those documents or information, the rate of them will be low and the value of previous experience of this person became very low. Therefore, the community agent can detect that there is a "fraudulent" member in the community.

In a previous version of the prototype, when a person introduced a document in the system, s/he did not indicate the quality value of it. In this case the previous experience was calculated only in base to the rate that agent gave. We have introduced this change because we have a reference value for each document that can be used to compare quality values that different users have given to the same document and detect if there are "fraudulent members" and also can be used to sort documents when we have not enough information, at least we will have a quality value given by the person who introduced the document in the community.

The three situations must be applied to each user agent depending on the situation. If a user agent makes a request to search for documents it will receive answers from different user agents and, depending upon the situation between the requester and the other agents, the requester must apply one of the three situation steps and not only one situation for all the agents that have answered. This is shown in Figure 8 where agent *A* makes a request to search for documents about topic T and agents *X*, *Y*, *Z* answer because they have documents about T. In this case agent *A* applies situation 1 to agent *X* because agent *X* is not a known agent, and situation 3 to agents *Y* and *Z* because it has already interacted with both agents on previous occasions.

EVALUATION OF THE PROTOTYPE

Once the prototype has been finished we have evaluated it. To do this, different approaches can be followed, from a multi-agent point of view or from a social one. First of all we have focused on the former and we are testing the most suitable number of agents advisable for a community. Therefore, several simulations have been performed. As result of them we found that:

- The maximum number of agents supported by the Community Manager Agent when it receives User Agents' evaluations is approximately 800. When we tried to work with 1000 agents for instance, the messages were not managed conveniently. However, we could see that the Manager Agent could support a high number of petitions, at least, using simpler behavior.

Figure 8. Mechanism through which to obtain trustworthy documents by using the model

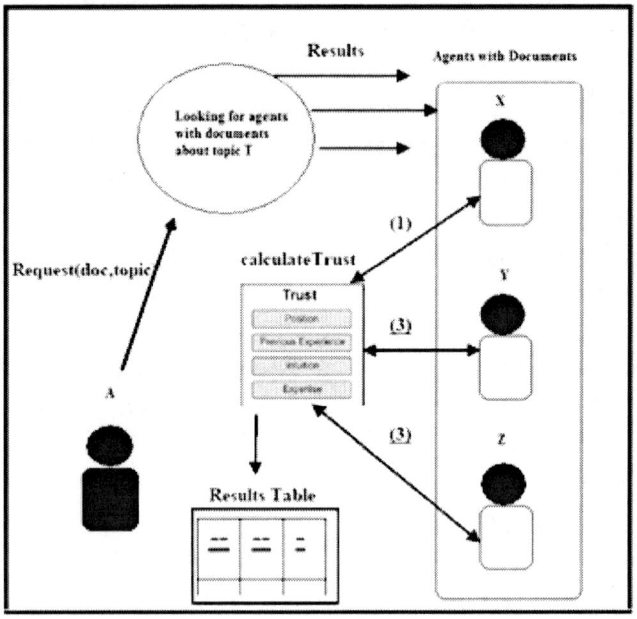

- On the other hand, if we have around 10 User Agents launched, they need about 20 o more interactions to know all agents of the community. If a User Agent has between 10 and 20 interactions with other members it is likely that it interacts with 90% of members of its community, which means that the agent is going to know almost all the members of the community. Therefore, after several trials we detected that the most suitable number of agents for one community was around 10 agents and they needed a average of 20 interactions to know (to have a contact with) all the members of the community, which is quite convenient in order to obtain its own value of reputation about other agent.

All these results are being used to detect whether the exchange of messages between the agents is suitable, and to see if the information that we propose to be taken into account to obtain a trustworthy value of the reputation of each agent is enough, or if more parameters should be considered. Once this validation is finished we need to carry out further research to answer one important and tricky question, which is how the usage of this prototype affects the performance of a community.

RELATED WORK

This research can be compared with other proposals that use agents and trust in knowledge exchange. With regard to trust, in models such as eBay (1995)(ebay, 1995) and Amazon (1996) (Amazon, 1996), which were proposed to resolve specific situations in online commerce, the ratings are stored centrally and the reputation value is computed as the sum of those ratings over six months. Thus, reputation in these models is a single global value. However, these models are too simple (in terms of their trust values and the way in which they are aggregated) to be applied in open multi-agent systems. For instance, in (Zacharia et al, 1999)(Zacharia, 1999) the authors present the Sporas model, a reputation mechanism for loosely connected online communities where, among other features, new users start with a minimum reputation value, the reputation value of a user never falls below the reputation of a new user and users with very high reputation values experience much smaller rating changes after each update. The problem with this approach is that when somebody has a high reputation value it is difficult to change this reputation, or the system needs a high amount of interactions. A further approach of the Sporas authors is Histos which is a more personalized system than Sporas and is orientated towards highly connected online communities. In (Sabater & Sierra, 2002)(Sabater, 2002) the authors present another reputation model called REGRET in which the reputation values depend on time: the most recent rates are more important than previous rates. Carbó et al (2003) (Carbó, 2003) presents the AFRAS model, which is based on Sporas but uses fuzzy logic. The authors present a complex computing reputation mechanism which handles reputation as a fuzzy set while decision making is inspired in a cognitive human-like approach. In (Abdul-Rahman & Hailes, 2000)(Abdul-Rahman, 2000) the authors propose a model which allows agents to decide which agents' opinions they trust more and to propose a protocol based on recommendations. This model is based on a reputation or word-of-mouth mechanism. The main problem with this approach is that every agent must maintain rather complex data structures which represent a kind of global knowledge about the whole network.

Barber and Kim (2004) present a multi-agent belief revision algorithm based on belief networks (Barber, 2004). In their model the agent is able to evaluate incoming information, to generate a consistent knowledge base, and to avoid fraudulent information from unreliable or deceptive information sources or agents. This work has a similar goal to ours. However, the means of attaining it are different. In Barber and Kim's case they define reputation as a probability measure, since the information source is assigned a reputation value of between 0 and 1. Moreover, every time a source sends knowledge, that source should indicate the certainty factor that the source has of that knowledge. In our case, the focus is very different since it is the receiver who evaluates the relevance of a piece of knowledge rather than the provider as in Barber and Kim's proposal. Some of these trust and reputation models are summarized in Table 1.

In (Huynh et al, 2004)(Huynh, 2004) the authors present a trust and reputation model which integrates a number of information sources in order to produce a comprehensive assessment of an agent's likely performance. In this case the model uses four parameters to calculate trust values: interaction trust, role-based trust, witness reputation and certified reputation. We use certified reputation when an agent wishes to join a new community and uses a trust value obtained in other communities, but in our case this certified reputation is made up of the four previously explained factors and is not only a single factor.

Also, works such as (Guizzardi et al, 2004) (Guizzardi, 2004) use the term 'Community' to support knowledge management but it is not used a specific trust model for communities.

The main differences between these reputation models (summarized in Table1) and our approach are that these models need an initial number of interactions to obtain a good reputation value and it is not possible to use them to discover whether or not a new user can be trusted. A further difference is that our approach is orientated towards collaboration between users in CoPs. Other approaches are more orientated towards competition, and most of them are tested in auctions.

CONCLUSION

Communities of practice have the potential to improve organizational performance and facilitate community work. Because of this we consider it important to model people's behavior within communities with the purpose of imitating the

Table 1. Other trust and reputation models

Model	Authors	Reputation Management	Features
ebay	-	Global values	Simple values obtained through interactions
Sporas	Zacharia	Global values	Reduces changes when reputation is very high Most recent reputation values are the most important
Histos	Zacharia	Pair wise ratings in the system as a directed graph	Divides Reputation into three dimensions: Individual, Social and Ontological
Regret	Sabater and Sierra	Decentralized values	Most recent reputation values are the most important Presents a witness reputation component
Afras	Carbó and Molina	Decentralized values	Based on BDI agents Based on Sporas model but using fuzzy logic Compares and combines fuzzy sets
Fire	T. Dong Huynh and Nicholas R. Jennings	Decentralized values	Four main components: interaction trust, role-based trust, witness reputation, and certified reputation

exchange of information that are produced in those communities. Therefore, we are attempting to encourage the sharing of information in organizations by using CoPs and knowledge bases. To do this we have designed a multi-agent three-layer architecture where the artificial agents use similar parameters to those of humans in order to evaluate knowledge and knowledge sources. These factors are: reputation, expertise, position, previous experience and even intuitions.

This approach implies several advantages for organizations as it permits them to identify the expertise of their employees and to measure the quality of their contributions. Therefore, it is expected a greater exchange and reuse of knowledge.

In addition, this work has illustrated how the architecture can be used to implement a prototype. The main functionalities of the prototype are:

- Controlling those employees who try to introduce valueless knowledge with the goal of obtaining some profit such as points, incentives, rewards, etc.
- Providing the most suitable knowledge for the employee's queries according to the employee features and needs.
- Detecting the expertise of the employees within an organization.

All these advantages provide organizations with a better control of their knowledge repositories which will have more trustworthy knowledge and it is consequently expected that employees will feel more willing to use it.

ACKNOWLEDGMENT

This work is partially supported by the MELISA (PAC08-0142-3315) and MECENAS (PBI06-0024) project, Junta de Comunidades de Castilla-La Mancha, Consejería de Educación y Ciencia, both in Spain. It is also supported by the ESFINGE project (TIN2006-15175-C05-05) Ministerio de Educación y Ciencia (Dirección General de Investigación)/ Fondos Europeos de Desarrollo Regional (FEDER) in Spain and CONACYT

(México) under Grant of the scholarship 206147 provided to the first author.

REFERENCES

Abdul-Rahman, A. and Hailes, S., (2000), "Supporting Trust in Virtual Communities". Proceedings of the 33rd Hawaii International Conference on Systems Sciences (HICSS'00), Vol. 6.

Ackoff, R., (1989), "From Data to Wisdom". Journal of Applies Systems Analysis. Vol. 16, pp. 3-9.

Allen, T., (1984), "Managing the Flow of Technology: Technology Transfer and the Dissemination of Technological Information within the R&D Organization", Cambridge, MA: MIT Press.

Amazon (1996). "URL: http://www.amazon.com".

Balasubramanian, S., Brennan, R., Norrie, D., (2001), "An Architecture for Metamorphic Control of Holonic Manufacturing Systems". Computers in Industry, Vol. 46(1), pp. 13-31.

Barber, K. and Kim, J., (2004), "Belief Revision Process Based on Trust: Simulation Experiments". 4th Workshop on Deception, Fraud and Trust in Agent Societies, Montreal Canada, pp. 1-12.

Carbó, J., Molina, M., Dávila, J., (2003), "Trust Management through Fuzzy Reputation". International Journal of Cooperative Information Systems. Vol. 12(1), pp. 135-155.

Davenport, P., (1998), "Working Knowledge: How Organizations Manage What They Know". Boston, MA, Project Management Institute, Harvard Business School Press.

Desouza, K., Awazu, Y., Baloh, P., (2006), "Managing Knowledge in Global Software Development Efforts: Issues and Practices". IEEE Software, pp. 30-37.

Dillenbourg, P., (1999), "Introduction: What Do You Mean By 'Collaborative Learning'?." Collaborative Learning Cognitive and Computational Approaches. Dillenbourg (Ed.). Elsevier Science.

eBay (1995). "URL: http://www.ebay.com".

Felder, R. and Silverman L, (1988), "Learning and Teaching Styles in Engineering Education". Engineering Education. Vol. 78(7), pp. 674-681.

Felder, R. M. (1996). "Matters of Style". ASEE Prism. Vol. 6(4), pp. 18-23.

Fuentes, R., Gómez-Sanz, J., Pavón, J. (2004). "A Social Framework for Multi-agent Systems Validation and Verification". Wang, S. et al Eds. ER Workshops, Springer Verlag, LNCS 3289, pp. 458-469.

Gebert, H., Geib, M., Kolbe, L., Brenner, W., (2003), "Knowledge-enabled Customer Relationship Management - Integrating Customer Relationship Management and Knowledge Management Concepts". Journal of Knowledge Management. Vol. 7(5), pp. 107-123.

Geib, M., Braun, C., Kolbe, L., Brenner, W., (2004). Measuring the Utilization of Collaboration Technology for Knowledge Development and Exchange in Virtual Communities. 37th Hawaii International Conference on System Sciences 2004 (HICSS-37), Big Island, Hawaii, IEEE Computer Society, Vol. 1, pp. 1-10.

Guizzardi, R., Perini, A., Dignum, V., (2004), «Providing Knowledge Management Support to Communities of Practice through Agent-Oriented Analysis». Proceedings of the 4th International Conference on Knowledge Management (I-KNOW), Granz, Austria.

Hillery, G., (1955), «Definitions of Community: Areas of Agreement», Rural Sociology, Vol. 20, pp. 118-125.

Huynh, T., Jennings, N., Shadbolt, N., (2004), «FIRE: An Integrated Trust and Reputation Model for Open Multi-agent Systems». Proceedings of the 16th European Conference on Artificial Intelligence (ECAI).

Imbert, R., and de Antonio, A., (2005), «When emotion does not mean loss of control». Lecture Notes in Computer Science, T. Panayiotopoulos, J. Gratch, R. Aylett, D. Ballin, P. Olivier, and T. Rist (Eds.), Springer-Verlag, London, pp.152-165.

Kakabadse, N., Kouzmin, A., Kakabadse, A., (2001), «From Tacit Knowledge to Knowledge Management: Leveraging Invisible Assets». Journal of Knowledge and Process Management, Vol. 8(3), pp. 137-154.

Kautz, H. (2004), «Knowledge Mapping: A Technique for Identifying Knowledge Flows in Software Organizations», EuroSPI, pp. 126-137.

Malhotra, Y. (2000), «Knowledge Management and Virtual Organizations», IDEA Group publishing, Hershey.

Millen, D., Fontaine, M., Muller, M., (2002), "Understanding the benefits and costs of communities of practice". Communications of the ACM., Vol. 45(4), pp. 69-73.

Mui, L., Halberstadt, A., Mohtashemi, M., (2002), "Notions of Reputation in Multi-Agents Systems: A Review". International Conference on Autonomous Agents and Multi-Agents Systems (AAMAS), pp. 280-287.

Nonaka, I. and Takeuchi, H., (1995), "The Knowledge Creation Company: How Japanese Companies Create the Dynamics of Innovation", Oxford University Press.

Sabater, J. and Sierra, C., (2002), "Social REGRET, a Reputation Model based on social relations", Proceedings of the Fifth International Conference on Autonomous Agents. Vol. 3(1), pp. 44-56.

Schafer, B. J., Konstan, A., J., Riedl, J. (1999). "Recommender Systems in E-Commerce", 1st ACM Conference on Electronic Conference (EC), pp. 158-166.

Ushida, H., Hirayama, Y., Nakajima, H., (1998), "Emotion Model for Life like Agent and its Evaluation". Proceedings of the Fifteenth National Conference on Artificial Intelligence and Tenth Innovative Applications of Artificial Intelligence Conference (AAAI / IAAI), Madison, Wisconsin, USA, pp. 8-37.

van-Elst, L., Dignum, V., Abecker, A., (2003), "Agent-Mediated Knowledge Management". International Simposium AMKM, Stanford, CA, USA, Springer, pp. 1-30.

Wang, Y., Vassileva, J., (2003), "Trust and Reputation Model in Peer-to-Peer Networks". Proceedings of the 3rd International Conference on Peer-to-Peer Computing.

Wasserman, S. and Glaskiewics, J., (1994), "Advances in Social Networks Analysis". Sage Publications.

Wenger, E., (1998), "Communities of Practice: Learning Meaning, and Identity". Cambridge U.K., Cambridge University Press.

Wenger, E., McDermott, R., Snyder, W., (2002), "Cultivating Communities of Practice", Harvard Business School Press.

Wooldridge, M., Ciancarini, P., (2001), Agent-Oriented Software Engineering: The State of the Art.

Zacharia, G., Moukas, A., Maes, P. (1999). "Collaborative Reputation Mechanisms in Electronic Marketplaces". In 32nd Annual Hawaii International Conference on System Science (HICSS-32).

This work was previously published in International Journal of Cognitive Informantics and Natural Intelligence, Volume 3, Issue 1, edited by Yingxu Wang, pp. 75-94, copyright 2009 by IGI Publishing (an imprint of IGI Global)

Chapter 6
Dynamic Negotiation Mechanism for Improving Service Quality for Replicas in Data Grids

Ghalem Belalem
University of Oran (Es Senia), Algeria

ABSTRACT

In order not to be limited in term of calculation, storage and communication, the concept of grid, which does not cease evolving, makes it possible to offer a practical operation of work unified as well as a great storage and computing power. To manage the division in the data grid, technical replication is used, but in spite of their advantages, the competitor access to the data could involve inconsistencies, from where the great challenge to ensure the consistency management between replicas of object. In this chapter, we describe model double-layered adapted to the applications on a large scale and which represents the support of the hybrid approach of consistency management of replicas based on pessimistic and optimistic approaches. This hybrid approach present an adapted mechanism based on the various negotiation forms between virtual consistency agents to be able to reduce the number of conflicts between replicas in data grids.

INTRODUCTION

Replication techniques are used to provide multiple critical copies and to maintain them. In coherent state, they improve the overall system availability and performance. In splitting of replication advantages, there are many problems that must be resolve like (Gray et al. 1996; Xu et al. 2002):

- How do we select and estimate the metrics for taking replication decisions?

DOI: 10.4018/978-1-60960-553-7.ch006

- When do we replicate a given object?
- Where do we place the replicas of a given object?
- How do we ensure consistency of all replicas of the same object?
- How do we route client requests to appropriate replicas?

Among these problems, the main critical concerns the consistency problem that needs to maintain the data consistency between a set of replicated data distributed among a set of computer. The main objective of a replica consistency approach is to avoid or even reduce the inconsistency between replicated data.

Many current applications can barely tolerate a certain degree of contradiction between replicas where the strong consistency is not a condition, for examples in the approximate readings from meteorological sensors often suffice when performing predictive modeling of weather conditions, the network security applications or in video conferencing applications (Olston and Widom, 2005). Our principal aim, in this paper, is to propose a hybrid mechanism of negotiation for the decision-making to the presence of the conflicts between the replicas in data grid (Foster and Kesselmann, 2004). This mechanism of negotiation is integrated in the hybrid consistency approach (Belalem and Slimani, 2007) inspired from the two pessimistic and optimistic traditional approaches. The structure of our present paper will as follows: the next section will describe the fundamental principles of pessimistic and optimistic consistency approaches.

Section 3, will dedicated to the description of the model used in our adapted negotiation mechanism. In section 4 section, we describe our mechanism of negotiation proposed for the decision-making to the meeting of the divergences between replicas which cannot be solved and present algorithms of our negotiation process. Section 5 is reserved for the characteristics of our proposed process. Section 6 presents some experiments to position and evaluate our approach compared to the other traditional ones. Finally section 7 will enclose this work by the presentation of the some future tracks.

APPROACHES OF CONSISTENCY MANAGEMENT

The Consistency is a relation which defines the degree of similarity between copies of a distributed entities. In the ideal case, this relation characterizes copies which have identical behaviors. Although in the real cases, even when the copies evolve in a different way, consistency defines the threshold of dissimilarity authorized between these copies. We hope of a consistency protocol which ensures the execution of the operations of users, the mutual consistency of copies in accordance with a behavior defined by a model of coherence. The consistency protocol gives an ideal view as if there is only one user and only one copy of the data in the system. Replica consistency management can be achieved, either synchronously, using the so-called pessimistic algorithms, or asynchronously, deploying optimistic ones (Belalem and Slimani, 2007; Saito and Shapiro 2005). Fundamental tussles between pessimistic and optimistic approach are those related to scalability and security. The execution of pessimistic consistency assures that any change in one replica is atomically notified to all other replicas. Therefore, there is an inherent guarantee that all replicas will have the same data all the time, making this approach indispensable in the mission of critical and sensitive applications like the distributed banking application. On the other hand, the optimistic approach is employed for applications (large scale systems, mobile environments and system weakly coupled), which evolves rapidly in terms of response time for example. So that we can say that, the pessimistic approach is interested in consistency more than availability, while the optimistic approach supports the availability more than the consistency (Belalem and Slimani, 2007; Saito and Shapiro 2005).

TECHNIQUES OF PESSIMISTIC CONSISTENCY

The technique of pessimistic consistency is interesting, since it guarantees a data consistency all the time. This approach gives users an illusion of having a single, highly available copy of data. However, the guarantee of the total maintenance of consistency involves a high communication cost (Pacitti et al. 1999; Yu and Vahdat, 2001). The pessimistic algorithms (Saito and Shapiro 2005) prohibit the access to a replica unless it is updating. The advantage of the pessimistic approach is to avoid the problems involved in the reconciliation. A basic protocol, called RAWA (Read Any Write All) (Zhoun et al. 2004) consists in obtaining an exclusive bolt on all the copies before to effect a writing (respectively reading) on one of the copies. The availability of the readings is improved with protocol ROWA (Read Once Write All) (Goel et al. 2005). The readings lock and access only one copy, while the write access mode continues to lock and modify all the copies. Nevertheless, this protocol is blocking in the event of breakdowns. An alternative ROWAA (Read Once Write All Available) (Zhoun et al. 2004) adapts this protocol to the cases of crashes by locking only the available copies. When a copy covers its availability, it must initially synchronize itself to execute the remaining updates. Another strategy of replication is proposed by the vote protocol family by Quorum (Amir and Wool, 1998; Rodrigues and Raynal 2003). The transactions are sent to a group of copies which vote (to decide which update is the most recent, writing or reading). These strategies are adapted to cases of unavailability of frequent nodes. Moreover, if a reproduction is inalienable (i.e. the cause is node failure), it can prevent other reproductions from being temporarily consulted until the failure of node is detected.

Several characteristics for pessimistic approach, we can be summerized these as follows (Belalem and Slimani, 2007; Saito and Shapiro 2005):

- Quality of Service (QoS) is very well, in pessimistic consistency;
- Very badly adapted to uncertain and unsteady environments (i.e. the mobile environments and data grid) with high rate of changes. But it performs well in local area networks, in which latencies are small and failures uncommon;
- It cannot bear the updating cost when the degree of replication is very high.

TECHNIQUES OF OPTIMISTIC CONSISTENCY

The techniques based on optimistic consistency promise higher availability and performance, but let replicas temporarily diverge and users see inconsistent data (Saito and Shapiro 2005). Also means the optimistic strategy allow users to reach any copy for the reading or the writing operations, even when there are breakdowns of network or when some copies are unavailable. It scales well in the front of a high number of replicas (Saito and Shapiro 2005). This also means that the approach can lead to replica inconsistency. On the other hand, the approach requires a follow-up phase to detect and then correct divergences between replicas by converging them toward a coherent state. Although this approach does not guarantee a high consistency with respect to the pessimistic one. One can also indicate some characteristics of the optimistic approach like (Belalem and Slimani, 2007; Saito and Shapiro 2005; Kuenning et al. 1998):

- Optimistic consistency improve availability, applications make progress even when network links and sites are unreliable;

- It is well adapted to the large scale systems, because it's require little synchronization among replicas, and mobile environments;
- In optimistic consistency, QoS is not very signifiant. Often it attached factor degree of quality. Because the states of copies can be temporarily mutually contradictory;
- An update can be applied to one copy without being synchronically applied to other copies, and there will can be even a substantial time since the application of an update in a copy until the propagation of the update to other copies. The concurrent updates with the various copies can cause conflicts. For example, in a system distributed of the reservation of the airline company, the use of optimistic coherence, two copies cannot accept a reservation for the same seat (Belalem and Slimani, 2007).

BI-LEVELS MODEL FOR CONSISTENCY APPROACH

For the consistency management of replicas in large scales systems, we proposed a process of consistency management which profit as well as possible from between the traditional approaches pessimistic and optimistic (Belalem and Slimani, 2007). This process uses a model of two levels: level 0 is physical and comprising the localization of replicas, for level 1 is logical one and represents the various agents, where each agent is responsible for part of level 0 (see Figure 1). In our work, we consider a grid as a collection of distributed collections of Computing Elements (CE's) and Storage Elements (SE's). These elements are linked together through a network to form a *Site* or a *Cluster*. Sites are in turn linked together to form a grid. Replicas are stored on Storage Elements and are accessible from Computing Elements. Our model is described in Figure 1 (Belalem et al. 2009).

Figure 1. Bi-Levels Model for Consistency Management with Virtual Consistency Agent

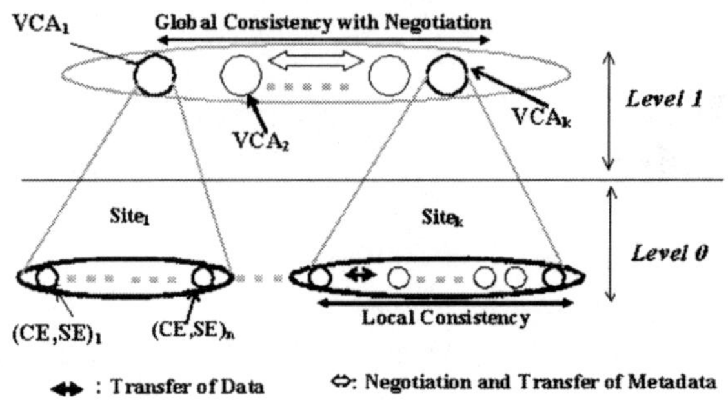

- **Level 0:** in this level we find sites that compose a grid. Each site contains a set of Computing Elements (CE's) and Storage Elements (SE's). Replicated data are stored on SE's and accessed from CE's via reading or writing operations. Each replica attached to additional information is called metadata. The latter gives a high description of replica and it can contain several additional information on the state of this replica, for example, number of versions, timestamp, origin of the update, factor of priority, etc.
- **Level 1:** in this level we define *k* Virtual Consistency Agents (VCA's) each one corresponding to each site of a grid. A virtual consistency agent VCA_i is responsible to manage replica consistency within a site S_i, that we call intra-site consistency. Then, each VCA_i cooperates with others VCA_j to ensure a replica consistency for the whole grid. In the presence of the critical situations (a significant conflict's factor of replicas). This cooperation is based on the principle of negotiation between the agents candidates with the conflicts resolution met. We refer this kind of replica consistency as an inter-sites consistency. In practice, the functionalities of a VCA_i are done by a Computing Element of site S_i. Hence, within a site S_i we suppose that we have

Figure 2. General process of our approach for reduction of divergent replicas in Data Grid

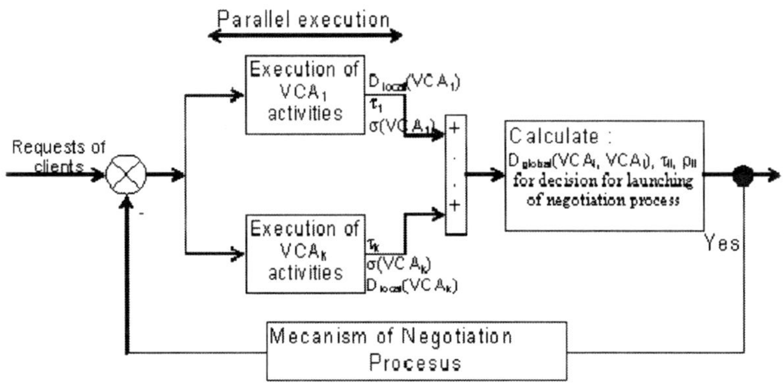

k-1 couples (CE,SE), that are able to store and access to replicated data, while the kth couple (CE,SE) represents the VCA_i and ensures replica consistency.

CONSISTENCY MANAGEMENT AND NEGOTIATION PROCESS

We will describe, in what follows, the core of the consistency process based on the negotiation between VCA (See Figure2). We will start by giving definitions of some basic elements, then we will present the two levels of service of consistency management and finally the mechanism of negotiation used for the conflicts resolution between the replicas.

Process Consistency Management Elements

The negotiation process consists of several elements, we can mention:

- Agent: An agent is a program that operates autonomously and accomplishes unique tasks without direct human supervision. It cooperates or competes with others agents to perform some set of tasks or satisfy some set of goals (Alda et al. 2004; Demazeau and Costa, 1996).
- Strategy of replication: Among the most used replication strategies (Chang and Chang, 2006; Pacitti and Ozsu, 2003; Saito and Shapiro 2005), we note:
  a. Single-Master: In single-master replication strategy (called leader-follower) one replica is selected to be the master. All clients then issue their update operations (e.g., insert, update, delete) to the master replica. The master will then propagate the update operations to the other replicas. Read operations can be issued to any replica (Chang and Chang, 2006; Saito and Shapiro 2005).
  b. Muti-Masters: In multi-master strategy, clients can issue their operations to more than one replica. When accepting operations on any replica in the system, it is possible to let the replica that receives the operation commit the change locally. This allows the client continue working based on that operation, and then propagate the change to the other replicas in the background (Pacitti and Ozsu, 2003; Saito and Shapiro 2005).
- Divergence: two replicas (r_i, r_j), of the same data, are known as divergent, if

Figure 3. VCA_i activities modeled in Petri net

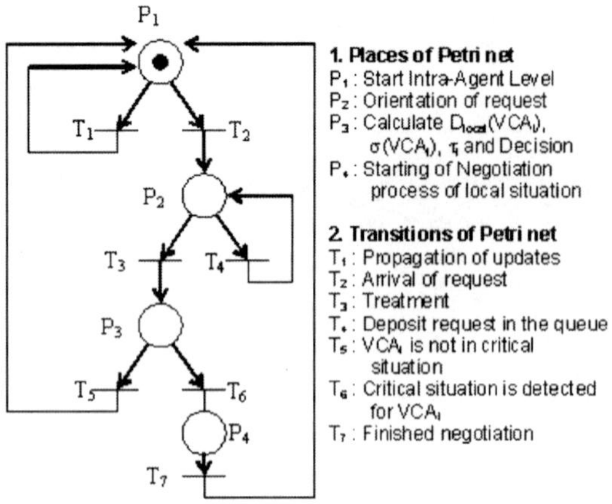

Metadata(r_i) ≠ Metadata(r_j) and *version(r_i) ≠ version(r_j)*, we speak about weak divergence.

- Conflict: two replicas (r_i, r_j), of the same data, are known as in conflict, if *Metadata(ri) ≠ Metadata(r_j)* and *version(r_i) = version (r_j)*, we speak about strong divergence. We associate, this definition of conflict, *Conflicts_Nbr* metric to indicate the number of conflicts of VCA.

Intra-Agent Level

The principal aim of the VCA is the control of process of local consistency. Its fundamental mission is to make converge the replicas towards the same local reference within the site (Figure3). This process can be announced according the following phases:

i. Phase of reception and treatment: according to the replication strategy of site, the VCA directs the customer's request towards the node of the free master so that it is treated, if not, it deposits it in the queue of its site.
ii. Phase of control of degree of divergence tolerated of a VCA: The objective of this phase consists in following the evolution of degree of divergence of replicas within a site by the VCA. To study the evolution of the divergence of replicas inside a site, we put forward the three following measures:

a. Measure rate of the number of conflicts per site (τ_i): this measurement makes it possible using VCA to know the rate of conflicts by the total number of replicas of the same data inside a site, and it is given by:

$$\tau_i = \frac{Conflicts\_Nbr(VCA_i)}{n_i}$$

Where :

VCA_i: is the identifier of ith VCA;

n_i: is the number of replicas in site of VCA_i of given object.

(1)

b. Measure distance within a site (D_{local}): we define Dlocal measurement between the versions maximum and minimal of replicas of the same object inside a site of a VCA. This measurement makes it possible to give a vision on the age of the replicas. It can be to calculate by:

$$D_{local}(VCA_i) = \overset{n_i}{\underset{t=1}{Max}}(V_{it}) - \overset{n_i}{\underset{t=1}{Min}}(V_{it}) \quad (2)$$

c. Measure dispersion of versions (σ): this measurement makes it possible to inform us on the manner of dispersion of the versions of the replicas around an average (3) of the same data inside a site. This measurement is given by the formula (4).

$$\bar{V}_i = 1 / n_i \sum_{t=1}^{n_i} V_{it} \quad (3)$$

Box 1.

```
Algorithm Local Critical_Situation
Input VCAi
Let τm, Dm-L be parameters of tolerance
/*τm : Rate of conflicts number tolerated */
/* Dm-L : Local distance tolerated*/
Begin
 Calculate : τi, σi; Dlocal for VCAi
 If (τi>τm) or (σi>ε) or (Dlocal> Dm-L) Then return (True)
                // existence of critical situation of VCAi
Else return (False)
End
```

$$\sigma(VCA_i) = \sqrt{1 / n_i \sum_{t=1}^{n_i} (V_{it} - \bar{V}_i)^2} \qquad (4)$$

We detect critical situations of one VCA to the meeting of one of the following cases:

a. τ_i > Rate of conflicts number tolerated;
b. $D_{local}(VCA_i)$>Distance tolerated;
c. $|\sigma(VCA_i)| > \varepsilon$; where $\varepsilon <<1$

The algorithm to check the existence or not of critical situation of VCA_i, is given in Box 1.

If VCA is in critical situation, i.e., that one of these events is detected, then it starts the local process of negotiation, see Box 2 (Algorithm Intra-Agent Level).

Box 2.

```
Algorithm Intra-Agent Level
Begin
/* Is a request arriving at the site controlled by VCAi */
Switch (Strategy) Do
Case Single Master: If Master = free Then treatment of the
request of the customer
          Else to deposit the request in the queue of the site
Case Multi Master: If ∃Master = free Then treatment of the
request of the customer
          Else to deposit the request in the queue of the site
EndSwitch
If Local Critical_situation(VCAi)
               Then Process of local negotiation
                        Else Propagate of updates.
End
```

Inter-Agent Level

In Inter-agent level two situations can be treated (As shown in Figure 4).

The first situation corresponds to the competitive negotiation and the second one represents cooperation negotiation.

1. The process of the competitive negotiation is started following the meeting of critical situation between agents. From a total point of view to follow the evolution of divergences of replicas between various VCA, we put forward the following measures:
    a. Measure rate of the number of conflicts (τ_{ij}): this measurement makes it possible to estimate the rate of conflicts

Figure 4. Stages of Inter-Agent Level

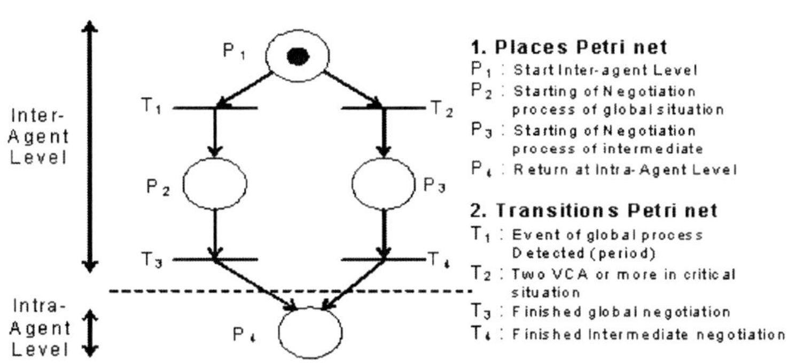

between agents. For example, the rate of conflicts between VCA_i and VCA_j are to give by formula (5).

$$\tau_{ij} = \frac{Conflicts\_Nbr(VCA_i) + Conflicts\_Nbr(VCA_j)}{n_i + n_j} \quad (5)$$

We can generalize the formula (5) to estimate the rate of conflicts in the system by:

$$\tau_* = \frac{\sum_{i=1}^{k} Conflicts\_Nbr(VCA_i)}{\sum_{i=1}^{k} n_i} \quad (6)$$

b. Measure global distance inter-agents (D_{global}): to study the distance between two VCA, we use the Euclidean distance. This measurement makes it possible to propose an estimate to us on the distance of the two VCA compared to the versions of the replicas of the same object. It is given by:

$$D_{global}(VCA_i, VCA_j) = \begin{cases} \sqrt{\sum_{t=1}^{n_i} \left|V_{it} - V_{jt}\right|^2} & If\ n_i \le n_j \\ \sqrt{\sum_{t=1}^{n_j} \left|V_{it} - V_{jt}\right|^2} & Else \end{cases} \quad (7)$$

c. Measure coefficient of correlation (ρ_{ij}): this measurement describes the degree of correlation between two VCA and it is represented according to covariance (formula 8).

$$Cov(VCA_i, VCA_j) = \begin{cases} 1/n_i \sum_{t=1}^{n_i} (V_{it} - \bar{V}_i)(V_{jt} - \bar{V}_j) & If\ n_i \le n_j \\ 1/n_j \sum_{t=1}^{n_j} (V_{it} - \bar{V}_i)(V_{jt} - \bar{V}_j) & Else \end{cases} \quad (8)$$

The coefficient of correlation makes it possible to form groups of the VCA where VCA_i inside the group are very close the ones with others. The coefficient of correlation between VCA_i and VCA_j is presented under the formula (9):

$$\rho_{ij} = \frac{Cov(VCA_i, VCA_j)}{\sigma(VCA_i) * \sigma(VCA_j)} \quad (9)$$

We associate δ_{ij} a degree of confidence at coefficient of correlation ρ_{ij}. δ_{ij} is an actual value to check the assumption of correlation ρ_{ij} between two VCA. A value δ_{ij} near to zero means that the assumption is false and that there is thus no correlation between VCA_i et VCA_j. A value δ_{ij} distant from zero (positive or negative) means that the assumption is checked and that there is thus correlation between VCA_i et VCA_j. It will be define by:

$$\delta ij = \begin{cases} \rho_{ij}\sqrt{(n_i - 2)(1 - \rho_{ij}^2)} & If\ n_i \le n_j \\ \rho_{ij}\sqrt{(n_j - 2)(1 - \rho_{ij}^2)} & Else \end{cases} \quad (10)$$

We define a critical situation between VCA_i and VCA_j if one of the following conditions is checked:

a. τ_{ij} > Rate of conflicts number tolerated
b. $D_{global}(VCA_i, VCA_j)$ > Distance tolerated
c. $|\rho_{ij}| < \varepsilon$; where $\varepsilon \ll 1$

The algorithm to check the existence or not of critical situation between VCA_i and VCA_j, is given in Box 3.

Box 3.

```
Algorithm Global Critical_Situation
Input VCAi, VCAl
Let D^(m-G) be parameters of tolerance
/*τ^m: Rate of conflicts number tolerated */
Begin
 Calculate : τil, σi; Dglobal for VCAi, VCAl
 If(τil>τ^m) or (σi>ε) or (Dglobal> D^(G-b))
    Then return (True)
    // existence of critical situation between VCAi and VCAl
    Else return (False)
End
```

If the event of the critical situation between VCA is detected, we carry out then the construction VCA groups in conflict and start of the process of negotiation intermediate by the algorithm in Box 4.

2. In the process of cooperative negotiation, the VCA try to reach the total commun utility of the system, which corresponds to consistency global by mechanism cooperation between VCA. This stage of inter-agent level is described by negotiation process of global situation.

Negotiation Process

As announced before, the mechanism of negotiation combines several cooperation forms for conflicts resolution according to the various situations met.

a. **Negotiation process of local situation:** In local situation, the VCA of the site in critical situation, acts like an initiator of negotiation, it announced its conflict situation (crisis plan) for the group of VCA on network by diffusion. The agents receive and evaluate this situation. The VCA which have the capacity to solve this crisis plan send to the initiator bids by indicating their capacities to carry out this crisis announced and whole information associated of correlation degree to this situation. At this time the initiator has a role of arbitration, it evaluates the bids and grants its crisis plan to the VCA the best suitable one also called the contractor. Lastly, the initiator and the contractor exchange information necessary during achieving of this critical situation.

Box 4.

```
Algorithm Competitive Group
Initialization
   Indice ={1,...,k}; Gr_comp = Ø; r=1;
/* k : total number of VCA of the system */
/* Gr_comp : group of agents in competition */
   While (Indice? Ø) and (r < k) Do
        Gr_inter = Indice - {r}; Gr_comp = {VCAr};
        While (Gr_inter ? Ø) Do
        Choose i of Gr_inter set such as the value i is smallest possible
           Gr_inter= Gr_inter – {i};
           If Global critical situation (VCAr, VCAi) Then
               Begin
                       Gr_comp = Gr_comp ∪ {VCAi}
                       Indice = Indice - {i}
               End
           EndIf
        EndWhile
        If card(Gr_comp) > 1 Then Negotiation process of intermediate situation for
        Gr_comp ;
   Repeat r = r +1 ; Until (r ∈ Indice) or (r > k);
EndWhile
```

Box 5.

```
Algorithm Intermediate Situation
For all elements of the group; research the most stable VCA
VCAs = most stable VCA of the group
// σ(VCAs) = Min {σ(VCAi)}   ∀ i∈ Group
Propagate informations of VCAs toward the other VCA of the group /* we speak about dominant strategy */
EndFor
```

Box 6.

```
Algorithm Dutch_auction
Begin
AGT_Max = Choose a VCA' which has largest version, it proposes its offer with the other VCA of the
system,
i :=1 ;
Repeat
If Majority_vote(AGT_Max) Then
        /* the bid is accepted by majority of VCA */
        AGT_Max propagates information of its metadata
        to the other VCA, as well as the data,
        Stop.
i :=i+1 ;
Choose VCA' having the i^th greatest version, it proposes its offer to the other VCA of the system;
AGT_Max = VCA et aller à
Until i>=k ; /* k : total number of VCA */
 /* No VCA could be accepted by majority */
English_auction. /* alternative 2 */
End Dutch_auction
```

b. **Negotiation process of intermediate situation:** In this stage of process, the VCA in conflict situation gather to find a agreement consensus to their critical situation, by the algorithm in Box 5.

c. **Negotiation process of global situation:** In this situation, global consistency is put in priority, the whole of VCA negotiate their information to arrive at consensus of agreement. The negotiation process used aims at converging the various replicas, of the same data in the complete system, towards a global reference replica according to one of the two following alternatives :

   A. One of the aims of QoS in the consistency management is obtained the most recent information (most up to date). Generally this information is associated at replicas versions. The principle of negotiation of downward biddings, or also called the Dutch auction (Belalem, 2008; Buyya and Vazhkudai, 2001), is used to arrive at a consensus of agreement (Dutch_auction algorithm), see Box 6.

   B. The second alternative is based on the principle of stability of the VCA, i.e. we support the VCA which has the smallest coefficient of divergence. We use in this alternative the principle of the English auction (see Algorithm English_auction).

The algorithm used for the majority vote is described in Box 7.

The algorithm to study the degree of the correlation between VCA_i and VCA_j is given in Box 8.

CHARACTERISTICS MODEL

From its structure the Bi-Levels model has several characteristics for examples:

Box 7.

```
Algorithm English_Auction
Begin
    AGT = Choose VCA wich has largest version
    /* Degree of stability from point of view the maximum version */
    For i=1 to k Do
      If (σi < σAGT) and (τi < τAGT) Then AGT = VCAi /* we favour the stability of VCA */
    EndFor
        AGT propagates information of its metadata to the other VCA, as well as the data
End English_Auction
```

i. Hybrid consistency management: to manage consistency among replicas, our proposal combines the pessimistic and optimistic approaches;
ii. Incremental management of consistency: inconsistency among replicas is first treated at level 0 (local consistency) then at level 1 (global consistency);
iii. Flexibility: the separation between couples (CE,SE) and their VCA's allows to diversify the strategies of replication at the site level.
iv. Scalability: thanks to the tree structure, our proposed approach is very adapted to scalability of grid: additing or removing (CE,SE), sites, increase or reduction of the number of replicas;
v. Reduced communication cost: the tree structure confers to our proposal a linear complexity. The communication cost to obtain consistencies intra and inter-sites are summarized in the following table.
vi. Adaptive negotiation: The mechanism of negotiation used combines several forms of interaction, for example: by arbitration, negotiation by compromise or integration according to presented situations;
vii. Simplicity: the hierarchical tree-based approach is solely composed by two levels and it is topology independent what ever the complexity of a grid;
viii. Transparency: a site is discerned by user as a single logical entity. Hence, it can use all capabilities of a site (computing, storage) without knowing its localization in the grid nor its composition.

Box 8.

```
Algorithm CorrelationDegree
Input VCAi, VCAi
Begin
 Calculate: ρii, δii
 If (||ρi|-1|<ε) or (δii > ε)
    Then return (False)
    Else return (True)
End
```

SIMULATION EXPERIMENTS AND RESULTS

Data is not really present, so, because this absence we have realized a simulator to implement different protocols and to compare results between them. Our goal is to study and to compare our protocol with the two traditional approaches (pessimistic and optimistic), we use two kinds of metrics:

- This category called also measurement of performance, allows the study of the behavior of our approach with that of pessimistic approach.
- The second category of metric called also measurements of qualities of service (QoS), allows studying the quality of the rendered service of our approach compared to the optimistic approach.

Figure 5. Average response time according to number of sites

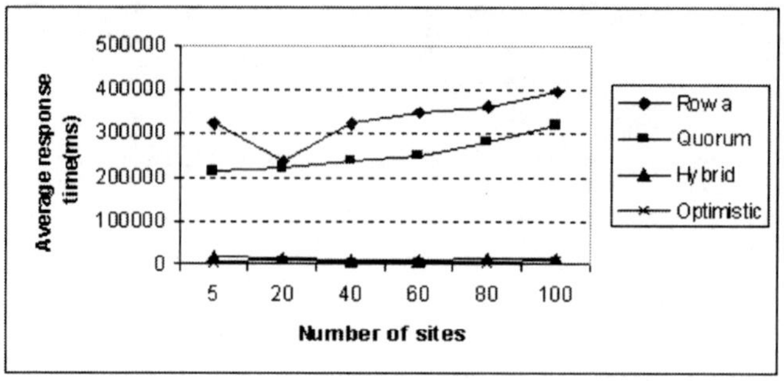

Figure 6. Profit Expressed as a percentage

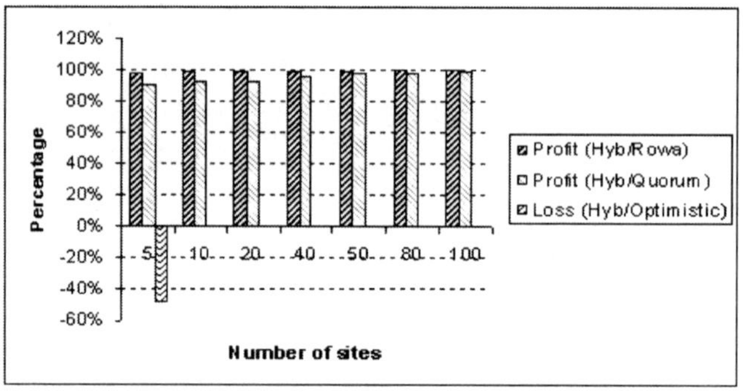

In the first simulation, to study the performance, we choose to compare our approach with the two protocols pessimistic: *ROWA (Read One Write All)* and *majority Quorum.*

The Figure 5 illustrates the evolution of the average response time according to the number of sites. The average response time of requests tends to increase with the increase the number of sites for the pessimistic approaches (ROWA and Quorum). This results obtained show that these approaches (ROWA and Quorum) are impracticable in data Grids.

The results of simulation as shown in Figure 6, showed the profits of our hybrid approach. The benefit to be gained is very significant compared to the pessimistic approaches (Rowa and Quorum).

The results of Figure 7 show that the number of divergences of data Grid by period in our approach is lower then the optimistic approach.

The conflict count of data Grid by period as shown in Figure 8, we observe that number of conflicts in our protocol is small compared with optimistic approach. In conclusion, our approach suggested makes it possible more quickly to solve the conflicts met, which more quickly reduces the number of conflicts met during simulation.

CONCLUSION AND FUTURE WORK

The main problem introduced by replication techniques is maintaining consistent replicas. In Data Grid environment, strong consistency is not adapted due to their prohibilitive cost. Weak consistency approaches can be used in these systems by tolerating divergences between replicas for at least some time period. In these divergence situations, reconciliation poses many problems and

Figure 7. Number of divergences of data Grid by period

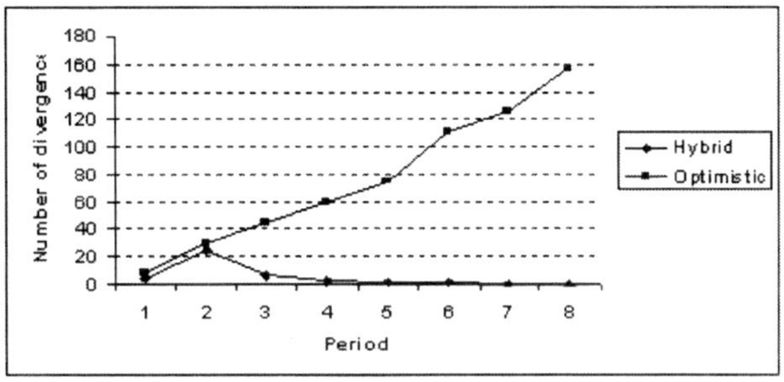

Figure 8. Count of conflicts of data Grid by period

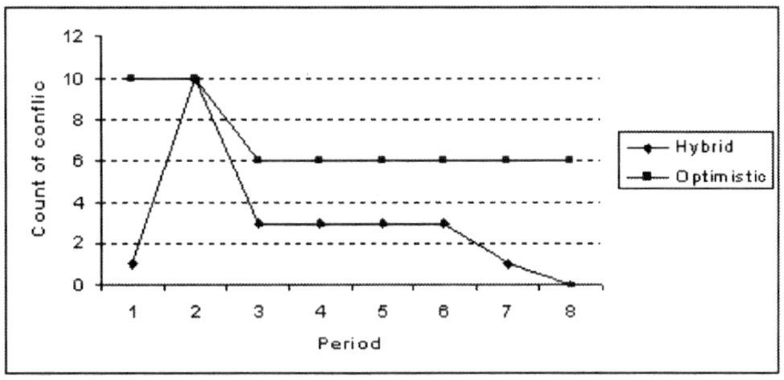

in particular in mechanisms of conflicts resolution between replicas. In this paper, we presented mechanism for convergence and resolution of conflicts between replicas that is suitable for data grid. This mechanism proposed is based on the various negotiation forms between virtual consistency agent (VCA). This proposed comprises two levels. In Intra-agent level, VCA checks and controls the degree of divergence tolerated. If this divergence is important, the VCA initiates the process of negotiation to regulate its local consistency, that we named negotiation process of local situation. The Inter-agent level controls the evolution of divergence degree between VCA. With an aim of converging very quickly towards an acceptable consistency, this level is divided of intermediate and global process. First is started if two VCA or more observe a degree of divergence not tolerated. Second is started periodically to converge the replicas of the same data to global reference replica. Our mechanism proposed is very promising in large scale environments. By its no blocking aspect of all requests (rarely), it allow to increase the performance by reducing the response time and by its aspect of reduction of the divergences between the replicas, it provides an improvement in the quality of service. Currently, the mechanism negotiation proposed articulated on the Bi-level model is in process of experimentation with Java environment.

There are a number of directions which we think are interesting and are worth further investigation. We can mention:

- Development of the Web service for consistency management of replicas: We propose to integrate our approach in the form of Web service in the Globus environment by using technology WSDL (Foster and Kesselmann, 2004);
- To propose the incorporation of the weight in the suggested plans by VCA candi-

dates with the resolution of critical situations, in function their stability of degree of consistency and the size their group of membership;

- To take into account the factor time during the phase of negotiation, i.e., to define intervals of time during which we must make to us decisions;
- In the current version of our approach, we have placed the replicas randomly. It is worthwhile to explore the possibility of making a static or dynamic placement to improve QoS in the data grid (Haddad and Slimani, 2007);
- Load balancing: From this point of view, and for improving even more performances and the quality of service of our approach, we propose to extend it by a service of load balancing (Yagoubi and Slimani, 2007; Li and Lan 2005*),* which allows to balance the requests on the various sites of Data Grid;
- Extend the proposed approach to consistency management of replicas for Cloud Computing environments (Belalem et al. 2010).

REFERENCES

Alda, S., Cramers, A. B., Bilek, J., & Hartmann, D. (2004), Support of Collaborative Structural Design Processes through the Integration of Peer-to-Peer and Multi-agent Architectures, *in Proceedings of the 10th International Conference on Computing in Civil and Building Engineering (ICCCDE-X)*, Weimar, Germany.

Amir, Y., & Wool, A. (1998). Optimal availability quorum systems: Theory and Practice. *Information Processing Letters*, *65*(5), 223–228. doi:10.1016/S0020-0190(98)00017-9

Belalem, G. (2008). Economic Model for Consistency Management of Replicas in Data Grids with OptorSim Simulator, *Networks for Grid Applications, Second International Conference - (GridNets 2008),* (pp. 121-129), Beijing, China, October 8-10, 2008, Lecture Notes of the Institute for Computer Sciences, Social Informatics and Telecommunications Engineering, Vol. 2, Springer.

Belalem G., Benotmane Z. & Benhallou K. (2009). Self Adjustable Negotiation Mechanism for Convergence and Conflict Resolution of Replicas in Data Grids, *International Journal of Cognitive Informatics and Natural Intelligence, (IJCINI*). 3(1), 95-110

Belalem, G., & Slimani, Y. (2007). A hybrid approach to replica management in data grids [IJWGS]. *International Journal Web and Grid Services*, *3*(1), 2–18. doi:10.1504/IJWGS.2007.012634

Belalem, G., Tayeb, F. Z., & Zaoui, W. (2010). Approaches to Improve the Resources Management in the Simulator CloudSim, *First International Conference Information Computing and Applications - (ICICA'2010),* (pp. 189-196), Tangshan, China, October 15-18, Lecture Notes in Computer Science, Vol. 6377, Springer.

Buyya, R., & Vazhkudai, S. (2001). Compute power market: Towards a market-oriented Grid, CCGRID'01, *First International Symposium on Cluster Computing and the Grid*, (pp. 574-581), Brisbane, Australia.

Chang, R.-S., & Chang, J.-S. (2006). Adaptable Replica Consistency Service for Data Grids. *Third International Conference on Information Technology: New Generations (ITNG'06)*, pp. 646-651, Las Vegas, Nevada, USA.

Demazeau, Y., & Costa, A. C. R. (1996). Populations and Organizations in Open Multi-Agent Systems, *in Symposium on Parallel and Distributed Artificial Intelligence (PDAI'96)*, Hyderabad, India.

Foster, I., & Kesselmann, C. (Eds.). (2004). *The Grid 2: Blueprint for a new computing infrastructure*. Elsevier Series in Grid Computing. Morgan Kaufmann Publishers.

Goel, S., Sharda, H., & Taniar, D. (2005). Replica synchronisation in grid databases. [IJWGS]. *International Journal Web and Grid Services*, *1*(1), 87–112. doi:10.1504/IJWGS.2005.007551

Gray, J., Helland, P., Neil, P. O., & Shasha, D. (1996), The dangers of replication and a solution. *In ACM SIGMOD International Conference on Management of Data*, (pp. 173-182), Montreal, Quebec, Canada, 4-5 June 1996. ACM Press.

Haddad, C., & Slimani, Y. (2007). Economic model for replicated database placement in Grid. *In Proceedings of* Seventh *IEEE International Symposium on Cluster Computing and the Grid (CCGrid'07)*, (pp. 283-292), Rio de Janeiro, Brazil.

Kermarrec, A.-M., Rowstron, A., Shapiro, M., & Druschel, P. (2001). The IceCube approach to the reconciliation of divergent replicas. *PODC '01: Proceedings of the twentieth annual ACM symposium on Principles of distributed computing*, (pp. 210-218), Newport, Rhode Island, USA.

Kistler, J. J., & Satyanarayanan, M. (1992). Disconnected operation in the coda file. *ACM Transactions on Computer Systems*, *10*(1), 3–25. doi:10.1145/146941.146942

Kuenning, G. H., Bagrodia, R., Gay, R. G., Popek, G. J., Reiher, P. L., & Wang, A.-I. (1998). Measuring the Quality of Service of Optimistic Replication, *ECOOP '98: Workshops on Object-Oriented Technology*, pp. 319-320, Brussels, Belgium.

Li, Y., & Lan, Z. (2005). *A survey of load balancing in grid computing. High Performance Computing and Algorithms, Lecture Notes in Computer Science* (*Vol. 3314*, pp. 280–285). LNCS.

Olston, C., & Widom, J. (2005). Efficient Monitoring and Querying of Distributed, Dynamic Data via approximate Replication. *IEEE Data Eng. Bull*, *28*(1), 11–18.

Pacitti, E., Minet, P., & Simon, E. (1999). Fast Algorithms for Maintaining Replica Consistency in Lazy Master Replicated Databases, *Int. Conf. on Very Large Databases*, Edinburgh, UK.

Pacitti, E., & Ozsu, M. T. (2003). Replica Consitency for Lazy Multi-Master Configurations in a Cluster of Autonomous Databases [Lyon, France.]. *DBA*, *03*, 318–327.

Petersen, K., Spreitzer, M., Terry, D., & Theimer, M. (1996). Bayou: replicated database services for world-wide applications, *EW 7: Proceedings of the 7th workshop on ACM SIGOPS European workshop*, (pp. 275-280), Connemara, Ireland.

Ranganathan, K., & Foster, I. (2001). *Identifying Dynamic Replication Strategies for a High-Performance Data Grid* (pp. 75–86). In GRID.

Rodrigues, L., & Raynal, M. (2003). Atomic broadcast in asynchronous crash-recovery distributed systems and its use in quorum-based replication. *IEEE Transactions on Knowledge and Data Engineering*, *15*(5), 1206–1217. doi:10.1109/TKDE.2003.1232273

Saito, Y., & Shapiro, M. (2005). Optimistic replication. *ACM Computing Surveys*, *37*(1), 42–81. doi:10.1145/1057977.1057980

Vidot, N., Cart, M., Ferrié, J., & Suleiman, M. (2000). Copies convergence in a distributed real-time collaborative environment, *CSCW '00: Proceedings of the 2000 ACM conference on Computer supported cooperative work*, (pp. 171-180), Philadelphia, Pennsylvania, USA.

Xu, J., Li, B., & Li, D. (2002). Placement problems for transparent data replication proxy services. *IEEE Journal on Selected Areas in Communications*, *7*, 1383–1398.

Yagoubi, B., & Slimani, Y. (2007). Task Load Balancing Strategy for Grid Computing. *Journal ofComputer Science*, *3*(3), 186–194. doi:10.3844/jcssp.2007.186.194

Yu, H., & Vahdat, A. (2001). The Costs and Limits of Availability for Replicated Services, *In SOSP '01: Proceedings of the Eighteenth ACM Symposium on Operating Systems Principles*, pp. 29-42, New York.

Zhoun, W., Wang, L., & Jia, W. (2004). An analysis of update ordering in distributed replication systems. *Future Generation Computer Systems*, *20*(4), 565–590. doi:10.1016/S0167-739X(03)00174-2

KEY TERMS AND DEFINITIONS

Data Grid: Data grid is a grid computing system that deals with data — the controlled sharing and management of large amounts of distributed data.

Dutch Auction: Dutch auction referred specifically to a type of auction that starts with a high price that keeps going down until the item sells. This is the opposite process to regular auctions, where an item starts at a minimum price and bidders wrestle over it by increasing their offers.

English Auction: Bidding starts with a low price, and is raised incrementally as progressively higher bids are solicited, until either the auction is closed or no higher bids are received.

Multi-Master Strategy: In this strategy, a system supporting several masters per object.

Quorum: In general allow writes to be recorded only at a subset (a write quorum) of the up nodes, so long as reads are made to query a subset (a read quorum) that is guaranteed to overlap the write quorum.

Single Master Strategy: In this strategy, a system supporting one master per object;

VCA: Virtual Consistency Agent.

Section 2

Chapter 7
Ambient Intelligence on the Dance Floor

Magy Seif El-Nasr
Penn State University, USA

Athanasios V. Vasilakos
University of Peloponnese, Greece

ABSTRACT

With the evolution of intelligent devices, sensors, and ambient intelligent systems, it is not surprising to see many research projects starting to explore the design of intelligent artifacts in the area of art and technology; these projects take the form of art exhibits, interactive performances, and multi-media installations. In this paper, we seek to propose a new architecture for an ambient intelligent dance performance space. Dance is an art form that seeks to explore the use of gesture and body as means of artistic expression. This paper proposes an extension to the medium of expression currently used in dance—we seek to explore the use of the dance environment itself, including the stage lighting and music, as a medium for artistic reflection and expression. To materialize this vision, the performance space will be augmented with several sensors: physiological sensors worn by the dancers, as well as pressure sensor mats installed on the floor to track dancers' movements. Data from these sensors will be passed into a three layered architecture: a layer analyzes sensor data collected from physiological and pressure sensors. Another layer intelligently adapts the lighting and music to portray the dancer's physiological state given artistic patterns authored through specifically developed tools; and, lastly, a layer for presenting the music and lighting changes in the physical dance environment.

INTRODUCTION

Ambient Intelligence (AmI) integrates concepts ranging from ubiquitous computing to Artificial Intelligence (AI) with the vision that technology will become invisible, embedded in our natural surroundings, present whenever we need it, attuned to the users' senses, and adaptive to users. In an Ambient Intelligent environment, people are surrounded with networks of embedded intelligent devices that can sense their state, anticipate, and perhaps adapt to their needs. One can imagine the implementation of such a vision within a dance environment where lighting, scenery, and audio change dynamically in the performance to reflect the dancers' movements and state. In this paper, we address such a vision. Particularly, we aim to address the design of a new ambient intelligent dance environment.

The project explores the visual depiction of the self and body in the light of the rising new technologies and media. The focus is the graphic representation of the dancer's creative expression in time. The artificial representation of the dancer is generated by transforming actual physiological signals from a dancer's body into visual and audio forms. Since the resulting form will represent the individual whose biological signals generate and sustain it, it will be a personal signature of that individual in digital space. The dance space will envelop its user via stage lighting and sound. By offering a new way of exploring the relationship between the dancer and her artificial reflection through the dance environment, this project will provoke profound and lasting aesthetic and reflective responses from its users/audience. The pursuit of this project is expected to establish a new area of creative inquiry in dance with several potential spin-offs and artistic collaborations.

To realize this vision, dancers will wear wireless physiological sensors that measure three functions: 1) skin conductance, 2) cardiac activity, and 3) body temperature. In addition, pressure sensors will be installed in the physical dance floor. Data from the sensors will be processed through two interface systems; one extracts physiological data, normalizes the signal, and interpolates any missing data, while the other collects pressure signals from all pressure mats and computes light IDs for lights affecting the dancer. These Light IDs and the physiological data will then be fed to two intelligent subsystems: an intelligent on-stage lighting system and an intelligent music system. These systems will adapt the lighting and music to reflect the dancer's movements and physiological state.

As with any artistic performance giving artists a language to identify the style and manner with which the intelligent systems can reflect dancer's state and movements is of utter importance. For this purpose, we have developed two tools: lighting and music tool. Both tools allow artists to identify the style and manner of reflection at a high level. The lighting tool allows designers to set several constraints; for example, she can set constraints indicating the shift of warmth of color in specific regions or level of contrast and shift in contrast between different regions as a function of the physiological state. Similar to the lighting tool, the intelligent music tool will allow artists to author constraints and patterns of music movements and shifts as a function of dancer's movements and physiological state.

The lighting and music intelligent systems use these artistic settings as constraints to adapt the lighting and music in real-time based on the given dancer's physiological state and movements. The intelligent lighting system uses non-constraint optimization to adjust each on-stage light color and angle reflecting the dancers' physiological state and movements while maintaining desired artistic style and patterns. The intelligent music system uses a rule-based system to dynamically and unobtrusively adapt the music to the dancer's physiological state given authored constraints and music patterns.

There have been numerous projects that integrated virtual imagery in performance, examples include (Crow & Csuri, 1985); Gruen, 1983; (Meador, Rogers, O'Neal, Kurt, & Cunningham, 2004). However, there is very little work that allows adaptation of on-stage lighting and music as an extension of the dancers' cognitive space. The area of Cognitive Informatics (Wang, 2006, 2007) fits our area of inquiry. In particular this is a field studies the mechanisms and process of natural processing and intelligence, including emotions, cognition, decision making, and its application to entertainment, engineering, educational, and other applications. We intend to explore the improvisational artistic ability of lighting designers and musicians. This is a particular area of research that has received very little attention due to its complexity integrating not only emotional and cognitive intelligence, but also artistic and adaptive intelligences. We seek to develop lighting and music intelligent tools that can extend the dancers' form and cognitive state. To our knowledge this particular focus has not been explored in previous literature exploring the use of lighting and music within performance and entertainment spaces.

The presented research has several contributions. It presents a new system that integrates intelligent systems and sensor technology in as a facilitator of self expression within a dance performance. In addition to its artistic contribution, the project also has several technical and design contributions, including tools that allow artists to author the style and patterns of lighting and music temporal variations in response to sensor data. While the music tool is similar to the tools used in the game industry, the lighting tool is original; it is developed based on Seif El-Nasr's previous work (Seif El-Nasr, 2005), which integrates tacit knowledge collected through two years of practical and theoretical training in the area of lighting design. By just adjusting the parameters supplied by the tools artists can create very different performances.

In this paper, we will discuss the ambient intelligent dance system in more detail. We will first discuss previous work in the area of interactive dance. We will follow with a detailed discussion of the proposed ambient intelligent system describing the sensors, tools, and intelligent systems developed. We will then describe a current prototype of the system. We will conclude by discussing limitations and future work.

PREVIOUS WORK

Numerous composers, choreographers, dancers, and theorists have explored the use of technology in theatre and dance. We do not intend to describe all the work that has been done in the realms of academic research, installations, or interactive productions here. However, we will discuss few examples that have influenced our work. Discussing these examples will situate our work; uncover its uniqueness, and its purpose.

One of the most influential and significant work that used animated figures for choreography is the work of Merce Cunningham. In his dance performance *Trackers*, he used a computer system called *Life Forms* devised by Tom Calvert to choreograph his dance movements. *Life Forms* is a piece of software designed to provide several stylized animated characters that allow users to create dance choreography or explore certain steps. In addition to using animation for choreography, Cunningham also developed a virtual dance installation in collaboration with Paul Kaiser and Shelley Eshkar, which was presented at Siggraph 1998. This installation was composed of a mental landscape in which motion-captured hand-drawn figures performed intricate choreography in 3D.

Besides the use of animated characters in a virtual performance, several performers have explored the use of animation within a real-life dance performance. For example, projected graphics have been used on backdrops in the San Francisco ballet *Pixellage*. In one of the scenes they used a virtual animated ball (projected on the screen behind the dancers) which dancers threw to each other. Another ballet performance called *The Catherine Wheel* used an animated character to represent the spiritual figure of Saint Catherine. By using an animated character, artists can easily represent the spiritual nature of the character as opposed to using real life effects or make-up.

Another example of the mix between virtual and real characters is depicted in the work of Meador et al.. They developed a collaborative production that mixes the use of virtual and real dancers within a dance stage. They used three different projectors within a dance performance; one of these projectors was used to project a virtual character that interacted with the dancers on stage. Their work was influenced by the work of Dan Saltz who directed *The Tempest* 2000 produced by the Interactive Performance Lab Group at the University of Georgia [http://dpa.ntu.ac.uk/dpa_search/result.php3?Project=136]. In this production of *The Tempest*, they projected the character *Ariel* as a virtual character. They used motion capture to animate the character in real-time. The use of a synthetic character for *Ariel* added to his magical quality, and thus enhanced the overall performance.

Another example of the use of technology in dance performances is the use of motion capture to inform changes in projected imagery. Troika Ranch, a Dance Company situated in New York City [http://www.troikaranch.org/], developed a motion capture system called *MidiDancer*, which uses several cameras to capture performers' movements. They explored the use of the *MidiDancer* as a method of dynamically synthesizing dancer's movements and using these synthesized movements to dynamically alter the projected video during the performance. Even though they presented several unique and interesting ideas, the use of motion capture within dance productions is still an area under research and exploration.

Ulyate and Bianciardi showed their work on the Interactive Dance Club in Siggraph 1998. The interactive dance club was composed of several zones where they experimented with several setups and sensors, including infra-red, pressure, and vision. They divided the dance floor into different zones which induced different interactivity paradigms. For example, in one zone they had a set of parallel light beams that detected when beams were broken. By breaking beams of light, participants could trigger 4-16 musical notes.

Similar to the Interactive Dance Club, Todd Winkler explored the use of space, gesture, and motion capture equipment for music composition. He focused on the use of dance and space to compose electronic music. He used the *Very Nervous System* (VNS) which is a system composed of one or two cameras that detect speed and location of dancers. He explored several methods of mapping the output data from VNS to music parameters, such as frequency, pitch, timber, etc.. He presented two productions in the late 1998 showing his work.

Beyond projection as a way to influence the dance space, Louis-Philippe Demers have explored the utility of adjusting physical stage lighting within an art installation. He developed a system that uses several sensors including, pas sensors, video sensors, optical and infrared sensors, sonar sensors, and 3D ultrasound devices to predict blocking and gather gesture information. Using these as input, he developed a system that manipulated on-stage lighting in terms of light brightness, color, and angle. He showed this system in several projects, including *The Shadow Project* and *Lost Referential*. Although it is clear that the artistic visions between the project presented in this paper and Demers' work is quite different, the two projects share the idea of dynamically adjusting lighting angles and colors. However, in our work we integrate a set of AI algorithms enabling change in light direction, color, and intensity based on the dancer's physiological state and movements as well as authored artistic styles and patterns which will be dictated through a lighting tool that we developed.

In addition to the use of technology in dance performances, there are several other research projects that integrated ambient intelligent systems in public installations and exhibitions. One example of such installations is the work of Tokushia et al.'s work on MYSQ (2006). MYSQ is a system that enables users to create video clips and share them with other users. MYSQ consists of a booth that integrates pressure sensors and several cameras to capture participants' dance movements or actions. Participants can create and enhance the video captured through adding cinematic effects and editing the video content through movements within the booth. Furthermore, participants can also collaborate during this content creation process. While this project, and others like it, may share the type of sensors that we use in our project, the aesthetic and artistic vision and the intelligent systems developed and presented in this paper are quite different. However, there are several sensor and vision technologies that these projects have used that we would like to further explore for the proposed project, including infrared vision systems.

DANCE SPACE

We envision a space similar to a proscenium theatre stage. Stage lights will be rigged on posts. We implant pressure sensors in the dance floor to track dancers' positions and movements. We also include a 3D surround sound system to play the music composed for the performance. The dancer wears an armband that collects physiological information while she freely moves around in space. Sensor information will be transmitted wirelessly through a local network to a computer that then analyzes this information and alters the music and on-stage lighting to express the dancer's physiological state.

Dancers wear the SenseWear® PRO_2 Armband [http://www.bodymedia.com/technology/index.jsp], which is a wearable body monitor that enables continuous collection of low-level physiological data, including heat flux, skin temperature, near body temperature, and galvanic skin response. It also includes a streaming program that we use to continuously and wirelessly stream the physiological data to a PC for processing while dancers freely move in space. The choice of this particular device was made due to our experience with it.

We adopt a pressure mat similar in design to others, such as ; an example can be seen in figure 1. As shown the device is interfaced to a micro controller. This pressure sensor mat is designed as an on-off switch, and thus is good for determining if a person stepped on the mat. We will use several mats to cover the dance floor. Signals from these sensors are sent directly to a host computer that assembles and identifies light IDs for lights on the stage affecting the dancer.

In addition to supplying music and sound content, artists will control the stylistic parameters for lighting and music adjustments. For this purpose, all the proposed intelligent systems include tools that allow artists to input stylistic constraints to direct the lighting and music changes.

Architecture

The architecture is shown in Figure 2. A *physiological signal interface* normalizes the physiological signals, interpolates missing data, and transfers the resultant data as a set of XML messages to two intelligent systems: *Intelligent on-stage Lighting System* and *Intelligent Music System*. The *Pressure Sensors Interface* analyzes the pressure sensor signals identifying lights relevant to the dancer's positions and sends these light IDs to the intelligent systems. The physiological state is stored in a structure called *Dancer Physiological State* represented in XML. The lights relevant to dancer's positions are stored as a list of light IDs that continuously changes as dancers move.

The intelligent Music System manipulates the music by dynamically substituting authored music segments and transitions based on the dancer's arousal level and authored rules. The *intelligent*

Figure 1. An example pressure pad

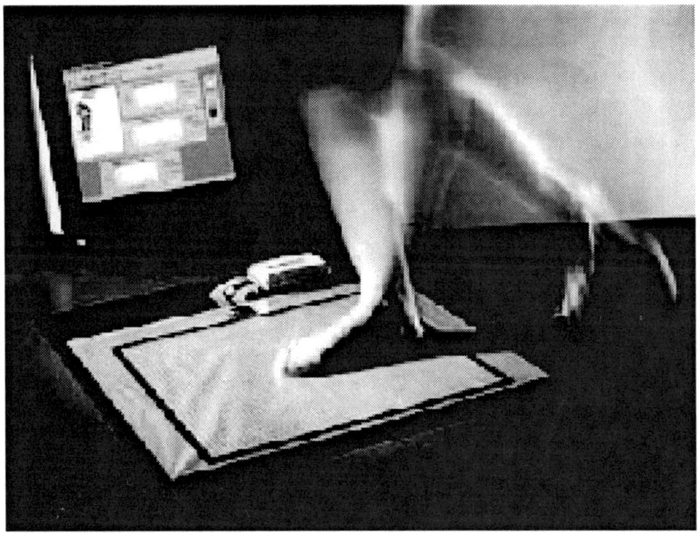

on-stage lighting system determines colors and angles for stage lights given the light IDs of lights affecting the dancer, the dancer's physiological state, and artistic constraints dictating lighting style and patterns. It categories lights on stage as: focus lights which are lights affecting the dancers given by the list of light IDs (output of *Sensor Analysis System*), and non-focus lights, all lights not in the list of light IDs. Based on this difference, it identifies, for each physical on-stage light, a color represented in RGB and an angle rotation. This information is then translated to light board hex code by the *On-stage Lighting Trans System*.

Physiological Signal Interface

Using the physiological sensors discussed in section 3, we collect GSR (Galvanic Skin Response) and body temperature. These signals are continuous numerical values. We pass these signals

Figure 2. Architecture of the System

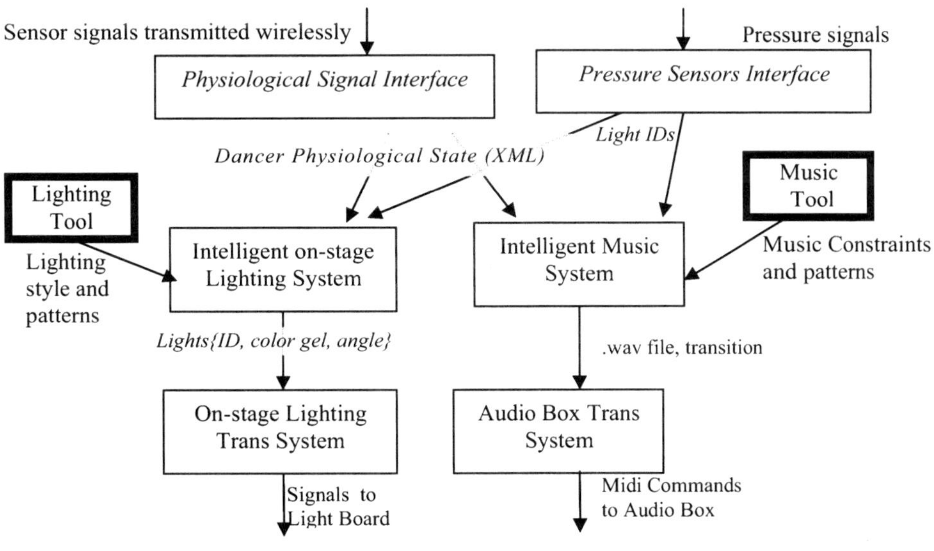

Figure 3. GSR reading while dancing for a 45-minute segment

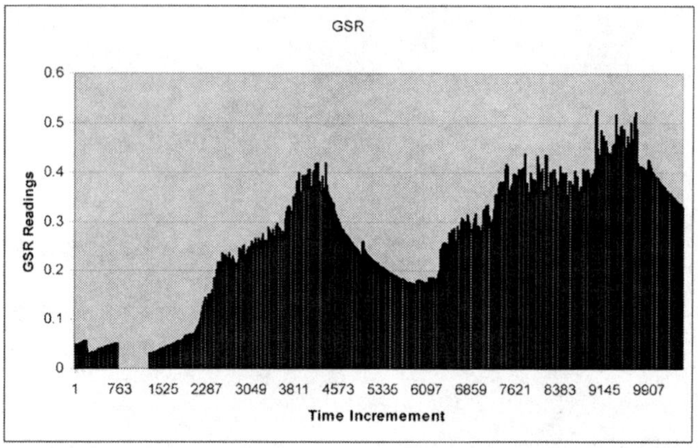

through a filter and synchronize their readings and sampling rates. The output of this system is a continuous function describing the physiological state in time increments, where the sampling rate is the *max* of all sampling rates of the used sensors.

An example GSR reading from a dance performance that lasted 45 minutes is shown in figure 3. As shown, the GSR readings follow the intensity in the performance. The peaks of the graph depict fast intense movements. This graph is then smoothened, where minor variations are ignored and missing data is filled by a simple interpolation algorithm given the known data and the missing data; the resultant data is passed to the intelligent systems.

Pressure Sensor Interface

Gathering blocking information is important to allow more intelligent lighting changes and setup, e.g. composing lights to focus on the dancer. At preproduction, we manually map lights to specific mat numbers. Receiving pressure signals from a specific mat indicates that a person has stepped on the mat. Therefore, instead of gathering or mapping 3D positions, we pass dancer position as a mat number(s). Using the lights-to-mat mapping, we determine which lights impact the dancer at any particular moment in time. The output of this system is a list of light IDs of lights affecting the dancer at the specific moment in time. Since this particular output is continuously changing, its output is buffered and fed to the next layer for processing as a process within the next layer becomes available.

Expressing Arousal through Lighting

Lighting Patterns for Expressing Arousal–Based on Film and Theatre Lighting Theory

Films and theatre productions use several color and lighting techniques to parallel and support the dramatic intensity expressed in the narrative. The specific effects or colors used for expressing emotions vary. For example, some shows use warm colors to signify positive emotions and cool colors to signify negative emotions; other shows use an opposite pattern. We believe that the actual link between emotions and color is ambiguous and may vary with culture.

In this section, we concentrate on discussing several contrast and affinity patterns that are used to evoke or parallel tension[1] (Almeida, 2005). We formulated these patterns based on a qualitative study of over thirty movies, including *Equilibrium*, *Shakespeare in Love*, *Citizen Kane*,

The Matrix, and *The Cook, The Thief, His Wife and Her Lover*. According to our study, the techniques used can be divided into shot-based color techniques: color techniques used in one shot, and scene-based color techniques: techniques used on a sequence of shots.

An example shot-based color technique is the use of high brightness contrast in one shot. Brightness contrast is a term used to denote the difference between brightness of different areas in the scene. High brightness contrast denotes high difference between brightness in one or two areas in a shot and the rest of the shot. This effect is not new; it was used in paintings during the Baroque era and was termed *Chiaroscuro* which is an Italian word meaning light and dark. An example composition can be seen in Giovanni Baglione's painting *Sacred love versus profane love* shown in figure 4. This kind of composition is used in many movies to increase arousal. Perhaps the most well known examples of movies that use this kind of effect are film noir movies (shown in figure 5), e.g. *Citizen Kane*, *The Shanghai Gesture*, *This Gun for Hire*. Another form of contrast used in movies is the contrast between warm and cool colors. Several movies use a high warm/cool color contrast composition, where contrast is defined as the difference between warm colored lights lighting the character and cool colored lights lighting the background. These kinds of patterns are usually used in peak moments in a movie, such as turning points. Lower contrast compositions often precede these heightened shots, thus developing another form of contrast, contrast between shots.

In addition to color and brightness contrast, filmmakers also used affinity of color, e.g. affinity of high saturated warm colors or unsaturated cold colors in one shot. An example movie that extensively used this technique is *The Cook, the thief, his wife, and her lover*. Other examples include *The English Patient*, which used affinity of de-saturated colors, and *Equilibrium*, which used affinity of cold unsaturated colors.

Figure 4. Chiaroscuro Technique used in Sacred love versus profane love Painting

Figure 5. Film Noir uses contrasts and shadows

The perception of contrast, saturation, and warmth of color of any shot within a continuous movie depends on colors used in the preceding shots. Also, the process by which color is used to evoke or project dramatic intensity depends on the sequence and temporal ordering of the effects discussed above. For this purpose, we define our patterns in terms of techniques spanning time over several shots.

The first technique we discuss is the use of affinity of saturated colors for a period of time. Movies, such as *The Cook, the thief, his wife, and her lover*, sustained affinity of highly saturated warm colors for a period of time. We believe that the temporal factor is key to the effect of this approach; this is due to the nature of the eye. The eye tries to balance the projected color to achieve white color. Hence, when projected with red color, the eye will try to compensate the red with cyan to achieve white color. This causes eye fatigue, which in turn affects the participant's stress level, thus affecting arousal.

In contrast to the use of affinity, several movies used contrast between shots to evoke tension. For instance, filmmakers used warm saturated colors in one shot then cool saturated colors in the other, thus forming a warm/cool color contrast between shots to reflect a decrease in dramatic intensity. Some designers use saturated colored shots then de-saturated colored shots creating a contrast in terms of saturation; example films that used this technique include *Equilibrium* and *The English Patient*.

Based on these observations, we identify the patterns shown in Table 1.

These patterns will be used by the intelligent lighting system to manipulate lighting in real-time to reflect a decrease or an increase in dancer's physiological state based on the current lighting state as will be discussed below.

Table 1.

Pattern No.	Description
I	Subjecting audience to affinity of high saturated colors (where high saturation ranges from 70% to 100%) for some time increases projected tension
II	Subjecting audience to contrast in terms of high saturated then low saturated colors (where saturation ranges from 100% to 10%) over a sequence of shots decrease projected tension
III	Subjecting audience to contrast in terms of low saturated then high saturated colors (where saturation ranges from 10% to 100%) over a sequence of shots increase projected tension
IV	Subjecting audience to contrast in terms of high brightness then low brightness (where brightness ranges from 100% to 10%) over a sequence of shots increase projected tension
V	Subjecting audience to contrast in terms of low brightness then high brightness (where brightness ranges from 10% to 100%) over a sequence of shots decrease projected tension
VI	Subjecting audience to contrast in terms of warmth then cool colors (where warmth ranges from 100% to 10%) over a sequence of shots decrease projected tension
VII	Subjecting audience to contrast in terms of cool then warm colors (where warmth ranges from 10% to 100%) over a sequence of shots increase projected tension
VIII	Subjecting audience to increase of brightness contrast subjected in a shot (where brightness contrast is measured in terms of difference between bright and dark spots in an image) over a sequence of shots increases projected tension
IX	Subjecting audience to decrease of brightness contrast subjected in a shot (where brightness contrast is measured in terms of difference between bright and dark spots in an image) over a sequence of shots decrease projected tension
X	Subjecting audience to increase of warmth/cool color contrast subjected in a shot (where contrast is measured in terms of difference between warm and cool spots in an image) over a sequence of shots increases projected tension
XI	Subjecting audience to decrease of warmth/cool color contrast subjected in a shot (where contrast is measured in terms of difference between warm and cool spots in an image) over a sequence of shots decreases projected tension

The Lighting Tool for Entering Artistic Constraints

There are several methods by which artists can author lighting changes. We have defined several levels of artistic control as follows.

High-Level Control

Based the patterns identified above, we defined a visual interface where artists can select the type of mapping of lighting reflection based on physiological signals, as one of the following patterns:

1. Physiological signal is mapped to brightness contrast increase/decrease, where contrast is established between focus and non-focus areas, i.e. difference in brightness between colors of lights lighting focus areas and others lighting non-focus areas.
2. Physiological signal is mapped to warm/cool color contrast increase/decrease, where contrast is established between focus and non-focus areas, i.e. difference in warm and cool colors of lights lighting focus areas and others lighting non-focus areas.
3. Physiological signal is mapped to saturation affinity increase/decrease, where brightness and hue is constant.
4. Physiological signal is mapped to warmth affinity increase/decrease, where brightness and hue is constant.

Depending on artistic input indicating which pattern or style he/she desires, the intelligent lighting system will adjust light colors based on the selected pattern.

Control on Stylistic Constraints

In addition, artists can also define several constraints that will affect how the lighting system adjusts colors and angles of light. These constraints are as follows:

- Importance of visual continuity
- Importance of projected depth through lighting
- Importance of projecting realistic motivation for lighting direction
- Importance of dancer's *modeling*[2] and visibility
- Desired mood angle on the dancer
- Type of transitions (e.g., gradual fade in, pulsing, or just a shift).

These constraints, similar to the patterns discussed above, were developed based on lighting design theories. Based on the values of these constraints, the lighting system will adjust colors and angles to reflect the desired effect while maintaining the desired constraints.

Lower-Level Control

We acknowledge that artists need additional control, and thus devised a language that will allow them to develop their own patterns given variations in the physiological state. The language is similar to game engine trigger languages or rule-based languages, where artists author rules in the following format:

```
trigger: (increased-by physiological-state(t)
    physiological-state(t-1) 5)
    ;; condition on an increase of 5 units between
    physiological states at times t and t-1
Action: (increase-by color-contrast 5)
```

Where the trigger part defines the condition by which the action becomes eligible, which in this case is an increase of 5 units in the physiological signal from time t-1 to time t. The action in this case is to increase color-contrast by 5.

We defined several trigger types that artists can use; these trigger types include increased-by or decreased-by on specific physiological states given time. We also defined several knobs that artists can use as action clauses. These action clauses include:

- Change in color-contrast: increase or decrease of color-contrast, where color contrast is defined as the contrast in terms of warm and cool colors between the dancer and the areas surrounding the dancer.
- Change in brightness-contrast: increase or decrease of brightness-contrast, where brightness contrast is defined as the contrast in terms of brightness between the dancer and the areas surrounding the dancer.
- Change in brightness of specific area, where areas defined as dancer area, non-action areas (areas surrounding the dancer), and background areas
- Change in Saturation of specific area, where areas defined as dancer area, non-action areas (areas surrounding the dancer), and background areas
- Change in Warmth of the color in a specific area, where areas defined as dancer area, non-action areas (areas surrounding the dancer), and background areas

These action clauses were developed based on the qualitative study discussed in section 4.3.1.

The Intelligent On-Stage Lighting System

Given the light IDs of lights affecting the dancer (the output of the sensor analysis system), the patterns and constraints authored through the tool described above, the *intelligent on-stage lighting system* computes angles and colors for each lights differentiating between lights affecting the dancer, and other lights.

The *intelligent on-stage lighting system* categorizes lights affecting the dancer as focus lights and other lights as non-focus lights. Given this categorization, it then calculates an angle and an RGB color values given the lights in the light ID list. In determining the angle of light on the dancer, the *intelligent on-stage lighting system* takes into account the quality of light and their influence in projecting stylistic constraints (defined above), including realistic motivation for the lighting direction, modeling, and desired mood. It uses a non-linear optimization system based on hill climbing to select an angle for each key light that minimizes the following function:

$$\lambda_v(1-V(k,s)) + \lambda_{\_}\left|k-k^-\right| + \lambda_m\left|k-m\right| + \lambda_l \min_i\left|k-l_i\right|, \quad (1)$$

where k and s are defined as the key light azimuth angle relative to the camera and the subject angle relative to the key light, respectively, as shown in figure 3, k^- is the key light azimuth angle from the previous frame, $\lambda_{\_}$ is the cost of changing the key light angle over time (enforcing *visual continuity*), λ_m is the cost of deviation from the desired mood angle (enforcing *mood*), m is the desired mood angle suggested by the artist, λ_l is the cost of azimuth angle deviation from a practical source direction (enforcing *realistic lighting direction*), l_i is the azimuth angle of light emitted by the practical source i, and λ_v is the cost of deviation from an orientation of light that establishes best visibility and modeling (enforcing *visibility and modeling*).

Based on Millerson's documented rules, we formulated the following equation to evaluate the visibility and modeling of a given key light azimuth angle:

$$V(k,s) = \sin(k)\cos(s). \quad (2)$$

The system then uses rules based on Millerson's guidelines to select fill and backlight azimuth angles depending on the value of the key light angle. According to Millerson's guidelines , fill light azimuth and elevation angles are calculated to be the mirror image of the key light angle. We define backlight azimuth angle as:

$$b = (k + \pi) \bmod 2\pi \quad (3)$$

These calculations only occur when dancers move and thus lights will adjust their angles to best show the dancer and model her, if modeling and visibility importance are set as high by the artist in the artistic constraints. The other lights on the stage are set to a default angle that creates a wash on the stage.

The interaction between colors assigned for each area composes the contrast and feeling of the entire scene. Similar to the angle optimization system, the lighting system uses non-linear optimization to search through a nine-dimensional space of RGB values for ideal RBG values given artistic constraints and patterns as well as given physiological state and dancer's location. It evaluates each RBG color value by using a multi-objective cost function, where each objective evaluates the color against the stylistic constraints and action clauses discussed above, including establishing depth, adhering to desired warmth, saturation, and lightness, and maintaining visual continuity. The cost function is defined as follows:

$$\lambda_d \left(D(c^t) - d\right)^2 + \lambda_c \left(\text{contrast}_\phi(c^t) - \delta\right)^2 + v(x) + \sum_{i \in \{f,n,b\}} P(c_i^t, c_i^{t-1}), \quad (4)$$

where

$$p(c_i^t, c_i^{t-1}) = \lambda_{s_i} \left(S(c_i^t) - s_i\right)^2 + \lambda_{h_i} \left(H(c_i^t) - h_i\right)^2 + \lambda_{l_i} \left(L(c_i^t) - l_i\right)^2 + \lambda_{w_i} \left(W(c_i^t) - w_i\right)^2 + \lambda_{ch} E(c_i^t, c_i^{t-1}), \quad (5)$$

where c^t is a vector of light colors for focus f, non-focus n, and background b, and areas at frame t. Color c_i^t is represented in RGB color space; S(c) denotes the saturation of color c; H(c) denotes the hue of color c; L(c) denotes lightness of color c (in RGB color space).

ELE uses a well-known formula for measuring color difference as follows:

$$E = \sqrt{\left(\frac{\Delta L}{k_L S_L}\right)^2 + \left(\frac{\Delta C}{k_C S_C}\right)^2 + \left(\frac{\Delta H}{k_H S_H}\right)^2 + \Delta R}, \quad (6)$$

where $\Delta R = R_T f(\Delta C\ \Delta H)$ and ΔL, ΔC, and ΔH are CIELAB metric lightness, chroma, and hue differences respectively; S_L, S_C, S_H are weighting functions for the lightness, chroma, and hue components; and k_L, k_C, k_H are parameters to be adjusted depending on model material information.

The depth, $D(c)$, of a color vector c is defined as the color difference between colors lighting the background areas and those lighting other areas, formulated as follows:

$$D(c) = \sum_{b \in B} \sum_{n \in NB} E(c_b, c_n), \quad (7)$$

where B are the indices for background lights; NB are the indices for non-background lights; and E is the color difference defined above.

Based on the results collected by Katra and Wooten described in , we used a multiple, linear regression method to formulate color warmth in RGB color space, as follows:

$$warmth\left(\begin{bmatrix} R \\ G \\ B \end{bmatrix}\right) = \begin{bmatrix} 0.008 \\ 0.0006 \\ -0.0105 \end{bmatrix}^T \begin{bmatrix} R \\ G \\ B \end{bmatrix} - 0.422. \quad (8)$$

The optimization problem discussed above is a constraint-based optimization problem, where the color, c, is constrained to a specific space of values defined by style (e.g., realistic style restricts the values of saturation or hue). The lighting system uses a boundary method to bind the feasible solutions using a barrier function $v(x)$, such that $v(x) \rightarrow \infty$ as x approaches the boundary defined by the feasibility region.

Although gradient descent has major drawbacks, including occurrence of oscillations and being easily stuck in a local minimum, the lighting system uses gradient descent for several reasons. First, it provides a fast and simple solution. Second, a local minimum in this case is preferable because it provides a solution closer to the older one, thus ensuring visual continuity. Third, alternative methods rely on the existence of a second derivative, which is not necessarily true in this case.

Expressing Physiological State through Music

Dynamically expressing tension in music (with the exception of some types of music) is difficult due to the melodic nature of music. Since our emphasis is on lighting, we will develop a simple system for improvising music; the system is similar to the work on adaptive music done by the game industry. The basic idea is that composers will compose the improvisational piece as a set of several pieces that can be interchangeable and that vary in their projected tension level. For example, composers will supply us with: piece1 which can be replaced with piece11 piece12, piece13, etc. where piece11, piece12, piece13 show different tension levels depending on the pattern or pieces already played. Artists will use XML to identify the pieces, as follows:

```
<Music Piece>
     <Name> Piece1 </Name>
     <FileName> piece1.wav </FileName>
</Music Piece>
<Music Piece>
     <Name> Piece11 </Name>
     <FileName> piece11.wav </FileName>
</Music Piece>
<Music Piece>
     <Name> Piece12 </Name>
     <FileName> piece12.wav </FileName>
</Music Piece>
     <Music Piece>
     <Name> Piece13 </Name>
<FileName> piece13.wav </FileName>
</Music Piece>
```

They will then use rules to denote the tension value of pieces given specific patterns, e.g.

```
(def-music-tension- rule
     (  (:TRUE  (played piece1))
        (:TRUE  (played piece2))
        (:TRUE (played piece3))
        (:TRUE  (candidate piece14)) )
(:where     (followed-directly piece2 piece1)
     )
(:action    (assert! (increasedby tension 10)))
     )
```

Where *played* is a symbolic predicate representing the fact that piece1 was played. This particular fact is placed in the rule-base database when piece1 is selected for playing and is played. The *where* part of the rule indicates some specific temporal or transitional constraints, such as followed-directly, which indicates some piece followed another piece directly, e.g. piece2 followed piece1 directly. The rule above denotes that if the piece14 is played after piece3, piece2, and piece1, and piece2 followed piece1 directly then the tension value would increase by 10 increments. Using these rules the *intelligent music system* will evaluate several candidate pieces given the increase/decrease of tension values and the dancer's physiological signal. For example, if the physiological signal dictates an increase of 5-7 increments in dancer's state then a closer increase to 5-7 increments would be the best match. The intelligent music system will evaluate all candidate pieces in terms of their tension value increase and will select an increase that is closer to the dancer's physiological state increase. If there are several appropriate candidates, then the system will choose one randomly. The system will select transitions in a similar manner.

In order for the intelligent music system to select a candidate piece given the authored pieces, it will use a rule-based system similar to the one discussed in. Again, artists will author several rules indicating when a particular piece becomes applicable for playing. For example, piece14 can be selected only if piece1 and piece2 have been played, this can be expressed as:

```
(def-music-selection-rule
        (   (:TRUE (played piece1))
            (:TRUE (played piece2)) )
        (:action  (assert!  (candidate
    piece14)))
```

The intelligent music system will then put piece14 in the candidate list if piece1 and piece2 have been played at some point in the past. The system then selects a piece to play from the candidate list by comparing their increase/decrease of tension value using the rules above. When a piece is selected to play, then the fact (play *?piece*), where *?piece* is the piece that has been selected, will be placed in the fact database.

Projecting Lighting and Music

On-Stage Lighting Trans System

The output of the intelligent on-stage lighting system is: a list of light IDs, and for each light ID, an angle, and a color in RGB color format. This output is passed to the *On-stage Lighting Trans system* which translates these commands to appropriate hex code commands used by the lighting board. The hex code will include routines to initiate light rotation or color commands for the appropriate lights given the output of the intelligent on-stage lighting system.

Projecting Music

As described above, the intelligent music system determines which wav files to play. Commands for switching between wav files will be sent to the audio box, which can dynamically switch between wav files. We assume all segments will use cross fade as a transition method. Cross fade is a method of transitioning between two music pieces where one is faded out and the other is faded in.

PROTOTYPE

We have implemented a system that takes in the physiological signal and selects lighting angles and colors to reflect the physiological state. At this stage, we have implemented the lighting system in a virtual environment, concentrating on three different capabilities (a) reading in physiological signals, evaluating the robustness of the sensors, (b) translating physiological sensor data to lighting changes using the patterns we have identified, (c) implementing the lighting tool, and (c) evaluating the types of variations that can be induced by the lighting tool we devised. We developed the virtual lighting system in WildTangent. The system accepts allows artists to author lighting patterns at the three levels discussed in section 4. Furthermore, it reflects the authored patterns and constraints through dynamic real-time change of lighting based on the physiological state.

We implemented the *Physiological System Interface* which streamed in a physiological signal collected through the Body Media device (e.g., shown in Figure 3). Figure 3 shows a signal that we used in our first experiment. This signal was collected wirelessly through the body media device while one of the authors was dancing. There were several missing data due to data lost in communication between the device and wireless base. These fields were filled in through a simple interpolation algorithm that we devised, which assumes monotonic linear increase or decrease based on the known data before and after the data loss.

We used pattern 1 from the list discussed in section 4.3. The system then improvised the lighting in the virtual space based on the physiological signal shown in figure 3 resulting in a continuous change in brightness contrast. Three screenshots of the resulting changes are displayed in Figure 6.

LIMITATIONS

We have described an ambient intelligent environment for a dance space. Our goal is to extend the current expression modes of dance by allowing lights and projected images to change and adapt depending on dancer's movements and physiological state. By changing lighting color and angle directly mapping the dancer's condition, we are presenting the dancer's state as a signature within the physical space. Such an interface will also allow dancers to use the environment as their expressive space and to project their self through the environment.

We have intentionally limited the technical design to only adapt to physiological state and not emotional state. This is due to the fact that extracting or predicting emotions is still a hard and open problem. It is especially problematic because most often dancers feel an amalgam of emotions and not one particular emotion. One possible way to predict emotional states is to use a high resolution image processing algorithms to analyze facial expressions and gestures. These techniques are still under research and are generally challenged by variation in lighting conditions. Therefore using them for this project is problematic. In addition, the lighting patterns we extracted from film techniques use light and color primarily to project tension rather than actual emotions. Hence, even if we can device a method for predicting emotional states, defining lighting design patterns that can universally represent emotional states is difficult.

While few interactive theatre productions used vision to capture on-stage motion, we decided to use pressure sensors. This decision was made for several reasons. First, most vision techniques are challenged by variations in the level of illumination within the captured images. This is due to the fact that most vision techniques use pixel colors to define edges and track movement. Since we propose a performance where lighting color and angle change dynamically to reflect the dancers' state, this environment, by definition, will constitute a challenge to any vision based system. Second, privacy of dancers may be an issue for our piece. Third, we need to establish a mapping between dancers' positions and lights on stage. While we could use vision techniques to track movements, determining 3D position and its relation to lights on stage is hard. However, there are several previous projects that effectively used camera systems, e.g. (Tokuhisa 2006); we intend to look at these projects for inspiration for future enhancements to the proposed architecture.

CONCLUSION

In this paper, we have proposed a new ambient intelligent architecture that expresses a dancer's

Figure 6. Linearly increasing brightness contrast (where center of room is the focus)

a) High Contrast   b) Moderate Contrast   c) Low Contrast

physiological state through manipulation of music and stage lighting around the dancers. The goal of the described architecture is to enable different modes of expression through a dance space, and provide a method that imprints a signature of the dancers' self in the physical dance space. The contribution of the paper is two fold: technical and aesthetic. The technical contribution is in the integration and the creation of several novel intelligent systems and tools that allow artists to dictate patterns of lighting and music movement in time as a reflection of dancer's physiological state. The aesthetic contribution is in extending the medium of dance expression beyond the body to integrate the environment itself as a reflection of the dancer's state. In this paper we described the proposed architecture detailing the intelligent lighting system and the intelligent music system as well as the tools that allow artists to dictate music and lighting changes as a function of dancer's physiological state and movements. We have also discussed our initial prototype of the system. In future work, we aim to continue development of the architecture described as well as evaluate its aesthetic utility within different dance forms.

REFERENCES

Almeida, P. (2005). *Identifying Low-level Visual Patterns that Stimulate Emotions and Moods in Movies and Video Games.* Master's Thesis. Penn State University.

Alton, J. (1995). *Painting with Light.* Berkeley: University of California Press.

Birn, J. (Ed.). (2000). *Digital Lighting & Rendering.* Indianapolis: New Riders.

Block, B. (2001). *The Visual Story: Seeing the Structure of Film, TV, and New Media.* New York: Focal Press.

Bordwell, D., & Thompson, K. (2001). *Film Art: An Introduction* (6th ed.). New York: Mc Graw Hill.

Brown, B. (1996). *Motion Picture and Video Lighting.* Boston: Focal Press. 23.

Calahan, S. (1996). *Storytelling through lighting: a computer graphics perspective.* Paper presented at the Siggraph Course Notes.

Calvert, T., & Mah, S. (1996). *Life Forms: an Application of Computer Graphics to Support Dance Choreography.* Paper presented at the Siggraph 96 Visual Proceedings: The art and interdisciplinary programs of Siggraph, New Orleans, LO.

Cheshire, D., & Knopf, A. (1979). *The Book of Movie Photography.* London: Alfred Knopf, Inc. Clark, A. (2001). Adaptive Music. *Gamasutra.com, May 15, 2001.*

Cooper, D. (1995). Very Nervous System. *Wired.*

Crawford, J., Schiphorst, T., Gotfritt, M., & Demers, L. P. (1993). *The Shadow Project.* Paper presented at the Symposium on Arts and Technology.

Crow, F., & Csuri, C. (1985). Music and Dance Join a Fine Artist and a Paint Machine. *IEEE Computer Graphics and Application*, 11-13.

Crowther, B. (1989). *Film Noir: Reflections in a Dark Mirror.* New York: Continuum.

Cunningham, M., Kaiser, P., & Eshkar, S. (1998). *Hand-drawn Spaces.* Paper presented at the Siggraph 1998, Special Sessions.

Demers, L. P. (1993). *Interactive and Live Accompaniment Light for Dance.* Paper presented at the Dance and Technology Conference, Simon Fraser University, Vancouver.

Demers, L. P., & Jean, P. (1997). *New Control Approaches on Lighting.* Paper presented at the Shadow Light '97, Flemish Opera House.

Demers, L. P., & Vorn, B. (Artist). (1998). *Lost Referential* [Exhibition]. Forbus, K., Kleer, J. De. (1993). *Building Problem Solvers.* Cambridge: MIT Press. Gillette, J. M. (1998).

Designing with Light (3rd. ed.). Mountain View, CA: Mayfield. Gruen, J. (1983). Dancevision. *Dance Magazine, 57,* 78-79. 24.

Hill, B. H., Roger, T. , and Vorhagen, F. W. (1997). Comparative Analysis of the Quantization of Color Spaces on the Basis of the CIELAB Color-Difference Formula. *ACM Transactions on Graphics, 16*(2), 109-154.

Katra, E. a. W., B. R. (1995). *Perceived lightness/ darkness and warmth/coolness in chromatic experience*. Unpublished manuscript.

Luo, M. R., Cul, G., and Rigg, B. (2000). *The Development of CIE 2000 Colour Difference Formula: CIEDE2000*, 2000, from http://www.ifra.com/Website/ifra.nsf/html/colorqualityclub.html

Magerkurth, C., Mandryk, R., Benford, S., Sanneblad, J. (2004). International Workshop on Gaming Applications in Pervasive Computing Environments. http://www.ipsi.fraunhofer.de/ambiente/pergames2005/ papers/PROCEEDINGS_PerGames_2004.pdf.

Meador, S., Rogers, T. J., O'Neal, K., Kurt, E., & Cunningham, C. (2004). Mixing Dance Realities: Collaborative Development of Live-Motion Capture in a Performing Arts Environment. *ACM Computers in Entertainment, 2*(2).

Miller, M. (1997). Producing Interactive Audio: Thoughts, Tools, and Techniques. *Gamasutra.com, October 15, 1997.*

Millerson, G. (1991). *The Technique of Lighting for Television and Film* (3rd ed.). Oxford: Focus Press. Patterson, S. (2001). Interactive Music Sequencer Design. *Gamasutra.com, May 15, 2001.*

Rokeby, D. (1986). *Very Nervous System*, from http://homepage.mac.com/davidrokeby/vns.html

Rokeby, D. (Artist). (1986-1990). *Very Nervous System*

Ross, R. (2001). Interactive Music et al., Audio. *Gamasutra.com, May 15, 2001.*

Srinivasan, P., Birchfield, D., Qian, G., & Kidane, A. (2005). *A Pressure Sensing Floor for Interactive Media Applications.* Paper presented at the ACM SIGCHI International Conference on Advances in Computer Entertainment Technology (ACE), Valencia, Spain.

Tokuhisa, S., Okubo, S., Suguro, K., Kotabe, T, and Inakage, M. (2006). MYSQ: An entertainment system based on content creation directly linked to communication, *ACM Computers In Entertainment (CIE)*, Vo. 4 , No. 3.

Ulyate, R., & Bianciardi, D. (1998). *Interactive Dance Club.* Paper presented at the Siggraph 98, Orlando, Florida. 25.

Ulyate, R., & Bianciardi, D. (2001). *The Interactive Dance Club: Avoiding Chaos in a Multi Participant Environment.* Paper presented at the CHI'01 Workshop New Interfaces for Musical Expression (NIME'01).

Vasilakos, A., & Pedrycz, W. (2006). *Ambient Intelligence, Wireless, Networking, Ubiquitous Computing.* MA, USA: Artech House Press.

Wang, Y. (2006), Keynote: Cognitive Informatics - Towards the Future Generation Computers that Think and Feel, *Proc. 5th IEEE International Conference on Cognitive Informatics* (ICCI'06), Beijing, China, IEEE CS Press, July, pp. 3-7.

Wang, Y. (2007), The Theoretical Framework of Cognitive Informatics, *International Journal of Cognitive Informatics and Natural Intelligence*, 1(1), Jan., pp. 1-27.

Wagner, M. G., & Carroll, S. (2001). *DeepWave: Visualizing Music with VRML.* Paper presented at the Proceedings of the Seventh International Conference on Virtual Systems and Multimedia (VSMM '01).*WinAmp.* from www.winamp.com.

Windows Media Player. from www.microsoft.com Winkler, T. (1995). *Making Motion Musical: Gesture Mapping Strategies for Interactive Com-*

puter Music. Paper presented at the Proceedings of International Computer Music Conference.

Winkler, T. (1997). *Creating Interactive Dance with the Very Nervous System.* Paper presented at the Proceedings of Connecticut College Symposium on Arts and Technology.

Winkler, T. (1998). *Motion-Sensing Music: Artistic and Technical Challenges in Two Works for Dance.* Paper presented at the Proceedings of the International Computer Music Conference.

ENDNOTES

1 Joint work done by the first author, Professor Seif El-Nasr, in collaboration with her master's student Priya Almeida.

2 Modeling is a term used by lighting designers to emphasis the number and angles of light affecting the actor. This will help shape the dancer's depth, and aid in bringing him out of the background

This work was previously published in International Journal of Cognitive Informatics and Natural Intelligence, Volume 3, Issue 2, edited by Yingxu Wang, pp. 1-17, copyright 2009 by IGI Publishing (an imprint of IGI Global).

Chapter 8
Kansei Experience:
Aesthetic, Emotions and Inner Balance

Ben Salem
Eindhoven University of Technology, The Netherlands

Ryohei Nakatsu
National University of Singapore, Singapore

Matthias Rauterberg
Eindhoven University of Technology, The Netherlands

ABSTRACT

Deliberate exploitation of natural resources and excessive use of environmentally abhorrent materials have resulted in environmental disruptions threatening the life support systems. A human centric approach of development has already damaged nature to a large extent. This has attracted the attention of environmental specialists and policy makers. It has also led to discussions at various national and international conventions. The objective of protecting natural resources cannot be achieved without the involvement of professionals from multidisciplinary areas. This chapter recommends a model for the creation of knowledge-based systems for natural resources management. Further, it describes making use of unique capabilities of remote sensing satellites for conserving natural resources and managing natural disasters. It is exclusively for the people who are not familiar with the technology and who are given the task of framing policies.

INTRODUCTION

We spend a significant amount of time interacting with Information and Communication Technology (ICT), using mobile phones, desktop computers, game consoles and so on. As a result, Human Computer Interaction (HCI) nowadays has shifted its approach from a focus on the computer, originally a scarce resource, to a focus on the user, and the merging of information and communication technologies. It has evolved into being about the user and what an ICT system can deliver to him/

her. We are convinced that now is the time to look further and assess the user gains, from ICT usage, in terms of experience and affects. Thanks to Cognitive Informatics (CI), there has already been some work in this area with the study of the user internal information processing of the brain (Wang, 2007). CI has focused on the relationship between information, computer science and mathematics on the one hand and neurobiology, cognitive science and psychology on the other hand (Wang, 2003). CI studies the way ICT users process internally information. However there is no strong emphasis on the experience user gain from such processing. We would like to go further, as we believe there is now a need to address the affects such process has on ICT users. There is a need for an experience that stimulates and triggers some cognitive functions with a strong affect their beholder. The most relevant cognitive functions, in this context, are: *reflexes*, *sensations*, *thoughts*, *dreams*, *emotions*, *moods*, and *drives*. These functions can be ordered according to their life-span (see Figure 2, note that this is a simplified description of these functions). The functions at the short end are triggered and running before we become aware of them. The functions at the other end are what make us ourselves and we are aware of them most of the time via introspection and retrospective analysis. As for the cognitive functions in the middle range, these are the functions that we are mostly aware of while they emerge in our mind and then disappear.

We have used this simplified list of functions during the implementation of a new interaction we have proposed called Kansei Mediated Interaction (KMI). That is because these cognitive functions are strongly associated with different systems in our body (brain, spinal cord, somatic system, autonomic system, endocrine system, and genetic system in Figure 3). In turn, these links help us design the right interaction (challenges, stimulus, body intakes, behaviour, sex in Figure 3) through various body parts and control systems. To achieve KMI, we have proposed to implement the user interface interaction using a combination of channels and medias exploring these links between cognitive functions and body systems (e.g. narrative, actuators, drinks, day long event in Figure 3). In this example we focus on the implementation of KMI within entertainment systems.

Our conviction of a need for ICT usage to be an enriching experience, an experience that yields positive affect, better feelings and enhanced inner

Figure 1. Three interaction modes: Explicit, Implicit and Kansei

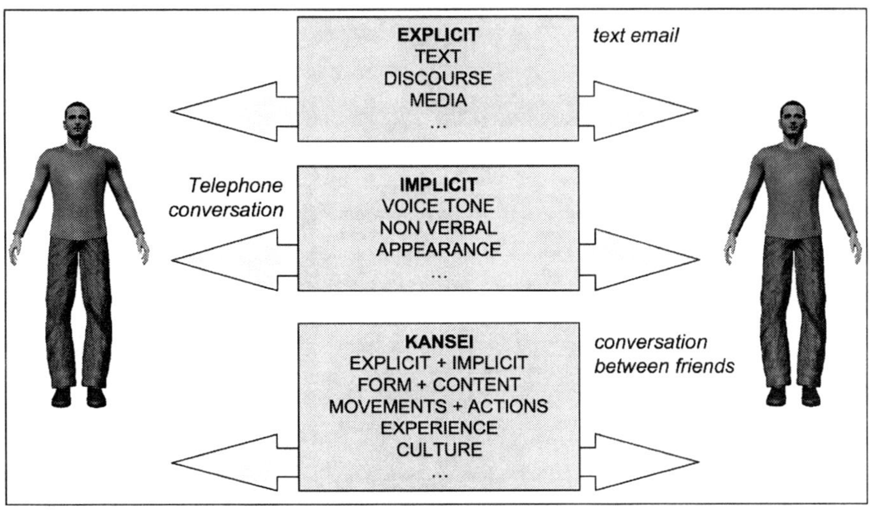

Figure 2. Familiar cognitive functions duration: from seconds to a lifetime

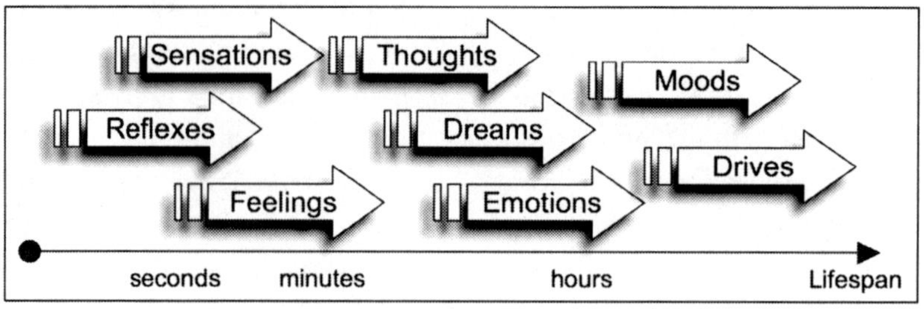

balance, implies the re-focus of HCI into the user experience and the fulfilment of his/her needs, requirements and desires. Within this view, Salem and Rauterberg (2004) have proposed a Needs Requirements and Desires (NRD) model, where the user is at the centre, and all needs requirements and desires radiate from it. It is an egocentric, egoistic and hedonistic model. In this model, needs relates to the *essentials*, requirements to the *necessary* and desires to the *optional*. The fulfilment of user needs have been addressed with ergonomics, and the fulfilment of user requirements with applications and interface designs. We believe the user desires has been somewhat poorly, if at all, addressed. A good start would be a different user interaction principle.

Novel User Interaction

Current understandings of users, as they have emerged from the office and the workplace (the initial domain of application for ICT), are not adequate and relevant anymore (Monk et al., 2002). A different approach seems necessary, to take into consideration the nowadays prevalence of ICT usage. There are many approaches that could be adopted, one is ubiquitous computing, addressing issues related to the growingly pervasive nature of ICT (Weiser, 1999). Another approach is ambient intelligence (Aarts, 2002), (Aarts & Marzano, 2003) and its attempt at predicting what a person may want (Maes, 2005).

The term "Ambient Intelligence" (AmI) reflects is the integration of: *Ubiquitous Computing*, *Ubiquitous Communication*, and *Intelligent*

Figure 3. From human control mechanisms to entertainment [CNS: Central Nervous System, PNS: Peripheral Nervous System] (from Salem et al., 2006).

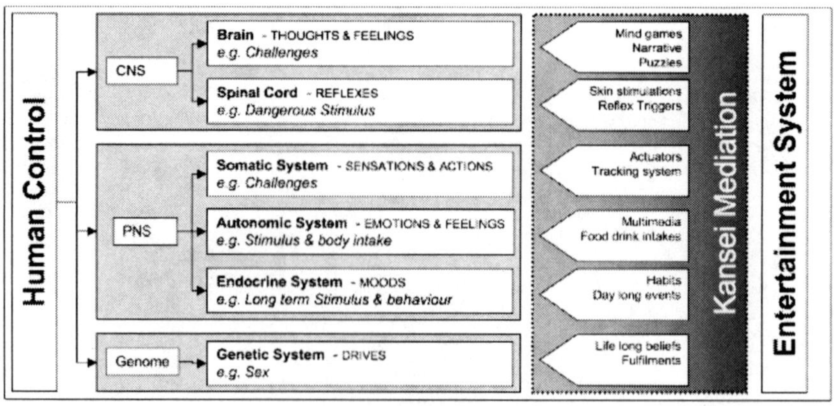

User Friendly Interfaces. The emphasis of AmI is on greater user-friendliness, more efficient services support, user empowerment and support for human interactions. An AmI environment is capable of recognising and responding to the presence of different individuals, working in a seamless, unobtrusive and often invisible way. Some key features characterise AmI, such as context awareness, personalisation and adaptation (Vasilakos & Pedrycz, 2006). Context awareness is achieved when location, identities of nearby people and objects, and changes to those objects are taken into account by the AmI system (Schilit & Theimer, 1994).

We are enthusiastic about the potentials of this approach, however, we are concerned with the proactive aspect of the interaction and believe in a balanced interaction that is neither fully proactive nor fully reactive to the user. In doing so, we have chosen to move away from a focus on the user skills, cognitive load, behaviours and abilities; to a focus on the user emotions, feelings, affections and experience. In effect, transforming ICT usage into an experience of the emotional and sensual aspects of interaction (McCarthy & Wright, 2004). To achieve this we would like to shift the purpose of the interaction design from solely aiming for a greater usability of the interface to seeking better addressing of a wider range of issues including in particular aesthetic and beauty. Aesthetic and beauty are known to give satisfaction to the mind (see various encyclopaedias definitions of the two concepts). Why shall there be a tension between usability and aesthetics? While usability can be easily measured and evaluated, aesthetic cannot, but there is no reason why these two should not be brought together, and a good design implies a balance between beauty and usability (Norman, 2002).

Human Computer Interaction (HCI) Trend

We propose the investigation of a computing paradigm that goes beyond user needs and requirements. We see Cultural Computing as potentially the paradigm that allows the user to experience the fulfilment of his/her desires. In this paradigm, emphasis is put on cultural values and artefacts, rather than on conventional interaction design aspects. Such an emphasis creates opportunities for Cultural Computing to develop as an HCI (Human Computer Interaction) paradigm. Through cultural computing, we are seeking a media that can deliver an experience of aesthetics, emotions and inner balance. To deliver an aesthetically successful interface, one has to rely on some rules and style. While proposing a model approach or some mechanisms to deliver aesthetics, MacLennan (1997) points out that mathematical modelling is always incomplete. Ngo, Teo and Byrne (2002) have focused on the understanding of UI Aesthetics as the visual organisation of the interface (i.e. layout). They have proposed 14 measures to assess the aesthetics of the UI. These are: (1) Balance, (2) Equilibrium, (3) Symmetry, (4) Sequence, (5) Cohesion, (6) Unity, (7) Proportion, (8) Simplicity, (9) Density, (10) Regularity, (11) Economy, (12) Homogeneity, (13) Rhythm and finally, (14) Order and Complexity. It is a rather complex set of measures with some overlapping. These measures are universal and intercultural, in the sense that all human appreciate them as aesthetic properties. This is an important, as otherwise the cultural computing paradigm could rely on other cultural values that are not necessarily universally shared.

The potential cultural dependency of the interaction developed is still however present (Nisbett et al., 2001). Although the cultural dependency is somewhat a drawback it has many advantages. It allows for a much richer experience to be rendered. This is thanks to the complexity and depth of the semantics involved and the user familiarity with them. There is also the advantage of higher bandwidth of information at the interface as symbolic meanings and implicit knowledge can be used. The interface is not limited to explicit messages and meanings anymore. However, there is a challenge in finding culturally rich media that could be used to deliver our proposed system.

KANSEI MEDIA

We propose the use of Kansei aesthetics and media as an approach to deliver effectively enrichment and positive affect. Kansei aesthetics deal with the aesthetic aspect of attribute and properties. It is about the aesthetics of feelings and impressions. In other words a combination of the different aesthetics described in the previous sections. Kansei media is the combination of communication channels that carry these aesthetic values.

Kansei (感性) is a multifaceted word with a context dependent meaning. Kansei deals with emotions, feelings and moods. It relates to attributes and properties that render feelings and impressions. Kansei is about subjective, personal and self-centered experiences. Originally, Kansei was associated with engineering and was defined as the translating of a consumer feelings and image of a product into some of the design elements used to create that product (Nagamachi, 1995). Kansei engineering relates user subjective perception and experience with product parameters and specifications. In this perspective, Kansei Media is a form of multimedia communication that carries non-verbal, emotional and Kansei information (Nakatsu et al., 2005a). Kansei Communication is about sharing implicit knowledge such as feelings, emotions and moods (see Figure 1). The media channels used to do so, are voice tone, non-verbal communication, appearance etc…

Kansei media is about exchanging cultural values in a seemingly natural and unconstrained way. As such, there is potentially a drawback in that Kansei media is culturally dependent and calling for a personalisation of the mediation to suit cultural particularities. However, this can be addressed by ensuring that Kansei media is about as much universal content as possible (e.g. not about local trends and contents but global beliefs and values). In Kansei media several new media are added to multimedia and multimodal communication, they relate to Kansei information, which is essentially aesthetics information. Such information will trigger feelings, impressions and emotions.

Kansei Media Interaction KMI

The integration of multiple, multimodal and Kansei Media can enable a type of interaction that is neither biased towards cognition, nor biased towards awareness. An interaction biased towards cognition would be one that only requires cognition from the user such as text command interface. While, an interaction biased towards awareness would be based on the user perception of the interface, for example a simple shooting game on a virtual environment. A balanced combination of cognition and awareness to deliver a positive experience is what we call Kansei Mediated Interaction (KMI). In this paper we propose KMI as the underlying interaction principle to support the cultural computing paradigm and to deliver a Kansei Experience. KMI goes beyond traditional HCI and aims to let the user experience aesthetics, emotions and inner balance. We hope KMI will enlighten the user, rather than just deliver an interface with high usability.

What are the design guidelines applicable to KMI? First, from the user interaction view the key features are: (1) natural and usable, (2) context aware and adapted, (3) automated capture and access, and (4) always available but not invasive. Second, from the user experience view the key dimensions are: (1) beauty and pleasantness, (3) emotions, (4) cultural values, (5) satisfactions, and (6) aesthetic experiences.

Furthermore, a Kansei Media Interface is designed for a person, not generic users. This means that the user interface should be so flexible and intelligent to tailor its features to the user. This would increase the usability of the system by enhancing and optimising our senses and enriching our experience. For this reason, personalisation is required and fundamental to KMI. A KMI system should be sensible to the user's feedback, and should be capable of modifying the interaction that has been or will be performed. In other words, KMI systems should be adaptive. Generally speaking, adaptive systems are designed to deal with changing environmental conditions whilst maintaining performance objectives. With KMI it is also about maintaining a positive and enriching experience, while being ware of the context of use. Context awareness and adaptation have the potential to yield an experience fusion between the reality, the culture and the society of the user is in and the ICT service s/he is enjoying. The system is aware of its own state and in relation with other systems notably those that relate to its user(s). At the same time a KMI must be aware of user intentions, state of mind and, desires. If the interaction is Kansei there is a feeling of harmony, because of the synergetic combination of the media the user is exposed to. If in addition the interaction is made natural and highly immersive, the user can have a Kansei Experience. In the sense that the user experiences a reality augmented with his/her experience of the KMI. It is an experience of environments fusion/juxtaposition, an experience that could be compared to enlightenment that we call Kansei Experience.

Kansei Experience should deliver an experience that is effective, efficient and satisfactory. These are similar to the guidelines of ISO 9241 (ISO, 1984). However, in that standard effectiveness is described as accuracy and completeness, efficiency relates to the expenditure of resources and satisfaction to the freedom from discomfort and the attitude towards usage. In our case we expend these definitions. Effectiveness relates to how good the system is at causing/achieving the desired results. Efficiency is about how minimal are the resources to achieve the desired results. As for user satisfaction, it relates to the user experience and its aesthetics (see Jordan, 1998). But how can Kansei Experience be implemented?

AESTHETICS

In Kansei media Experience the quality of the experience and the aesthetic of the experience should be used as a measurement of successful implementation. Nakatsu, Rauterberg and Vorderer (2005) pointed out that aesthetics should be systematically connected to entertainment theory. Aesthetics could be defined, explained and explored in many ways and still remains as evasive and unknown. What is aesthetics and why as humans we have sought and are seeking aesthetics? In general, aesthetic can be associated with the concept of beauty, it can also be the canon of beauty or its experience. In one sense, it is the measurement of beauty. Such a measure is associated with pleasure. Thus featuring beauty yields aesthetic pleasantness (Berlyne, 1960). Mentioning pleasantness implies a key fact about aesthetic, it is experienced by the beholder of the experience (Matravers, 2003). This makes the concept rather subjective and difficult to quantify. Furthermore, beauty needs not and should not, within this context, be limited to visual beauty. Any of our senses could be involved as well (Suzuki, 1959; Servomaa, 2001). So to redefine the concept in a more encompassing way, it is better to say that aesthetic is a subjective assessment of the beauty of an experience (Arcilla, 2002). The beauty of an experience can be translated into the pleasantness of the experience (e.g., 'tea ceremony' discussed by Ekuan, 1998, p. 28ff). In turn the pleasantness of an experience relates to the emotions triggered and their intensity. The emotional outcome of an experience is therefore an essential part of its aesthetic assessment, since emotions have a hedonistic bias in their occurrence (Cupchik, 1994,). The aesthetic experience is ultimately about a satisfaction resulting from the experience. Cupchik (1994) discusses two principles from pragmatic and emotional processing in everyday life that are generalised to the aesthetic realm. In everyday processing, important experiences are linked with bodily feelings of pleasure and arousal (Berlyne, 1960). In addition, meanings, which are contingent on specific contexts, are associated with blends of primary emotions (Millis, 2001). Aesthetics has therefore any combination of the following four key components: (1) beauty, (2) pleasantness, (3) emotions, and (4) satisfaction. Aesthetics could result from exposure to a perceivable form (e.g., physical, acoustic, olfactorial), the performance of an action (e.g., body expression, motor activity, etc.) or simply a mental experience (e.g., reading a book, meditation, etc.); see also the concept of *integrated presence* in Nakatsu, Rauterberg and Vorderer (2005).

Aesthetic can be described as the combination of beauty, pleasantness, emotions and satisfaction. It can be experienced from the perception of a form, the performance of an action or from a mental experience (Salem and Rauterberg 2005b). Such perceptions include a combination of any of sensory, synesthetic (combination and mixture of senses), cognitive, autonomic and motor experience (see Figure 6). Furthermore another type of aesthetics exists, namely social aesthetic. It relates to one's culture and is represented by attitudes, norms, values and beliefs. In general, the artifacts produced by that society also represent the social aesthetic.

Aesthetics of the Form and Content

The classical understanding of aesthetics relates it to the perception of physical beauty and balance. It is about the perception of proportions, symmetry, harmony and appearance (e.g., Locher et al., 1998). This kind of aesthetics is relevant to the interface design in the sense of the forms, their arrangement and the timing of their rendering. It relates to the visual form and content as well as to other forms and contents. Sounds, for example and other modalities are involved as well. In effect any of our senses let us experience aesthetics of this kind. And thus a synergetic combination of sense in a multimedia multimodal experience will yield a stronger experience of aesthetics. This would occur when one modality reinforces the perception resulting from another, e.g. sound and image render the same message (Karat et. al., 2002). This is achievable because human perception can take several forms that can be used in this context (see Figure 6). The experience can be sensory (how we perceive), synesthetic (how we combine perception), cognitive (how we think), autonomic (how our 'gut feeling' tells us we feel), and motor (how we act and do). This is of particular relevance for the developer of aesthetic pleasing experience.

Aesthetics of the Movement and Action

The aesthetics of the movement and action is an understanding of aesthetics inspired from the world of performing art such as mime theatre and dance. This aspect of aesthetics is related to the perceived quality of performed movements and actions, such as the strength shown by an athlete while running the 100mtrs, the lightness of a ballet dancer and the caricatured gestures of a mime performer. The aesthetics of the movements can also be related to one's movements and actions. Yamamoto (1999) relates it to the Japanese concept of *iki* which includes even to daily behaviour. Moen (2005) investigates experience from the perspective of dance and body movement philosophy. He proposes full-body movement interaction called kinaesthetic movement interaction. It is an interaction where the body movements are used both as Input and output, exploiting human kinaesthetic sense. Moen sees dance as the basis for movement interaction, and has developed a device: Bodybug that can be interacted with by dancing. The vision being of users choreographing body movement interactions to achieve aesthetics, pleasure, functional satisfaction while respecting personal differences (Moen, 2005).

In the context of user interaction, the performance of a movement or an action can be assessed them in terms of effect on Needs, Requirements and Desires (NRD). The fulfilment of one of the

NRD would yield a positive aesthetic experience of the movement or the action, and ultimately of the interaction. An example of this is the performance of an action that will result in a comfortable relaxing body posture.

Aesthetic of the Experience

An aesthetics experience can have either of three origins: (1) the aesthetics of the perception (AoP, as discussed in the section before last), (2) the aesthetics of the cognition (AoC; Cupchik, 1994), or (3) the aesthetics of the action (AoA, as discussed in the last section). In all three cases aesthetics relate to the experience one has and his/her assessment of it. Aesthetics are related to subjective, personal, changing and sometimes-irrational aspects of one's life. This could be for example the way one feels after watching a movie (AoP), listening to some music (AoP), looking as some painting (AoP), having a break-through idea or a deep insight (AoC), or highly immersed dancing (AoA).

Cultural Aesthetics

While so far we have explained aesthetics from a personal point of view, there is another aspect of aesthetics relating to more than a person, that is of relevance. Some examples of cultural aesthetics would be related to trends, culture or religions. Within the context of entertainment we wish to focus on social aesthetics related to culture (see also Rauterberg, 2004). Culture has long been associated with arts, leisure and entertainment. It seems therefore a good starting point that we wish to develop further with Kansei aesthetics.

Kansei, Inner Balance and Enlightenment

Ultimately the fulfilment of all of one's desires is achieved when one becomes freed from them. Attachment to desires can only yield disillusion. True fulfilment can be reach with enlightenment (see the teaching of Buddha as an example). However is enlightenment a definable concept that can be applied to computing? And, what is enlightenment? (To stay within the scope of this paper, we will only briefly address this topic).

The peoples of three distinct regions of the world created the religious and philosophical traditions that have continued to nourish humanity into the present day: Confucianism and Daoism in China; Hinduism and Buddhism in India; monotheism and philosophical rationalism in The Mediterranean and the Middle-East. 'Monotheism' and 'philosophical rationalism' are the religious and philosophical foundations of the Western world. In this perspective there are three monotheist religions in the West (in chronological order): Judaism, Christianity and Islam.

Religions are sets of beliefs and rituals made up of both personal practices and systems of values. In a simplified sense, religions are a way of life and a coherent set of beliefs that lead to a better spirituality. For the sake of simplification, and within the context of this paper, religious lead to a detachment from earthly concerns, and enlightenment. We could state that enlightenment has been defined both in the West and in the East. Kant (1784) gave a first answer to the question, "What is Enlightenment?" He indicated that the 'way out' that characterises Enlightenment in the West is a process that releases us from the status of 'immaturity'; and by 'immaturity,' he meant a certain state of our will that makes us accept someone else's authority to lead us in areas where the use of reason is called for. Enlightenment is defined by a modification of the pre-existing relation linking will, authority, and the use of reason. In the Hindu religion, enlightenment is compared to Moksha. It is achieved when one reach higher awareness than consciousness, time, space and causation. A level when being is no more different from experiencing. In Buddhism, Bodhi is the state reached by the Buddha and his disciples. It is reached on the complete understanding of the true nature of the universe. Zen Buddhism uses two concepts Kencho (見性) and Satori (悟り) to describe enlightenment. Kensho refers to the first perception of true-nature. Satori on the other hand refers to a more permanent perception. Kensho is the initial awakening experience, while Satori is the more lasting experience of enlightenment. Both are however transitional compared with the Nirvana, which is the state of great inner peace, balance and contentment.

How can we make use of an understanding of enlightenment in the context of ICT usage? We can understand that enlightenment is a state of awareness and experience. The beholder of which has a deeper understanding of him/herself and is free from many desires. Desires that otherwise would be detrimental to his/her experience and thus existence. We postulate that enlightenment could be achieved by an enjoyable, enriching and engaging experience.

Enlightenment is therefore of great interest within the perspective of the paradigm of cultural computing. Within this paradigm, it is our vision that Kansei Experience can deliver enlightenment to the user by helping him/her attain a deeper and richer experiencing and understanding of reality. To do so, Kansei Experience needs to be enjoyable in the sense of entertaining and stimulating. It also needs to be enriching, by helping reach a better and deeper understanding of the instant and it needs to be engaging by being immersive and helping the user focus and detach him/herself from reality. We advocate that these qualities will make Kansei Experience an aesthetically pleasant experience, an experience that trigger and involve emotions and an experience that yield an inner balance to its user.

METHODOLOGY: IMPLEMENTING KANSEI EXPERIENCE

Inspiration from a Precursor

There are many early systems we could have selected as inspiration. Virtual Reality Systems and Game consoles are some examples. However, we

Figure 4. The Sensorama Simulator (from Heiling, 1962)

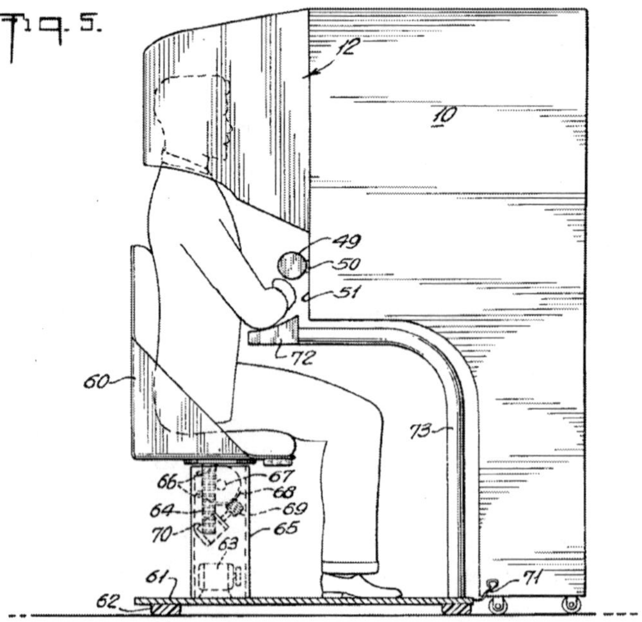

Figure 5. Maiko Girls heading to a function in Gion, Kyoto, Japan (photo B. Salem)

feel that a relevant precursor to our proposal is the Sensorama system. It was designed to stimulate the senses of the user with the aim of simulating a realistic experience. Various applications were proposed for this system. It was developed to be an arcade game, a training and educational tool. the system let the user experience a ride on a motorbike around a city. Rightfully, Sensorama is considered as the precursor of many interaction developments such as multi-media and Virtual Reality. Sensorama implemented a range of media with the intent of simulating a realistic experience. It relied on the synesthetic perception of combined media, such as smells, wind, noises associated with a bike ride in a town (Heiling, 1962).

The aim of the Sensorama system was to render a high-fidelity realistic experience, similar to a music system playing a high-fidelity recording. We aim to go beyond this and render a Kansei Experience that relates to reality, but enhances it to enlighten the user.

Another source of inspiration is the Maiko girls in Kyoto, Japan. They are apprentice Geishas and are highly sophisticated and skilled entertainers. Although not actually an ICT system, Maiko girls "deliver a service to users". In essence, Maiko job and skills are to make their client feel good and pleased. This is achieved by combining art performance, entertainment, game play and educated and cultured conversations. It is also thanks to a highly subtle and perfectly timed body language and facial expressions in combination with a complex accoutrement.

What is most relevant about these two examples apart from their intrinsic values, is that while the first is essentially recreating a reality the second is about rendering an experience. Whit KMI we are interested in a combination of both. While it is straightforward to understand the rendering of a reality in similar fashion to a Virtual Reality system, the rendering of an experience is a more challenging task that requires the mastering of some aesthetics and emotion triggering principles.

Interaction Experience

Our aim is about allowing the user to experience an interaction that is closely related to the core aspects of his/her culture. In a way that let him/her engage with the interface using the values and attributes of his/her own culture. As such it is important to understand one's cultural values and

Figure 6. Several forms of human perception.

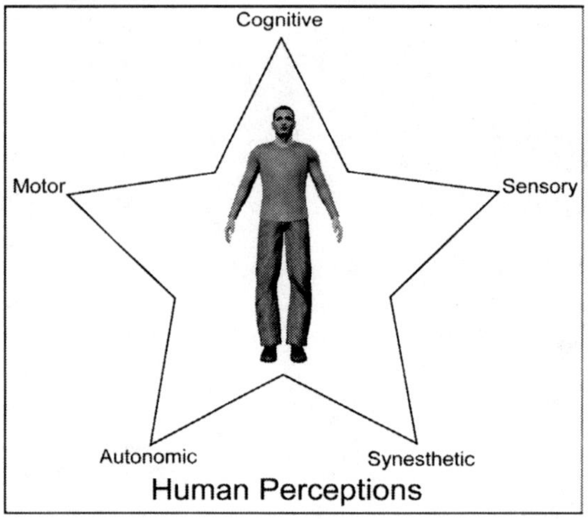

how to use and render them during the interaction. As there are many cultures that we could investigate, there is a need to select a subset of cultures. In this paper we will focus on two representative sub-cultures, one from the Western world prevailing in Britain, and one from the Eastern World prevailing in Japan.

To understand both cultures in a way that would fit the objectives of our interests, we have sought illustrative examples of both. Thus, we have investigated illustrative stories that are well known, accessible, classical in their respective cultures and relevant from the point of view of cultural computing. We also looked for stories that would be helpful in the understanding of the essential aspects of both Western and Eastern cultures. Our approach is to create an interaction based on the cultural values highlighted in these stories.

Illustrative Stories

The origin of stories from the Western culture (i.e. the Mediterranean) can be traced back to Greek mythology. An example of Greek mythology is the Illiad and the Odyssey written by Homer in the 8th Century BC. Illiad tells the story of the siege of the city of Illium during the Trojan War. The Odyssey describes the heroic journey back home from the Trojan war of Odysseus king of Ithaca. As for the Eastern Culture of Japan, the original works are the two mythological books Kojiki and Nihonshoki., completed in 712 and 720 respectively. Kojiki describes various myths and legends of Japan. It starts at the beginning of the world and ends at the reign of Empress Suiko. Nihonshoki, on the other hand, is a compilation of the chronicles of Japan. Kojiki emphasise the mythical while Nihonshoki is more factual.

The historical mythologies could have been used, but within the context of our work they are rather complex in content, narrative and plot. We have selected less predominant stories but relevant nonetheless. For the Eastern culture it is the story of The Ox Herding attributed to a Ch'an master (circa 1200s). Although the story is of Chinese origin, as with many other cultural elements, it has been adopted into Japanese culture. For the Western culture we have selected Alice in Wonderland by Lewis Carroll (1865). These stories help understand the underlying cultural value or question it. For the Eastern culture, the value dealt with is enlightenment, while it is orderly rational reason for the Western culture. In the next section we present an overview of these stories, and describe the introductory phase of the experience of these stories.

Figure 7. Various representations of the Ox Herding Steps

7.1. Step 3	7.2. Step 4	7.3. Step 5	7.4. Step 6	7.5. Step 10

Eastern Culture: Story of the 'Ox Herding'

This short story has ten steps:

1. Seeking the ox: Starting the journey, while being unaware that a dualist approach cannot lead to the understanding of the true nature of mind. There should not be a mind/body separation.
2. Seeing the ox tracks: Although the ox is still not seen, the tracks confirm its existence, and leads to it. Through self-discipline and training, it is possible to rediscover one's true self.
3. Seeing the Ox: The path to enlightenment has been seen. It is about finding one's true self, through trial and errors (see Figure 7.1).
4. Catching the Ox: Although the ox has been seen, the difficulty now is to catch it (see Figure 7.2).
5. Herding the Ox: Kencho is finally obtained after a long period of disciplinary training. However, the Kencho attained is only a stepping stone towards Satori (see Figure 7.3).
6. Coming Home on the Ox Back: The efforts paid off. The ox and the herder move together effortlessly. This shows the state in which one completely finds one's true self, that is, the state in which one obtains Satori (see Figure 7.4).
7. Ox is forgotten: The ox and the herder become one. Dualism has been overcome and the herder has no worldly attachments any more.
8. Both Ox and self-forgotten: The separation of reality from the mind is gone. Enlightenment is experienced and the mind has escaped.
9. Returning to the source: Back to square one. The world carries on as always.
10. Returning to help others: The enlightened has renounced all to selflessly help others (see Figure 7.5).

The ten Ox Herding Pictures (Figure 7) are an imagery of an illusion to be negated before a seeker can experience enlightenment. In these pictures, the ox symbolise the mind, while the herder the seeker. The illusion being that reality is separate from the mind (Buddhanet, 2006). These metaphorical steps help one achieve Kensho and at a later stage Satori.

Implementing the Ox-Herding Experience

The user, the Ox-Herder, is in a beautiful surrounding. There is a refreshing stream, beautiful trees and so forth. However, the Ox-Herder is still not satisfied. S/He is looking for something else, inner peace and contentment. S/He therefore embarks in a spiritual journey, unaware that the true nature of the mind cannot be find in a dualistic view of the

Figure 8. Various representations of the Ox Herding Steps (drawings by Ian Young)

8.1 Setting the scene	8.2 Looking around

8.3 Seeking the ox	8.4 Seeing the ox track

world. As a metaphor, the herder is searching for the ox, it is a spiritual search leading to a change of lifestyle and the eradication of bad habits. This path is difficult to find. The Ox-Herder is a little lost, little confused, running here and there. S/He is searching for something but he is not even sure what he is looking for. The spiritual path has not started yet, but the Ox-Herder feel somewhat uncomfortable and unsatisfied. S/He realise that material things are not sufficient for long-lasting happiness. This is the inspirational motivation to search and seek the ox.

Figure 8 shows the storyline of the early stage of the ox-herding experience as we are implementing it in a videogame. In our proposed implementation, physiological parameters (heart rate, galvanic skin conductance, breathing rate) are used alongside a conventional game controller. These parameters are used to ensure a transfer of user inner state into the game character progress through the narrative.

Western Culture: Story of 'Alice in Wonderland'

Western culture is based on Cartesian logic, analytical reasoning and a linear and constant flow of time. Western culture in general is based on Monotheist religions (Judaism, Christianity, and Islam) that are concerned with certainty and absolutism: in the sense of absolute certain truth.

In this context, Alice's adventures are the antithesis of Western Culture. Her adventures happen in a world of paradox, the absurd and the

Figure 9. Alice Adventures in Wonderland (from Wikipedia)

9.1. Alice 9.2. The Caterpillar 9.3. Cheshire Cat 9.4. Tea Party

Figure 10. Inside our Alice installation (from Hu et al., 2008)

10.1 Looking down the rabbit hole

10.2 Drink me and Eat me

10.3 In dialogue with the caterpillar

10.4 Talking to the Cheshire cat

improbable (Wikipedia, 2006). The key aspects of Alice in Wonderland can be resumed in the following points: (1) a non-linear non-constant time flow; (2) a distortion of space and people; (3) a counter-intuitive, common sense defying heuristics.

As a children's story, Alice in Wonderland can be used to give interesting examples of many of the basic concepts of adolescent psychology. Her experiences can be seen as symbolic depictions of important aspects of adolescent development, such as initiation, identity formation, and physical, cognitive, moral, and social development (Lough, 1983). Alice's adventures are directly challenging the strongly held belief of a linear, single track and sequential reality.

Implementing the Alice Experience

We have built a large-scale installation to re-create six stages of the Alice in wonderland adventures. The stages are: Down the rabbit hole, eat me and drink me, swimming in the sea of tears, talking to a caterpillar, and in discussion with the Cheshire

Table 1. Comparing Ox Herder and Alice Adventures in Wonderland stories

	Ox Herder	**Alice in wonderland**
Characters	The herder and the ox	Alice and others, each possessing its own paradox
Plot	Successive steps to attain higher experience	A succession of paradox and illogical events
Timeline	Could be up to lifetime	Non-linear
Location	Delocalised	Various locations somehow "connected"
Artefacts	None are essential	Cakes, mushroom, tea and various familiar objects.
Thoughts	Spiritual and holistic	Analytical, logical and linear
Value	Harmony	Personal agency
Narrative	3rd person	1st person

cat. The installation is designed for individual exploration. Through the various stages, users experience boredom and curiosity (waiting and then following a white rabbit), spatial disorientation (going down the rabbit hole), size disorientation (thanks to eat me and drink me), re-genesis (swimming in sea of one's own tears), the self (in a dialogue with the caterpillar) and finally sanity (in a dialogue with the Cheshire cat).

Figure 10 shows the Alice Installation we have built for investigation of the Alice experience.

Discussion: Guidelines for Kansei Mediated Experience

Both stories are closely related to their respective cultures and are different in many aspects that we list in table 1. These differences reinforce a certain cultural dependency of Kansei Experience systems.

The challenges of implementing a successful Kansei Mediated Experience require the adoption of a different computing paradigm. Cultural Computing is one paradigm that allows for the rendering of a user experience. Furthermore, we suggest Kansei Media, as it relates to cultural values, emotions and aesthetics. To deliver a Kansei Experience, we propose the following issues as the main challenges to address at the implementation of the system: (1) The Kansei Media to use, (2) the cultural values to base the Kansei Experience on, (3) the affects on the user we wish to render and (4) the aesthetics canons to follow.

The Kansei media can be based on a variety of contents, for example to seek calmness, introspection or engagement. Is the Kansei Experience about strengthening personal agency and will or reinforcing harmony and holistic balance. The cultural values could focus on the system of thoughts like linear analytical thinking, or on other cultural value like morals or beliefs. The affects on the user can be about self questioning, attaining a deeper perception. Finally, the aesthetic canons can rely on music, visual atmosphere and other aspects of the experience to deliver beauty to the user.

However, because of the complexity of the experience proposed, the same user could understand and appreciate the implementation of an interaction inspired from 'Alice in Wonderland Adventures', while be puzzled if s/he was presented with an interaction inspired from the 'Ox Herder story'. Similarly, another user would appreciate the second implementation and be confused by the first.

CONCLUSION

In this paper we have attempted to define and explain a novel concept we call Kansei Experience using Kansei Media. In our opinion Kansei Experience is a promising development because it allows for the delivery of both explicitly and implicitly rich experiences. Kansei Experience is also of relevance nowadays with the acknowledgment of emotions and feelings as important component of our cognitive functions. For a possible implementation of Kansei Experience we have proposed entertainment as the application domain. Our vision is of an environment where the interaction is based on a story that is strongly related to the culture of the user. The drawback of this vision is the cultural dependency of Kansei Experience.

Finally, we hope to have raised the awareness of Kansei media and communication in the field of HCI and demonstrated through our proposed Kansei Experience of the significant potential a Kansei approach could have on ICT usage and entertainment.

ACKNOWLEDGMENT

This paper is an extension and a rewriting of the paper by the authors on Kansei Mediated Entertainment (Salem et al., 2006).

REFERENCES

Aarts, E. (2002). Ambient Intelligence – Experience Technology, Ambient Intelligence in HomeLab, Eindhoven, NL, Philips.

Aarts, E., Marzano, S. (2003) (eds.). The New Everyday – Views on Ambient Intelligence, Rotterdam, NL, OIO Publishers.

Arcilla, R.V. (2002). Modernising media or modernist medium? The struggle for liberal learning in our information age. Journal of Philosophy of Education, 36(3), pp. 457-465.

Berlyne, D.E. (1960). Conflict, Arousal, and Curiosity. New York, NY, USA, McGraw Hill.

Buddhanet, (2006). The Ten Ox Herding Pictures, see www.buddhanet.net/oxherd1.htm

CCP (2004). Cultural Computing Program see www.culturalcomputing.uiuc.edu

Cupchik, G.C. (1994). Emotion in aesthetics: reactive and reflective models. Poetics, 23, pp. 177-188.

Erickson T.D.(1989). Interfaces for Cooperative Work: An Eclectic look at CSCW '88, ACM SIGCHI Bulletin, Volume 21, Issue1, July 1989, pp. 56-64.

Ekuan K. (1998). The aesthetics of the Japanese lunchbox. Cambridge, MA, USA, MIT Press.

Gaines, B.R. (1985). From ergonomics to the fifth generation: 30 years of human-computer interaction studies, Computer Compacts, Volume 2, Issue 5-6, November 1984 – January 1985, pp. 158-161.

Harslem, E., Nelson, L. E.(1982). A retrospective on the development of Star, Proceedings of the 6th international conference on Software engineering, Tokyo, Japan Pages: 377 - 383

Heilig M. (1962). Sensorama Simulator, US Patent 3,050,870.

Hiltz S.R. (1984), Online Communities: A case study of the office of the future, Norwood, NJ, USA: Ablex Publishing.

Hu, J., Bartneck, C., Salem, B., Ratuerberg, M. (2008), Alice's adventures in cultural computing, International Journal of Arts and technology, Volume 1, Number 1, pp. 102-118.

ISO (1984), Ergonomic requirements for office work with visual display terminals (VDTs) - Part 11 : Guidance on usability, ISO 9241-11:1998(E).

Jones, P.F. (1978). Four principles of man-computer dialogue, Computer-Aided Design, Volume 10, Issue 3, (May 1978), pp.197-202.

Jordan, P.W. (1998). Human Factors for Pleasure in Product Use, Applied Ergonomics, Vol29, No. 1, pp. 25-33.

Kant I. (1784). Beantwortung der Frage: Was ist Aufklärung? Berlinische Monatschrift, vol. 2, pp. 481-494.

Karat, C.M., Karat, J., Vergo, J., Pinhanez, C., Riecken, D., Cofino, T. (2002). That's Entertainment! designing Streaming, multimedia Web experience. International Journal of Human Computer Interaction, 14, 3-4, pp. 369-384

Locher, P.J., Stappers, P.J., Overbeeke, K.C. (1998). The role of balance as an organizing design principle underlying adults' compositional strategies for creating visual displays, Acta Psychologica, 99, pp. 141-161.

Lough GC, (1983). Alice in Wonderland and cognitive development: teaching with examples. Journal of Adolescence. 1983 Dec; 6(4): 305-15.

MacLennan, B.J. (1997). "Who Cares About Elegance?" The Role of Aesthetics in Programming Language Design, ACM SIGPLAN Notices, 32(3), March 1997, pp. 33-37.

Maes, P. (2005). Attentive Objects: Enriching People's natural interaction with everyday objects, Interactions, July + August 2005, pp.45-48.

Matravers, D. (2003). The aesthetic experience. The British Journal of Aesthetics, 43(2), pp. 158-174.

McCarthy, J., Wright, P. (2004). Technology as Experience, Interactions, September + October 2004, pp. 42-43.

Millis, K. (2001). Making meaning brings pleasure: the influence of the titles on aesthetic experiences. Emotion, 1(3), pp. 320-329.

Moen, J. (2005) Towards People Based Movement Interaction and KinAesthetic Interaction Experience, In Proc. AARHUS'05, 21-25 August 2005, Arhus, Denmark, pp. 121-124.

Monk, A., Hassenzahl, M., Blythe, M., Reed, D. b(2002). Funology: Designing Enjoyment, In Proc. CHI 2002, 20-25 April 2002, Minneapolis, MN, USA, pp. 924-925.

Nagamachi M., Kansei Engineering: A New Ergonomic Consumer-Oriented Technology for Product Development, International Journal of Industrial Ergonomics, 15, 1995, pp. 3-11.

Nakatsu R, Rauterberg M, Salem B (2005a). Forms and Theories of communication: from multimedia to Kansei mediation. Multimedia Systems, 11(3), pp. 304-312.

Nakatsu R, Rauterberg M, Vorderer P (2005b). A new framework for entertainment computing: from passive to active experience. Lecture Notes in Computer Science, vol. 3711, pp. 1 – 12.

Nisbett, E.R., K. Peng, I. Choi, A. Norenzayan (2001). Culture and Systems of Thoughts: Holistic Versus Analytic Cognition, Psychological Review, April 2001, 108, 2, 291-310.

Ngo, D.C.L., Teo, L.S., Byrne, J.G. (2002) Evaluating Interface Esthetics (sic), Knowledge and Information Systems, 2002, 4, pp. 46-79.

Norman, D.A. (2002) Emotions & Design: Attractive things Work Better, Interactions, July + August 2002, pp.36-42.

Pierce JS, Pausch R, Sturgill CB, Christiansen KD (1999). Designing a successful HMD-based experience. Presence, 8(4), pp. 469–473.

Rauterberg M (2004). Positive effects of entertainment technology on human behaviour. In: R. Jacquart (ed.), Building the Information Society (pp. 51-58). IFIP, Kluwer Academic Press.

Rheingold H. (1991). Virtual reality, New York, NY, USA, Simon & Schuster.

Salem B., M. Rauterberg (2004). Multiple User Profile Merging: Key Challenges for Aware Environments, Proceedings EUSAI- European Symposium on Ambient Intelligence 2004, Lecture Notes in Computer Science, Vol. 4161, pp. 103-116.

Salem B (2005). Commedia Virtuale: from theatre to avatars. Digital Creativity, 16(3), pp. 129-139.

Salem B, Rauterberg M (2005a). Power, Death and Love: a trilogy for entertainment. Lecture Notes in Computer Science, vol. 3711, pp. 279 – 290.

Salem B, Rauterberg M (2005b). Aesthetics as a key dimension for designing ubiquitous entertainment systems. In: M. Minoh & N. Tosa (eds.) The 2nd International Workshop on Ubiquitous Home—ubiquitous society and entertainment. (pp. 85-94) NICT Keihanna and Kyoto.

Salem B, Rauterberg M., Nakatsu R. (2006). Kansei Mediated Entertainment, In Proceedings ICEC2006, Lecture Notes in Computer Science, 4161, pp. 103-116.

Schilit B., Theimer M. (1994). Disseminating Active Map Information to Mobile Hosts, IEEE Network, Vol. 8, No. 5, pp. 22-32.

Servomaa, S. (2001). Aesthetics of the art of flowers: ikebana. In: G. Marchiano & R. Milani (eds.), Proc. Intercontinental Conference 'Frontiers of Transculturality in Contemporary Aesthetics' (pp. 367-377), Turin, Italy.

Suzuki, D.T. (1959). Zen and Japanese Culture. Princeton, NJ, USA, Princeton University Press.

Stewart T.F.M. (1976). Displays and the software interface, Applied Ergonomics, Volume 7, Issue 3, September 1976, pp. 137-146.

Sutherland, I.E. (1964). Sketch pad a man-machine graphical communication system, Proceedings of SHARE design automation workshop 1964, pp. 6.329-6.346.

Tosa N, Matsuoka S, Ellis B, Ueda H, Nakatsu R (2005). Cultural Computing with Context-Aware Application: ZENetic Computer. Lecture Notes in Computer Science, vol. 3711, pp. 13 – 23.

Thacker C.P., McCreight E.M., Lampson B.W., Sproull R.F. and Boggs D.R. (1982). "Alto: A Personal Computer," Computer Structures: Principles and Examples, D. Siewiorek, D.G. Bell and A. Newell, editors, McGraw-Hill.

Vasilakos A., Pedrycz W. (2006), Ambient Intelligence, Wireless Networking, Ubiquitous Computing, Boston, MA, USA, Artech House.

Yershov, A.P. (1965). One View of Man-Machine Interaction, Journal of the ACM, Vol, 12, issue 3 (july 1965), pp. 315-325.

Wang Y. (2003). Cognitive Informatics: A New Transdisciplinary Research Field, Brain and Mind, Vol. 4, No. 2, pp. 115-127.

Wang Y. (2007). The Theoretical Framework of Cognitive Informatics, International Journal of Cognitive Informatics and Natural Intelligence, Vol.1, No. 1, pp. 1-27.

Weiser M. (1991). The Computer of the Twenty-First Century, Scientific American, September 1991, pp. 94-10.

Weiser M. (1999). Some computer science issues in ubiquitous computing, ACM SIGMODIBLE Mobile Computing and Communication Review, VOl. 3, No. 3, pp. 3-11.

Weizenbaum, J. (1966). ELIZA – A computer program for the study of natural language communication between man and machine, Communications of the ACM, Volume 9, Number 1 (January 1966), pp. 35-36.

Yamamoto, Y. (1999). An aesthetics of everyday life. M.Sc. Thesis, University of Chicago, USA.

The Ox Herding Pictures were Taken from the Following Websites

Figure 3: http://www.sacred-texts.com/bud/mzb/oxherd.htm,

Figure 4: http://www.hsuyun.org/Dharma/zbohy/VisualArts/OxHerdingPictures/oxherding2.html

Figure 5: http://oaks.nvg.org/wm2ra4.html

Figure 6: http://www.donmeh-west.com/tenox.shtml

Figure 10: http://www.buddhanet.net/oxherd10.htm

This work was previously published in International Journal of Cognitive Informantics and Natural Intelligence, Volume 3, Issue 2, edited by Yingxu Wang, pp. 18-36, copyright 2009 by IGI Publishing (an imprint of IGI Global)

Chapter 9
IPML:
Structuring Distributed Multimedia Presentations in Ambient Intelligent Environments

Jun Hu
Eindhoven University of Technology, The Netherlands

Loe Feijs
Eindhoven University of Technology, The Netherlands

ABSTRACT

This paper addresses issues of distributing multimedia presentations in an ambient intelligent environment, examines the existing technologies and proposes IPML, a markup language that extends SMIL for distributed settings. It uses a metaphor of play, with which the timing and mapping issues in distributed presentations are covered in a natural way. A generic architecture for playback systems is also presented, which covers the timing and mapping issues of presenting an IPML script in heterogeneous ambient intelligent environments.

INTRODUCTION

Ambient Intelligence (AmI) is introduced by Philips Research as a new paradigm in how people interact with technology. It envisions digital environments to be sensitive, adaptive, and responsive to the presence of people, and AmI environments to change the way people use multimedia services (Aarts, 2004). The environments, which include many devices, will play interactive multimedia to engage people in a more immersive experience than just watching television shows. People will interact not only with the environment itself, but also with the interactive multimedia through the environment.

For many years, the research and development of multimedia technologies have increasingly

DOI: 10.4018/978-1-60960-553-7.ch009

Figure 1. The mapping problem caused by variations in playback system architecture

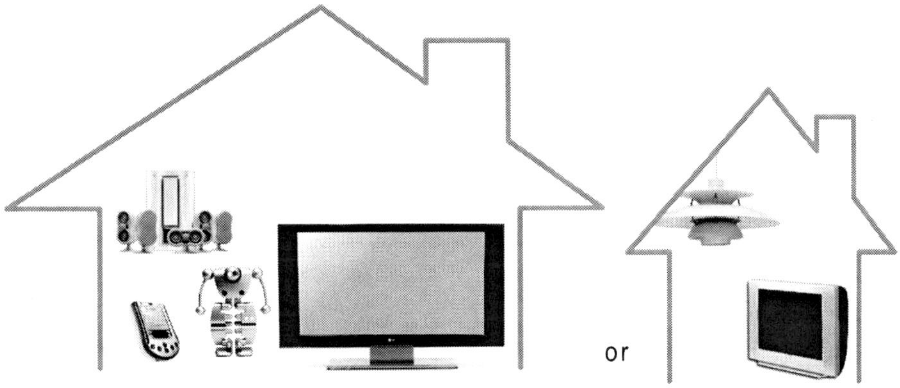

focused on models for distributed applications, but the focus was mainly on the distribution of the media sources. Within the context of AmI, not only are the media sources distributed, the presentation of and the interaction with the media will also be distributed across interface devices. This paper focuses on the design of the structure of multimedia content, believing that the user experience of multimedia in a distributed environment can be enriched by structuring both the media content at the production side and the playback system architecture at the user side in a proper way. We refer to the adaptation at the user side as the mapping problem. One important aspect of the mapping problem is sketched in Figure 1. The content source and the script should be independent from the question which specific devices are available at the user's side. This may vary from a sophisticated home theater with interactive robots (left) to a simple family home with a television-like device and a lamp (right). There is no a priori limit to the type of devices, for example PDAs and controllable lights are possible as well. The playback environment need not even be a home; it could be a professional theater or a dedicated installation. The structure should enable both the media presentation and the user interaction to be distributed and synchronized over the networked devices in the user environment. The presentation and interaction should be adaptive to the profiles and preferences of the users, and the dynamic configurations of the environment.

As El-Nasr and Vasilakos (2006) point out, there is very little work that allows the adaptation of the real environment configuration to the cognitive spaces of the artists, in our example, the authors of the content and the script. The area of Cognitive Informatics (Wang, 2006, 2007) provides interesting insights into this issue. In particular this is a field studies the mechanisms and process of natural processing and intelligence, including emotions, cognition, decision making, and its application to entertainment, engineering, educational, and other applications. On the one hand, the users and the authors should not be bothered by the complexity hidden behind the surface of the ambient intelligence; on the other hand the ambient intelligent environment should be able to interpret the user's needs in interaction and to adapt to the author's requirements in presentation. The common part that the users and the authors share is not a particular user's environment, but only the media content. The media content should be structured in such a way, that the requirements from the both sides can meet. To structure the media content, the following issues need to be addressed:

1. By what means will the authors compose the content for many different environments?

The authors have to be able to specify the following with minimized knowledge of the environments: (a) Desired environment configurations; (b) Interactive content specification for this environment.

2. How can the system play the interactive media with the cooperation of the user(s) in a way that: (a) makes the best use of the physical environment to match the desired environment on the fly; (b) enables context dependent presentation and interaction. Here the term "context" means the environment configuration, the application context, the user preferences, and other presentation circumstances; (c) synchronizes the media and interaction in the environment according to the script.

This paper first examines existing open standards for synchronized media. Then the notion of "play" is introduced as a unifying concept, first in an informal way, later formalized through the design of the language. The language is developed in two steps: first an existing scripting language and then the language IPML which takes full advantage of the notion of play and addresses all of the aforementioned issues. The latter language is based on a generic architecture for the playback system that covers the timing and mapping problems.

Then we discuss the three main architectural design elements which are needed to bring the plays, written in this language, to live: distributed agents, an action synchronization engine, and an IPML mapper. It is through this design that we validate the concepts and thus prove the feasibility of IPML and demonstrate its value.

OPEN STANDARDS FOR SYNCHRONIZED MEDIA

SMIL and MPEG-4 are contemporary technologies in the area of synchronized multimedia (Battista, Casalino, & Lande, 1999, 2000). SMIL focuses on Internet applications and enables simple authoring of interactive audiovisual presentations, whereas MPEG-4 is a superset of technologies building on the proven success in digital television, interactive graphics applications and also interactive multimedia for the Web. Both were the most versatile open standards available at moment of starting the design trajectory.

But both were challenged by the requirement for distributed interactions. It requires that the technology is first of all able to describe the distribution of the interaction and the media objects over multiple devices. The BIFS in MPEG-4 emphasizes the composition of media objects on one rendering device. It doesn't take multiple devices into account, nor does it have a notation for it.

SMIL 2.0 introduces the *MultiWindowLayout* module, which contains elements and attributes providing for creation and control of multiple top level windows (Rutledge, 2001). This is very promising and comes closer to the requirements of distributed content interaction. Although these top level windows are supposed to be on the same rendering device, they can to some extent, be recognized as software interface components which have the same capability.

To enable multimedia presentations over multiple interface devices, StoryML was proposed (Hu, 2003). It models the interactive media presentation as an interactive *Story* presented in a desired environment (called a *Theater*). The story consists of several *Storylines* and a definition of the possible user *Interaction* during the story. User interaction can result in switching between storylines, or changes within a storyline. *Dialogues* make up the interaction. A dialogue is a linear conversation between the system and the user, which in turn consists of *Feed-forward* objects, and the *Feedback* objects that depend on the user's response. The environment may have several Interactors. The interactors render the media objects. And finally, the story is rendered in a Theater.

One problem of StoryML is that it uses a mixed set of terms. "Story" and "storylines" are from narratives, "media objects" are from computer science, whereas interactors are from human computer interaction. Scripting an interactive story requires various types of background knowledge. It is questionable whether StoryML has succeeded in both keeping the scripting language at a high level and let the script authors only focus on the interactive content. "Movies did not flourish until the engineers lost control to artists – or more precisely, to the communications craftsmen." (Heckel, 1991)

StoryML uses storytelling as a metaphor for weaving the interactive media objects together to present the content as an "interactive story". This metaphor made it difficult to apply StoryML to other applications when there are no explicit storylines or narratives. Moreover, StoryML can only deal with linear structure and use only a storyline switching mechanism for interaction.

Reflecting on our experiences with StoryML, it is necessary to design a script language that has a more generic metaphor, that supports both linear and nonlinear structures and that can deal with complex synchronization and interaction scenarios. Next we introduce the metaphor of "play", for the design of the new scripting language, IPML.

PLAY

Instead of storytelling, Interactive Play Markup Language (IPML) uses the more powerful metaphor of *play*. A play is a common literary form, referring both to the written works of dramatists and to the complete theatrical *performance* of such. Plays are generally performed in a *theater* by *actors*. To better communicate a unified interpretation of the text in question, productions are usually overseen by a *director*, who often puts his or her own unique interpretation on the production by providing the actors and other stage people with a *script*. A script is a written set of directions that tell each actor what to say (lines) or do (actions) and when to say or do it (timing). If a play is to be performed by the actors without a director and a script from the director, the results are unpredictable, if not chaotic. It is not the intention of this paper to give a definitive and extensive definition of the term "play", nor to reproduce all elements of such a rich art form. Only the necessary parts are taken for easier understanding and communication when composing a markup script. Here we use the word "play" for both its written form of a script, and the stage performance form of this script.

Timing in a Play

Timing in a play is very important whether it be when an actor delivers a specific line, or when a certain character enters or exits a scene. It is important for the playwright to take all of these into consideration. The following is an example from *Alice in Wonderland* (Carroll & Chorpenning, 1958):

ALICE: Please! Mind what you're doing!

DUCHESS (tossing ALICE the baby): Here... you may nurse it if you like. I've got to get ready to play croquet with the Queen in the garden. (She turns at the door.) Bring in the soup. The house will be going any minute! (As the DUCHESS speaks, the house starts moving. The COOK snatches up her pot and dashes into the house.)

COOK (to the FROG): Tidy up, and catch us! (The FROG leaps about, picking up the vegetables, plate, etc.)

ALICE (as the FROG works): She said "in the garden." Will you please tell me –

FROG: There's no sort of use asking me. I'm not in the mood to talk about gardens.

ALICE: I must ask some one. What sort of people live around here?

A few roles are involved in this part of the play. Their lines and actions are presented by the playwright in a sequential manner, and these lines and actions are by default to be played in sequence. However, these sequential lines and actions are often not necessarily to happen immediately one after another. For example, it is not clear in the written play how much of time the duchess should take to perform the action "*tossing Alice the baby*" after Alice says "*Mind what you're doing*" and before the duchess says "*Here... you may nurse it if you like*". The director must supervise the timing of these lines and actions for the actors to ensure the performance is right in rhythm and pace. Furthermore, things may happen in parallel – For example, the house starts moving as the duchess speaks, and Alice talks as the frog works. Parallel behaviors are often described without precise timing for performing. It is up to the directors to decide the exact timing based on their interpretation of the play. For example, the director may interpret "*As the DUCHESS speaks, the house starts moving*" as "*at the moment of the duchess start saying 'The house will be going in any minute', the house starts moving*".

Mapping: Assigning Roles to Actors

Actors play the roles that are described in the script. One of the important tasks of the director is to define the cast – assign the roles to actors. This is often done by studying the type of a role and the type of an actor, and finding a good match between them. This is also exactly the problem for distributed presentations: determining which device or component to present certain type of media objects. It can be very hard for a computer to carry out this task, unless these types are indicated in some way otherwise.

In some traditional art of play, these types are even formalized so that a play can be easily performed with a different cast. We found a perfect source of inspiration in Beijing Opera. The character roles in Beijing Opera are divided into four main types according to the sex, age, social status, and profession of the character: male roles (Sheng, Figure 2(a)); female roles (Dan, Figure 2(b)); the roles with painted faces (Jing, Figure 2(c)) who are usually warriors, heroes, statesmen, or even demons; and clown (Chou, Figure 2(d)), a comic character that can be recognized at first sight for his special make-up (a patch of white paint on his nose). These types are then divided into more delicate subtypes, for example Dan is divided into the following subtypes: Qing Yi is a woman with a strict moral code; Hua Dan is a vivacious young woman; Wu Dan is a woman with martial skills and Lao Dan is an elderly woman. In a script of Beijing Opera, roles are defined according to these types. An actor of Beijing Opera is often only specialized in very few subtypes. Given the types of the roles and the types of the actors, the task of assigning roles to actors becomes an easy matching game.

Figure 2. Role types in Beijing opera

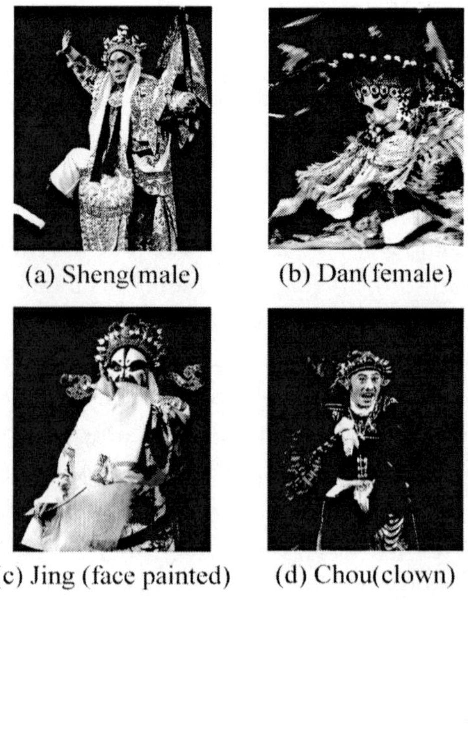

(a) Sheng(male) (b) Dan(female) (c) Jing (face painted) (d) Chou(clown)

Interactive Play

Plays can be interactive in many ways. The actors may decide their form of speech, gestures and movements according to the responses from the audience. This is the case in Beijing opera, which sometimes can still be been seen today, and which may be performed in the street (Figure 3) or in a tea house, where the actors and the audience are mixed – the actors and the audience share the stage. The movements of the actors must be adapted to the locations of the audience, and the close distance between the audience and the actors stimulates the interaction. An example of such interaction is that the characters often strike a pose on the stage, and the audience is supposed to cheer with enthusiasm. The time span of such a pose depends on the reactions of the audience. Although this is often not written in the script, such an interactive behavior is by default incorporated in every play of Beijing opera.

Other interactive plays allow the audience to modify the course of actions in the performance of the play, and even allow the audience to participate in the performance as actors. Thus in these plays the audience has an active role. However, this does not mean that the readers of a novel or the members of audience in the theater are passive: they are quite active, but their activity remains internal.

The written text of the play is much less than the event of the play. It contains only the dialog (the words that the characters actually say), and some stage directions (the actions performed by the characters). The play as written by the playwright is merely a scenario which guides the director and actors. The phenomenon of theater is experienced in real-time. It is alive and ephemeral – unlike reading a play, experiencing a play in action is of the moment – here today, and gone tomorrow.

To prepare for the formalization in the next section, we fix some terms. The word *performance* is used to refer to the artifact the audience and the participants experience during the course of performing a *script* by preferred *actors*, monitored and instructed by a *director*. The script is the underlying content representation perceived by the authors as a composite unit, defining the temporal aspects of the performance, and containing the *actions* which are depicted by the content elements or the references to these elements. Traditional multimedia systems use a different set of terms

Figure 3. 19th century drawing of Beijing opera

which are comparable to the terms above; they are in many cases similar, but should not be confused.

In the next section we review the language elements of SMIL (Ayars et al., 2005), later taking them as a starting point for the design of our IPML, preserving the good ingredients and developing extensions that are necessary.

SMIL

Synchronized Multimedia Integration Language (SMIL) is an XML-based language for writing interactive multimedia presentations (Ayars et al., 2005). It has easy to use timing modules for synchronizing many different media types in a presentation. SMIL 2.0 has a set of markup modules.

Without attempting to list all the elements in these modules, we show an object-oriented view of some basic elements in Figure 4: *Par*, and *Seq* from the timing and synchronization module, Layout, *RootLayout*, *TopLayout* and *Region* from the layout module, *Area* from the linking module, *MediaObject* from the media object module, *Meta* from the meta information modules and *Head*, *Body* from the structure module. Details about the corresponding language elements can be found in the SMIL 2.0 specification (Ayars et al., 2005).

The *Region* element provides the basics for screen placement of visual media objects. The specific region element that refers to the whole presentation is the *RootLayout*.

Common attributes, methods and relations for these two elements are placed in the super-class named the *Layout*.

SMIL 2.0 introduces a *MultiWindowLayout* module over SMIL 1.0, with which the top level presentation region can also be declared with the *TopLayout* element in a manner similar to the SMIL 1.0 root-layout window, except that multiple instances of the *TopLayout* element may occur within a single *Layout* element.

Each presentation can have *Head* and *Body* elements. In the *Head* element one can describe common data for the presentation as whole, such as *Meta* data and *Layout*. All *Region* elements are connected to the Head.

Figure 4. SMIL in UML

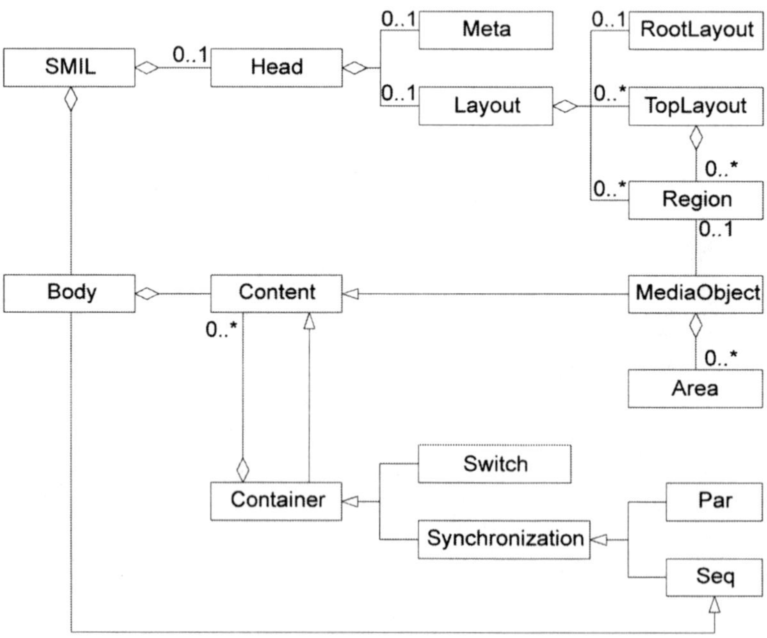

The *MediaObject* is the basic building block of a presentation. It can have its own intrinsic duration, for example if it is a video clip or an audio fragment. The media element needs not refer to a complete video file, but may be a part of it. The *Content*, *Container*, and *Synchronization* elements are classes introduced solely for a more detailed explanation of the semantics of the *Par*, *Seq*, *Switch* and *MediaObject*, and their mutual relations.

Par and *Seq* are synchronization elements for grouping more than one *Content* element. If the synchronization container is Par, it means that direct sub-elements can be presented simultaneously. If the synchronization container is *Seq*, it means that direct sub-elements can be presented only in sequence, one at a time. The Body element is also a *Seq* container.

The connection between *Content* and *Container* viewed as an aggregation has a different meaning for the *Synchronization* element and for the *Switch* element. If the *Container* element is *Switch*, which means that only one sub-element from a set of alternative elements should be chosen at the presentation time depending on the settings of the player.

With the *Area* element, a spatial portion of a visual object can be selected to trigger the appearance of the link's destination. The *Area* element also provides for linking from non-spatial portions of the media object's display. It allows breaking up an object into temporal subparts, using attributes *begin* and *end*.

IPML

SMIL seems to have the ingredients for mapping and timing:

- Its timing and synchronization module provides versatile means to describe the time dependencies, which can be directly used in the IPML design without any change.
- The SMIL linking module enables non-linear structures by linking to another part in the same script or to another script. Although the *Area* element can only be attached to visual objects, this limitation can be easily solved by lifting the concept up to a level that covers all elements that need to have a linking mechanism.
- The SMIL layout module seems to be very close to the need of distribution and mapping. The concept of separating mapping and timing issues into two different parts, i.e. *Head* and *Body*, makes SMIL very flexible for different layouts – if a presentation has to be presented to a different layout setting, only the layout part must be adapted and the timing relations remain intact, no matter whether this change happens before the presentation in authoring time, or during the presentation in run time.

Upon first investigation, SMIL appears not directly applicable for the distributed and interactive storytelling: it does not support a notion of multiple devices. However later we found that we went one step too far – the StoryML does incorporate the concept of multiple actors, but it is its linear timing model and narrative structure which limit its applicability.

What needs to be done is to pick up SMIL again as the basis for the design, extending it with the metaphor of theater play, and bringing in the lessons we learnt from StoryML. Figure 5 shows the final IPML extension (marked gray) to SMIL. The Document Type Definition (DTD) of IPML can be found in (Hu, 2006).

Note that in Figure 5, if all gray extensions are removed, the remaining structure is exactly the same as the SMIL structure as illustrated in figure 5. This is an intentional design decision: IPML is designed as an extension of SMIL without overriding any original SMIL components and features, so that the compatibility is maximized. Any SMIL script should be able to be

Figure 5. IPML in UML

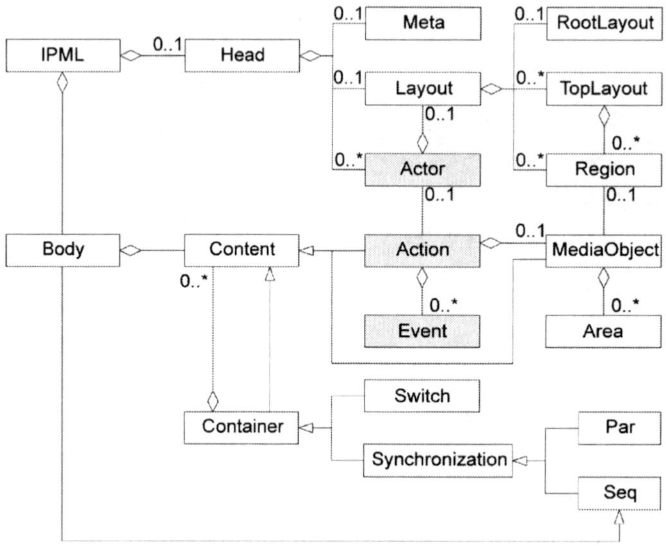

presented by an IPML player without any change. An IPML script can also be presented by a SMIL player, although the extended elements will be silently ignored. The compatibility is important, because it can reduce the cost of designing and implementing a new IPML player – the industry may pick up the IPML design and build an IPML player on top of a existing SMIL player so that most of the technologies and implementations in the SMIL player can be reused.

Actor

The *Head* part of an IPML script may contain multiple *Actor* elements which describe the preferred cast of actors. Each *Actor* has a *type* attribute which defines the requirements of what this actor should be able to perform. The *type* attribute has a value of URI, which points to the definition of the actor type. Such a definition can be specified using for example RDF (McBride, 2004) and its extension OWL. RDF is a language for representing information about resources in the World Wide Web. It is particularly intended for representing metadata about Web resources. However, by generalizing the concept of a "Web resource", RDF can also be used to represent information about things that can be identified on the Web, even when they cannot be directly retrieved on the Web. OWL adds more vocabulary for describing properties and classes: among others, relations between classes (e.g. disjointness), cardinality (e.g. "exactly one"), equality, richer typing of properties and characteristics of properties (e.g. symmetry), and enumerated classes. The "thing" we need to describe is the type of the actor.

During the performance time, the real actors present to the theater to form a real cast. Each actor then needs to report to the director about what he can perform, i.e. his actor "type". The "type" of a real actor is defined by the actor manufacturers (well, if an actor can be manufactured). The real actor's type can again be described using an RDF or OWL specification. The director then needs to find out which real actor fits the preferred type best. The mapping game becomes a task of reasoning about these two RDF or OWL described "types". First of all the user's preferences should be considered, even if the user prefers a "naughty boy" to perform a "gentleman". Otherwise, a reasoning process should be conducted by the director, to see whether there is exactly an actor

Figure 6. Mapping Alice and Duchess to large screen and Frog to robot

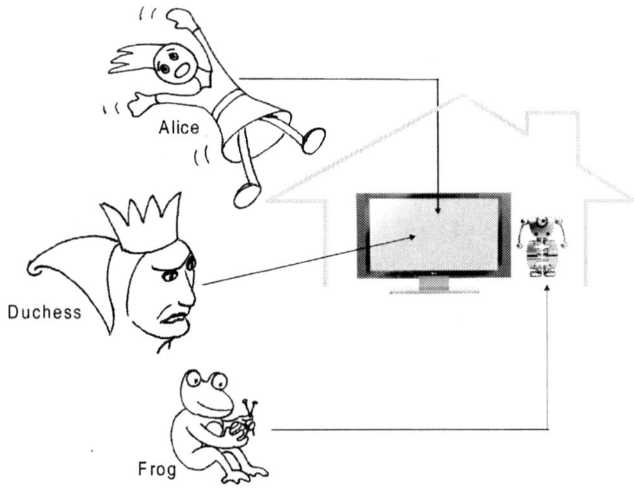

has a type that "equals to" the "gentleman", or to find an "English man" that indeed always "is a" "gentleman", or at least to find a "polite man" that "can be" a "gentleman" and that matches "better than" a "naughty boy", etc. This reasoning process can be supported by a variety of Semantic Web (Berners-Lee & Fischetti, 1999) tools, such as Closed World Machine (CWM) (Berners-Lee, Hawke, & Connolly, 2004) and Jena (McBride, 2001) just for example.

Although *Alice in Wonderland* would be a difficult play to map, we can use it to illustrate some ideas again. For example, Alice could be played on several devices. But the Duchess is supposed to appear impressive and with dominance, so a close-up on a large screen serves that purpose best. The Frog preferably is active by playing with physical objects, so a robotic device would be best. Moreover, when played by a robot, it can jump and run around. One possible mapping, taking these simple constraints into account, is shown in Figure 6.

Action

The *Action* element is similar to the *MediaObject* element in SMIL. However, Action can be applied to any type of content element which is not explicitly defined using different media objects such as *Img*, *Video* and *Animation* in SMIL. The *Action* element has an attribute *src* giving the URI of the content element and its type either implicitly defined by the file name extension in the URI if there is one, or explicitly defined in another attribute *type*. The *type* attribute defines the type of a content element as the *type* attribute of *Actor* defines the actor type, using a URI referring to a definition.

Action may have an attribute *actor* to specify the preferred actor to perform it. If it is not specified, the type of the content element may also influence the actor mapping process: the director needs to decide which actor is the best candidate to perform this "type" of action. Again, the user preference should be taken into account first; otherwise a reasoning process should be conducted to find the "gentleman" who can nicely "open the door for the ladies".

In addition, the Action element may have an *observe* attribute which specifies the interested events. This attribute is designed for an actor to observe the events that are of interest during the course of performing a specific action. For example, when an actor is performing an action to present a 3D object, it may be interested in the controlling events for rotating the object. This actor can then "observe" these events and react on it. Note that these observed events have no influence on the timing behavior. It will not start nor stop presenting this 3D object unless they are included in timing attributes, i.e., *begin* and *end*. Events that are not listed will not be passed by the director to the actor during this action, thus the event propagation overhead can be reduced.

However, some actors may be interested in the events that are not related to certain actions. To accommodate this and not to change the original SMIL structure, we require these Actors to perform an action of the type *null*, specified using a special URI scheme "*null:*", which allows events to be "observed" during an action of "doing nothing".

Event

The third extension of IPML to SMIL is event based linking using the *Event* element. *Event* elements in an *Action* element are similar to *Area* elements in a visual *MediaObject* element in SMIL, with the exceptions that it does not require the parent *Action* element to have a visual content to present, and that the events are not limited to the activation events (clicking on an image, for example) on visual objects. An *Event* has an attribute *enable* to include all interested events during an action, including all possible timing events and user interaction events. Once one of the specified events happens, the linking target specified using the attribute *href* is triggered. Similar to the *Area* element, the *Event* element may also have *begin*, *end* and *dur* attributes to activate the *Event* only during a specified interval. Event based linking makes IPML very flexible and powerful in constructing non-linear narratives, especially for the situations where the user interaction decides the narrative directions during the performance.

Again with *Alice in Wonderland*

To show what an IPML script would look like in practice, we again use the example from Alice in Wonderland. Since we can't embed multimedia content elements in this printed paper and we only have printed lines and action instructions, we introduce two exotic URI schemes: "*say*:" for the lines and "*do*:" for the action instructions, just for the fun of it.

Now that we have a scripting language that can be used for describing a distributed presentation, a playback system is needed to turn a written "play" to a performance. Next the structure of such a playback system is presented.

ACTORS: DISTRIBUTED PAC AGENTS

The actors are in the system not only to perform their actions to present the multimedia objects, but also to provide the interface for the users to interact with the system. Many interactive architecture structures have been developed along the lines of the object-oriented and the event driven paradigms. Model-View-Controller (MVC) (Krasner & Pope, 1988) and Presentation-Abstraction-Control (PAC) (Coutaz, 1987) are the most popular and often used ones (Buschmann, Meunier, Rohnert, Sommerlad, & Stal, 1996).

The MVC model divides an interactive agent into three components: model, view and controller, which respectively denotes processing, output and input. The model component encapsulates core data and functionality. View components display information to the user. A View obtains the data from the model. There can be multiple views, each of which has an associated controller component. Controllers receive input, usually as events that encode hardware signals from input devices.

Coutaz (1987) proposed a structure called Presentation-Abstraction-Control, which maps roughly to the notions of View-Controller pair,

Box 1.

```
<ipml>
<head>
    <actor id="ALICE"   type="http://alice.wonderland.eu/lovelygirl" />
    <actor id="DUCHESS" type="http://alice.wonderland.eu/seriouswoman" />
    <actor id="COOK"    type="http://alice.wonderland.eu/cook" />
    <actor id="FROG"    type="http://alice.wonderland.eu/frog" />
    <actor id="HOUSE"   type="http://alice.wonderland.eu/woodenhouse" />
</head>
<body>
    <!----->
    <action id="As_1" actor="ALICE" src="say:Please! Mind what you're doing!"
/>
    <par>
        <action id="Dd_2" actor="DUCHESS" src="do:tossing Alice the baby" />
        <action id="Ds_2" actor="DUCHESS" src="say:Here...you may nurse it if
you like, I've got to get ready to play croquet with the Queen in the garden."
/>
        <action id="Ad_2" actor="ALICE"   src="do:receiving the baby"
begin="Dd_2.babytossed"/>
    </par>
    <action id="Dd_3" actor="DUCHESS" src="do:turns at the door" />
    <action id="Ds_3" actor="DUCHESS" src="say:Bring in the soup." />
    <!----->
    <par>
        <action actor="HOUSE" src="do:moving" />
        <seq>
            <par>
                <action actor="DUCHESS" src="say:The house will be going any
minute!" />
                <action actor="COOK"    src="do:snatches up her pot and dashes
into the house" />
            </par>
            <action actor="COOK" src="do:turns to the FROG" />
            <action actor="COOK" src="say:Tidy up, and catch us!"
            <par>
                <action actor="FROG"  src="do:leaps about" />
                <action actor="FROG"  src="do:picking up the vegetables,
plates, etc." />
                <action actor="ALICE" src="say:She said 'in the garden', will
you please tell me -" />
            </par>
```

continued on following page

Box 1. Continued

```
            <action actor="FROG"  src="say:There's no sort of reason asking
me, I'm not in the mood to talk about gardens." />
            <action actor="ALICE" src="say:I must ask some one. What sort of
people live around here?" />
        </seq>
    </par>
</body>
</ipml>
```

Model, and the Mediator pattern (Gamma, Helm, Johnson, & Vlissides, 1995). It is referenced and organized in a pattern form by Buschmann et al. (1996): the PAC pattern "defines a structure for interactive software systems in the form of a hierarchy of cooperating agents. Every agent is responsible for a specific aspect of the application's functionality and consists of three components: presentation, abstraction, and control. This subdivision separates the human-computer interaction of the agent from its functional core and its communication with other agents."

In the design of the IPML player, PAC is selected as the overall system architecture, and the actors are implemented as PAC agents that are managed by the scheduling and mapping agents in a PAC hierarchy, connected with the channels, and performing the actions. Hu (2006) argues in detail why PAC is preferred to MVC for the IPML player.

Distributed PAC

This structure separates the user interface from the application logic and data with both top-down and bottom-up approaches (Figure 7). The entire system is regarded as a top-level agent and it is first decomposed into three components: an Abstraction component that defines the system function core and maintains the system data repository, a Presentation component that presents the system level interface to the user and accepts the user input, and in between, a Control component that mediates the abstract component and the presentation component. All the communications among them have to be done through the control components.

At the bottom-level of a PAC architecture are the smallest self-contained units which the user can interact with and perform operations on. Such a unit maintains its local states with its own Abstraction component, and presents its own state and certain aspects of the system state with a

Figure 7. Distributed PAC in a hierarchy

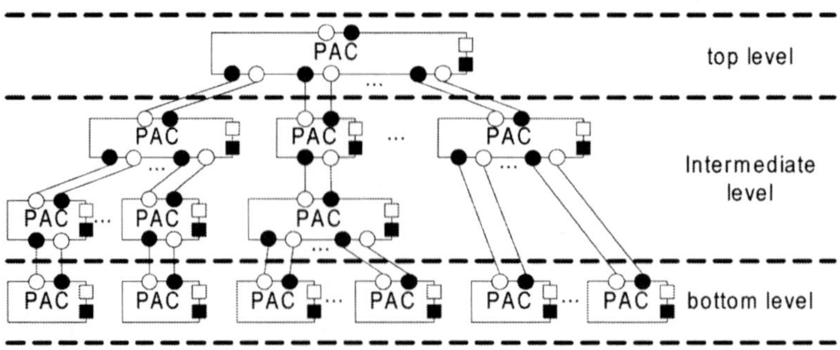

Presentation component. The communication between the presentation and the abstraction components is again through a dedicated Control component.

Between the top-level and bottom level agents are intermediate-level agents. These agents combine or coordinate lower level agents, for example, arranging them into a certain layout, or synchronizing their presentations if they are about the same data. The intermediate-level may also have its interface Presentation to allow the user to operate the combination and coordination, and have an Abstraction component to maintain the state of these operations. Again, with the same structure, there is a control component in between to mediate the presentation and the abstraction.

The entire system is then built up as a PAC hierarchy: the higher-level agents coordinate the lower level agents through their Control components; the lower level agents provide input and get the state information and data from the higher level agents again through the Control components. This approach is believed more suitable for distributed applications and has better stability than MVC, and it has been used in many distributed systems and applications, such as CSCW (Calvary, Coutaz, & Nigay, 1997), distributed real-time systems (Niemelä & Marjeta, 1998), web-based applications (Illmann, Weber, Martens, & Seitz, 2000; Zhao & Kearney, 2003), mobile robotics (Khamis, Rivero, Rodriguez, & Salichs, 2003; Khamis, Rodriguez, & Salichs, 2003), distributed co-design platforms (Fougeres, 2004) and wireless services (Niemelä, Kalaoja, & Lago, 2005).

To a large degree the PAC agents are self-contained. The user interface component (Presentation), the processing logic (Abstraction) and the component for communication and mediation (Control) are tightly coupled together, acting as one. Separations of these components are possible, but these distributed components would then be regarded as PAC agents completed with minimum implementation of the missing parts. Thus the distribution boundaries remain only among the PAC agents instead of composing components.

Based on this observation, each component is formally described by Hu (2006), modeling the communication among PAC agents with push style channels from the Channel pattern. In Figure 1, a "●" indicates a data supplier component; a "○" indicates a data consumer component and a connecting line in between indicate the channel. The symbol "■" indicates that the attaching component has a function of physically presenting data to the user, and the symbol "□" indicates the function of capturing input from the user interface or the environment.

Actor: A PAC Agent

After this detailed comparison with MVC we are ready to harvest the fruits of the PAC style: the PAC agents are perfectly suited to implement the notion of "play", that is, the central notion in IPML scripts.

An actor is basically a PAC agent. It reacts on the user input events and scheduling commands, and takes actions to present media objects. Hu (2006) describes an example implementation of an actor based on the Distributed PAC pattern and other patterns including Synchronizable Object, Channel and Action described therein.

IPML Player: An IPML Actor

The final IPML system is simply an IPML actor, or in other words, an Actor implementation that is capable of presenting the IPML scripts. IPML is a presentation description language that extends SMIL, describing the temporal and spatial relations among distributed actions on synchronizable content elements.

Note that IPML is an extension of SMIL, and a SMIL document is actually a composite content element by itself. Hence an IPML actor is first of all a SMIL player and it may present the

Figure 8. IPML system: an IPML actor

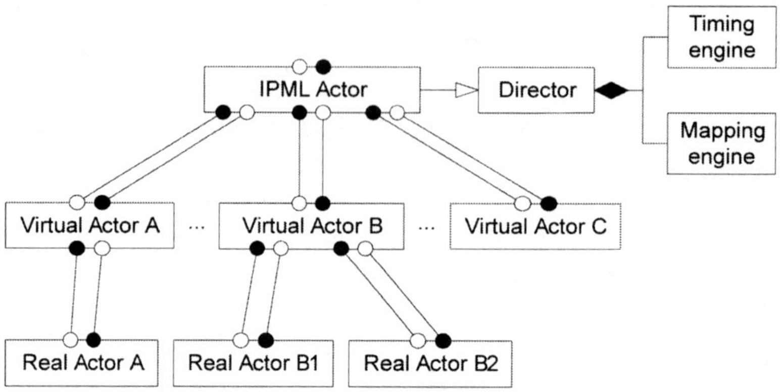

contained content elements to its own Presentation component. What makes IPML superior to a SMIL player is the capability of distributed presentation, interaction and synchronization: It can delegate content presentations to other actors, synchronize the presentation actions of these actors, and propagate distributed user interaction events among these actors.

The IPML actor implements the role of a Director, which has a mapping engine, creates, manages and connects the virtual actors, and has a timing engine which schedules the timed actions for the delegating virtual actors (Figure 8). Depending on the physical configuration of the "theater" – the presenting environment, the mapping engine may also connect appropriate "real actors" to virtual actors, where the virtual actors keep the role of software drivers for the "real actors". The mapping engine may make use of distributed lookup and registration services such as UPNP (Michael Jeronimo, 2003) and JINI (Edwards, 2000) to locate and maintain a list of "real actors", but this architecture leaves these possibilities open to the implementation of the mapping engine.

The timing and mapping engines are essential parts of the system. In the flowing two sections they are presented in more detail.

ACTION SYNCHRONIZATION ENGINE

An interactive play has been defined as a cooperative activity of multiple actors that take actions during certain periods of time to present content elements. An action, as the basic component of such an activity, has a time aspect per se. That is, a timing mechanism is needed to decide when the actor should commence the action, how long the action should take, and how the actions are related to each other in time.

IPML has been presented as the scripting language, in which the SMIL timing model is used for describing the time relations between actions. A runtime synchronization engine is presented here. It provides a powerful, flexible and extensible framework for synchronizing the actions. This engine is used by the director to schedule the actions for the actors, no matter whether the actors are distributed over the network.

ASE Model

ASE is a runtime Action Synchronization Engine that takes the timing and synchronization relations defined in an IPML script as input, and creates an object-oriented representation based on an extended version of the Object Composition Petri Net (OCPN) (Little & Ghafoor, 1990).

Figure 9.

$G_{ASE} = (PL, TR, AR, MA, DU, RE, PP, TC, CT)$,

where,

$PL = \{pl_1, pl_2, \cdots, pl_n\}$, where $n \geq 0$	[*places*]
$TR = \{tr_1, tr_2, \cdots, tr_m\}$, where $m \geq 0 \wedge PL \cap TR = \varnothing$	[*transitions*]
$AR : I \cup O$, where [	directed arcs]
$I = PL \leftrightarrow TR$	[*input* arcs]
$O = TR \leftrightarrow PL$	[*output* arcs]
$MA : PL \rightarrow \mathbb{N}$	[*tokens* mapping *places* to natural numbers]
$DU : PL \rightarrow \{null\} \cup \mathbb{R}$	[*durations*]
$RE : PL \rightarrow \{null, re_1, re_2 \cdots, re_k\}$ where $k \geq 0$	[*resources*]
PP: PL	[*priority places* a subset of PL
$TC = \{tc_1, tc_2 \cdots, tc_q\}$ where $q > 0$	[runtime *transition controllers*]
CT: $(TR \times TC \times TR)$	[*TC* controlled transition pairs]

An ASE model is a nine tuple that extends OCPN, see Figure 9.

The behavior of the Petri net is governed by a set of firing rules that allows the tokens to move from one place to another. The inclusion of a null value in the ranges of the functions *DU* and *RE* means that there are places without a pre-determined duration, and there are places not related to any content resources.

The ASE model distinguishes priority places from other places. Special firing rules will be used for these priority places to implement the IPML *endsync* semantics and to cope with nondeterministic durations and interaction events. A priority place is drawn as a circle in an ASE graph like other places, but using a special (thicker) circle to emphasize its priority.

The added transition controllers *TC* make it possible to change the structure between two transitions in run time. It may fire another linked transition instead of the current enabled transition, which can be used to repeat or skip the structure between the controlled transition pairs. The controller may use a counter to control the number of repeat iterations, and may add and remove timer places in the structure to control the total duration for repeat. This mechanism is useful when dealing with IPML *restart*, *repeatCount* and *repeatDur* semantics. In an ASE graph, a box represents a transition controller, and dashed lines connect the controlled transition pairs (Figure 10).

As already mentioned, there are places that do not have a pre-determined duration. The actual duration of these places can only be determined

Figure 10. Transition controller

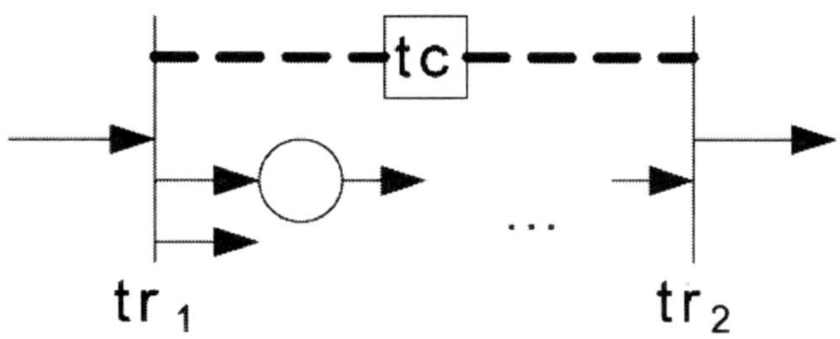

after the actions at these places have been carried out. These places are called *nondeterministic places*:

$NP = \{pl: PL|DU(pl) = null\}$.

Non-deterministic places in an ASE graph are circles marked with a question mark.

Some nondeterministic places are not related to any content resources. These places are used in an ASE model to represent the actions that need to be taken by the engine itself to check certain conditions, to detect user interaction events, or simply to block the process etc. These places are called *auxiliary places*:

$AP = \{pl: NP|RE(pl) = null\}$.

There are also places that do have duration, but do not have a content element attached to it. These places are used by the ASE model to include an arbitrary interval to construct temporal relations. These places are called *timer places*:

$TP = \{pl: PL|DU\{pl\} \neq null \wedge RE(PL) = null\}$.

Timer places are indicated with a clock icon with the hands pointing to 9:00am. To construct an ASE graph from an IPML script structurally, it is sometimes necessary to connect two transitions. Since an arc can only be the link between a transition and a place, a zero-duration timer place can be inserted to maintain the consistency. These zero-duration timer places are called *connecting places*, indicated with a clock icon with its hands pointing to 0:00pm, and marked with an anchor link.

Table 1 shows the graph representations of the different ASE places and their priority versions. To show how an ASE would look like, the example of Alice in Wonderland is again used. The script fragment between two "<!---->" lines in the example given in the section "Again with *Alice in Wonderland*" are converted to a temporal structure as shown in Figure 11. The connecting places are not visible in this structure, but they are essential in the process of converting an IPML script to an ASE model. Every temporal element in an IPML is firstly formally mapped to an ASE model that utilizes connecting places. The purpose is to keep it always possible to embed sub-models into this model, which corresponds to the hierarchical temporal structure of the IPML elements. After the entire IPML is converted, the connecting places in the model are removed whenever it is possible to simplify the final ASE model.

The firing rules of ASE and the formal process of translating an IPML script into a simplified ASE model are described by Hu (2006). More comprehensive examples are also given therein.

Table 1. Places in ASE

Place	Normal	Priority
Normal	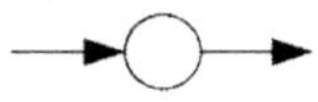	
Nondeterministic	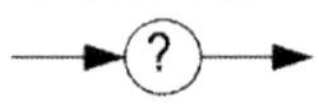	
Timer	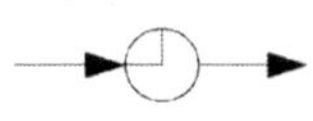	
Connecting	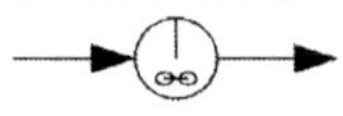	

Figure 11. An example of the ASE model

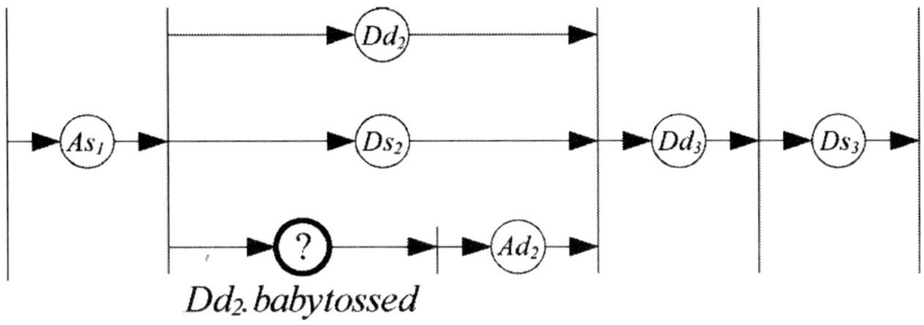

Object-Oriented Implementation of ASE

The ASE in the IPML system is implemented in an object-oriented manner (Figure 12): Places and transitions are objects with input and output references that realize the arcs; transition enabling and firing are simply event-driven invocations. The Observer pattern can be used to implement the structure, where the transitions observe the token states of the connected places. Transition controllers are also objects with references to and from two related transitions. If the different types of the places are omitted from Figure 12, the remaining static structure is rather simple. The dynamic behavior of these objects is driven by the firing rules and the implementation of the dynamic behavior is straightforward. The remaining design problem now is how to convert an IPML timing structure to an ASE model.

Get Ready Just-in-Time

For an action to be immediately taken at the scheduled time, actors need to get ready prior to that time. For media objects, enough data needs to be prefetched; For robotic behaviors the mechanical system needs to be at a ready position for the next move. Two extreme strategies could be adopted by the director. First, the director informs all actors to get prepared for all possible actions before the performance is started; second, the director never requests the actor to get ready before any action. The first strategy guarantees the

Figure 12. Object-oriented implementation of ASE

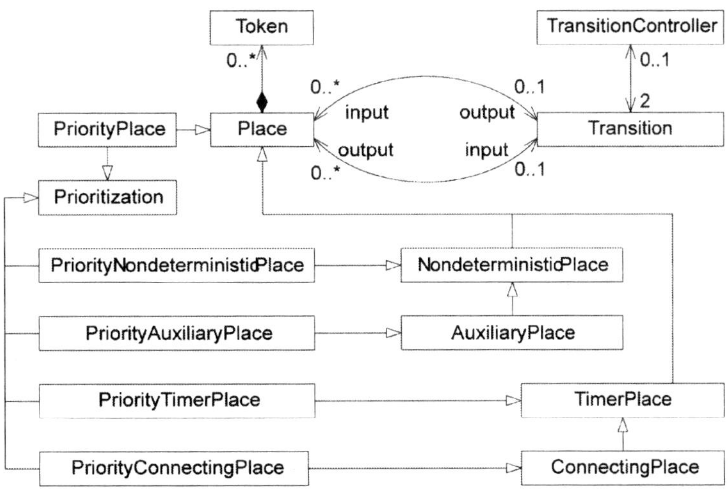

smooth transitions between actions, and manages nondeterministic timing and user interaction well. However the cost is also obvious: it may result in a long initial delay and for media objects, and every actor needs a large buffer for prefetching all media objects in advance. The second strategy minimizes the initial start delay and the buffer requirement, but every transition between two actions will take time because the next action only starts to be prepared after the previous one stops. Hence smooth transitions between action places are not possible, unless the actions do not need to be prepared, which is rare in multimedia presentations. The nondeterministic user interactions make the situation even worse – Users may experience a long delay between their input actions and the system reactions. A different approach is needed for the IPML system.

Just-in-Time Approach

The director in the IPML system uses a "*just-in-time*" approach, in which the action preparation process is required to be completed just before the action time. With this strategy, the director informs the actor to prepare an action before the action time with the necessary preparation time taken into account. This strategy therefore requires less use of data buffers and facilitates more efficient use of network bandwidth.

In an ideal situation, i.e. the action time of all actions can be determined in advance, the start-preparation time for each action, that is, when an actor should start preparing an action, can be calculated based on its playback time, its QOS request, and the estimation of the available network bandwidth.

However for an IPML performance, the accurate action time often can not be determined before the performance takes place, because of the nondeterministic action durations and user interaction events. The best an ASE can do is to predict the earliest action time for each action as if the non-mechanistic events would happen at the earliest possible moments. This can be done before the performance starts, as long as the ASE model has been established. During the performance, a dynamic error compensation mechanism can be used to adjust the estimate of the action time for each action and the start-reparation time as well.

Action Time Prediction

Once an ASE is converted and simplified from a given IPML, the director estimates the earliest possible action time for each action in the ASE by first assigning the duration of all nondeterministic places to zero and then traversing the ASE. The action time of an action is the firing time of its starting transition. There are two possible cases for a transition in the ASE: 1. it has no priority input place, or 2. it has at least one priority input place. For case 1, the firing time of the transition is the firing time of the preceding transition plus the maximum duration of the input places. For case 2, the firing time of the transition is instead the preceding transition plus the smaller one between the minimum duration of the priority places, and the maximum duration of the non priority places. Note that for time independent actions, such as presenting images and text, if the duration is not explicitly given, it is considered nondeterministic and its duration is considered as zero for prediction. For time-dependent actions, like presenting audio and video media objects, if the duration is not defined explicitly, the duration of the place is the implicit duration of the object if it can be determined from the server in advance.

During this traversal process of predicting the earliest possible firing time of each transition, it is also necessary to deal with the transition controllers to get more accurate values. The restart controller deals with events that could restart an element during the active duration of the element, so the earliest case for its ending transition would be that there is no restart at all. Thus, the restart controller is ignored during prediction. The repeat controller deals with the *repeatDur* attribute as well as the *repeatCount* attribute. The *repeatDur* attribute sets the duration of repeating an element, so the firing time of the ending transition should be the end of the repeat duration. The *repeatCount* attribute specifies the number of times to repeat, thus the firing time of the ending transition is extended as many times as specified. If any of them is set to be "*indefinite*", the duration is considered nondeterministic hence a value of zero.

Dynamic Adjustment

Obviously, the actual action time of every action will not be earlier than the prediction made prior to the performance. The differences between the actual action time of the actions that have already been taken and their predicted earliest times can be used to adjust the predicted action time of the actions that have not yet been performed. The predicted action time can then be updated for those yet to happen. The updated prediction of the action time can then be used to update the start-preparation time. Note that start-preparation time should not be updated if the preparation request has already been sent to the actor, since the actor may have already started preparation and an ongoing preparation process should not be interrupted. Nevertheless an ASE action time prediction with this dynamic adjustment mechanism does make the predicted action time of later actions closer to the actual action time, hence seems more intelligent than without.

Distributed Time

Since they inhabit on different hardware platforms, actors may have time systems that are different from the directors. In order to get everything synchronized, the actors must use the director's time, or at least agree on the time difference. A simple approach to get actors have the same time as the director's, is to use clock synchronization mechanisms to synchronize the clocks of the underlying platforms.

Clock Synchronization

An actor may perceive data skews due to asynchrony of its local clock with respect to the clock of the director, which may arise due to network delays and/or drifts in the clocks. In the absence of synchronized clocks, the time interval of an actor may have drifted to a value bigger or smaller than that of the director.

Clocks can be synchronized using an asynchronous protocol between the transport level entities in the presence of network delays compounded by clock drifts. Most clock synchronization protocols require the entities to asynchronously exchange their local clock values through the network and agree on a common clock value. These protocols use knowledge of the network delays in reaching agreement. For instance, the NTP requires the entities to receive their clock values from a central time server that maintains a highly stable and accurate clock and to correct the received clock values by offsetting the network delays. For clock synchronization protocols to function correctly, it is desirable that the network delay is deterministic, i.e., the degree of randomness in the delay is small and the average delay does not change significantly during execution of synchronization protocol. Accordingly, the transport protocol may create a deterministic channel with high loss and delay sensitivity to exchange clock control information. Clock synchronization is a complex topic of its own, and details of such protocols are outside the scope of this paper.

Software Clocks

The actors are not the only ones who inhabit a hardware platform. There may be other hardware or software components sharing the same platform clock. Applying clock synchronization mechanisms to the shared clock may result in unexpected consequences on the components that are not under the supervision of the play director but have other time critical tasks of their own.

To avoid this side effect, the IPML system requires every actor to implement a software clock. The actor's clock is then synchronized with the director's clock using NTP according to the director's time on a regular interval basis. During the update interval, the actor's clock ticks ahead according to the local platform time.

"Action!" Delayed

The director issues action scheduling commands over the network to the actors. "Action!" the director yells and expects the actor immediately starts the action. In real performance, these directing commands travel at the speed of sound and will be heard by the actor almost "immediately". However in the IPML system, these commands are not the only data traveling through the network. A command may need to be cut into pieces, packaged and queued at the director's side waiting for the network service to move it over. Once the packages arrive at the actor's side, they are again put in the queue for the network service to retrieve them. Once retrieved, depending on the network protocol, the data packages might need to be verified and confirmed before the command is reassembled and handed over to the actor. All these take time. Depending on the protocol and the bandwidth, it varies from few milliseconds to hundreds of milliseconds, or even more. The command will eventually be heard late. Since a particular network protocol is not assumed for transporting the commands, it is necessary to handle the delay at the architecture level.

Several strategies are adopted in the IPML system. First of all, all scheduling commands from the director are tagged with a time stamp that indicates when exactly the command is issued. Upon receipt, the actor retrieves the time stamp and compares it to its local software clock. Since the actor's clock is synchronized with the director's clock, the traveling time of the command can be calculated. If the command is not to start an action to present time-dependent content, the traveling time of the command is ignored. Otherwise if the traveling time is bigger than a QOS threshold, the actor will skip a fragment of time-dependent content that should have been performed right after the command was issued and before the command is received. Thus the distributed content elements can always be synchronized over the network, at the price of a small portion of the content being dropped at the beginning. If the network has enough bandwidth, the dropped content is hardly noticeable by the user.

IPML MAPPING

As learnt from the formal study on the mapping issues (Feijs & Hu, 2004), a mapping process is a set of controlling commands, to be sent through control channels to the components that are capable of copying and combining streams from input channels to output channels. All the actors in the IPML system are all connected through channels in a PAC hierarchy and controlling commands can be sent though these connecting channels. How the mapping can be handled in a dynamic setting is briefly described next.

Virtual Actors

Virtual actors are required in the IPML system architecture (Figure 8) as an essential layer of software PAC agents for dynamic mapping. These agents can be provided by the vendors of the real actors as a software driver, or by the content producers as a "recommended" actor if there is no real actor available. These virtual actors can be provided by an installation package which requires the user to install it in advance, or for example an Internet resource identified by a URL such that the virtual actor can be downloaded and installed automatically. Here one shall not try to cover the security and privacy consequences of this automatic downloading and installation process, since it has been an issue for all Internet applications and should be taken care of by dedicated protocols and subsystems.

Once the virtual actors are available to the IPML system, it is then registered and maintained by the mapping engine of the director.

Channel Resources

The system also provides and maintains a distributed channel service over the connected devices. Here it benefits from the design of the channel patterns (Hu, 2006): all channels between actors and the director are distributed objects managed by a channel service, hence the network resources can be easily monitored and allocated with QOS and load balancing taken into account. The director may query the channel service so that the communication conditions can be taken into account during the mapping process.

Mapping Heuristics

The IPML director has a list of available virtual actors together with their types given. The director also has access to the channel service to query the channel resources to find out whether a virtual actor is connected to a real actor. Given an actor type as the requirement, the IPML uses the following heuristics to map the required actor type to a virtual actor:

1. The user preference has top priority.
2. If after 1 multiple virtual actors can be selected, the ones having the "closest" type have the priority over the others.
3. If after 2 multiple virtual actors can be selected, the ones with a real actor connection have priority over those without.
4. If after 3 multiple virtual actors can be selected, the one that has been selected most recently for this type is again selected. If none of them have ever been selected, the director randomly selects one from these virtual actors.
5. If none of the virtual actors can be selected, the director creates a "dummy" virtual actor for this type. The "dummy" virtual actor will do nothing but ignore all requests.

In step 2, how to decide an actor is the "closest" to another among the others is not clearly described. It depends on how the types are defined. In practice, one may leave it to an ontology reasoning system for example a semantic web tool for RDF or OWL type descriptions.

During the action time, these heuristic conditions may change, for example, the real actors may connect and disconnect from the "theater" at any time, and users may change their minds at any time to have a "gentleman" instead of a "naughty boy" to be the actor or vice versa. To dynamically update the mapping relations, the director needs to repeat this mapping process on a regular interval basis.

Actor/Director Discovery

The problem now is how the virtual actors, the real actors and the director can find each other for registration and connection. This is actually a well-known device/service discovery problem and many middleware standards (for example JINI, HAVI, OSGi and UPnP) have a solution for it. So one may simply leave the discovery task of registering virtual actors to the director, and leave the task of connecting virtual actors and the real actors to these middleware infrastructures.

CONCLUSION

On top of existing network technologies and platform architectures, a generic architecture has been designed to enable playing IPML in a networked environment with user preference and dynamic configurations taken into account. The architecture has been implemented and tested in Java, and several demonstrators have been built upon this architecture (Feijs & Hu, 2004; Hu & Bartneck, 2005; Hu & Feijs, 2003; Hu, Janse, & Kong, 2005; Janse, van der Stok, & Hu, 2005). It has been applied in various projects, from big projects funded by the Information Society Technologies program of the European Commission (NexTV, IST-1999-11288; ICE-CREAM, IST-2000-28298), to small educational projects at the Department of Industrial design, Eindhoven University of Technology. The users of the architecture range from the professionals inside Philips Research, to undergraduate industrial design students.

In this design, the metaphor of play was an essential design decision. The scripting language and the architecture of the playback system are designed around this metaphor. The concept of mapping and timing are well covered in the architecture and proven to be easily understandable by both the system designers and the scriptwriters. In this era of digitalization, this might be yet another example that we still have much to learn from the traditional arts such as play.

REFERENCES

Aarts, E. (2004). Ambient intelligence: a multimedia perspective. *IEEE MultiMedia*, *11*(1), 12–19. doi:10.1109/MMUL.2004.1261101

Ayars, J., Bulterman, D., Cohen, A., Day, K., Hodge, E., Hoschka, P., et al. (2005). *Synchronized Multimedia Integration Language (SMIL 2.0) - [Second Edition]* (W3C Recommendation).

Battista, S., Casalino, F., & Lande, C. (1999). MPEG-4: A Multimedia Standard for the Third Millennium, Part 1. *IEEE MultiMedia*, *6*(4), 74–83. doi:10.1109/93.809236

Battista, S., Casalino, F., & Lande, C. (2000). MPEG-4: A Multimedia Standard for the Third Millennium, Part 2. *IEEE MultiMedia*, *7*(1), 76–84. doi:10.1109/93.839314

Berners-Lee, T., & Fischetti, M. (1999). *Weaving the Web: The Original Design and Ultimate Destiny of the World Wide Web by Its Inventor*. Harper San Francisco.

Berners-Lee, T., Hawke, S., & Connolly, D. (2004). *Semantic Web Tutorial Using N3*. Turorial.

Buschmann, F., Meunier, R., Rohnert, H., Sommerlad, P., & Stal, M. (1996). *Pattern-Oriented Software Architecture, Volume 1: A System of Patterns*: John Wiley & Sons, Inc.

Calvary, G., Coutaz, J., & Nigay, L. (1997). From single-user architectural design to PAC*: a generic software architecture model for CSCW. *CHI'97 Conference*, 242-249.

Carroll, L., & Chorpenning, C. B. (1958). *Alice in Wonderland*. Dramatic Publishing Co., Woodstock.

Coutaz, J. (1987). PAC, an Implemention Model for Dialog Design. *Interact*, *87*, 431–436.

Edwards, W. K. (2000). *Core JINI*. Prentice Hall PTR.

El-Nasr, M. S., & Vasilakos, T. (2006). DigitalBeing: An Ambient Intelligent Dance Space. *Fuzzy Systems, 2006 IEEE International Conference on*, 907-914.

Feijs, L. M. G., & Hu, J. (2004). Component-wise Mapping of Media-needs to a Distributed Presentation Environment. *The 28th Annual International Computer Software and Applications Conference (COMPSAC 2004)*, 250-257.

Fougeres, A.-J. (2004). Agents to cooperate in distributed design. *IEEE International Conference on Systems, Man and Cybernetics, 3*, 2629-2634.

Gamma, E., Helm, R., Johnson, R., & Vlissides, J. (1995). *Design Patterns - Elements of Reusable Object-oriented Software*. Addison-Wesley.

Heckel, P. (1991). *The Elements of Friendly Software Design*.

Hu, J. (2003). StoryML: Enabling Distributed Interfaces for Interactive Media. *The Twelfth International World Wide Web Conference*.

Hu, J. (2006). *Design of a Distributed Architecture for Enriching Media Experience in Home Theaters*. Technische Universiteit Eindhoven.

Hu, J., & Bartneck, C. (2005). Culture Matters - A Study on Presence in an Interactive Movie. *PRESENCE 2005, The 8th Annual International Workshop on Presence*, 153-159.

Hu, J., & Feijs, L. M. G. (2003). An Adaptive Architecture for Presenting Interactive Media onto Distributed Interfaces. *The 21st IASTED International Conference on Applied Informatics (AI 2003)*, 899-904.

Hu, J., Janse, M. D., & Kong, H. (2005). User Evaluation on a Distributed Interactive Movie. *HCI International 2005, 3 - Human-Computer Interfaces: Concepts, New Ideas, Better Usability, and Applications*, 735.731-710.

Illmann, T., Weber, M., Martens, A., & Seitz, A. (2000). A Pattern-Oriented Design of a Web-Based and Case Oriented Multimedia Training System in Medicine. *The 4th World Conference on Integrated Design and Process Technology*.

Janse, M. D., van der Stok, P., & Hu, J. (2005). Distributing Multimedia Elements to Multiple Networked Devices. *User Experience Design for Pervasive Computing, Pervasive 2005*.

Khamis, A., Rivero, D. M., Rodriguez, F., & Salichs, M. (2003). Pattern-based Architecture for Building Mobile Robotics Remote Laboratories. *IEEE International Conference on Robotics and Automation (ICRA'03), 3*, 3284-3289.

Khamis, A., Rodriguez, F. J., & Salichs, M. A. (2003). Remote Interaction with Mobile Robots. *Autonomous Robots, 15*(3). doi:10.1023/A:1026268504593

Krasner, G. E., & Pope, S. T. (1988). A cookbook for using the model-view controller user interface paradigm in Smalltalk-80. *Journal of Object Oriented Program, 1*(3), 26–49.

Little, T. D. C., & Ghafoor, A. (1990). Synchronization and Storage Models for Multimedia Objects. *IEEE Journal on Selected Areas in Communications, 8*(3), 413–427. doi:10.1109/49.53017

McBride, B. (2001). Jena: Implementing the RDF Model and Syntax Specification. *Semantic Web Workshop, WWW2001*.

McBride, B. (2004). *RDF Primer* (W3C Recommendation).

Michael Jeronimo, J. W. (2003). *UPnP Design by Example: A Software Developer's Guide to Universal Plug and Play*. Intel Press.

Niemelä, E., Kalaoja, J., & Lago, P. (2005). Toward an Architectural Knowledge Base for Wireless Service Engineering. *IEEE Transactions on Software Engineering, 31*(5), 361–379. doi:10.1109/TSE.2005.60

Niemelä, E., & Marjeta, J. (1998). Dynamic Configuration of Distributed Software Components. *ECOOP '98: Workshop ion on Object-Oriented Technology*, 149-150.

Rutledge, L. (2001). SMIL 2.0: XML for Web Multimedia. *IEEE Internet Computing, 5*(5), 78–84. doi:10.1109/4236.957898

Wang, Y. (2006). Cognitive Informatics - Towards the Future Generation Computers that Think and Feel, Keynote, *Proc. 5th IEEE International Conference on Cognitive Informatics* (ICCI'06), Beijing, China, IEEE CS Press, July, pp. 3-7.

Wang, Y. (2007). The Theoretical Framework of Cognitive Informatics. *International Journal of Cognitive Informatics and Natural Intelligence, 1*(1), 1–27. doi:10.4018/jcini.2007010101

Zhao, W., & Kearney, D. (2003). Deriving Architectures of Web-Based Applications. *Lecture Notes in Computer Science*, *2642*, 301–312. doi:10.1007/3-540-36901-5_31

This work was previously published in International Journal of Cognitive Informantics and Natural Intelligence, Volume 3, Issue 2, edited by Yingxu Wang, pp. 37-60, copyright 2009 by IGI Publishing (an imprint of IGI Global)

Chapter 10
Adaptive Multiplayer Ubiquitous Games: Design Principles and an Implementation Framework

Chen Yan
Game School of the Jilin Animation Institute, China

Stéphane Natkin
Centre d'Etude et de Recherche en Informatique du Conservatoire National des Arts et Métiers, France

ABSTRACT

One of the goals of ubiquitous computing technologies is to provide an adaptable and personal content at any time and in any context. As a consequence a user-centered design is required. The goal of this research is to develop new gameplays and new narration principles for Multiplayer Ubiquitous Game. We aim to formalize a narrative mechanism to generate events which can stimulate the user's physical actions with the real world, and social communications with other players. Based on the analysis of the relationship between the real world and the virtual world, a narration adaptive to the user's profile is proposed. A prototype using these principles has been developed using off the shell services available on location-based mobile phones.

ADAPTIVE NARRATION IN MULTIPLAYER UBIQUITOUS GAMES

An increasing complexity of relationships between the real world and the virtual world is arising in the next generation games (Björk, Holopainen, Ljungstrand, & Åkesson, 2002). The new types of interaction experimented in Massively Multiplayer Online Games (MMOG) like "World of Warcraft" (Blizzard, 2004), geolocalized games like "Botfighter2" (AliveMobile, 2000) or "Mogi" (Newtgame, 2003), Mixed Reality games like

DOI: 10.4018/978-1-60960-553-7.ch010

"Age Invaders" (Khoo & Cheok, 2006) or relying on the real time political events of the real world like "Geo-Political Simulator" (Eversim, 2004) and Internet and mail based adventure games like "In Memoriam" (Lexis Numérique, 2003) have one or several of the following properties:

- Pervasive: the game interacts with the player's life at uncontrolled times through email and phone calls, for example.
- Social: the game leads to social interactions between the players and more generally between people.
- Ubiquitous: The game relies on a ubiquitous computer system using all of the daily objects as interface and is aware of the user's context and needs.
- Mobile: the gameplay relies on the player's physical mobility.
- There is no general analysis of the type of entertainment which relies on mixed reality interactive media and, of course, no underlying narrative theory. In this paper, we present a method to develop Multiplayer Ubiquitous Games (MUG). Our goal is to define a model of mixed reality interactive narration which is able to:
- Define the global principle of the game: goal of the game, why the user is interested to play and what type of interactions are involved.
- Define the ludic and narrative principles, the objects in the real and the virtual worlds and their semantic relations, and the user model.
- Define the learning process of the user model and the decision process of the ludo-narrative system.

The research project relies on four steps presented in this paper. The first step is to classify and clarify some concepts used in the analysis of the possible interaction between Virtual Worlds (VW) and Real Worlds (RW) for entertainment applications. In the first section we recall a general model of the relationship between RW and VW and state a terminology. It leads to a classification of applications and seven criteria with their definitions and possible values. The second step is to specify a relation scheme between the information related to the player behavior and possible narration schemes. According to the information available, we consider three possible levels of the user model: generic, localized and personalized. Considering the model of the user as a key element of the game system, we propose three types of narration scheme: global, context-oriented and character-based. In the third step we defined a model of the user, implemented in the game, which allows an adaptive driving of the game evolution according to the user's preferences. The functional architecture of this feedback loop between the RW and the VW is presented. The last step is to validate our approach through an experimental MUG game.

REAL AND VIRTUAL WORLDS

Basic Concepts

In this section we will define the main components of mixed realty ubiquitous systems we are dealing with. In the RW there are one or several people who know that their actions may interact with the VW. We will call these people the users of the system. This means that the user has a representation in the virtual world whose behavior is perceptible to him. The identification of a user in the virtual space is known as "avatar", which is an anonymous and dynamic character put in charge to explore the VW, and sometimes may be partly autonomous without control of the user.

The part of the RW which is concerned by this study is the user's physical environment when he is involved in the dedicated applications. It contains all of the contextual information needed to interpret the meaning of the virtual world within the user's physical and social context. We consider three kinds of real objects able to interact with the system:

- Explicitly represented Real Object (ERO). An ERO is a natural RW object explicitly represented in the system, such as the user represented as an avatar, the user is an ERO. It can also be some physical variables, like the location of users, or the users' emotional states.
- Implicitly represented Real Object (IRO). An IRO may be some implicit hypothesis about the state of objects in the RW. For example, the game is designed for the oriental knight culture; players should know and respect this knight system to play the game.
- Unpredicted interacting Real Object (URO). A URO appears when the system generates some unpredicted interactions between the RW and the VW. For example, virtual objects can be sold in the RW through the Internet, which may be not an explicit feature of the game, but will have an impact on the RW economy.

We call, at a given time the current value of the state of these objects, the RW state or context.

A virtual world (VW) is an imaginary space composed of virtual objects governed by simulated physical, economical and social laws. The user is represented in this world and experiments with the sense of immersion and presence. We define two types of virtual objects:

- An Image Virtual Object (IVO) is the image of a real object. For example, the identification of the user could be a cursor, a car or a character, which is an IVO. It may also include some estimates of real world variables, like the weather.
- A Purely Virtual Object (PVO) is an object that has no match in the RW. For example, it could be Non Player Characters (NPC), virtual items and locations that exist only in the game.

We call, at a given time the current value of the state of these objects, the VW state or context.

In our context, the notion "mixed reality" has the following meaning: the real objects and virtual objects can co-exist and interact in the physical and virtual gaming environments. An experimental example is Human Pacman (Adrian David Cheok, Siew Wan Fong, Kok Hwee Goh, Xubo Yang, Wei Liu, Farzam Farzbiz & Yu Li 2003). It's an outdoor mixed reality role-playing game augmented by computers. In this game system, the RW includes two players and their physical movements, the physical objects as cookie's ingredients and their locations, and the geographical characteristics of the outdoor area, which are all EROs. The VW context is composed of the players' representations as "pacman", "ghost" and "helper" (IVO) plus their locations, and a fantasy VW map (PVO).

The "ubiquitous" feature, considered by this study, denotes the ability of a system to spread the application in the time and physical space using the integration of communication networks, mobile services, multiple platforms and various distributions of smart objects in RW (Natkin, 2006). The location-based mobile multiplayer game "Botfighters2" is an experimental example in ubiquitous gaming. The aim of the game is to chase virtual objects in a town and to fight against the opponent team. The RW context contains the physical positions of the players in the city and the physical geography of the city (EROs). The VW context contains players' avatars, the virtual map of the city (IVOs) and virtual weapons and strongholds (PVOs). The main interface of the game is the mobile phone, but TV broadcast is occasionally used to launch seasonal missions. The player attacks other players using the SMS service on his mobile phone and builds the active community to exchange information through the internet. Socialization is one of the core elements

Figure 1. Relation between two worlds: a reactive system

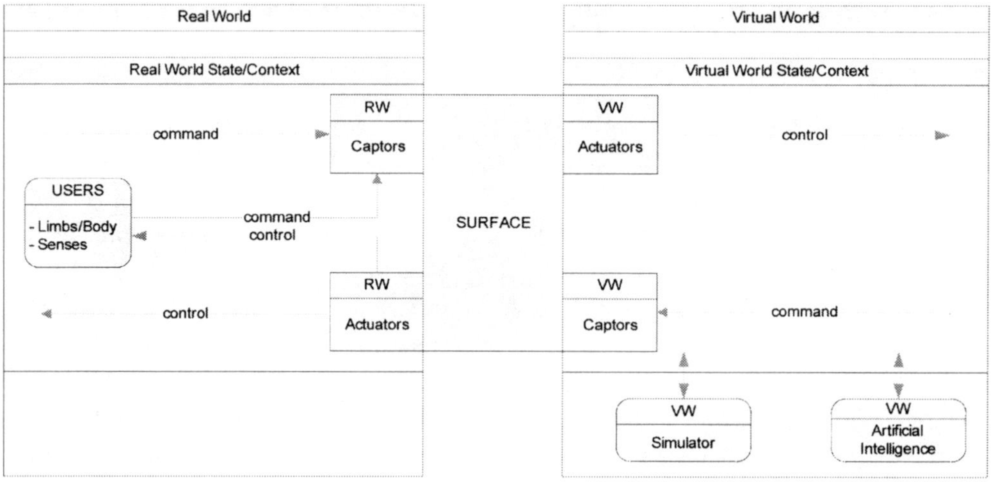

of the gameplay, as the mission assigned to each player leads them to meet other players. The consequences of this socialization, from love affair to business, are unpredictable (URO).

Correspondence between RW/VW

Most of the interactions between the RW and the VW are known and are generally a consequence of the design of a reactive system: the VW is continuously modified by the action of the users and, at least from the user's cognitive point of view; the RW is altered by the events from the VW. We illustrate this relation between RW and VW in Figure 1.

At the boundary of the VW and the RW is the interface. In the RW the interface starts with the user's senses and limbs acting as sensors and actuators (user commands). The RW may also use data from other RW phenomenon such as the weather, the location of the user's car or the results of a football championship. Some of this information is measured by dedicated captors; others are taken from general information systems through, for example, the Internet. In a symmetrical way the VW gathers all of the required information to generate some effects in the VW or to produce some feedback on the RW or on the user through some real actuators. Following (Coutaz, Lachenal, Berard & Barralon, 2002) we call the Interactive Surface the physical materialization of the information which flows from the RW to the VW and vice-versa. In other words, the surface is the outside of a physical entity which makes information observable to the user and to the system. For example, in interaction based on the user's body movement, the displacement of the body is regarded as the command information for the system. The immediate feedback, the perception by the user of his displacement, could be the movement of an avatar seen through the screen of a PDA. In the case of ubiquitous computing, the interface between the RW and the VW is fully distributed in terms of time and space, leading to the perception of a VW omnipresent and diffused through the whole reality (see Figure 1).

Mixed Reality Ubiquitous Applications

Game and more generally digital entertainment can be considered as a foreseen of the evolution of mass media in relation to general usage of inter-personal communications (Natkin, 2006).

This is particularly true for ubiquitous multi-user's applications. An analysis and a classification of these applications allow us to understand the current trends in terms of possible interactions in mixed reality and ubiquitous design, and especially to extract some useful criteria. The nature of the application depends on the objective of the user's interactions with the system. We are mainly interested in applications designed for the general public, so professional ubiquitous and VR systems dedicated to a specific sector are out of the scope of our study. We are also only interested in reactive systems with aware users. Six types of considered applications are briefly recalled here and presented in detail in (Natkin & Yan, 2005) and (Yan, 2007).

Type 1: Art Installations

Interactive art, that is to say interactive plastic art or musical pieces, has been extensively developed during the last century. In relation to our definition of a user, we are only dealing with installations where the spectator is an aware user. In many cases, he feels that he is one of the designers of the artwork. Several examples of this type of installations are given in (Codognet, 1998). A sound installation is presented in (Leprado, 2007). "Bubbles" is a plastic art piece presented in (Wolfgang & Kiyoshi, 2000).

Type 2: Entertainment

We consider three subclasses of Interactive Digital Entertainment (IDE).

Type 2-1: Games

Our work is focused on computer games which gameplay relies either on complex physical activities or social interactions. We consider, for example the pioneer Augmented Reality Game "ARQuake" (Thomas, Close, Donoghue, Squires, De Bondi, Morris & Piekarski, 2000), the Ubigames ("Ubiquitous games") experiments started by "Pirates!" (Bjork, & Ljungstrand, 2001) and many experimental mixed reality games carried out in Mixed Reality Lab (NUS) since 2003.

Type 2-2: Toys

Games have a goal and a reward system. However the relationships between RW and VW can also be found in toys. Tamagotchi (Maita, 1997) and more general simulations of ecological systems are good examples of toys which include the management of a VW. Toys can also be used as input or output devices for different platforms.

Type 2-3: Interactive Media Show

The possible applications of interactive television are still currently under investigation. The concept of interactive show has opened some research issues related to the relationship between RW and VW (Natkin & Yan, 2005). A commercial example is "Sofia's Diary" (beActive, Portugal, 2003), a daily interactive TV soap opera, in which each user can build his own heroine, "Sofia", control her everyday life wherever and whenever with his mobile phone. It is broadcasted on radio stations and on TV shows, published in magazines, in newspapers or on the web...

Type 3: Virtual Assistance

The most common virtual assistance is the GPS electronic map which shows the best path to a given location. In some sophisticated applications or services, an intelligent agent could be used to give advice, as in the project of "AURA" (Garlan, Siewiorek & Steenkiste, 2002).

Type 4: Training

Computer Supported Collaborative Learning (CSCL) is a subset of Computer Supported Collaborative Work (CSCW) in the goal of acquiring skills or knowledge. Only a part of CSCW is relevant to our work as most of the CSCL environment does not include a representation of the user in the virtual training space.

Type 5: Access and Sharing Data

In our context we only consider applications which define a VW as a means to represent, access, retrieve and share complex information. A typical example is a 3D interface for digital libraries (Cubaud, Dupire & Topol, 2005).

Type 6: Space

The goal of the relationship between the RW and the VW can be to define a virtual or physical meeting area for various social purposes.

Type 6-1: Virtual Agora
A virtual agora is a virtual place of virtual interaction and virtual socialization. "There" (Makena Technologies, Inc., 2001) or "Second Life" (Linden lab, 2003) are famous virtual 3D agora on the web which offer a virtual world that enables users to connect and collaborate just as they would in the real world.

Type 6-2: Intelligent Space
Space here means a physical place for people with natural and physical interactions towards the VW. A Smart Ambient is a closed or open area where computers, smart objects, or multimodal sensors are embedded. The goal is to allow users to work or live "more efficiently", as showed in "Easyliving" (Brumitt, Meyers, Krumm, Kern & Shafer, 2000).

Classification Criteria and Values

As a consequence of the analysis of numerous systems from the previous types (Yan, 2007), we have defined a set of criteria to identify the system from the application designer's point of view.

Criterion 1: Social Feature

The system may be designed for a single user's application or a multi users' application. A single user system has one user at a time, but even in this case the system can be designed to partly rely on a social goal. For example, the solo game The "Sims2" (EA, 2004) which has an explicit social goal, based on the exchange of Sims' goods, as opposed to the solo game "Metal Gear Solid" (Konami, 1998) which does not have a social goal. In Multi Users systems, several users can share information and resources. One of the major system design goals is to meet social needs. The relationship between users can be cooperative, where users find appropriate partners to build a group to communicate and negotiate among themselves to achieve their objectives. The relation can also be competitive where users either compete in groups as in "Counter Strike" (Valve software, 2004), or choose their opponents to compete individually.

Criteria 2: Duration of a Turn

A turn (game) of an application is defined as the sequence to reach, from the beginning to an end-point objective. At the end, the user can restart a new turn, but there is no relationship in the system between the two turns. The mean time of a turn is called the duration of the application. Of course, an application with a finite duration may have a short duration (a few minutes like in arcade games), or a long duration (an adventure game) which needs several hours to be completed. Some applications have no forecasted ends (Blogs, MMOGs), some can end but this ending is not a goal, such as the death of a Tamagotchi.

Criteria 3: Types of Real Objects

The application design defines which objects from the RW will have an image in the system and what kinds of characteristics are measured to trigger the system decisions. We identify four possible sources of real contextual information. The first source is from the user. The user's intentional physical actions will have an explainable effect on the VW, and his involuntary actions could also be recorded by the system, for example the user's biological information is interpreted as emotional states. The second source comes from other people, who are not aware of their influence on the VW (who are not users according to our terminology). For example the number of people in a given area can affect the system decision. The third source of information comes from the physical world, such as the time of day or the geographical data. The physical social world is the fourth source of information. Social, economic or political information like stock exchange quotations can be used by the system.

Criterion 4: Representation of VW

Some VWs are virtual instruments. By virtual instruments, we mean that the system is a simple memoryless loop which reacts in real time to the user actions. Most of digital instrument systems do not include a representation of the user and are out of the scope of our study.

VW systems can be persistent or not. Persistence means that the VW seems to evolve even when the user is not interacting with it. This evolution may be real, like in MMOG, or simulated like in a computer simulated aquarium or a game like "Animal Crossing" (Nintendo, 2006). A classical typology considers the nature of virtual objects (IVO and PVO) and the type of relationship between the objects in these worlds. For example we can consider VW based on abstract objects and semantic navigation, like a large hypermedia system or VW based on RW simulation such as a virtual reality or augmented reality system. In a given application, several classes of objects, relationships and navigation schemes may be used.

Criterion 5: Feedback of VW on RW

We distinguish three levels of feedback. Open systems have no physical feedback. In our context, this is the case when ERO just triggers the meaningful reactions of their counterparts, IVO. The second level includes systems which have a predictable feedback between PVO and ERO or IRO. For example, in some augmented reality games, the augmented computer vision enables the user to perform physical actions in real time, like the user's body movement. And, PVO may have some expected influence on the real world (IRO). This is the goal of a propaganda or education game. But, as we have already mentioned, the VW may also generate some unpredicted effects on the RW. This is the third level of feedback when the PVO affects the URO.

Criterion 6: Surface

Surfaces between RW and VW may be considered according to four points of view. The first one is to consider the surface location. The surface could be localized in one place or delocalized in several fixed places or could even be mobile. Another viewpoint is to observe the instance of each interaction, chosen or proactive. In most games the time and the place to play are always chosen by the player. The proactive action generates events which can interfere with the user's life at non chosen times like "In Memoriam" or "Majestic" (Electronic Art, 2001). Thirdly, from a system architecture viewpoint, surface can be considered according to the local intelligence of devices used in the interface. It could be just local captors/actuators, simple reactive devices or it may be autonomous smart objects. The last viewpoint comes from ergonomics and is related to nature of the interface devices. Considering shape and usage, the surface can be a general purpose interface, a dedicated object, or a modification of an everyday life object. Consider, for example, the interface of a PC, the interface of a car game using a toy steering wheel, or the driver interface in a real car, including the steering wheel, where control is acquired through a GPS.

Criteria 7: Captor / Actuator

Many applications use a combination of multiple various sensors to obtain the required situation information and carry out the actions of the system. Here we consider only the RW captors and actuators, which are always physical sensor objects.

The sensors can be dedicated to the application or not dedicated to the application. The dedicated captors may be a keyboard, a mouse or a joystick for capturing the user's actions, a microphone for capturing the voice, a data glove for gestures and mobile phone for location... The dedicated actuators could be a HMD, 3D goggles for the user's perceptions in order to have an augmented vision and headphones for sound effects... Captors out of the user's control could be a camera or infrared sensors installed in a space. Dedicated actuators out of the user's control could be a projector, a screen, a printer, a loudspeaker...

The sensors which are not dedicated to the system are devices used in our daily life and diverted from their uses for the VW application. For instance, the large screen in a city square, the TV, the telephone... could be the actuators; radars, weather forecast captors, the internet...can capture the required RW information for the system.

Use of These Criteria

Our main goal, through this analysis and the classification of new digital entertainment applications, was to understand schemes which relate multiplayer games, ubiquitous computing and narration. The most interesting applications related to the social aspects are Massively Multiplayer Online Games and Virtual Agora. "Second Life" or "Habbo Hotel" (Sulake Corporation, 2000) are a typical mix of these domains based on social goals. The use of narration as a social mechanism is mainly deployed in on line games, leading to scientific analysis and the understanding of game design principles (Bartle, 2005) (Genvo, 2006). Experiments on ubiquitous computer games rely on very a simple narration scheme with only one type of quest for all players. The use of this scheme as a social mechanism is still poor and experimental. In Augmented Reality Games, like "Age Invaders" (Khoo & Cheok, 2006), the social goal is the core of the design but, in counterpart, the narration scheme is inexistent or poor. The social relation is based on physical meetings, like in arcade games, or on the support of a team, like in sports. In all cases, the idea of "ubiquitous" computing, relying on a system aware of the player context and needs is at a very basic stage of development. The system knowledge of the player is limited to his age and gender, his location and his involvement in the quest in progress. Our analysis also leads us to find some good principles to build MUG, combining ideas already used in these different fields.

Coming back to the goal of our study, which is to define an adaptive MUG relying on social relations with complex relationships between the RW and the VW, and taking into account the classification criteria, we came to the following simple conclusions.

As a MUG is a game it must have a challenge and a reward scheme. The social goal leads us to consider a cooperative game, but, as it may be very difficult to build a ubiquitous game where all participants play against the computer, it must also be a competitive game (criterion 1). Therefore, it must be a team competitive game. As it must be played in the real world and according to real life, it must have some rather short turns, from a few hours to a day (criterion 2). The preceding points lead to designing a kind of collective sport. However, as we want to use some narrative scheme to build stable social relations between players like in MMORPG, it must be, in some way, an adventure game and, in another way, a persistent mix of RW and VW. Online pro-active games like "Majestic" or "In Memoriam 2" (Lexis Numerique, 2006) have these properties but are mainly played through standard computer interfaces with limited physical interactions between the RW and the VW. Therefore, the MUG must rely on physical objects which have an immediate and foreseeable impact on the player (criterion 3): a detailed knowledge of the player itself (civil status, habits and relatives, his practices as a player, his current location and the location of his friends…), simple RW objects (the town itself, weather, traffic conditions…) and, eventually, some important RW events which can be used to diversify game plots (election days, sport events…). As the player is involved in an adventure game, the PVO objects (criterion 4) must include some "personal virtual belongings" that the player can chase, conquer or buy to progress in the game. Virtual locations and events, able to modify the PVO (disasters, for example) can be easily used to create dynamic quests. To avoid a too complex and unreliable architecture, the use of dedicated smart objects should be limited (Gaming places like in "Age Invaders"). In counterpart the use of existing information systems (large LCDs on walls, TV screens in public places), can be widely used. The mobile phone is the most affordable ubiquitous and geo localized technology. It should be the core of our system, combined with a standard Web interface. A more ambitious experiment should also try to take advantage from standard broadcast media (radio, TV) as a way to spread information in real time to the players (criterion 6 and7).

The feedback (criterion 5) is the core of our study: the MUG must be based on an adaptive narrative system, to be truly context-aware or location-aware and to take decisions according to the player's motivation. The following section presents an adaptive narrative engine for MUG.

NARRATION MODEL BASED ON A USER'S MODEL

During the studies on narration for mixed reality ubiquitous systems, we found that the possible actions of the user in both universes (VW and RW) are uncontrolled and variable in a mixed reality mode. Understanding and modeling the user experiences during the system design is necessary and will be helpful and give instructions or guidance to the narration generations of MUG. Thus our method for solving the problematic of an adaptive narration is to employ and integrate a user's model in the system. The user model will not only take into account the knowledge of the user's states or behaviours in classical online gaming situations, but will also consider the user's actions in augmented outdoor or mobile gaming environments. Thus, the game narration mechanism can be responsive to this complex user model in considering the real world context.

Related Works

User modeling is one of the major subjects of HCI (Human-Computer-Interaction) research. It generally concentrates on two axes: one is the adaptation to user's interaction and the other is the personalization of services (or information). The goal of the first type of research is the adaptation of the user interface according to its context of usage; the second type of work tries to personalize a given application to the user's preferences and needs. The challenge of user modelling is to find a way to collect and represent the set of user's information needed and to predict user's behaviours under a set of possible stimulation.

In digital game research, the idea of a player model came from the analysis of player behaviors in Massively Multiplayer Online Role-Playing Games (MMORPGs). Players with similar motivations can be grouped into the same "Player types" (Bartle, 1996): achievers, explorers, socializers and killers. This classification of players' roles allow game designers to understand the players' motivations and to define quests types fitting players' needs. However, in MMOG's design, users out-game information or his personality traits are poorly taken into account by the system. As we can see, the user's activity space embedded with computing and information systems becomes ubiquitous and proactive. Our point of view is to consider the interaction between the real and the virtual world in a mixed reality mode, and the possible user actions in both universes. In this mode, the player may change between types over time and he may also be influenced by the real world. Thus, a user model will contain more evaluations, not only about his online information, but also about his physical activities, and some psychological and social parameters deduced from existing cognitive and sociological models…

Using such a user model, a system can provide the user with services or information suitable for his specific mood and generate interactions and adaptive interfaces compatible with the user's specific context. In the field of MUG, the user model has not been studied yet. Related works can be found in the fields of education and in interactive narration (Natkin & Yan, 2006). In IDtension (Szilas, Rety & Marty, 2003), a user model is used to create conflicts according to the classical dramaturgy theory. Interactive Drama Architecture (IDA) (Magerko & Laird, 2003) uses a real time director agent guided by a user model. It anticipates the user's behavior and determines the plots. Some authors also use the character-based model as a key point of a narrative scheme: (Cavazza Charles & Mead 2002; Mateas & Stern, 2003) use the classical model of character conflicts to generate a dramaturgy; (Peinado & Gervás 2004) use Bartle's players' classification as a key element of an automatized role-play game manager.

Considering MUG, the complex relationships between player and the RW, between player and the VW and also between the RW and the VW, leads to elaborate a user model with more criteria. We must also define an adaptive narration mechanism based on it, in order to improve the gameplay in a mixed reality and ubiquitous mode.

User's Model

In our context, the user model and the stimulations are used by a narration engine to control the evolution of the MUG system. The user model depends on a set of parameters that can be either statically defined by the game designer or dynamically adjusted by the changes in the user's physical states or even the user's social features and personality. This leads to three levels of parameters in the user model: generic, localized or personalized.

A generic parameter is a general hypothesis about the player, a location of all users on the game map and statistics of players' actions. It does not distinguish one player from another. For example, it is a statistic about the players' community (e.g. 65% Male, 35% Female, 34-years-old mean age, occidental).

A second level of parameters includes some real-time data related to the user's location. In this case, the user has an identifier and his current location is a user-state variable of the model. Most location-based mobile games, like "Mogi" and "Botfighter2" implicitly use a user location model to detect the distance between players and virtual objects in order to manage personal exchanges between players according to their proximity.

A third level of a user's model contains personalized parameters which define state variables about each user. Some of these variables are already used in classical games: for example, various kinds of challenges according to the skill level of the user. However, in mixed reality environments much more detailed data can be used: civil status, personal habits, social relationships...

According to (Natkin & Yan, 2006), a user model only based on generic parameters is called a Generic model (G). A model that includes both generic and localized parameters is called a Localized model (L), and if the model includes at least one personalized parameter it is called a Personalized model (P). In the following, we will describe which kinds of player information is collected and identified. We have to consider the player's knowledge from several different points of view. It is important to understand that the behavior of the user in his everyday life is different from his behavior as a player. For example, from a classical psychological point of view, a human personality changes infrequently, but the player's behavior in the game (as he is playing) may change from one turn to the next. There is not a simple mapping between the player's characteristics and his avatar's feature or its evolution. Table 1 shows a classification of this data and its position in the level of the user model.

These different types of information are not used in the same way. Statistical information about the user himself can be correlated to very general social models. For example the proportion between "fighting quests" and "socialization quests" may be bound to the ratio between males

Table 1. Structure of the user model

Subclass	Static exact	Statistical	Real Time Exact	Real Time Estimated	Example	Comments
User himself						
Civil status	P	G			Age	Give indication on what kind of gameplay may interest the user
	P	G			Gender	
	P	G			Profession	
	P	G			Country/ State/Town	
Social Relation	P	G			Friend of other player	Useful to create collaborative or competing teams based on real world relationships
	P	G			Wife or Husband of player player	
	P	G			Profile of the related player	
Preferences	P	G			Leisure	To upgrade user's interest and familiarity
	P	G			Consumption habits…	
User as player						
Accounting		P or G			Time since first account	Information needed to know the user and his playing habits. Quest can be chosen according to these habits
		P or G			Time spent playing	
		P or G			Frequencies of play times	
		P or G			Mean duration of a turn	
Choice Selected by the player	P	G			Type of account	Idem
	P	G			Type of Avatar	
Statistics about Physical Interactions		L			Distribution of the duration of play in each location	Allows the system to personalize user's quest according to his practices.
		L			Type of location visited during the game	
		P or G			Distribution of the type of interface used	
		P or G			Ergonomics data (Interface used)	
	P	G			Trace of physical interactions	

continued on following page

Table 1. Continued

Subclass	Static exact	Statistical	Real Time Exact	Real Time Estimated	Example	Comments
Real Time Interactions			P	G	Playing or not	Real time control of the player's state.
			L		Localisation	
			P	G	Type of interactive device used/available	
			L		Relative position to other players	
			P or L		Interaction with other players	
User as an Avatar						
Standard information	P				Name	Same usage as in MMORPG.
	P				Graphic representation	
	P				Level	
	P				Abilities	
			P		Location in the VW	
			P		Status in the game	
			P		Trace in the game	
Social model in the game				P or G	Sociability (group, guild...)	Used to compute user's profile and to generate adaptive quests
				P or G	**Motivation (FFM profile)**	

and females or to the distribution of the players' ages. However, if this information is known on a personal basis, it can be used in a much more subtle way. It has been observed that the relationships between players in the game have an influence on the relation in real life (Gustavo & Talmud, 2006). And the converse is also true, the choice of a competitive or cooperative quest can be interpreted in more depth if the system knows that the players involved are friends or are married.

The data about the user as a player can be used, for example, to adapt quests either to general preferences (global data) or to individual choices. According to the value of the corresponding parameters, the system may construct a game level composed of a great number of complicated, long quests or short mini-games. The physical location of these quests may also be chosen according to the same process.

The User's Social Model

In the sequel of this section we discus mainly the social model of the player in the game. Our needs, in terms of a model, are related to the trend of player's social features but are more focused on the automatic generation of narration and gameplay schemes. These needs can be summarized as follow:

- The goal of the model is not to understand in a wide way the personality of the user but to deduce his interests as a player and therefore offer him an adapted scenario scheme. In particular, in contrast to the real personality of the user, its virtual personality in the game may evolve very quickly (Bartle, 2005).
- The user sociability model must not only be descriptive but also operational: we must be able to deduce some simple rules

to decide which type of quest can relates the social state of the player to the global narration needs.

- The model must be adaptive: we must be able to build a feedback loop that improves the model according to the player choices and actions.
- The model must be sufficiently flexible and responsiveness to classify users in smooth and changing categories.

As the model will monitor a narration system according to supposed player's need, we must allow the user to change his mood and alter the sociability model. For example, using Bartle's classification, the system considers that he is mainly a "socialiser" so generally offers him some "social" quest. But this week he wants to be a warrior…

The simplest way to reach these goals would be to define the user's model as a set of weighted possible types of quest. For example in an MMORPG the following types could be considered: Search for NPC or other players, Hire some efficient teams, Fight against enemies or monsters, Conquer areas, Explore new territories, Discover characteristics of an object or a place… Each time the player chooses a quest in a given set, the weight of this set increase. The quests are randomly chosen according to the distribution of weights. But this model relies too much on the definition of quest, which may vary even in a given game. We may also consider building a player model based on Bartle's classification and relates the classification to the type of quest. For example Killers should prefer to conquer area or kill monster rather than to hire a team. But in all these cases the user model, from our point of view, is related to just one type of game (MMORPG).

We are not trying to analyze the personality of the user. We want to find a way adaptive to measure the adequacy of quests that are offered at his/her interest. We are therefore oriented to the quantified representations of the personality. Such a model should interpret the motivations that promote communication and social interaction between players. We have therefore chosen the theory "Five Factor Model (FFM)" (Costa, 1992) that qualifies and quantifies the human personality into five factors. The five factors (Pierce & Jane, 2003] are: The Need for stability (N), The Extraversion (E), The Originality (O), The Accommodation (A), The Consolidation (C). The personality of a human is then described and identified by five normal distributions of these factors considered separately. A simple approximation uses three possible values for each factor -, =, + (Table 2). For example, a personality can be defined by a quintuplet. From our point of view, the main interest is its operation ability: it has been used in several fields, for example for personnel's profession selection. The personality profile is determined by using a survey based on a questionnaire.

To simplify the algorithm to adapt the model of motivation we chose a continuous representation of the factors. At a given time the social model of the player, called the User Profile, is defined as a vector M of 5 frequencies:

$M=(MN, ME, MO, MA, MC)$ where $-1 \leq M_i \leq 1$

The initial values are computed using the FFM test principle. The player is invited to fill out a form which is used to set the initial values of the user model parameter and which includes a FFM test. The profile changes according to a feedback loop related to the player choices in the game. The technical aspects of this process are introduced in the next section. The goal is to relate the profile of the user to the profile of game quests. It does not matter if the psychological profile is correct and if the nature of the quest is fully translated. The only goal is to suggest interesting quests to a player. However, we want the player to not only remain in control of the interpretation of his actions during the game, but also be able to directly alter his profile through a simple editing interface.

Table 2. Factors of the personality in FFM

	Level		
Factors	-	=	+
Neuroticism (N)	Resilient (N-)	Reactive (N=)	Nervous (N+)
Extraversion (E)	Introvert (E-)	Ambivert (E=)	Extravert (E+)
Originality (O)	Preserver (O-)	Moderator (O=)	Explorer (O+)
Accommodation (A)	Challenger (A-)	Negotiator (A=)	Adapter (A+)
Conscience (C)	Flexible (C-)	Equilibrist (C=)	Focused (C+)

This allows him to choose a type of behavior in the game and to change this behavior according his progression.

THE NARRATION MODEL

Main Principle

We define an interactive narration model as a mechanism which generates narrative schemes correlated to the user's profile. As guidance model, these narrative schemes orient the definite quests pre-scripted by the game designer which can be translated into actions in the real world. In the context of mixed reality, it means that the narrative emerges from the user's physical actions or natural activities in the gaming space. The ultimate goal of this mechanism is to give an interest (event) to different players either in their individual experience or their social experience during the game. This model must have the following properties:

- Coherent with a narrative scheme
- Adaptable to the evolution of the state of the real world
- Controllable in time and space
- Able to be carried out using the ubiquitous and pro-active computing environment of the game
- Adaptable to the player characteristics and motivation
- Providing a variable and renewable experience to the player

Quest System

The narration model generates quests. Quest system is a traditional storytelling technique used in various genres of games. According to a terminology in (Guardiola, 2000), a quest is defined by three main characteristics:

- A goal, for example, find a secrete code.
- Obstacles, which are opposed to achieving the goal, such as the existence of a secret passage to access codes.
- A resolution method, which makes it possible to cross the obstacles, for example, the activation of a mechanism that opens a secret passage.

MMORPG usually uses a quest system, which is an extensive database of pre-scripted quests. The set of quests is designed to fulfill some narrative structure but also to develop the structure of the virtual world providing each player with the kind of experience he is interested in. Generally in MMORPG, all of the available quests are open to each player who decides to choose one of them, according to his experience and environment. In MUG the same principles can not be used directly, for several reasons: the player is involved in the real world leading to time and space constraints, he has to physically meet other players, the duration of a turn is limited, he may use mobile devices as interface which changes the ability to control the execution of the quest … Therefore, a MUG like "Botfighter2" uses a very simple and linear narration scheme using the same quest (or type of quest) for all players at a given type. We try to define a quest system which has the same properties as the MMORPG quest database and which can be used in MUG. These quests can be generated according to the current users' states at the running time and to some social goals defined by the game designer. For example, the narration model can decide at a given time to send an SMS to all users to meet together for a special quest, which induces desired social behaviors of the players. Another possible decision could be creating virtual objects that generate certain plots, such as a new virtual treasure which might induce a battle. The great advantage of these approaches is that by controlling the phenomena, the game designer combines the narrative aspect with freedom of the players in reaction to the phenomena and with the capacity to develop collective impulses.

Levels of Narration

We propose three levels of narration to respond to the user model in different situations according to different needs. We call them a Global Narration, a Context-oriented Narration or a Character-based Narration.

A Global Narration generates quests according to a storyline that does not distinguish players' identities, contexts or histories. This storyline relies on some standard structures of narrative and a global sociological or cognitive model. Its objective is mainly to attract and stimulate players to play the game. For example, a quest is set to be globally announced to all the players. The result of these generated events provokes a social phenomenon, a public cause, for the whole group of players.

According to the requirement of the real environment surrounding the user (place, physical environment, users' movements, temperature, or even political or economic information), the narrative system needs a Context-oriented Narration to make appropriate responsive reactions. The Context-oriented narration distinguishes different circumstances in order to trigger different scenarios. Actually the games based on real world information are mostly location-based mobile multiplayer games, such as "Mogi". The scenario of "Mogi" is to seek and collect virtual objects with a mobile phone in different places in the real city. The goal is to complete collections of these virtual objects. In this game, some virtual objects will only appear somewhere in the city at a fixed time, which enables the player to move to obtain these objects. The Context-oriented narration uses the physical positions of the players in the landscape and the landscape itself to create the virtual objects that will become quest goals.

The system could also identify and generate a personalized storyline and be pertinent for each individual user. The goal of such a mechanism, which may be very difficult to put into practice, is to generate much more efficient stimulation according to an individualized cognitive or social model. As a consequence, the user may have a stronger experience while playing the game. A well known example of Character-based Narration is the use of Bartle's classification of players' behaviors in MMORPG (Peinado & Gervás, 2004). In this case, the system tries to develop quests according to the role and the power of the players involved at a given time. This is a powerful driving force of dramaturgy and personal interest for the game.

Figure 2. Relations between the user model and the narration types

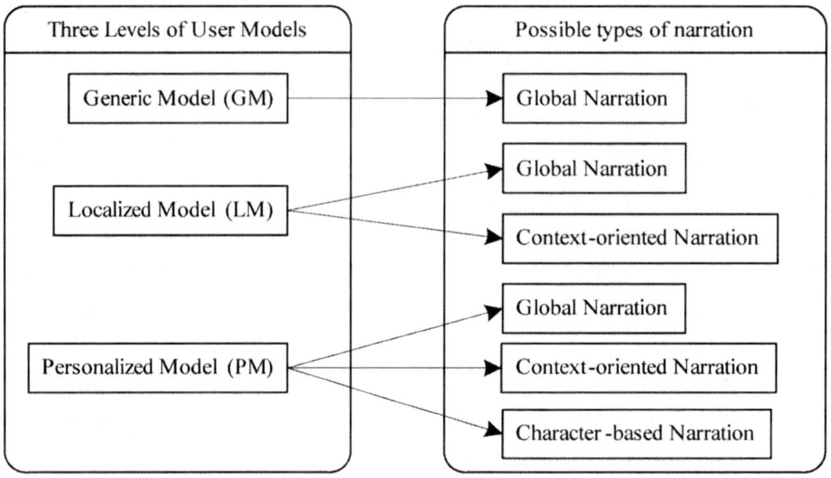

RELATION BETWEEN THE NARRATION AND THE USER MODELS

The level of narration that can be generated depends on the accuracy of the information of the user possessed by the system. It is clear that a game managing a PM for each user has enough information to generate plots at three levels of narration. The system may induce his/her special needs and specific interest, and supply him/her with an individual narration, by considering the coherence with the context-oriented narration and the global narration. An LM allows Global or Context-oriented narration where the system may give local "surprises" according to his/her local context. A GM only leads to a Global narration in which the system determines a global information or service for all of the users. The possibility of mixing the various narrative levels is a way to maintain the balance between the interaction of the user and the dramatic control of the system. However, the ability to manage a PM depends on various aspects of the game system (technical complexity, social and environmental contexts, privacy constraints). These relationships are schematized in Figure 2.

Each game system only uses one type of user model. Most systems contain an implicit GM; location-based mobile games use a geo-localized LM. In MUG, a PM can be used so that the scenario could be interpreted in a degraded way with some predefined forms, which are ranging from a personal event to a global one. The ability to use a PM depends on many factors such as the technical complexity, the social and environmental contexts, the constraints of private lives, etc. On the other hand, the possibility to mix different levels of narration keeps a balance between the user interaction and the control of dramaturgy in the mode of mixed reality.

NARRATIVE STRUCTURE

Our research goal, on game narration, is to define a narrative structure has the same properties as those of MMORPGs in term of socialization. It should also yield to the time and spatial constraints. Finally, the narrative structures should be more complex and comparable with those found in a live action role-playing game or a proactive game like In Mémoriam (Ubisoft, 2004). On the basis of the seven criteria mentioned above, we propose a method to build a narrative scheme for the MugNSRC which relies on three layers of content: Episode, Mission and Quest. The principles of the narrative structure we describe here construct the essential gameplay elements of MugNSRC.

The first layer, the episode, is "the story" by itself that must be told and which depends on the achievement of certain "social events". The episode is the macroscopic component of the narrative structure; it determines the space and temporal settings for the game. It can be interesting to find here Aristotle's three-act structure in which the paroxysm is an outstanding social event: a duel, a festival, an exposure…in the real world.

The second layer is the missions, which discriminates the episode in terms of time and space by identifying the roles which must be held by the players. The mission determines a global transformation of the real and virtual world state. It is itself a dramaturgic structure which is based on a type of conflict and drives a group of player to a clearly definite objective (from which comes the term of "mission"). The specification of the missions and their relation order constitute the scenario of an episode. The arrangement of the missions follows the traditional narrative structure: initial situation - transformation - final situation. Contrary to the pure fictional narration, the situations relate partly to the state of real world objects.

The third layer is the quest, which is assigned by the game system to each player in order to carry out the mission. It is the unit of intention and of action perceived by the player. A quest has the same attributes as a mission plus the specification of the way in which the quest is proposed and presented to the player, i.e., the type of interface. Each quest is defined by three main characteristics: a goal, obstacles and solution methods. If given pre-conditions are satisfied, a set of quests will be activated or proposed to the player. These preconditions rely on the real or virtual state of the player, the real world context, and the progression order of the game scenario.

In the narrative structure, the execution of the narration, i.e., the fact of playing an episode is neither deterministic, nor sequential, nor linear. The succession of the missions (and, on a precise level, the quests) is not completely fixed; several missions can be held in parallel, thus, a great number of narrations can be generated by the game and the players. The influence from the real world conditions can also force the system to give up the execution of certain missions or quests or even to carry out the alternatives in a "degraded" mode. For example, according to the knowledge of the user on a general, localized or personalized level, the system can adjust the narration from an individualized plot to a contextual or global scenario (Yan, 2007). In other words, controlling the execution of the narration in a MUG is equivalent to the control of a distributed real-time system that tolerates failures, whose principal elements are the players and for which the measurement of the effectiveness, is the interest of these players.

Quest Generation

In this paper we do not discuss the formal representation of the narration model. It follows the maze technique of level design (Natkin, 2006): virtual objects have to be distributed in the space according to chosen path and goals. Quests are submitted to the players at given points of the game evolution. We have proposed a timed Petri Net model which can be adapted to this use (Grünvogel et al. 2004).

We assume that the game is composed of rounds or levels which have to be executed in a given time. "Botfighter2" uses quests which must be played in less than one week, "Majestic" uses one month episodes.

A round is generated according to the following steps:

a. Choice by the designers of a narrative structure. This structure is a set of events which is supposed to occur according to time and

Figure 3. Feedback loop of a MUG system

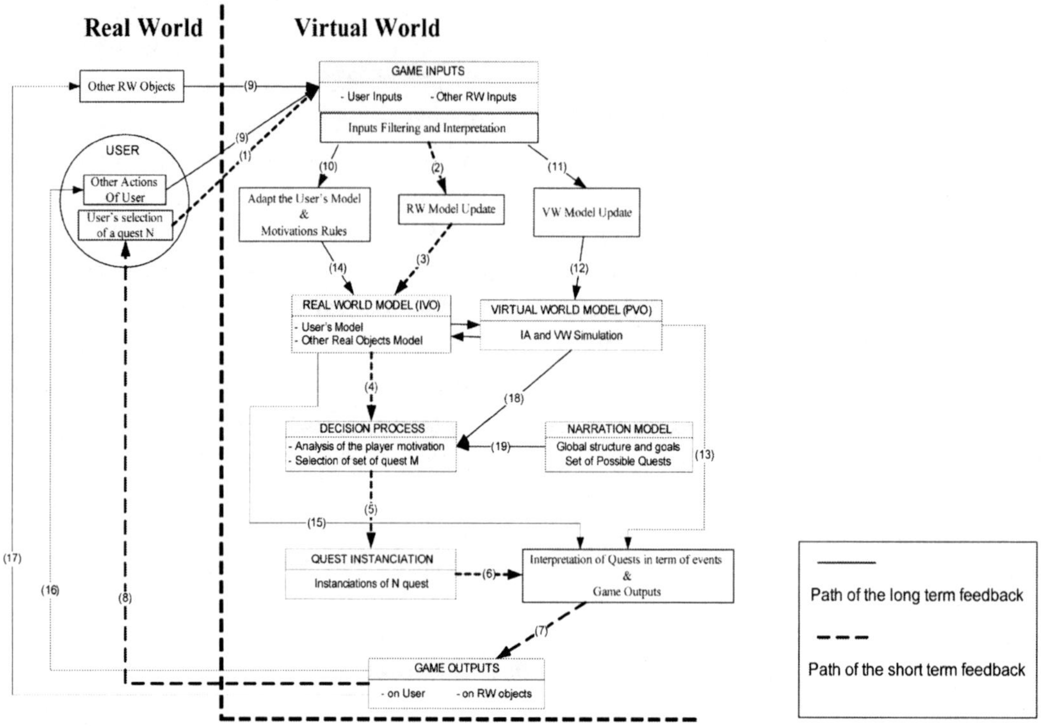

space constraints. This narrative structure is associated with a set of classes of quests.

b. Proposition of a subset of quest classes to each player according to his model and the current progression of the game.
c. Choice made by the player of his class of quest.
d. Instantiation of the class of quest according to the RW and VW situations.
e. Execution of the quest as a sequence of command to game output devices.

The Feedback Loops

In this section, we present the whole computing system with the integration of the user model and narration model in three kinds of feedback loops. Figure 3 shows three main feedback loops.

The main game loop is represented by the following sequence of entities:

- 1, 9: Inputs from the player and the RW
- 2, 3, 10, 14, 11, 12: Updating the RW and VW model
- 13, 15, 7, 16, 17: Output and the corresponding interactions from the game to the RW

The short term feedback loop selects a set of quests M and recommends them to the player, who chooses one of them. The selection of the potential quests M is based on the User's Model and the corresponding motivation scheme. It is represented by the dotted line sequence of edges (1 to 8).

The long term feedback loop reflects the adaptive nature of the User's Model: According to the quest selected by the user, the model is dynamically adapted to reflect the user's current motivations. This is represented by the sequence of entities 1, 10, 14, 15, 7, 8.

The Adaptive Scheme

The allocation of quests and the instantiations of quests rely on the user's model. Each class of quest is associated with a FFM profile, a vector $Q=(QN,QE,QO,QA,QC)$ where $-1 \leq Q_i \leq 1$. A distance $D(x,y)$ in R^5 is used; it can be either the Cartesian distance or a distance function used for classification (such as the Malahobis distance). The game selects a set of classes in the neighborhood of the user profile (according to d) and another quest class is randomly chosen. This second proposition is comparable to a mutant in genetic algorithms: it allows the player's choice to be diversified. It can also be combined with narrative constraint to ensure certain qualities of the gameplay. Then each selected class is instantiated using the user model: choice of the location, of the group of player involved in a given mission or with who they will probably meet during the game…

The list of the constructed quests is proposed to the player. He selects one quest and starts playing. Let Q be the profile associated with the chosen quest, the new profile of the player M' is given by an exponential smoothing:

$$M'=\alpha M+(1-\alpha)Q \text{ where } 0 \leq \alpha \leq 1$$

The choice α of determines the speed of adaptation. If $\alpha=1$, the user profile is not changed, if $\alpha=0$, the user profile becomes equal to the quest class profile.

A GAME PROTOTYPE

In this section, we present the principles of a prototype developed at France Telecom. The goal of this prototype is to demonstrate the feasibility, the usefulness of our concepts and the adaptive narrative mechanism. We take advantage of an arcade action game "Nippon Salary Racing Championship" (NSRC, ENJMIN school project, Gallais, Henry, Saphores, Rapine, Guillebon, Roubinet, Muller, Camous & Deschamps 2007) as a core for a MUG system: MugNSRC.

The original game NSRC is based on cartoon type races of wheelchair in the office at the virtual Japanese Company, the Bananamoto Ltd. Winners progress in a hierarchy of the company, and get better and higher offices in the Bananamoto building. The game is based on the platform Xbox360 for a maximum of 4 players. The game design includes several types of races from friendly to highly competitive ones. Game results allow them to progress in the company hierarchy and to win virtual money. The money is needed to buy new wheelchair pieces. These pieces are useful to increase the performance and to personalize each player's wheelchair. The design includes the management of the community of players, which can exchange pieces and wheelchairs. A ranking of the players is published. The map of the Bananamoto building, with a periodic design of new floors including new race circuits is also included in the community management. Finally, a player can suggest race or exchange meetings to other players. The designers of NSRC have chosen graphic, music and gameplay to encourage the creation of a player's community on the web: the nostalgic of old 8 bits games. This idea was the starting point of our design (see Figure 4).

The core of our proposal is to keep the principle of the race and the pieces exchange as a goal

Figure 4. Hyperspace acceleration of wheelchairs in NSRC

for physical meetings, the community is the core of the social aspects of the game. The prototype is designed to work in a small physical area (a town), but the principles can be extended to a world wide game, comparable to virtual communities, such as "Habbo Hotel".

Design Principles on 7 Criteria for Creating Real-World Gaming-System Interaction

Based on our 7 criterions, we transform NSRC into the MugNSRC as follows:

1. The game multiplayer competitive side, instantaneous racing, and player ranking in the Bananamoto, are of course maintained. However, the exchange principle is not enough to develop a cooperative side. Thus, several extensions are proposed:
   - Rules of progression in the Bananamoto hierarchy are altered to promote coalitions.
   - Players must cooperate in physical rallies to win special bonuses, which end in particularly important races where every player must compete in the same festive place. In these places real wheelchair races can also be organized.
   - The community is encouraged to create and exchange player's creations, videos and photos of races, machinima and new wheelchair pieces. Competitions between these user's creations are organized…
2. The game is turn-based in term of wheelchair racing. The physical rally is also based on short turns (1 to 3 hours). The community management is at the core of the socialisation. From this point of view, the game is persistent.
3. The main types of real objects used in the game are:
   - The area where the game takes place including gaming festive zones, where players meet and compete;
   - The player's home position, the current location of the players, the proximity between the players;
   - The weather and the time of the day: race conditions in the virtual world are altered by these parameters;
   - Some goodies (T-shirts, special real wheelchairs…) manufactured as wining prices of wheelchair games.
4. The PVO includes all objects of NSRC (Bananamoto building, wheelchairs…). Each user has an avatar which has a particular status in the company and owns virtual objects and money. Some virtual goods, rewards and clues are also created to be chased in the RW. Player's creations can be also considered as PVOs. IVOs are mainly composed by the map of the gaming area and the images of the locations in this area.
5 and 7. The player interacts with the virtual world during races on the X360 platform, but also using the mobile phone for rally turns to exchange data through SMS and to capture images and video. Effectively, captors and actuators are both the interface of the X360 and the one included in cell phones.
6. As the game combines features of virtual worlds, ubiquitous computing and electronic sports, the feedback is of the third type. If the game is a success, the evolution of the game and the community will probably reach beyond the fore-casted controls of the designers.

Adaptive Narration for MUG

To explain how the adaptive mechanism works, we use our Episode-Mission-Quests method as mentioned in the previous section.

One episode of the game is built on a social event for all players. The result of an episode must evolve significantly the state of the player community of Bananamoto. The type and size of the gaming area and duration of a game turn is fundamental to define an episode. MugNSRC is a festive game in which a great part of its gameplay is carried out in a limited zone with the scale from a campus to a whole city. As we wish that, during one episode, to imply the player in several missions and as we want to maintain a real social relationship during this phase, we established the duration of an episode to one day. E.g., players fulfill missions such as searching wheelchair pieces, solving puzzles or taking photos to gain game points or to maintain the Bananamoto's memory, at the end of the day, all of the players meet together to participate a festive evening party.

The episode is accomplished in terms of time and space by the fulfillment of missions. In our prototype, there are two types of missions: Competition and Creation. The Competition Missions are created to develop the potential conflict related to the competition in NSRC and to integrate the rare virtual resources. It can be, for example, to search unique pieces of given wheelchairs, or to obtain bonuses that a player needs to participate in certain races. The purpose of Creation Missions is to create events or objects taking part in the memory of the community. Searching for physical elements used in the organization of the races, and creating a video of the NSRC festival are typical examples. The relation order between missions which determines a narrative scheme is extremely complex especially in the case of MUGs. Our prototype deals only with the partial order set by the designer in the narrative scheme. For example, certain mission can start only if another mission is started; two missions may be held in different time and different real places, but both will be completed at the same time and the same place. In term of gameplay, players discover the development of the mission by carrying out the quests, by playing different roles such as competitor, creator, collector or organizer. These roles are implicit for players but the actions of these roles are essential to the progress of each mission.

The quest is the action unit proposed to the player. Here are examples of 4 pre-scripted quests:

Q1. find game goodies in area B and combine it to your wheel chair according to the instruction given somewhere… on the Web.
Q2. Collect ten virtual objects in area A for updating wheelchairs.
Q3. Find other players to exchange virtual pieces.
Q4. Find a team of players to produce an artwork (a race video, for example). Present the work in the festive racing area and gain some levels in the Bananamoto Company's virtual Hall of Fame.

Each quest's FFM profile and their values are defined by the designer according to the nature of the actions needed to carry out the quest. The possible quest profiles are given in Table 3.

If certain given pre-conditions in the real world are satisfied, a set of quests will be activated or proposed to the player. However, the post-condition to evaluate if a quest is accomplished or not can be very variable and specific. This is related to two problems. First, in a MUG system, the

Table 3. Quests qualified according to the five factors

	N	E	O	A	C
Q1: Item searching	0	+1	+1	0	-1
Q2: Exploration	-1	+1	+1	-1	0
Q3: Race	+1	+1	0	0	-1
Q4: Video Creation	+1	0	+1	0	0

conditions of the success, the failure or the abort of a quest depends on the state of the real world which, contrarily to the state of the virtual world, is difficult to be exactly evaluated. The second problem is related to real time constraints and the sequencing of the quests execution. For example some quest must be performed before a given hour to finish the first mission in the morning and the episode in a day. Therefore, in our prototype, we decided to verify three simple conditions: condition of location, condition of time and condition of loading or deposit. At certain moments, m quests may be proposed to the player who can accept n<m quests and the player may complete them in any order. The analysis of the state progression and the proposition of quests are executed by the system periodically. At the same time, certain quests can be exclusive, either because their achievements in parallel are physically impossible (e.g. they take place in too distant places), or because the quest contradict with certain aspects of the narrative scheme (e.g. two quests must be executed by two distinct players as we wish that they meet).

Assume that Alice and Bob start playing in zone A. Initially, according to the progression of the interpretation of the narrative structure and the state of the real world, a certain number of quests are selected. Secondly, a given number of quests are proposed to each player according to the state of his model and his personality profile. Assume the profile of Bob is *MBob*=(0.3,1,0,0,-0.8). Using the Cartesian distance, we get:

d(Q1,M)=1.06 d(Q2,M)=2.08

d(Q3,M)=0.73 d(Q4,M)=1.59

Therefore the system proposes the "Q3: Find other players to exchange virtual pieces." to Bob and proposes also "Q2: Collect ten virtual objects in area A for updating wheelchairs.". Q2 is chosen randomly but Alice is chosen as Bob's opponent, for she is in the same zone and she knows Bob. This decision of the narration model is a personalized quest (Charater-based Narration) as the system knows Bob well through his personality profile.

Assume that α=0.2. If Bob chooses Q3 his profile becomes:

MBob=aM+(1-a)Q2=(-0.74,1,-0.8,0.8,-0.16)

Later, two players Celine and David joined in the game in zone B. The system recognizes two players just coming online, so it will send a SMS to inform Alice and Bob to recommend a group race to them. This will lead to a global quest (Global Narration) Q4 for all 4 players. However, as it is a rainy afternoon the virtual circuit in the Bananamoto building is drenched. A Q1 quest is also proposed: maybe some of the players will have to find a new wheel with a better grip.

CONCLUSION

Our study is dedicated to some basic problems in the design of MUG: the complex relationships between the RW and the VW, the types of information covered by the notion of the player's model, and the method to correlate the player's model to the gameplay in general and to the narrative structure in particular. For each of these problems we propose an operational answer. A model of development for an adaptive MUG is also proposed. We have chosen to experiment our scheme in the framework of a game developed by students at the Graduate School of Games and Interactive Media (ENJMIN, France) to demonstrate our concepts with some concrete mixed reality ubiquitous gaming situations. A prototype and a small experiment were carried on by Orange in the city of Rennes (Yan, 2007). According to the observation of players while they play the MugNSRC, the game is capable of proposing adaptable narrative schemes, and offer different types of missions and quests to the players. It is also able to adapt to the hazard of the real world and the preferences of the players. A full experimentation of the game, like the Momentum experimentation, during several days and with at least fifty players is necessary to fully validate our proposals. This requires some heavy technical and financial means, not available at this time. Nevertheless, many problems are still open, in particular, the coherence between the adaptive scheme and the narration. We believe that there is no general answer to this question, which must be considered in the scope of each game or each type of game.

REFERENCES

Bartle, R. A. (1996), Hearts, Clubs, Diamonds, Spades: Players Who Suit MUDs, from http://mud.co.uk/richard/hcds.htm

Bartle, R. A. (2005). In Media, C. R. (Ed.), *Why People Play, Massively Multiplayer Game Development 2, Thor Alexander* (pp. 3–18). Hingham, MA: Virtual Worlds.

Bjork, S., Hansson, J., & Ljungstrand, P. (2001), Pirates! - Using the Physical World as a Game Board, Proceedings of INTERACT IFIP TC.13 Conference on Human-Computer Interaction, 2001.

Björk, S., Holopainen, J., Ljungstrand, P., & Åkesson, K.-P. (2002), Designing Ubiquitous Computing Games - A Report from a Workshop Exploring Ubiquitous Computing Entertainment, Personal and Ubiquitous Computing, January '02, Volume 6, Issue 5-6, pp. 443-458.

Brumitt, B., Meyers, J., & Krumm, A. Kern and Shafer, S. (2000), EasyLiving: Technologies for Intelligent Environments, Proceedings of the International Conference on Handheld and Ubiquitous Computing, Springer, 2000, pp.12-29.

Cavazza, M., Charles, F., & Mead, S. J. (2002), Character-based Interactive Storytelling. In IEEE Intelligent Systems, special issue on AI in Interactive Entertainment, pp. 17-24.

Cheok, A. D., et al. (2003), Human Pacman: A mobile entertainment system with ubiquitous computing and tangible interaction over a wide outdoor area, Proceedings of the 17th Annual Human Computer Interaction Conference, England, Sept. 2003, Springer-Verlag LNCS press.

Codognet, P. (1998), "Artificial Nature and Natural Artifice". Presented at the Art & Technology Conference, Tate Gallery, Liverpool, and at ARCO'02, Madrid, panel discussion on "Art and New Media".

Costa, P. T. Jr, & McCrae, R. R. (1992). *The NEO-PI-R: Professional manual*. Odessa, FL: Psychological Assessment Resources.

Coutaz, J., Lachenal, C., Berard, F., & Barralon, N. (2002). Quand les Surfaces Deviennent Interactives, Les Cahiers du Numérique. *Lavoisier*, *3*(4), 101–126.

Cubaud, P., Dupire, J., & Topol, A. (2005), Digitization and 3D Modeling of Movable Books, ACM-IEEE Joint Conference on Digital Libraries, Denver, USA, June, 2005.

Gallais, Henry, Saphores, Rapine, Guillebon, Roubinet, et al (2007), NSRC gamedoc, available on request at: contact@enjmin.fr

Garlan, D., Siewiorek, A. and Steenkiste, P. (2002), Project Aura: Toward Distraction-Free Pervasive Computing, IEEE Pervasive Computing.

Genvo, S. (2006), Le game design de jeux video, approche communicationnelle et interculturelle, PhD thesis, University of Metz, October 2006. Available at: http://www.omnsh.org/article.php3?id_article=97

Grünvogel, S. M., Vega, L., & Natkin, S. (2004), A new Methodology for Spatiotemporal Game Design, Proc of the Fifth Game-On International Conference on Computer Games: Artificial Intelligence, Design and Education CGAIDE'2004, pp. 109-113.

Guardiola, E. (2000). *Ecrire pour le jeu: Techniques scénaristiques du jeu informatique et vidéo*. Ed. Dixit.

Gustavo, M., & Talmud, I. (2006). The Quality of Online and Offline Relationships, the role of multiplexity and duration. *The Information Society*, 2006.

Khoo, E. T., & Cheok, A. D. (2006). Age Invaders: Inter-generational Mixed Reality Family Game. *The International Journal of Virtual Reality*, *5*(2), 45–50.

Le Prado, C., & Natkin, S. (2007), "Listen Lisboa: scripting languages for interactive musical installations". Sound and Music Computing Conference, SMC'07, Lefkada Greece.

Magerko, B., & Laird, J. E. (2003), Building an Interactive Drama Architecture, Proc of First International Conference on Technologies for Interactive Digital Storytelling and Entertainment, TIDSE'03. Darmstadt, Germany, pp. 226-237.

Mateas, M., & Stern, A. (2003), Integrating Plot, Character and Natural Language Processing in the Interactive Drama Facade, Proceedings of the TIDSE'03, Darmstadt, Germany, Fraunhofer IRB Verlag.

Natkin, S. (2006), Video Games & Interactive Media, A glimpse at new Digital Entertainment, AK Peters Ed, Wesley MA, USA, March, 2006.

Natkin, S., & Yan, C. (2005), Analysis of Correspondences between Real and Virtual Worlds in General Public Applications, Proc of Computer Graphics, Imaging and Visualization (CGIV05), Beijing, July 25-28, 2005, IEEE, 2005, pp. 223-231.

Natkin, S., & Yan, C. (2006), User Model in Multiplayer Mixed Reality Entertainment Applications, Proc of International Conference on Advances in Computer Entertainment Technology ACE'06, California, USA, June, 2006, ACM SIGCHI press.

Peinado, F., & Gervás, P. (2004), Transferring Game Mastering Laws to Interactive Digital Storytelling, Proceedings of the 2nd International Conference on Technologies for Interactive Digital Storytelling and Entertainment (TIDSE'04), 24-26 June, Darmstadt, Germany, LNCS, 3105, 2004, pp. 48-54.

Pierce, J. H., & Jane, M. H. (2003), The Five Factor Model: An Introduction to the Five-Factor Model of Personality, Center for Applied Cognitive Studies (CentACS), Charlotte, North Carolina, 2003. Available: http://www.childrenofmillennium.org/eugenics/pages/articles/bigfive.htm

Szilas, N., Rety, J. H., & Marty, O. (2003), Authoring highly generative Interactive Drama, Proc of International Conference of Virtual Storytelling (ICVS), Toulouse (France), November, 2003.

Thomas, B., Close, B., Donoghue, J., Squires, J., De Bondi, P., Morris, M., & Piekarski, W. (2000), ARQuake: An Outdoor/Indoor Augmented Reality First Person Application, Proc of 4th International Symposium on Wearable Computers, Atlanta, GA, Oct 2000, pp 139-146.

Wolfgang Münch and Kiyoshi Furukawa, (2000), Bubbles, Prototype at Schloss Wahn, Theaterwissenschaftliche Sammlung, Universität Köln, July 2000.

Yan, C. (2007), Jeux Vidéo Multijoueurs Ubiquitaires: principes de conception et architecture d'exécution, PhD dissertation, CNAM, Paris, December 2007.

Chapter 11
Formal Descriptions of Cognitive Processes of Perceptions on Spatiality, Time, and Motion

Yingxu Wang
University of Calgary, Canada

ABSTRACT

Recent researches in both cognitive informatics and computational intelligence are interested in the human perceptual senses of spatiality, time, and motion, which are fundamental cognitive life functions according to the Layered Reference Model of the Brain (LRMB). This paper presents the cognitive process of human perceptual senses on spatiality, time, and motion. The sense of spatiality is investigated into the coordinate system, orientations, and cognitive maps, followed by the development of the mathematical model and the cognitive process of human spatial senses. The sense of time with the biological clocks, cognitive clocks, and their mathematical models are analyzed in order to explain the cognitive process of human time sense. On the basis of the formal models of senses of spatiality and time, the sense of motion is modeled as a complex sense incorporating both of spatiality and time. Then, the cognitive, mathematical, and process models of the sense of motion are rigorously established. This work provides a theoretical framework for the rigorous implementation of the intelligent behaviors of cognitive computers, autonomous agent systems, and robots in cognitive informatics and computational intelligence.

INTRODUCTION

The human perceptual sense of spatiality, time, and motion are fundamental subjects studied in cognitive informatics, physics, cognitive psychology, and computational intelligence (Smith, 1993; Gray, 1994; Pinel, 1997; Matlin, 1998; Westen, 1999; Reisberg, 2001; Wilson and Keil, 1999; Wang and Wang, 2006, 2008; Wang et al., 2006). Cognitive informatics (Wang, 2002a, 2003,

DOI: 10.4018/978-1-60960-553-7.ch011

2007b) is an emerging discipline that studies the internal information processing mechanisms and processes of the natural intelligence - human brains and minds. In cognitive informatics, the human perceptual senses of spatiality, time, and motion are fundamental cognitive life functions according to the Layered Reference Model of the Brain (LRMB) (Wang et al., 2006), which reveals that the brain is functioning with 39 fundamental cognitive processes at 7 layers known as the *sensation, memory, perception, action, meta-cognitive, meta-inference*, and *higher cognitive* layers from the bottom-up.

The cognitive functions of the perception layer of LRMB may be considered as the thinking engine of the brain with a 7-layer hierarchical structure, known as the seven *basic perceptual senses*, supplementary to the five *external sensations* of vision, auditory, smell, tactility, and tastes, which implements self consciousness inside the abstract memories of the brain (Smith, 1993; Pinel, 1997; Matlin, 1998; Westen, 1999; Reisberg, 2001; Wang et al., 2006).

Definition 1. The perception layer is a subconscious layer of life functions of the brain for maintaining conscious life functions and browsing internal abstract memories in the cognitive models of the brain.

The perception layer of LRMB is a part of the subconscious life functions. The perception layer is the internal sensory layer that encompasses consciousness, stimuli, emotions, motivations, attitudes, sense of behaviors, sense of spatiality, sense of time, and sense of motion. This article puts emphases on the senses of spatiality, time, and motion, while the other perceptual processes at the perception layer of LRMB may be referred to (Wang, 2007c; Wang and Wang, 2008).

The entire human perceptual senses can be described in the 7-layer hierarchical model as shown in Figure 1, where the 7-layers perceptional senses are: L1 – stimuli; L2 – subconsciousness; L3 – consciousness; L4 – spatiality; L5 – time; L6 – motion; and L7 - behaviors from the bottom-up.

Figure 1. The hierarchical model of human perceptual senses

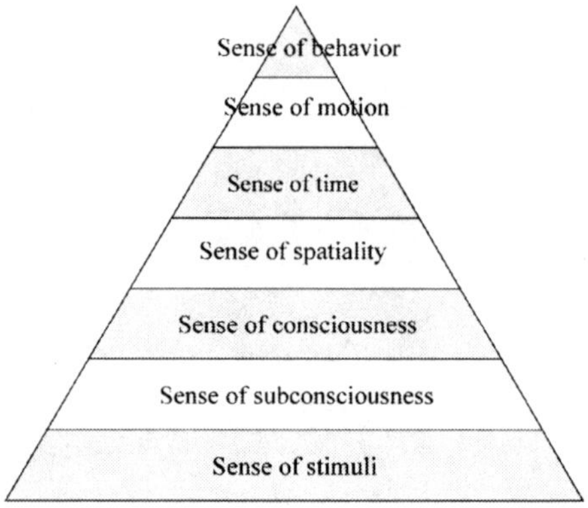

This article formally presents the cognitive process of perceptions on spatiality, time, and motion. The human senses of spatiality such as the coordinate system, orientations, and cognitive maps are investigated, which leads to the establishment of the mathematical model and the cognitive process of human spatial sense. Then, the sense of time from the aspects of biological clocks, cognitive clocks, and their mathematical models are analyzed in order to explain the cognitive process of time. Based on the formal models developed in the preceding sections, the senses of space and time, the sense of motion as a complex sense incorporating both of spatiality and time is elaborated with a set of cognitive, mathematical, and process models.

THE SENSE OF SPATIALITY

The sense of spatiality is not only studied in physics, but also interested in cognitive informatics, psychology, and computational intelligence.

Definition 2. The sense of spatiality is the most fundamental awareness and perception of the surrounding environment of a person, which encompasses the coordinate system, orientations, and cognitive maps.

Although, the spatial sense seems intuitive to adults, they are not easy for young children to acquire, and also are nontrivial capabilities for robots and machine intelligent systems (Jazar, 2007; Martin, 1998).

The Cognitive Foundation of Human Spatial Sense

Although the sense of spatiality of the brain is usually established during the early phase of children development, the concept or logical model of it, is acquired until 3 year-old, even as old as the age attending elementary schools.

The Coordinate Model of the Physical Space

The prominent senses of space are the three-dimensional (3-D) coordinates, i.e. *above/below* (*up/down*), *ahead/behind* (*front/back*), and *left/right*. The comprehensive complexity and difficulty in obtaining these pairs of sensational concepts is in an increasing order as given above (Matlin, 1998; Tversky et al., 1999). That is, the vertical senses up and down are most easily acquired because of the universal presence of the gravity effect, followed by front and back because it is intuitive and physical. However, some children may have difficulty to acquire the concepts of left and right until school time, because the pair of left and right is purely an abstract or logical sense.

The cognitive complexity of the coordinative spatial sense can be explained by a set of comparative analyses in Table 1. As given in Table 1, the pair of concepts, up and down, is absolute, which never change in any case, with any person, or toward any target property (objects or human beings) in the gravity system. However, the senses of left/right are relative, which are dependent on the targets if they are an object or human, and if they are perceived in the physical world or virtual world reflected in a mirror. Between the two extreme dimensions, the dimension of front and back is always opposite from the points of views of the observer and the target. The varying cognitive complexities of different coordinates, particularly those of the left/right dimension, explain why the mirror effect is always confusing people in everyday life.

Table 1. Analyses of coordinative spatial senses

Reference System	Target	Senses of Spatiality		
		Up/ Down	Front/ Back	Left/Right
In real world	Object	Same	Opposite	Same
	Human	Same	Opposite	Crossed
In mirror	Object	Same	Opposite	Crossed
	Human	Same	Opposite	Same
Cognitive complexity		*Low*	*Medium*	*High*

It is noteworthy that human spatial behaviors are almost always a closed loop, i.e., from home to a working place and then back home. Inside this primary loop, there are more inner loops in daily work and behavior. Further, for each inner loop of a certain behavior, there are routine and iterative procedures. Therefore, the spatial traces of human behaviors can be modeled as a series of embedded loops, each of them can be described by a behavioral cognitive process (Wang, 2008b, 2008d).

The Absolute and Relative Models of Orientations

The second category of human spatial senses is orientations that are an abstract sense about the spatiality.

Definition 3. An orientation is the directional angle of a vector in the physical space.

The orientations can be classified as *absolute* and *relative* orientations. The former is a set of directional angles in the physical space identifies by its 3-D coordinates, which may be simplified as the four directions of north, south, east, and west in the plane cases. The latter refers to the spatial relations between an object and an observer, which may be described as up/down, front/back, and left/right.

The cognitive senses of orientations of individuals are originally created and connected to the home of a person of childhood. The primary physical sense of vision and gravity are closely related to the logical senses of spatiality and orientations. Without them, these logical senses cannot be mapped onto the physical space in order to allow their cognitive comprehension.

The Abstract Space of Cognitive Maps

The third category of human spatial senses is the cognitive maps, which are a virtual analog representation and/or abstract semantic images of the real world geographical and spatial layouts in the long-term memory of the brain (Matlin, 1998).

Definition 4. The cognitive map is an internal analog or abstract map that visually and/or semantically represents a person's geographical knowledge about spaces, locations, positions, and orientations in the real-world.

The cognitive map is our geographical spatial knowledge about a city, a street, a building, or a room. The scopes of cognitive maps can be zoomed from the entire universe to a single object as small as a unit pixel of vision. The generation of a cognitive map costs certain mental power and memory. It also requires multiple rounds of explorations, learning, and rehearsals.

The scope of cognitive maps representing the physical world can be varying from small-scale local areas to large-scale global areas. Although the scope of cognitive maps is varying, it is constrained by the fundamental size of a unit visual frame of humans. The product of the scope and resolution of a cognitive map, or its amount of pixels, is relatively constant, because it is limited by the size of a vision frame (Wang, 2008c). That is, the larger the scope, the lower the resolution; and vice versa.

Theorem 1. The conservation of cognitive maps states that the area of a scope A is inversely proportional to the resolution R of a given sized cognitive map S_{CM}, subject to the size of the visual frame S_{vf} is a constant, i.e.:

$$\begin{aligned} S_{CM} &\triangleq S_{vf} = A \bullet R \\ &= 2,363\ [bit] \end{aligned} \quad (1)$$

In Theorem 1, the size of the visual frame S_{vf} is determined as 2,363 bits according to the experiments reported in (Wang, 2008c).

The identification and establishment of the relationship between an internal cognitive map, a map on paper, and the real-world geography is a typical and complex spatial recognition capability of human beings.

Mathematical Models of the Physical Space

On the basis of preceding discussions, the mathematical model of the sense of spatiality can be formally developed in this subsection.

Definition 5. The physical space $\vec{S}$ of the universe is a 3-dimensional Cartesian product, i.e.:

$$\begin{aligned} \vec{S} &\triangleq f(x, y, z) \\ &= \vec{X} \times \vec{Y} \times \vec{Z} \end{aligned} \quad (2)$$

where $x \in \vec{X}, y \in \vec{Y}, z \in \vec{Z}$, and $\vec{X}, \vec{Y}$, and $\vec{Z}$ and are three perpendicularly finite or infinite sets of coordinates.

It is noteworthy that the physical space is modeled as a vector structure, where each dimension is a vector. The 3-D vector model of spatiality may also explain the sense of orientation.

The axiom of the abstract universe is that it is infinitive in both space and time. The infinitive nature of the former helps to avoid the paradox that queries what the things are outside the universe. The infinitive nature of the latter avoids the paradox that queries when the universe originated and when it will cease to exist.

Definition 6. The position of an object p in the physical space $\vec{S}$ is a triple of specific 3-D coordinates, i.e.:

$$p \triangleq (x, y, z), x \in \vec{X}, y \in \vec{Y}, z \in \vec{Z}, p \sqsubseteq \vec{S} \quad (3)$$

On the mathematical models of the physical space and positions, cognitive orientations can be formally modeled below.

Definition 7. An absolute orientation ψ of a space vector is its directional angles within the 3-D physical space, i.e.:

$$\begin{cases} \psi_x \triangleq \cos^{-1} \dfrac{x}{\sqrt{x^2 + y^2 + z^2}} \\ \psi_y \triangleq \cos^{-1} \dfrac{y}{\sqrt{x^2 + y^2 + z^2}} \\ \psi_z \triangleq \cos^{-1} \dfrac{z}{\sqrt{x^2 + y^2 + z^2}} \end{cases} \quad (4)$$

The absolute orientation is usually denoted by a projection to the four planar directions of *north, south, east, and west*.

Definition 8. The relative orientation $\overrightarrow{\psi_r}$ of an object is its relative position p_j with the observer p_o, i.e.:

$$\begin{aligned} \overrightarrow{\psi_r} &\triangleq p_o \mapsto p_j \\ &= (x_o \mapsto x_j, y_o \mapsto y_j, z_o \mapsto z_j) \end{aligned} \quad (5)$$

where a denotes a determination of the relative direction of the vector between two positions in the physical space.

The relative orientation is usually denoted by *up/down, front/back, and left/right*, from the point of the observer's view.

Definition 9. The cognitive map CM of an individual is an abstract semantic memory as a 4-tuple, i.e.:

$$CM \triangleq (O, P, \Psi, R) \quad (6)$$

where

- O is a nonempty set of visual objects, $O \triangleq \{o_1, o_2, \ldots, o_n\}$
- P is a nonempty set of positions of the objects, $P \triangleq \{p_1, p_2, \ldots, p_n\}$
- Ψ is a nonempty set of orientations of the objects, $\Psi \triangleq \{\psi_1, \psi_2, \ldots, \psi_n\}$
- R is a nonempty set of relations between the objects and the physical space, $\vec{R} \triangleq O \times \vec{S}$

The Cognitive Process of Sense of Spatiality

Based on the mathematical model of spatial senses as described in Definitions 5, 6, and 8, a formal description of the cognitive process of spatial sense in Real-Time Process Algebra (RTPA) (Wang, 2002b, 2007a, 2008a, 2008b) is presented in Figure 2. The spatial sense process is divided into three steps known as: (i) Form a specific cognitive map (sCM); ii) Update the entire cognitive map (CM); and (ii) Memorization. In this process, two other meta processes, *VisionSensary*ST and *Memorization*ST, are invoked. The memorization step may be skipped when the spatial sense process is composed with or called by other processes.

Figure 2. The cognitive process of spatial sense in RTPA

```
The Process of Sense of Spatiality

SpatialSenseProcess (I:: oS, CMST, OARST;
                     O:: CM'ST, OAR'ST)ST ≙
{ I. Form a specific cognitive map (sCM)
  ↣ VisionSensation (I:: oS;  O:: PSET, ΨSET, RSET)ST
  → pST := (xoP, yoP, zoP)ST
  → ψS := ψS | ψS ∈ ΨSET = {up, down, front, back, left, right}
                           // Determine relative orientation
  → rST := oS × pST
  → sCMST := (oS, pST, ψST, rST)ST

  II. Update the entire cognitive map (CM)
  → CM'ST := CMST ⊎ sCMST

  III. Memorization
  → CM'ST := CMST ⊎ sCMST
  → OAR'ST := OARST ⊎ CMST
  ↣ Memorization (OAR'ST) ST
}
```

THE SENSE OF TIME

Time is a relative measure of moments, durations, sequence of events, changes, actions, and growth. It is also the 4th dimension of the physical universe, or *space-time*, which is logically infinitive without origin and end. Time is always a logic dimension of the physical and living systems, where existence can never be reversed. This is why a philosophical thought states that one cannot cross the same river for twice.

Definition 10. The sense of time is a conscious awareness of the continued progress of existence in the past, present, and future in the forms of moments, durations, and orders of occurrence of events.

Cognitive Perception of Time

Albert Einstein viewed that "the distinction between past, present, and future is only an illusion, however persistent (Einstein and Besso, 1972)." The sense of time is needed for most crucial human life functions such as activity planning, observing sequence of events, execution of dynamic behaviors, coordination of actions, predicating motions, determining priority of activities, and synchronization in groups and societies.

Figure 3. Sense of time by internal and external clocks

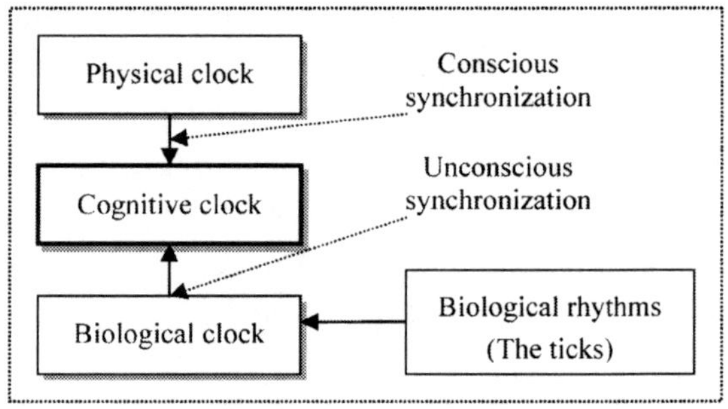

The sense of time can be classified by the external and internal clocks known as the biological, cognitive, and physical clocks from the bottom-up, as shown in Figure 3, on the basis of the biological rhythms.

The Biological Clock

Definition 11. The biological clock is an unconscious and subjective perception of time based on biological and physiological rhythms of human bodies.

The sources of time as an abstract concept originated from sun light or the cyclic geometrical relations between the earth and sun. Time is perceived as an infinitive sequence of points of moments, where the present is a dynamic moment in the flow of time with both of its future and past as infinitive.

The source of the biological clock, as the original but subconscious ticks, is the biological rhythms of human bodies. Typical pacemakers of the biological rhythms are sleep-awake cycle, heartbeats, breathing, hormonal/metabolic activities, pulse, blood pressure, and temperature of the body (Matlin, 1998). The suprachiasmatic nucleus located in the hypothalamus is proposed as responsible for the circadian rhythm (Rusak and Zucker, 1979).

Many physical, biological, and physiological factors may influence the sense of time. For instance, the perception of the biological pace is proportional to the temperature of human body as observed in (Coaen et al., 1993).

The Cognitive Clock

Definition 12. The cognitive clock is a conscious and subjective perception of time based on both the biological clock and the external physical clock.

As shown in Figure 3, the human sense of time in the form of the cognitive clock is both a subconscious (subjective) perception of the biological clock and a conscious (relative) perception of the external physical clock.

Corollary 1. The brain is an asynchronous system, because there is no obvious central clock at the conscious and subconscious psychological levels, or the physiological level.

There is only a relative clock (refers to external time), which the brain adopts, to synchronize daily activities with the external world. An experiment in a group of people who were living in isolated deep-underground cave, and without any clue of external time and the day-night cycle of the earth, has shown that the relative cognitive clock based on the physiological rhythmic cycle tends to synchronize at a 25-hours cycle per day rather than 24 hours (Wever, 1979; Aschoff, 1984). This observation might indicate that human beings still keep an ancient rhythmic cycle which formed several hundred million years ago that was slower than the modern time. In other words, the prehistory daily cycle of the earth must has been longer than what we currently have (Wang and Wang, 2006).

The Mathematical Model of Time

Axiom 1. Time is an abstract notion of the transitions of nature.

Axiom 2. The same metric of time can be used in different frames or reference systems independent from different observers.

Definition 13. The metric of time as an extended calendar, t, can be logically modeled with scopes of millisecond (ms), second (ss), minute (mi), hour (hh), day (dd), month (mm), and year (yyyy) from bottom-up, i.e.:

$$t \triangleq (yyyy,\ mm,\ dd,\ hh,\ mi,\ ss,\ ms) \quad (7)$$

The minimum interval of human sense is identified as within the scope of 25ms to 150ms (Coaen et al., 1993).

Theorem 2. Time $\vec{T}$ is a vector that is always monotonically ascending, i.e.:

$$\forall i, j \in \mathbf{N}, i < j \Rightarrow t_i < t_j, \quad t_i, t_j \in \vec{T} \quad (8)$$

Theorem 2 represents the logical causality for a sequence of events. Because any cause must always occur before its effect(s). Theorem 2 also ensures that effects always follow the cause over time.

Corollary 2. The lifecycle of a living system is unidirectional. Because any latest step of the growth is based on the preceding one(s).

The Cognitive Process of Time

Based on the mathematical model of time as described in Definition 13 and Theorem 2, a formal description of the cognitive process of sense of time in RTPA is presented in Figure 4. The sense of time process is divided into three steps known as: (i) Maintain absolute and relative time; (ii) Aware of duration; and (iii) Memorization.

The cognitive sense of time *t*yyyy:mm:dd:hh:mi:ss:ms and duration ΔtN is based on an external reference clock §*t*yyyy:mm:dd:hh:mi:ss:ms. Although both *t*ST and ΔtN are updated subjectively by frequent

Figure 4. The cognitive process of time sense in RTPA

```
The Process of Sense of Time

TimeSenseProcess (I:: §tyyyy:mm:dd:hh:mi:ss:ms;
                  O:: OAR'ST,
                      cₜ(tyyyy:mm:dd:hh:mi:ss:ms, -, -)ST,
                      cτ(τN, -, -)ST, cΔt(ΔtN, -, -)ST)ST ≙
{ I. Maintain absolute and relative time
  tyyyy:mm:dd:hh:mi:ss:ms := §tyyyy:mm:dd:hh:mi:ss:ms
                     // Syncronise to a system or public colock
  ↯ @1ms_Tick⊙ ↗
    (// Absolute clock
      → ↑ (tyyyy:mm:dd:hh:mi:ss:ms)
          mod 9999:12:(31/30/29/28):24:59:59:999
     // Relative clock
      → ( ◆ τhh:mi:ss:ms ? 23hh:59mi:59ss:999ms
              → ↑ (τN)
          | ◆~
              → τN := 0
        )
    )
  ↘ ⊙

  II. Aware of duration
   → τᵢN := τN
   →   ...
   → τⱼN := τN
   → ΔtN := τⱼN - τᵢN, j > i

  III. Memorization
   → cₜ(tyyyy:mm:dd:hh:mi:ss:ms, -, -)ST
   → cτ(τN, -, -)ST
   → cΔt(ΔtN, -, -)ST
   → OAR'ST := OARST ⊎ {cₜ(tyyyy:mm:dd:hh:mi:ss:ms, -, -)ST,
                         cτ(τN, -, -)ST, cΔt(ΔtN, -, -)ST}
   → Memorization (OAR'ST)ST
}
```

synchronizations to the external clock, a rigorous updating of the them can be modeled as a process as given in Step (i) at the 1ms interrupt level. In the sense of time process, the meta-processes of *Memorization*ST is invoked, which may be skipped when the sense of time process is composed with or called by other processes.

THE SENSE OF MOTION

Definition 14. The sense of motion is an important and complex awareness and perception about human behaviors, external animate or moving objects, the surrounding environment, and their real-time interactions.

Motions are particularly useful in modeling human behaviors and machine intelligence. It is noteworthy that the sense of motion is a complex sense based on both space and time. There is no motion without taking time or without a displacement in space. In "*Principles of Mathematics*," Russell perceived that "Motion is the occupation, by one entity, of a continuous series of places at a continuous series of times (Russell, 1901)." Hermann Minkowski (1864-1909) proposed the notion of space-time (Minkowski, 1908), which is intensively investigated by Einstein in his *special theory of relativity* (Einstein, 1905, 1916, 1995).

Taxonomy of Motions

Definition 15. Motion is a relative relation between an object and an observer in space over a series of continuous or discrete timing points.

It is noteworthy that motions are a relative and dual sense that depends on if any or both the object and observer are moving or at rest. When the reference frames are different, a motion may become relative rather than absolute.

According to the relationships of the reference systems for both the object and observer in space, motions can be classified into eight forms known as

absolute rest,
absolute observer motion,
absolute object motion,
absolute relative motion;
Relative rest,
relative observer motion,
relative object motion,
relative absolute motion

As shown in Table 2. In Table 2, $\overrightarrow{M}_X$ denotes a motion in reference frame *X*, and $\overrightarrow{M}_X$ indicates no motion related to frame *X*.

The contradictive sense in complex motions between the physical sense and the logical/cogni-

Table 2. Taxonomy of motions

Mode	Reference Systems	Object	Observer	Form of Motion
1	Identical (X)	$\overline{M}_X$	$\overline{M}_X$	Absolute rest
2		$\overline{M}_X$	$\vec{M}_X$	Absolute observer motion
3		$\vec{M}_X$	$\overline{M}_X$	Absolute object motion
4		$\vec{M}_X$	$\vec{M}_X$	Absolute relative motion
5	Different (X and Y)	$\overline{M}_X$	$\overline{M}_Y$	Relative rest
6		$\overline{M}_X$	$\vec{M}_Y$	Relative observer motion
7		$\vec{M}_X$	$\overline{M}_Y$	Relative object motion
8		$\vec{M}_X$	$\vec{M}_Y$	Relative absolute motion

tive sense of motions may result in significant physiological effects and kinetic uncomfortable such as seasickness, airsickness, and carsickness.

The most important property of motions is its velocity, which is a relation between the displacement of an object and the time spent for it (Cutnell and Johnson, 1998).

Definition 16. The instantaneous velocity $\vec{v}$ is a limit of the displacement $\vec{d}$ of an object over time t, i.e.:

$$\vec{v} = \lim_{\Delta t \to 0} \frac{\Delta \vec{d}}{\Delta t} \ [m / s] \tag{9}$$

Definition 17. The average velocity $\overline{\overline{v}}$ is the displacement of an object $\vec{d}$ per unit time t, i.e.:

$$\overline{\overline{v}} = \frac{\vec{d}}{t} \ [m / s] \tag{10}$$

The relative velocity of an observed object is dependent on the mode of its movement as illustrated in Figure 5 and Table 2.

Figure 5. The relative light speed in a spacecraft and observed on earth

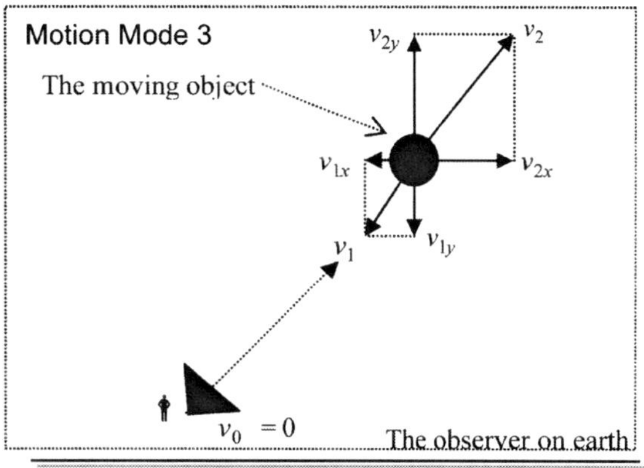

Theorem 3. The *relative velocity* $\vec{v}'$ of a moving object with a velocity $\overrightarrow{v_s'}$ in a moving frame with a velocity as observed in another reference frame *X* outside the given frame *S* is:

$$\overrightarrow{v_X'} = \overrightarrow{v_S} + \overrightarrow{s_X} \tag{11}$$

When the two frames *X* and *S* are the same, i.e., $\overrightarrow{s_X} = 0$, then $\overrightarrow{v_X'} = \overrightarrow{v_S}$

Mathematical Models of Motions

The model of motions is based on the physical and perceptual models of space-time as proposed by Hermann Minkowski (1864-1909).

Definition 18. The space-time $\vec{\mathfrak{S}}$ of motion in the universe is 4-dimensional known as the dimensions of the 3-D physical space $\vec{X}, \vec{Y}, \vec{Z}$ and $\vec{T}$,and, i.e.:

$$\begin{aligned} \vec{\mathfrak{S}} &\triangleq f(x, y, z, t) \\ &= \vec{X} \times \vec{Y} \times \vec{Z} \times \vec{T} \end{aligned} \tag{12}$$

where $x \in \vec{X}, y \in \vec{Y}, z \in \vec{Z}$ and $t \in \vec{T}$.

Figure 6. The 4-D space-time $\vec{\mathfrak{S}}$ and a series of movements

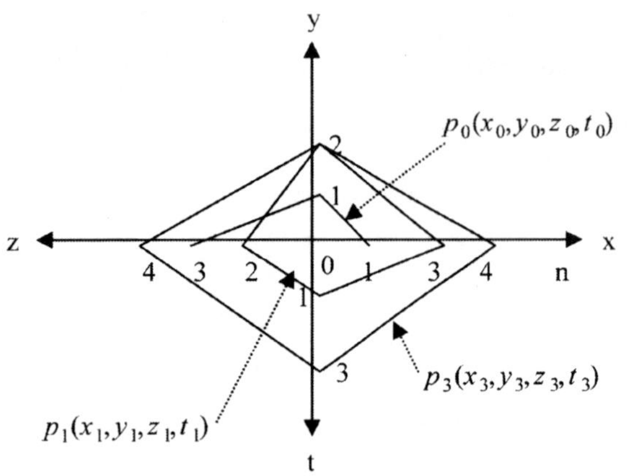

According to Definition 18, the space-time $\vec{\mathfrak{S}}$ can be modeled as shown in Figure 6. It is noteworthy that is a vector model, where each dimension $\vec{X}, \vec{Y}, \vec{Z}$ or $\vec{T}$ *or* is a vector.

The mathematical model of motion can be defined as a single or a series of transition(s) of an object's positions in the space over time.

Definition 19. An abstract motion $\vec{m}$ is a series of n transitions of positions $\vec{p}_i(x_i, y_i, z_i, t_i)$ over time $\vec{T}$ in $\vec{\mathfrak{S}}$ i.e.:

$$\vec{m}(x, y, z, \Delta t) \triangleq \mathop{\mathrm{R}}_{t=0}^{n-1}(\vec{p_t}(x_t, y_t, z_t, t_t) \rightarrow \overrightarrow{p_{t+1}}(x_{t+1}, y_{t+1}, z_{t+1}, t_{t+1})), n > 1 \tag{13}$$

Therefore, a single movement of an object is a motion in unit time as given below.

Definition 20. An space (3-D) displacement $\vec{d}$ of an object in $\vec{\mathfrak{S}}$ is a motion over a unit time period Δt, i.e.:

$$\begin{aligned}\vec{d}(x, y, z, \Delta t) &\triangleq \Delta\vec{p}(x, y, z, \Delta t) \\ &= \vec{p}_2(x_2, y_2, z_2, t_2) - \vec{p}_1(x_1, y_1, z_1, t_1)\end{aligned} \tag{14}$$

where $\Delta t = t_2 - t_1 = 1$ is a unit time in terms of yyyy, mm, dd, hh, mi, ss, or ms.

As illustrated in Figure 6, the perception of motion, in terms of the trace of a series of displacements can be described as:

$$(\vec{p}_0(1,1,3,0) \rightarrow \vec{p}_1(3,2,2,1) \rightarrow \ldots \rightarrow \vec{p}_3(4,2,4,3) \rightarrow \ldots)$$

or shortly

$$(\vec{p}_0(1,1,3) \rightarrow \vec{p}_1(3,2,2) \rightarrow \ldots \rightarrow \vec{p}_3(4,2,4) \rightarrow \ldots)$$

It is noteworthy in Definition 20 that a *space displacement* $\vec{d}(x, y, z, \Delta t)$ is a complex vector. Simple cases of displacement may be derived as partial differentials in each dimension or each pair of dimensions. The former is known as the linear displacement and the latter the planar displacement.

Definition 21. The linear displacement in each single dimension of the 3-D physical space $\vec{S}(x, y, z)$ is the rate of transitions in a specific dimension over time, i.e.:

$$\begin{cases}\vec{d}_x(x, y, z, \Delta t) \triangleq \dfrac{\partial\vec{d}(x, y, z, \Delta t)}{\partial x} = \vec{p}_{x2}(x_2) - \vec{p}_{x1}(x_1) \\ \vec{d}_y(x, y, z, \Delta t) \triangleq \dfrac{\partial\vec{d}(x, y, z, \Delta t)}{\partial y} = \vec{p}_{y2}(y_2) - \vec{p}_{y1}(y_1) \\ \vec{d}_z(x, y, z, \Delta t) \triangleq \dfrac{\partial\vec{d}(x, y, z, \Delta t)}{\partial z} = \vec{p}_{z2}(z_2) - \vec{p}_{z1}(z_1)\end{cases} \tag{15}$$

Definition 22. The planar displacement in each plane of the 3-D space $\vec{S}(x, y, z)$ is the rate of transitions in a specific pair of dimensions over time, i.e.:

$$\begin{cases}\vec{d}_{xy}(x, y, z, \Delta t) \triangleq \dfrac{\partial^2\vec{d}(x, y, z, \Delta t)}{\partial x \partial y} = \vec{p}_{xy2}(x_2, y_2) - \vec{p}_{xy1}(x_1, y_1) \\ \vec{d}_{yz}(x, y, z, \Delta t) \triangleq \dfrac{\partial^2\vec{d}(x, y, z, \Delta t)}{\partial y \partial z} = \vec{p}_{yz2}(y_2, z_2) - \vec{p}_{yz1}(y_1, z_1) \\ \vec{d}_{xz}(x, y, z, \Delta t) \triangleq \dfrac{\partial^2\vec{d}(x, y, z, \Delta t)}{\partial x \partial z} = \vec{p}_{xz2}(x_2, z_2) - \vec{p}_{xz1}(x_1, z_1)\end{cases} \tag{16}$$

A human behavior can be described by the same motion or displacement models as given in Definitions 15 through 22.

On the basis of the 4-D space-time $\vec{\mathfrak{S}}$, the space velocity of a moving object can be described below.

Definition 23. The space velocity of a moving object $\vec{v}(x, y, z, t)$ is the average rate of its transition of positions in the 3-D space, $\Delta\vec{p}(x, y, z)$ over a unit time Δt, i.e.:

$$\begin{aligned}\vec{v}(x,y,z,t) &\triangleq \lim_{\Delta t \to 0} \frac{\vec{d}(x,y,z,t)}{\Delta t} \\ &= \lim_{\Delta t \to 0} \frac{\Delta\vec{p}(x,y,z,t)}{\Delta t} \\ &= \lim_{\Delta t \to 0} \frac{\vec{p}_2(x_2,y_2,z_2,z_2) - \vec{p}_1(x_1,y_1,z_1,z_1)}{\Delta t}\end{aligned} \tag{17}$$

According to Definition 23, the linear and planar velocity can be derived as follows.

Definition 24. The linear velocity of a moving object along a single dimension, $\vec{v}_x, \vec{v}_y, or\ \vec{v}_z$ is a projected velocity on each dimension in $\vec{\mathfrak{S}}$, i.e.:

$$\begin{cases} \vec{s}_x(x,y,z,t) \triangleq \dfrac{\partial\vec{d}(x,y,z,t)}{\partial x} = \lim\limits_{\Delta t \to 0} \dfrac{\vec{p}_{x2}(x_2) - \vec{p}_{x1}(x_1)}{\Delta t} \\ \vec{s}_y(x,y,z,t) \triangleq \dfrac{\partial\vec{d}(x,y,z,t)}{\partial y} = \lim\limits_{\Delta t \to 0} \dfrac{\vec{p}_{y2}(y_2) - \vec{p}_{y1}(y_1)}{\Delta t} \\ \vec{s}_z(x,y,z,t) \triangleq \dfrac{\partial\vec{d}(x,y,z,t)}{\partial z} = \lim\limits_{\Delta t \to 0} \dfrac{\vec{p}_{z2}(z_2) - \vec{p}_{z1}(z_1)}{\Delta t} \end{cases} \tag{18}$$

Definition 25. The planar velocity of a moving object in a plane $\vec{v}_{xy}, \vec{v}_{yz}, or\ \vec{v}_{xz}$ is a projected velocity on each pair of dimensions in $\vec{\mathfrak{S}}$, i.e.:

$$\begin{cases} \vec{s}_{xy}(x,y,z,t) \triangleq \dfrac{\partial^2\vec{d}(x,y,z,t)}{\partial x \partial y} = \lim\limits_{\Delta t \to 0} \dfrac{\vec{p}_{xy2}(x_2,y_2) - \vec{p}_{xy1}(x_1,y_1)}{\Delta t} \\ \vec{s}_{yz}(x,y,z,t) \triangleq \dfrac{\partial^2\vec{d}(x,y,z,t)}{\partial y \partial z} = \lim\limits_{\Delta t \to 0} \dfrac{\vec{p}_{yz2}(y_2,z_2) - \vec{p}_{yz1}(y_1,z_1)}{\Delta t} \\ \vec{s}_{xz}(x,y,z,t) \triangleq \dfrac{\partial^2\vec{d}(x,y,z,t)}{\partial x \partial z} = \lim\limits_{\Delta t \to 0} \dfrac{\vec{p}_{xz2}(x_2,z_2) - \vec{p}_{xz1}(x_1,z_1)}{\Delta t} \end{cases} \tag{19}$$

The Cognitive Process of Sense of Motions

Based on the mathematical model of motions, as well as those of space and time, as described in Definitions 18, 19, 20, and 23, a formal description of the cognitive process of motion in RTPA is presented in Figure 7. The sense of motion process is divided into four steps known as: (i) Determine space positions and displacements; (ii) Identify motion in space velocity; (iii) Identify motion in linear velocity; and (iv) Memorization. In the sense of motion process, three other meta-processes, *SpaceSense*ST, *TimeSense*ST, and *Memorization*ST, are invoked. The memorization step may be skipped when the motion sense process is composed with or called by other processes.

Figure 7. The cognitive process of motion sense in RTPA

The Process of Sense of Motion

MotionSenseProcess (**I**:: o**S**; **O**:: v**ST**, OAR'**ST**) $\triangleq$

{ I. Determine space positions and displacements

→ TimeSenseProcess (**I**:: o**S**; **O**:: t_1**SS**)**ST**

↣ SpaceSenseProcess (**I**:: o**S**; **O**:: $p_1(x_1, y_1, z_1, t_1)$)**ST**

↣ TimeSenseProcess (**I**:: o**S**; **O**:: t_2**SS**)**ST**

↣ SpaceSenseProcess (**I**:: o**S**; **O**:: $p_2(x_2, y_2, z_2, t_2)$)**ST**

II. Identify motion in space velocity

$$\rightarrow \vec{v}(x,y,z,\Delta t) = \frac{\vec{p_2}(x_2,y_2,z_2,t_2) - \vec{p_1}(x_1,y_1,z_1,t_1)}{t_2 - t_1}$$

III. Identify motion in linear velocity

$$\rightarrow \begin{cases} \vec{v_x}(x,y,z,\Delta t) = \dfrac{\vec{p_2}(x_2,t_2) - \vec{p_1}(x_1,t_1)}{t_2 - t_1} \\ \vec{v_y}(x,y,z,\Delta t) = \dfrac{\vec{p_2}(y_2,t_2) - \vec{p_1}(y_1,t_1)}{t_2 - t_1} \\ \vec{v_z}(x,y,z,\Delta t) = \dfrac{\vec{p_2}(z_2,t_2) - \vec{p_1}(z_1,t_1)}{t_2 - t_1} \end{cases}$$

IV. Memorization

→ sOAR**ST** := {c($v(x, y, z, t)$, -, -)**ST**, c($v_x(x, y, z, t)$, -, -)**ST**, c($v_y(x, y, z, t)$, -, -)**ST**, c($v_z(x, y, z, t)$, -, -)**ST**}

→ OAR'**ST** := OAR**ST** ⊎ sOAR**ST**

↣ Memorization (OAR'**ST**)

}

CONCLUSION

This article has presented the cognitive processes of perceptions on spatiality, time, and motion, which have been recognized as fundamental cognitive life functions of human beings. The senses of spatiality of humans such as the coordinate system, orientations, and cognitive maps, have been rigorously explored, which have resulted in the mathematical model and the cognitive process of human spatial sense. The sense of time from the aspects of biological clocks, cognitive clocks, and their mathematical models has been formally studied, which have led to the explanation of the cognitive process of time. The sense of motion as a complex sense incorporating both of spatiality and time has been described by the cognitive, mathematical, and process models. The cognitive process models of the senses of space, time, and motion, as well as their mathematical models, have revealed a coherent theory of the hierarchical model of human perceptual senses of LRMB.

ACKNOWLEDGMENT

This work is partially sponsored by the Natural Sciences and Engineering Research Council of Canada (NSERC). The author would like to thank the anonymous reviewers for their valuable suggestions and comments on this work.

REFERENCES

Aschoff, J. (1984). Circadian Timing. *Annals of the New York Academy of Sciences*, *423*, 442–468. doi:10.1111/j.1749-6632.1984.tb23452.x

Coaen, S., Waad, L. M., & Enns, J. T. (1993). *Sensation and Perception* (4th ed.). Fort Worth, USA: Harcourt Brace College Publishers.

Cutnell, J. C., & Johnson, K. W. (1998). *Physics* (4th ed.). NY: John Wiley & Sons.

Einstein, A. (1905), On the Electrodynamics of Moving Bodies, *Annalen der Physik,* 17(891), June, (English translation in 1922).

Einstein, A. (1916), The Foundation of the General Theory of Relativity, *Annalen der Physik,* 49.

Einstein, A. (1995). *Relativity: The Special and the General Theory*. Reprint, Three Rivers Press.

Einstein, A., & Besso, M. (1972). *Correspondence, 1903–1955, Translated by P. Speziali from French.* Paris: Hermann.

Gray, P. (1994). *Psychology* (2nd ed.). New York: Worth Publishers, Inc.

Jazar, R. N. (2007). *Theory of Applied Robotics: Kinematics, Dynamics, and Control*. Berlin: Springer.

Matlin, M. W. (1998). *Cognition* (4th ed.). Orlando, FL: Harcourt Brace College Publishers.

Minkowski, H. (1908), *Space and Time*, Address, 80th Assembly of German Natural Scientists and Physicians, Cologne, Sept.

Pinel, J. P. J. (1997). *Biopsychology* (3rd ed.). Needham Heights, MA: Allyn and Bacon.

Reisberg, D. (2001), Cognition, second edition, Exploring the science of the mind, W.W. Norton & Company, Inc.

Rusak, B., & Zucker, I. (1979). Neural Regulation of Circadian Rhythms. *Physiological Reviews*, *59*, 449–526.

Russell, B. (1901), Is Position in Time and Space Absolute or Relative? *Mind*, July, London.

Smith, R. E. (1993). *Psychology*. St. Paul, MN: West Publishing Co.

Tversky, B., N. Franklin, H.A. Taylor, D.J. Bryant (1999), Spatial mental models from descriptions, *Journal of the American Society for Information Science*, Jan., 45(9), *pp. 656 – 668.*

Wang, Y. (2002a), Keynote, On Cognitive Informatics, *Proc. 1st IEEE International Conference on Cognitive Informatics* (ICCI'02), Calgary, Canada, IEEE CS Press, August, pp. 34-42.

Wang, Y. (2002b). The Real-Time Process Algebra (RTPA), *Annals of Software Engineering: An International Journal. Baltzer Science Publishers, Oxford, 14*(Oct), 235–274.

Wang, Y. (2003), On Cognitive Informatics, *Brain and Mind: A Transdisciplinary Journal of Neuroscience and Neurophilosophy*, Kluwer Academic Publishers, August, 4(3), pp. 151-167.

Wang, Y. (2007a), *Software Engineering Foundations: A Software Science Perspective*, CRC Book Series in Software Engineering, Vol. II, Auerbach Publications, NY., USA, July.

Wang, Y. (2007b). The Theoretical Framework of Cognitive Informatics, *International Journal of Cognitive Informatics and Natural Intelligence. IGI Publishing, USA, 1*(1), 1–27.

Wang, Y. (2007c). On The Cognitive Processes of Perception with Emotions, Motivations, and Attitudes, *International Journal of Cognitive Informatics and Natural Intelligence. IGI Publishing, USA, 1*(4), 1–13.

Wang, Y. (2008a), On Contemporary Denotational Mathematics for Computational Intelligence, *Transactions on Computational Science, 2*, Springer, Sept., pp. 6-29.

Wang, Y. (2008b). RTPA: A Denotational Mathematics for Manipulating Intelligent and Computational Behaviors, *International Journal of Cognitive Informatics and Natural Intelligence. IGI Publishing, USA, 2*(2), 44–62.

Wang, Y. (2008c), A Cognitive Informatics Theory for Visual Information Processing, *Proc. 7th International Conference on Cognitive Informatics* (ICCI'08), IEEE CS Press, Stanford University, CA., Aug., pp.317-323.

Wang, Y. (2008d). On the Big-R Notation for Describing Iterative and Recursive Behaviors, *International Journal of Cognitive Informatics and Natural Intelligence. IGI Publishing, USA, 2*(1), 17–28.

Wang, Y. (2009), On Abstract Intelligence: Toward a Unified Theory of Natural, Artificial, Machinable, and Computational Intelligence, *International Journal of Software Science and Computational Intelligence,* IGI, USA, Jan., 1(1), pp. 1-17.

Wang, Y., & Wang, Y. (2006). Cognitive Informatics Models of the Brain [C]. *IEEE Transactions on Systems, Man, and Cybernetics, 36*(2), 203–207. doi:10.1109/TSMCC.2006.871151

Wang, Y., & Wang, Y. (2008), The Cognitive Processes of Consciousness and Attention, *Proc. 7th International Conference on* Cognitive Informatics (ICCI'08), IEEE CS Press, Stanford University, CA., Aug, 30-39.

Wang, Y., Wang, Y., Patel, S., & Patel, D. (2006). A Layered Reference Model of the Brain (LRMB) [C]. *IEEE Transactions on Systems, Man, and Cybernetics, 36*(2), 124–133. doi:10.1109/TSMCC.2006.871126

Westen, D. (1999). *Psychology: Mind, Brain, and Culture* (2nd ed.). NY: John Wiley & Sons, Inc.

Wever, R. A. (1979). *The Circadian System of Man: Results of Experiments Under Temporal Isolation.* NY: Springer.

Wilson, R. A., & Keil, F. C. (Eds.). (1999). *The MIT Encyclopedia of the Cognitive Sciences.* Cambridge, MA: The MIT Press.

This work was previously published in International Journal of Cognitive Informantics and Natural Intelligence, Volume 3, Issue 2, edited by Yingxu Wang, pp. 84-98, copyright 2009 by IGI Publishing (an imprint of IGI Global)

Section 3

Chapter 12
The Cognitive Informatics Theory and Mathematical Models of Visual Information Processing in the Brain

Yingxu Wang
University of Calgary, Canada

ABSTRACT

It is recognized that the internal mechanisms for visual information processing are based on semantic inferences where visual information is represented and processed as visual semantic objects rather than direct images or episode pictures in the long-term memory. This article presents a cognitive informatics theory of visual information and knowledge processing in the brain. A set of cognitive principles of visual perception is reviewed particularly the classic gestalt principles, the cognitive informatics principles, and the hypercolumn theory. A visual frame theory is developed to explain the visual information processing mechanisms of human vision, where the size of a unit visual frame is tested and calibrated based on vision experiments. The framework of human visual information processing is established in order to elaborate mechanisms of visual information processing and the compatibility of internal representations between visual and abstract information and knowledge in the brain.

INTRODUCTION

It is recognized that, although over 90% information receptors of the brain are in the visual form, the internal processing mechanisms for the visual information are based on semantic or symbolic inferences rather than graphical reasoning (Hubel and Wiesel, 1959; Matlin, 1998; Payne and Wenger, 1998; Pinel, 1997;, Westen, 1999; Wilson, 2001). In other words, the brain carries out thinking, reasoning, and inference on visual stimuli and image information in an abstract ap-

DOI: 10.4018/978-1-60960-553-7.ch012

proach, and all visual information is represented and processed as visual semantic objects rather than direct images or episode pictures in long-term memory.

A fundamental question about the mechanisms of the brain is what the form of internal representations of visual information is in long-term memory (Glickstein, 1988; Goldstein, 1999; Wang, 2009b; Wang and Wang, 2006). Early studies perceived that visual information is stored as pictures and the eyes work as cameras (Gray, 1994; Smith, 1993). Contemporary studies reveal that it may be true only in Sensory Buffer Memory (SBM) and Short-Term Memory (STM), but images retained and recognized in LTM are in the form of abstract visual semantics or symbolic concepts (Coaen et al., 1994; Hubel and Wiesel, 1959; Wang, 2009b). Therefore, the mechanisms of visual knowledge processing are based on abstract semantic analyses and syntheses.

This article presents the cognitive informatics foundations of visual information processing in the brain and their applications in knowledge engineering and computational intelligence. In the remainder of this article, fundamental principles of visual perceptions such as the gestalt principles, the cognitive informatics principles, and the hypercolumn theory, are described. The visual information processing mechanisms are explained by the visual frame theory and the calibration of the size of a unit visual frame. The framework of human visual information processing is developed to elaborate the fundamental mechanisms of visual information processing in the brain for visual knowledge representation and manipulation.

COGNITIVE FOUNDATIONS OF VISUAL INFORMATION PROCESSING

The mechanisms of visual information representation, processing, recognition, and comprehension, as well as their relationships to those of abstract information processing, are a set of fundamental questions in explaining the nature of human vision. This section presents the classic gestalt (holistic) principles and the cognitive informatics principles of visual information processing. Hubel and Wiesel's hypercolumn theory for visual information processing in the visual cortex is introduced, which reveals the important mechanism of internal image information representation, interpretation, and processing.

The Holistic Principles

The classic *gestalt principles* of visual perception are developed in Germany based on experiments conducted in the 1920s and 1930s, where the term *gestalt* means an organized whole that is related to the philosophical doctrine of holism (Gray, 1994; Westen, 1999). The gestalt or holistic philosophy states that the whole is greater than the sum of its parts, which is inherited by modern system science.

In system algebra (Wang, 2008b), Wang creates a mathematical model of the holistic system principle that reveals the mechanism of abstract systems gains known as incremental union.

Definition 1. An incremental union of two sets of relations R_1 and R_2, denoted by *(*, are a union of R_1 and R_2 plus a newly generated incremental set of relations ΔR_{12}, i.e.:

$$R_1 \boxplus R_2 \triangleq R_1 \cup R_2 \cup \Delta R_{12} \qquad (1)$$

where $\Delta R_{12} \not\subseteq R_1 \wedge \Delta R_{12} \not\subseteq R_2$ and $\Delta R_{12} = 2(\#C_1 \bullet \#C_2) \subseteq R_1 \boxplus R_2$.

The incremental union operation on abstract systems is a new denotational mathematical structure, which provides a generic mathematical model for revealing the fusion principle and system gains during system unions and compositions.

Six gestalt principles for visual object and pattern perception are identified (Kanizsa, 1979) such as *similarity, proximity, good continuation,*

Table 1. The Gestalt Principles of Visual Perceptions

No.	Principle	Description
1	Similarity	The tendency to see resemble objects and patterns belong to a same group.
2	Proximity	The tendency to partite discrete image into groups.
3	Good continuation	The tendency to see intersected curves and images continued smoothly.
4	Simplicity	The tendency to see an image in the simplest way by analysis.
5	Closure	The tendency to see a completely enclosed border by ignoring gaps or cloaks.
6	Background contrast	The tendency to identify larger and dark objects in an image as the ground; and the smaller and brighter objects as the front-end figure.

simplicity, closure, and *background contrast,* as summarized in Table 1.

The classic gestalt principles reveal a set of important natural tendencies and fundamental mechanisms of human visual perceptions. However, they are inadequate in rigorousness in order to form a theory of visual information processing of the brain.

The Cognitive Informatics Principles

In cognitive informatics (Wang, 2002, 2003, 2007b, 2009a; Wang et al., 2009), a set of cognitive principles for visual information perceptions is identified as summarized in Table 2. The cognitive principles for visual perception may be considered as the aesthetic principles, which people intend to apply in perception and identification of perfect and coherent human figures, physical objects, and natural surroundings.

It is noteworthy that many of the visual perception principles in Table 2 have also identified in mathematics and philosophy. This is an interesting finding on the relationship between humanity, aesthetics, cognitive informatics, mathematics, and philosophy. Particularly, the perfection principle elicits an important human tendency to reconstruct the whole when only a portion of an image or picture can be seen.

The Hypercolumn Theory of the Visual Cortex

It is recognized there are 1.5 million axons that link each Lateral Geniculate Nucleus (LGN) cells to the visual cortex known as the *striate cortex* (Glickstein, 1988; Goldstein, 1999). Hubel and Wiesel, Nobel Prize laureates in Physiology and Medicine in 1981, discovered the special orientation selectivity of visual neurons for barlike stimuli with specific orientations (Hubel and

Table 2. The Cognitive Informatics Principles of Visual Perceptions

No.	Principle	Description
1	Association	The tendency to find links and relations among individual objects and images.
2	Symmetry	The tendency to identify a symmetry in images.
3	Perfection	The tendency to perceive a perfect image from the given partial information.
4	Abstraction	The tendency to use a semantic label to denote an image.
5	Categorization	The tendency to classify similar images into a group.
6	Analysis	The tendency to identify common meta- shapes or meta-figures in images.
7	Appreciation	The tendency to be sensitive on borders, intersections, changing points, or differences in images.

Wiesel; 1959, 1979). Hubel and Wiesel revealed that the basic structure of vision cells known as the hypercolumns.

Definition 2. A hypercolumn (HC) is a structured visual processing module in the visual cortex, corresponding to a unit area of the retina, which is capable to process a basic unit of visual information in the three aspects of ocular dominance, location on the retina, and orientation of the stimuli.

Tovee reported that there are over 2,500 hypercolumns in the striate cortex (Tovee, 1996), which is equivalent to the size of a visual frame as determined in the experiment described in the next section on the visual frame theory.

Based on Definition 2 and the number of HCs in a visual frame, a mathematical model of HC is developed by the author in order to formalize the discovery of Hubel and Wiesel (1959).

Definition 3. The formal model of an HC is a 3-tuple, i.e:

$$\begin{aligned} HC &\triangleq (E, P, O) \\ &= (\{L_e, R_e\}, \\ &\quad \{X, Y\}, \\ &\quad \{0^\circ, 5^\circ, 10^\circ, \ldots, 90^\circ, \ldots, 175^\circ, 180^\circ\} \\ &\quad) \end{aligned} \tag{2}$$

where E represents the ocular dominance (the left eye (L_e) or the right eye (R_e)), P the location on the retina determined by the coordinates (X, Y), and O the orientation of a stimulus in the scope of 0° through 180°, i.e.:

$$E = \{L_e, R_e\} \tag{3}$$

$$\begin{aligned} P &= X \times Y = [0 \ldots 50] \times [0 \ldots 50]^T \\ &= \begin{pmatrix} (0,0) & \cdots & (0,50) \\ \vdots & \ddots & \vdots \\ (50,0) & \cdots & (50,50) \end{pmatrix} \end{aligned} \tag{4}$$

$$\begin{aligned} O &= \{O_0, O_1, O_2, \ldots, O_{34}, O_{35}\} \\ &= \{0^\circ, 5^\circ, 10^\circ, \ldots, 175^\circ, 180^\circ\}) \end{aligned} \tag{5}$$

For instance, the following HCs represent that the left eye detects a 45° bar at the coordinates (2, 49) in the retina, and the right eye detects a 135° bar at the coordinates (32, 6), respectively.

$$\begin{aligned} HC_1 &= (E_1, P_1, O_1) = (L_e, (2, 49), 45^\circ); \\ HC_2 &= (E_2, P_2, O_2) = (R_e, (32, 6), 135^\circ) \end{aligned} \tag{6}$$

Complex shapes and images can be represented by multiple HCs corresponding to the given pattern of the stimuli.

Based on Hubel and Wiesel's, as well as Tovee's, experiments and Definitions 2 and 3, the following theorem and corollary can be derived.

Theorem 1. The symbolic representation mechanism of vision states that the basic unit of vision is a barlike area modeled by the HC rather than a simple dot.

Proof: Hubel and Wiesel's experiments and the HC layout determine that the semantics of internal image representation is an abstract symbolic structure as shown in Equation. 2, i.e., HC = (E, P, O).

Corollary 1. An image frame is represented by a set of 50 × 50 HCs.

Proof: Directly based on Definitions 2 and 3, Corollary 1 can be proven.

The advantage of the HC mechanisms in visual image detection and representation is that it is reliable, fault-tolerant, and anti-noisy. Any noisy HC in an image frame may be easily identified and corrected in later-phase cognitive processing. According to the hypercolumn theory and Theorem 1, the most sensitive shapes of traffic signs to the brain would be in the forms of bars rather than circles.

Figure 1. Calibration of the invariant size of a visual frame

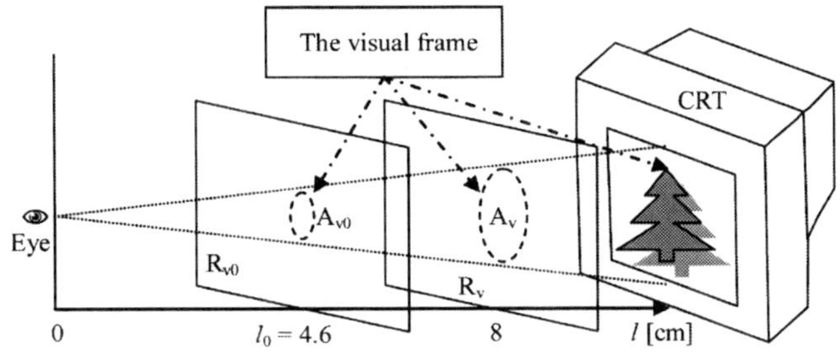

THE VISUAL FRAME THEORY OF HUMAN VISIONS

Definition 4. A visual frame is an invariant area of eyes with a certain sensorial resolution in number of pixels or bits.

The following psychological experiment is designed to test the typical size and resolution of a unit visual frame of human vision.

The Experimental Test of Human Visual Frames

Experiment 1. The layout of the experimental test is illustrated in Figure 1, where a CRT is adopted with length 28.0cm, width 20.6cm, and its resolution is 1,024 • 1,024 = 1Mb pixels. The minimum area of the visual frame of the eye is represented by A_{v0} where the minimum pixels can be visible at the nearest distance, $l_0 \approx$ 4.6cm. The diameter of the visual frame is tested as d ≈ 0.9cm.

In the test, the basic area of the visual frame A_{v0}, the area of the CRT A_{CRT}, and the resolution of the CRT R_{CRT} are obtained as follows:

$$\begin{aligned} A_{v0} &= 2 \bullet \frac{1}{4}\pi d^2 = 0.5 \bullet 3.14 \bullet 0.9^2 = 1.3 \ [\text{cm}^2] \\ A_{CRT} &= 28.0 \bullet 20.6 = 576.8 \ [\text{cm}^2] \\ R_{CRT} &= 1{,}024 \bullet 1{,}024 = 1{,}048{,}576 \ [\text{bit}] \end{aligned} \tag{7}$$

On the basis of the above testing results, the resolution of a unit vision frame of humans, R_{v0}, can be calibrated with that of the CRT by proportional equivalency, i.e., A_{v0}: R_{v0} = A_{CRT}: R_{CRT}. This leads to the finding of the unit size of human visual frame in the following theorem.

Theorem 2. The maximum resolution of a vision frame is a constant that is proportional to the number of visual sensation nerves, or the number of visible pixels in the visual frame, i.e.:

$$\begin{aligned} R_{v0} &= \frac{A_{v0} \bullet R_{CRT}}{A_{CRT}} \\ &= \frac{1.3 \bullet 1{,}048{,}576}{576.8} \\ &= 2{,}363.3 \ [\text{bit}] \end{aligned} \tag{8}$$

This result conforms well with the number of hypercolumns, #HC ≈ 2,500, according to Tovee (1996). Theorem 2 indicates that, although there are about 5 million cones and 120 million rods in the retina as the array of light receptors (Goldstein, 1999), the size of visual frame or the resulted image pixels of each eye is much smaller and it is invariant from the distance between the eyes and the visual object.

Properties of Visual Frames

Theorem 2 indicates that the size of a human visual frame is invariant in term of number of pixels within the visual field. According to Theorem 2, the closer the object, the higher the resolution in the visual frame; and vice versa. The maximum information that can be obtained by eyes without saccade is R_{v0} = 2,363 bits (pixels), no matter how large is the external image.

Therefore, large-frame information must be scanned by multiple units of visual frames via saccades. This explains why no one may read more than three or four words in a line without move one's eyes.

Corollary 2. The visual resolution of eyes R_v is inversely proportional to the distance of the object l, i.e.:

$$R_v \leq R_{v0} | \; l \geq l_0 = 4.6 \text{ cm} \quad (9)$$

According to Corollary 2, the visual resolution of the eyes is decreasing when the distance of the object in the visual frame is increasing.

When the maximum sample rate of vision s_v and the sample rate of each HC s_{HC} are known, i.e., s_v = 50 frame/s (Tuker, 1997), and s_{HC} = 50bit/s, the processing speed of visual information can be determined as follows.

Corollary 3. The maximum rate of human visual information processing S_v is a product of the maximum sample rate s_v and the maximum resolution of the vision frame R_{v0}, i.e.:

$$\begin{aligned} S_v &\triangleq s_v \bullet R_{v0} \\ &= 50 \text{ [frame/s]} \bullet 2{,}363 \text{ [bit/frame]} \\ &= 118{,}150 \text{ [bit/s]} \end{aligned} \quad (10)$$

Corollary 3 indicates that the maximum visual information transformation rate between sensory-buffer memory and short-term memory is equivalent to approximately 118.2kbps, which forms the upper bound of visual information processing of the human brain.

THE FRAMEWORK OF HUMAN VISUAL INFORMATION PROCESSING SYSTEM

The Framework of Visual Information Processing (FVIP) of the brain is shown in Figure 2, where three forms of memories, known as the Sensory Buffer Memory (SBM), *Short-Term* Memory (STM), and Long-Term Memory (LTM), are involved under the control of the Perception Engine (PE) of the brain (Wang and Wang, 2006; Wang, 2007a). The visual information is stored, retrieved, and manipulated in the three memories in different forms. SBM temporally stores the analog visual information as a direct image of the external object, which is transferred into STM as an analog visual frame. Except the part of abstract or symbolic information, the visual information retained in LTM can be classified into three types, namely the *basic image base*, the *semantic image base*, and the episodic image base.

The FVIP model reveals that the major forms of visual information represented in LTM is non-analog or non-photonic. Instead, it is symbolic, semantic, and denotational, except a small part of the visual information in the semantic image base such as common and simple shapes and solid figures (Wang, 2009b), or the episodic image base such as image of family members, home, highly impressed scenes of events, highly familiar places, and very frequently used facilities or tools.

Theorem 3. Acquired visual information is represented in symbolic or semantic form in LTM.

Proof: This theorem can be proven by Theorem 1 and related experiments.

Theorem 3 is supported by many observations and psychological experiments. For instance, Reed and his colleagues reported that the mental images in LTM are in propositional codes (Reed, 1972; Reed et al., 1974) or in the form of semantics.

Experiment 2. Novel and fancy images that have never been seen in one's experience may be perceived in subconscious dreams or during

Figure 2. The Framework of Visual Information Processing (FVIP) in the brain

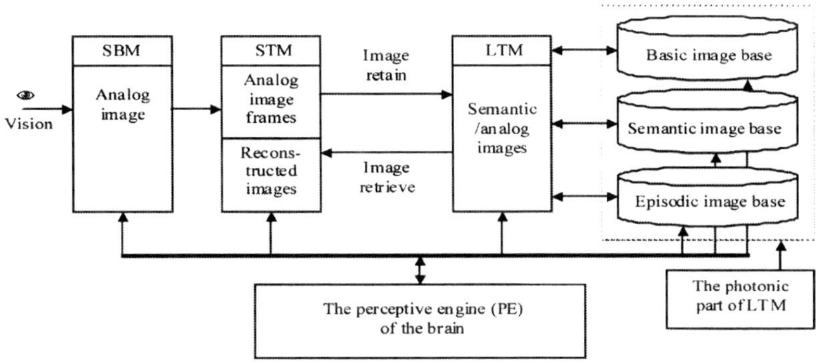

conscious imageries. This mental phenomenon indicates that the LTM does not retain photonic images from the real world, and images one sees during thinking is reconstructed in STM by the perceptual engine rather than retrieved by searching from LTM.

An advantage of the abstract internal representation of visual information is that a category of equivalent images may be treated by the same semantic images as shown in Figure 3. Therefore, highly efficient, flexible, and adaptive visual information processing and comprehension mechanisms may be naturally implemented (Wang, 2008a, 2009b).

It is noteworthy that the STM is the space for both analogy image processing acquired from SBM and internal image reconstruction retrieved from LTM. In other words, STM is the space of image information processing, coding, decoding, and reconstruction. Therefore, what the mind sees during visual information and knowledge processing are restructured images located in STM, particularly in the visual cortex.

Figure 3. The semantic image representation by abstract concept

Semantic symbol	Category of objects
Cup	
Lamp	

THE MECHANISMS OF VISUAL INFORMATION PROCESSING

The FVIP model developed in preceding section explains the fundamental mechanisms of human vision, visual information processing, imagery, perception, image reconstruction, and pattern recognition. On the basis of FVIP, the following fundamental questions about human vision can be answered:

- Why human long-tern memories of images are always blurred and vague?
- Why it is easy to compare two photos in STM or in both STM and SBM, but is not easy to compare those in LTM?
- How internal visual information is represented?

Corollary 4. The human tendency in visual information processing is to perform abstract or semantic visual inferences rather than direct diagram-based visual inference.

The above corollary is supported by the following experiment.

Experiment 3. The brain cannot carry out image inferences without looking at the real

Table 3. The Role of Abstraction in Human Inferences

	Abstract (Concept)	Analog (Image)	Description
Internal inference	√	×	Reasoning based on internal abstract concepts or images
External inference	√	√	Reasoning based on external abstract or visual objects

Figure 4. The semantic representation of concrete images

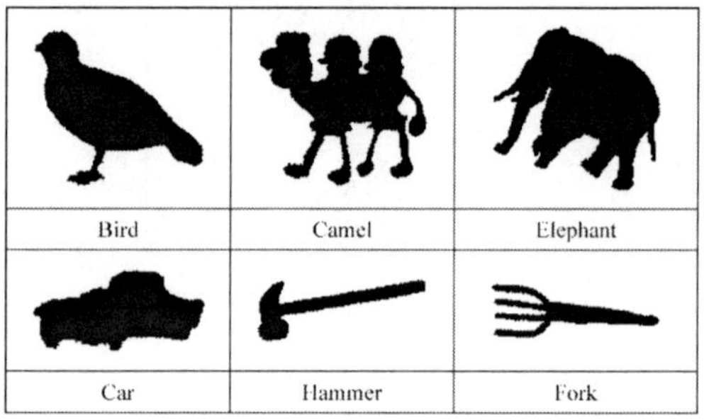

world images, e.g. the images and their relations in Figure 2 or Figure 3, because this cognitive process requires too large memory space beyond the capacity of STM in the brain for supporting complex inferences.

Based on Experiment 3, the role of abstraction in human inference can be observed and contrasted in Table 3, where an internal analog inference cannot be carried out based only on analogy images. Table 3 provides another evidence to support Theorem 3 and the FVIP model. That is, the internal visual representation in LTM is abstract semantic objects rather than image objects.

For example, typical semantic objects of basic shapes and images can be represented as shown in Figure 4, where 6 semantic objects are given to represent 6 elementary images adopted from the Silhouette database at http://www.lems.brown.edu/vision/software/216shapes.tar.gz, where 216 visual objects have been created.

CONCLUSION

This article has presented the cognitive informatics foundations of visual information and knowledge processing in the brain. A set of cognitive principles of visual perception, such as the Gestalt principles, the cognitive informatics principles, and the hypercolumn theory, has been elaborated. The visual information processing mechanisms of human vision have been explained by the invariant visual frame theory, where the size of a unit visual frame has been determined. The framework of human visual information processing has been developed to elaborate the mechanisms of human visual information processing. Based on it, the mechanisms of visual information processing and their compatibility with abstract information processing have been analyzed and contrasted.

It has been revealed that the visual information is represented in symbolic or semantic forms. The basic unit of vision has been identified as a barlike area known as Hubel and Wiesel's hypercolumns, and an image frame is represented by a set of 50×50 hypercolumns. The size of a visual frame has been calibrated as about 2,363 pixels with the property of an invariant resolution inversely proportional to the distance of visual objects. One of the major findings reported in this article has been that the human tendency in visual information processing is to perform abstract semantic visual inferences rather than direct diagram-based visual inferences. In other words, the internal representation of visual and abstract knowledge shares a unified and coherent form known as concept networks. According to the Hierarchical Abstraction Model (HAM), the internal representation of both visual and symbolic knowledge are in abstract forms, and only a higher-level abstract means is precise and adequate to express an object at a given level of abstraction in the HAM model.

ACKNOWLEDGMENT

This work is partially sponsored by the Natural Sciences and Engineering Research Council of Canada (NSERC). The authors would like to thank the anonymous reviewers for their valuable suggestions and comments on this work.

REFERENCES

Coaen, S., Ward, L. M., & Enns, J. T. (1994). *Sensation and Perception* (4th ed.). NY: Harcourt Brace College Pub.

Glickstein, M. (1988). The Discovery of the Visual Cortex. *Scientific American*, *259*, 118–127. doi:10.1038/scientificamerican0988-118

Goldstein, E. B. (1999). *Sensation and Perception*, 5th ed. NY: Brooks/Cole Publishing Co., ITP.

Gray, P. (1994). *Psychology* (2nd ed.). New York: Worth Publishers, Inc.

Hubel, D., & Wiesel, T. N. (1959). Receptive Fields of Single Neurons in the Cat's Visual Cortex. *The Journal of Physiology*, *148*, 574–591.

Hubel, D., & Wiesel, T. N. (1979). Brain Mechanisms of Vision. *Scientific American*, *82*, 84–97.

Kanizsa, G. (1979). *Organization in Vision: Essays on Gestalt Perception*. NY: Praeger.

Matlin, M. W. (1998). *Cognition* (4th ed.). Orlando, FL: Harcourt Brace College Publishers.

Payne, D. G., & Wenger, M. J. (1998). *Cognitive Psychology*. Boston: Houghton Mifflin Co.

Pinel, J. P. J. (1997). *Biopsychology* (3rd ed.). Needham Heights, MA: Allyn and Bacon.

Reed, S. (1972). Pattern Recognition and Categorization. *Cognitive Psychology*, *3*, 383–407. doi:10.1016/0010-0285(72)90014-X

Reed, S., Ernst, G., & Banerji, R. (1974). The Role of Analogy in Transfer between Similar Problem States. *Cognitive Psychology*, *6*, 436–450. doi:10.1016/0010-0285(74)90020-6

Smith, R. E. (1993). *Psychology*. St. Paul, MN: West Publishing Co.

Tovee, M. J. (1996). *An Introduction to the Visual System*. Cambridge, UK: Cambridge, University Press.

Tucker, A. B. Jr., (Ed.). (1997). *The Computer Science and Engineering Handbook*. FL: CRC Press.

Wang, Y. (2002). Keynote: On Cognitive Informatics. *Proc. 1st IEEE International Conference on Cognitive Informatics* (ICCI'02), Calgary, Canada, IEEE CS Press, August, (pp. 34-42).

Wang, Y. (2003). On Cognitive Informatics. *Brain and Mind: A Transdisciplinary Journal of Neuroscience and Neurophilosophy*, Springer, August, *4*(3), 151-167.

Wang, Y. (2007a). *Software Engineering Foundations: A Software Science Perspective. CRC Book Series in Software Engineering* (*Vol. II*). NY, USA: Auerbach Publications.

Wang, Y. (2007b). The Theoretical Framework of Cognitive Informatics. [IGI Publishing, USA.]. *International Journal of Cognitive Informatics and Natural Intelligence*, *1*(1), 1–27. doi:10.4018/jcini.2007010101

Wang, Y. (2008a). On Contemporary Denotational Mathematics for Computational Intelligence [Springer.]. *Transactions of Computational Science*, *2*, 6–29. doi:10.1007/978-3-540-87563-5_2

Wang, Y. (2008b). On System Algebra: A Denotational Mathematical Structure for Abstract System modeling. [IGI Publishing, USA.]. *International Journal of Cognitive Informatics and Natural Intelligence*, *2*(2), 20–42. doi:10.4018/jcini.2008040102

Wang, Y. (2009a). On Abstract Intelligence: Toward a Unified Theory of Natural, Artificial, Machinable, and Computational Intelligence. [IGI, USA, Jan.]. *International Journal of Software Science and Computational Intelligence*, *1*(1), 1–17. doi:10.4018/jssci.2009010101

Wang, Y. (2009b). On Visual Semantic Algebra (VSA): A Denotational Mathematical Structure for Modeling and Manipulating Visual Objects and Patterns. *International Journal of Software Science and Computational Intelligence*, *1*(4), 1–18. doi:10.4018/jssci.2009062501

Wang, Y., Kinsner, W., & Zhang, D. (2009). Contemporary Cybernetics and its Facets of Cognitive Informatics and Computational Intelligence. [B]. *IEEE Transactions on Systems, Man, and Cybernetics*, *39*(2), 1–11.

Wang, Y., & Wang, Y. (2006). Cognitive Informatics Models of the Brain. [C]. *IEEE Transactions on Systems, Man, and Cybernetics*, *36*(2), 203–207. doi:10.1109/TSMCC.2006.871151

Westen, D. (1999). *Psychology: Mind, Brain, and Culture* (2nd ed.). NY: John Wiley & Sons, Inc.

Wilson, R. A., & Keil, F. C. (2001). *The MIT Encyclopedia of the Cognitive Sciences*. MIT Press.

Chapter 13
Comparing Learning Methods

Mercedes Hidalgo-Herrero
Universidad Complutense de Madrid, Spain

Ismael Rodríguez
Universidad Complutense de Madrid, Spain

Fernando Rubio
Universidad Complutense de Madrid, Spain

ABSTRACT

In this article we perform some experiments to study how an automatic system learns a set of rules from its interaction with an artificial environment. In particular, we are interested in comparing these capabilities to the skills shown by humans to learn the same rules in similar conditions. We perform this analysis by conducting two experiments. On the one hand, we observe the evolution of the automatic learning system in terms of its performance along time. At the beginning, the system does not know the rules, but it can observe the positive/negative results of its decisions. As its knowledge about the environment becomes more precise, its performance improves. On the other hand, seventy students faced the same artificial environment in the same conditions, though this time the experiment was presented as a game. The objective of the game consists in gaining points, but the rules of the game are not known a priori. So, there is a clear incentive for finding them out. We use these experiments to compare the learning curves of both humans and automatic systems, and we use this information to analyze the similarities/ differences between both learning processes. In particular, we are interested in assessing how close the automatic system is from passing the Turing test.

COMPARING LEARNING METHODS

It is well known that the inspiration for the scientific research is frequently found in the borders between scientific fields. In these borders, researchers from different fields find useful knowledge that belongs to the standard background of a community but is unknown for researchers of other communities. This is the case of Cognitive Informatics (Wang, 2002), which puts different areas in contact: On the one hand, Computer Science and, on the other hand, Neurology, Psychology, and other Sciences related to the human brain. Interesting results about this research area can be found in previous issues

DOI: 10.4018/978-1-60960-553-7.ch013

of this journal. Among them, we could highlight (Kinsner, 2007; Flax, 2007; Rajlich & Xu, 2007; López, Núñez & Pelayo, 2007; Encina, Hidalgo-Herrero, Rabanal, Rodríguez & Rubio, 2008), as they cover several different aspects of the most recent research in the area.

One of the main concerns of Computer Science in general and Cognitive Informatics in particular is the relation between artificial and human cognitive processes. This issue has attracted the attention of Artificial Intelligence (AI) researchers for decades. One of the first approaches proposed to study this relation is a criterium that has become classical: The Turing test (Turing, 1950; Saygin, Cicekli, & Akman, 2000). Alan Turing proposed that a machine should be considered intelligent if its behavior is indistinguishable from that of a human being. This claim implicitly assumes that the human mind is a kind of optimal form of intelligence. The reason is not that we assume that the intelligence of humans is perfect, but that we do not have any other model to make this comparison. So, systems under assessment are compared with the only thing we know (almost by definition) that is intelligent.

Let us note that, during its evolution, the AI field has adapted itself to several practical problems. In the optimistic beginning, researchers tried to imitate the global human behavior in a wide sense. After their failure, the AI limited dramatically its scope to that it has currently, which consists in producing intelligent behaviors in very limited and restricted knowledge domains. As the goals of AI have changed along time, we think that those concepts proposed formerly to relate and compare human intelligence with artificial intelligence should be revisited and updated as well. In spite of the fact that achieving an artificial intelligent behavior in the *wide* sense (that is, with no domain restriction) is a hard task, creating machines whose behavior could be considered intelligent is not difficult when a specific domain is chosen, specially when the set of rules governing that domain is relatively simple.

However, it is not clear that a simple *preprogrammed* system should be considered intelligent, regardless of whether it shows a deep knowledge about a specific topic. Since most current AI systems are constrained to work under specific domains, we postulate that nowadays it is not reasonable to compare the behaviors of humans and machines in a specific context, at least *after* they have finished their *formative* period. Let us note that showing a human behavior requires, in particular, to show the capability to *learn*. Hence, a machine that was endowed with the rules governing a domain but no learning skills could not be considered intelligent if, regardless of its high performance from the beginning, it does not learn in the long term and improves its performance. That is, it should *not* pass the Turing test. It is worth to point out that learning skills are considered by AI researchers to be one of the cornerstones of any intelligent system. Hence, we postulate that an additional condition should be added to the classical Turing test: The behavior of the machine under test should show a kind of learning skills in the long term (which could require performing a very long test). Let us note that in small domains it is specially clear that an AI system could show a great knowledge of a topic but still be non intelligent. Let us consider that a system is developed to show its knowledge in the topic *"is 11 multiplied by 11 equal to 121? "*. A system that answers *"yes"* to the only question fitting into that domain would be as accurate as any informed human. In most programming languages, the effort needed to develop that system is negligible. If we consider other more complex domains, a system could show a total or partial accurate knowledge because it could have been preprogrammed with a set of solutions or a set of rules allowing to infer them. Nevertheless, these systems could still be considered as non intelligent. Let us note that our reason to claim that thing is not related to the classical argument of the Chinese room (Searle, 1980; Searle 1990; Hauser, 1997), whose argument is as follows: A person who does not speak Chinese stays inside a room with a Chinese dictionary. People from the outside write Chinese symbols in a paper and put them into the room. Then, the person inside the room answers by matching that symbol in the dictionary. The argument says that the room does *not* know Chinese because the person inside does not. The typical reply is that the room, as a *whole*, does know Chinese. In contrast, our reason is that, even under a pure *behaviorist* point of view, such a system provided with a full knowledge from scratch would not pass the Turing test, since it does not show the capability to *learn*, which is a key feature of the human mind.

Moreover, endowing an AI system with a learning mechanism could not be enough to guarantee a human-like behavior that would pass the Turing test. If a system includes a learning mechanism, then its knowledge will be constrained to its initial knowledge as well as the subsequent refinement of it due to its *fix mechanism* to acquire new knowledge. If that mechanism were not close to that of humans, the system behavior would diverge from that of humans in the long term. In particular, if the acquisition of new knowledge is too fast, it could be a symptom of the fact that the learning mechanism was (partial or completely) designed by the developers to converge to some preconceived model. In this case, that learning mechanism would not be versatile or human. If the preconceived model is actually *good*, the system could produce the fake feeling that the learning mechanism is good, while the role of the mechanism was just to conduct the system to the preconceived model. Let us note that this mechanism is actually slower and less efficient than providing the system with that model from scratch. Hence, assessing the intelligence of an AI system does not only consist in comparing the final performance of it with that of humans, but it also requires comparing the learning processes. Moreover, since a high performance may be a symptom of a preconceived model, we will consider that successes and failures during the learning process are equally important to compare machines and human beings. That is, a machine must both *succeed* and *fail* as a human being does.

Modified Turing Test

Taking into account the previous ideas, we get the following new formulation of the Turing test for the concrete case of learning: A machine is intelligent if the evolution of its learning process is similar to that of a human being. Let us note that comparing the learning processes of both humans and machines is hard because it requires comparing the cognitive theories each of them builds and refines along time, which enables the comparison of the levels of accurateness of each of them along time. Unfortunately, we lack a systematic way to do that. In spite of the fact that we can be provided with a formal model of the theories developed by the IA system, a human being is endowed with the full power of her *language* to develop and maintain her own theories. However, as we know, our language is not capable to reason systematically about itself up to that abstraction level. Our way to put in practice those ideas will be less ambitious but more practical, and it will partially lie on a behaviorist perspective: We will compare the human *performance* with that of the machine along time during the learning process. In this context, the performance will be just the *ratio* between successful and failed decisions, weighted to take into account the relevance of each success or failure. This comparison will provide us with a diagnosis result to decide whether the evolution of both learning processes match each other. In addition to that, we will also analyze the behaviors from a constructivist point of view, but restricting the type of anaylisis we would be able to perform.

We will present our modified interpretation of the Turing test, and we will apply it to a case study. In particular, we will apply that test to the AI system *Troglodytes*, presented in (López, Rodríguez & Rubio, 2003). In brief, this system consists of a population of agents moving around an environment. As a result of their movements, agents can achieve successes or failures. The rules governing the conditions that provide each of them in the environment are fix, but they are unknown for the agents. Agents are provided with a simple learning mechanism, which allow them to develop their own theories about the environment. If these theories are accurate, they will be able to profit from them to obtain the successes and to avoid the failures. We will compare the observed behavior of agents with that the human beings would show under the same conditions. We will make a set of human beings (specifically, students) to face the same computational environment governed by the same set of hidden rules. The experiment will be proposed as a *game* where successes will add points while failures will subtract them. The goal of the players will be to maximize their scores. This will give an incentive to guess the hidden rules of the game. If they do, they will be able to use them to get the successes and avoid the failures.

Next, after the results of both humans and agents have been collected, we will apply our modified version of the Turing test by comparing the evolution of the average performance of humans along time with that of agents, where the average performance is the mean of points gained/subtracted each turn. If both functions are similar, we will be able to say that the learning process of the Troglodytes system is intelligent, and, by extension, that the global behavior of that system is so. That is, if during a long enough experiment the evolution of humans and agents is close enough, then it would be reasonable to *extrapolate* these results beyond the actual times considered in the experiment. Hence, we can conclude that the system learns like a human. On the other hand, if both functions are clearly different, then we will say that the learning mechanism of the system is not that of a human being.

The rest of this article is structured as follows. In the following section we review some basic learning theories. Then, in Section 3 we present the running example we will use along the article. Next, Section 4 describes the experiment we have conducted to study how human beings learn how to deal with our running example. Afterwards, Section 5 presents the automatic environment designed to learn how to deal with our running example, while in Section 6 we summarize the results obtained with such an automatic system. Once both experiments have been conducted, in Section 7 we present our conclusions by comparing the results obtained with human beings and the results obtained with the automatic system.

LEARNING THEORIES

Before explaining in detail our experiments, let us start by describing some basic concepts. Our interest in an experiment based on the deduction of rules is justified by Gagne (Gagne, 1985) and Haygood and Bourne's (Haygood & Bourne, 1965) hierarchies of learning. Gagne's theory states that there are several types of learning depending on the different types of instruction that it requires. These levels of learning are summarized as follows: (i) Verbal information; (ii) intellectual skills; (iii) cognitive strategies; (iv) motor skills; and (v) attitudes.

For example, in order to learn attitudes the instruction should include a credible role model that will be presented to the learner.

Our experiment focuses on the intellectual skills, and the learning tasks for achieving them were ordered by Gagne from the simplest to the most complex, giving place to the following types of tasks:

- **Stimulus recognition:** If a stimulus is repeated many times it produces an effect. Even though the stimulus has no effect, the response takes place. It represents the stimulus/response association.
- **Response generation:** It differs from the precedent type in that the external stimu-

lus is combined with an internal one. For example, a trainer teaches to a dog to greet with its paw by using repeatedly two stimulus: (i) the phrase "up your paw", and (ii) to take its paw and put it up; both steps together with reinforcements.

- **Procedure following:** Several learnings of type 2 are concatenated.
- **Use of terminology:** This is a subclass of the precedent type but the response is a word.
- **Learning of discriminations:** Some objects (or nouns) that have been learned by relationships are linked. The relationships may be the result of personal methods that are not stated explicitly.
- **Concept formation:** Until now we have learned to make movements, to identify an object with its name, to name new objects, but we want to learn concepts. This task has to do with the internal representation of the external environment, and the aim is to predict what happens without changing it. For example, to deduce what may happen if a car crashes we do not make it collide with a wall, we just imagine the situation. This task involves identifying and abstracting the relevant properties of the stimulus: color, form, position, and so on.
- **Rule application:** A rule is a concatenation of two or more concepts. It can be formulated by an implication ($A \Rightarrow B$). The verbal reproduction of a rule does not mean having learnt it; a rule has been learnt when the learner can apply it to significant examples. There are two styles of learning rules: (i) To give explicitly the rule and ask for its use in several examples; (ii) The inductive process: many examples from which the learner can deduce the rule (discovery). This is the mechanism that we have adopted in our experiment.
- **Problem solving:** The humans can solve problems after having learnt rules. It consists in combining lower level rules to solve a problem never encountered by the learner before. It may involve generating new rules which receive trial and error use until the one that solves the problem is found. In Gagne's classification, problem solving is the higher intellectual skill.

On the other hand, Haygood and Bourne (Haygood & Bourne 1965) state that a concept has two components: the defining attributes and the rule which combines or relates them in order to form the concept. Consequently, the conceptual tasks are four:

- **Attribute learning:** The learner isolates the attributes that characterize an object and make it different from other entities. Hence, this task has to do with perception. Besides, she associates a name to it, so the task also involves designation.
- **Attribute use:** The subject has already differentiated some attributes of an object that she uses to classify it; the learner makes the object belong to a category.
- **Rule learning:** By means of her experience, the subject learns rules which are common to all the situations of a certain class.
- **Rule use:** Rules are used as tools of conceptual behaviors such as problem solving.

Both hierarchies consider that rule learning and use is a human characteristic. Besides, the treatment of rules is located on top of both of them. Since our aim in this article is to compare how humans learn with our program way of learning, we are going to follow the hierarchies presented above and base our comparison on the rules learning and use.

Once we have decided what we want to study, we must establish the tasks that will be carried out by the humans and the computer, and the method that we are going to use to observe the behavior and responses of both of them.

We will base our observations and the players tasks on the three main theories of learning: behaviorism, cognitivism and constructivism. Before explaining how we use these models let us explain them:

- **Behaviorism:** The aim of this theory is to use experimental methods to observe the behavior of the subject. With respect to learning, the premise above leads to consider that learning happens just when a correct response is given after the presentation of a stimulus, and the trainer can detect whether the subject has learnt or not by observing her behavior over a period of time. Since the subject internal or mental processes cannot be observed directly by an experimental method, this theory does not worry about them. The main tool to motivate the learning is the use of reinforcements of learned behaviors.
- **Cognitivism:** Its basic principle is to consider the learning as a change of the knowledge state. The changes occur when more knowledge is acquired, and this acquisition takes place via processes of codification and structuring developed by the learner. These two tasks also guide the trainers to designate the learning situations. Consequently, the learning depends on the work of the teacher, the situations that she designes, but also on the way the learner processes the information. It is in this theory that Gagne, Haygood and Bourne works are encompassed by cognitivism: they define with their hierarchies the changes of the knowledge state.
- **Constructivism:** This model states that learning takes place when the subject experiments and interacts with the environment. She learns via the action (Piaget, 1973). Thus, knowledge is embedded in the meaningful tasks that the teacher poses to the learner. The latter assembles her knowledge by composing and modifying it when she has to solve the posed problem.

With respect to the way of instruction, behaviorism transmits or transfers behaviors representing knowledge, while cognitivism worries about the transmission of knowledge in the most effective and efficient way. However, constructivism focuses on the building of knowledge from the experience. That is, from a constructivist point of view, learning is a process of building rather than an acquisition of knowledge.

Our interest in this experiment concentrates on the rules that the troglodytes program and the humans will deduce after having played with some boards. Both of them will base their deductions on the experiences and the interactions with the environment, i.e. the boards, because of the feedback provided by the environment. Therefore, the subjects achieve learning by building the rules from the experience, that is, by using constructivism.

However, apart from the final deduced rules, we can obtain information from the players' responses throughout the game. Thus, we observe their progress by using behaviorist techniques.

AN EXAMPLE

Moving in a Bidimensional World

In this section we introduce the running example we will use along the article. Let us consider a simple environment where troglodytes, palmtrees, lions, and dinosaurs are living in a forest. We will split those elements into two groups, intelligent and non-intelligent elements. For the sake of simplicity, the only intelligent elements will be the troglodytes, while the rest of elements of this example will be non-intelligent.

The conditions defining the behavior of the whole system are the following:

When a troglodyte arrives at a palm-tree she will achieve a profit by eating its bananas, if there are not lions near it. Otherwise, the lion will get angry because it is *its palm-tree*, and it will bite the troglodyte. This fact will make the troglodyte to achieve a failure.

When a troglodyte arrives at a place occupied by a dinosaur, and only one more troglodyte is near the dinosaur, both of them will coordinate to hunt it, and therefore both of them will obtain a profit by eating it. Otherwise, if a single troglodyte meets a dinosaur, this one will hit him, and the troglodyte will obtain a failure.

The same mechanism works when a troglodyte arrives at a place occupied by a lion, but the profit when hunting it will be minor, as lions are smaller than dinosaurs and carnivores are less nutritive than herbivores.

Palm-trees, dinosaurs, and lions will be periodically introduced in the environment while others will disappear due to the actions of the troglodytes. Let us remark that we need the forest to be constantly populated by these elements, because otherwise the mechanism of prizes and punishments would be lost. If the forest turned empty then there would no longer exist any strength forcing troglodytes to overcome themselves.

It is worth to point out that the previous rules will be unknown a priori by the troglodytes (and also by the humans of our experiment!). Thus, they will have to infer the rules by using their own intelligence. It is clear that the probability of troglodytes to succeed will directly depend on their capability to infer them, because if they do it then they will be able to get the prizes and avoid the punishments. Let us also note that in our simple system the designer knows the rules of the environment prior to using the agents, but it will not be the case in general. In fact, our general framework (see (López, Rodríguez & Rubio, 2003)) allows a designer to use intelligent agents to infer some properties of the environment that she actually does not know a priori.

Before describing the experiments, we will introduce some basic concepts that will be necessary to represent the environments where our intelligent agents will be introduced. The *environment* of the system is the variable containing all of the significant characteristics of a system. An environment will consist of the *elements* of a system and the *locations* of these elements. A set of *arcs* among locations specifies the system topology as a graph. For the sake of simplicity, we will assume that we are using a *rectangular environment*. That is, all locations are placed like in a *chessboard* in such a way that each of them is linked to those a *king* could move to in a chess game (that is, cardinal directions and diagonals). So, any arc is *reversible*. Let us note that in our rectangular environments the locations are joined through a *Moore neighborhood* (Moore, 1962), in contrast to the *Turing neighborhood* where diagonals are not considered.

Regarding the troglodytes, each of them have a counter representing its current *health*. Initially, each troglodyte has 100 health points. Afterwards, in each step of the experiment, the counter will be increased or decreased depending on the obtained successes and failures.

MEASURING HUMAN LEARNING CURVES

In order to be able to check whether our troglodytes learn the rules governing its environment in a human-like way, we have conducted an experiment with 70 people trying to learn the same rules as the troglodytes. The experiment was done in the Education Faculty of the Complutense University of Madrid, with 70 students who volunteer to be part of a *learning experiment*. To avoid interferences, the students did not know that their results were going to be compared with those of a machine.

In fact, they knew nothing at all about the nature of the experiment.

As our program has no idea about the meaning of the words *lion*, *dinosaur*, *palm-tree* or *troglodyte*, a fair comparison between the program and the students requires that the students do not know that these items are surrounding them. Otherwise, they would assume that lions are dangerous and palm-trees are good candidates to profit from them. That is, their previous knowledge (not related to the experiment) would allow them to infer part of the rules even without analyzing a single situation.

Due to the previous reason, each time a situation was presented to the student the environment was shown as the one it is depicted in Figure 1. It is a bidimensional board of size 7x7, because troglodytes can only see things at distance less than or equal to three, and the student/troglodyte is supposed to be located in the center of the board. Symbols of playing cards were used instead of figures more directly related to its actual meaning. Thus, lions were represented by hearts, dinosaurs by diamods, palm-trees by spades, and other troglodytes by clubs. It is important to remark that the students did not know that they were clubs. That is, the symbol representing them in the environment (©) was different. The reason is again the same: the program does not know it either!

We showed a list of possible situations to each of the students, and for each situation the student was asked to select what she wanted to do. Notice that nine possible answers were available in each situation, corresponding with the eight possible directions plus remaining in the same location. After the student selects her movement, we inform her about the results she has obtained. We do not only say to her whether she obtained a profit or a failure, but we also inform her about the numerical value of such profit/failure. After that, the student has 20 seconds to think about what happened, i.e. to try to infer the rules, and then she has to go on with the next situation.

Figure 1. Example of environment shown to the students

			♣			♦
♠	♦					
			♦			♦
♥		♦	©			♦
♠	♠	♦	♣	♠	♥	
♦	♦					

At the beginning, the students have no idea about the rules, and they behave somehow randomly. Then, they start to assemble/build the rules governing the experiment. From learning theories point of view, the knowledge about how the environment works is generated constructively. In Figure 2 we show the evolution of the average results of the students. The top part of the figure shows the evolution of the mean number of points obtained in each turn by the students, while the bottom part of the figure shows the mean of the cumulative points they have obtained. As it can be seen, the students received information about 450 movements. Although we were interested in having a larger number of movements analyzed, most of the students had time restrictions. Thus, we only showed them 450 movements. However, to compensate this reduction we posed to them test cases more representative, removing dummy situations. In fact, the 450 movements were selected from a set of size 900.

Let us briefly comment on the kind of rules the students inferred. It is interesting to note that the only rule that all the students discovered was that empty positions were neither successes nor failures. The rest of the rules were much more complex to discover for them. However, all the rules were discovered by at least one student. This fact is quite interesting, as it proves that no rule was too complex, although finding all of them was hard. In fact, some students found false rules, that is, rules that were not used in the experiment (although they were somehow related to others that were really used). However, it was funny to

Figure 2. Experiments with humans

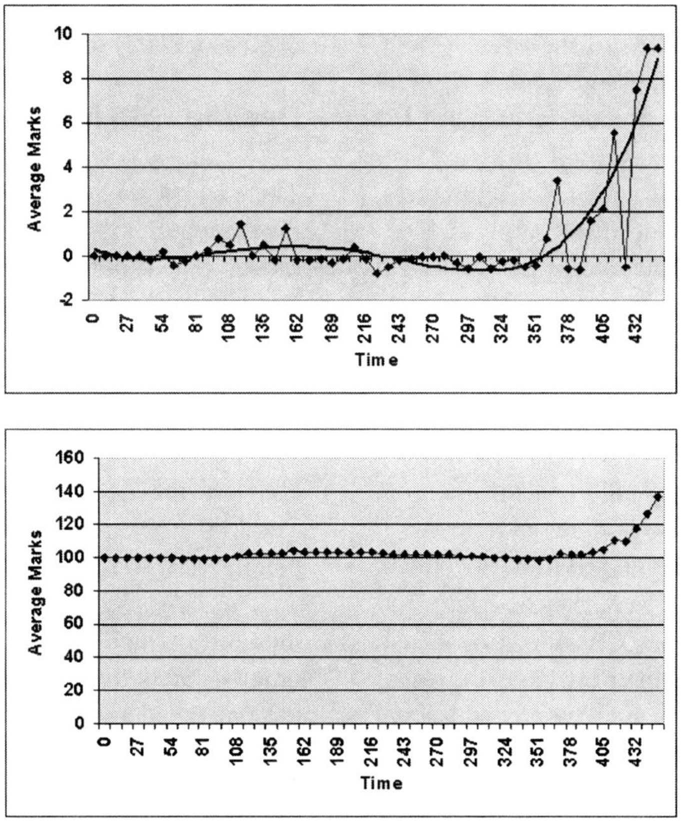

see that the use of those rules was profitable for them, as they were able to obtain profit by using them (although not in all the cases).

The rules built by the students are described below. Nearly everybody discovered that moving to a place occupied by a club (troglodytes) was neither dangerous nor profitable. Many students discovered that hearts (lions) were dangerous, while spades (palm-trees) used to be profitable. However, most of them did not discover why spades were dangerous in some cases (when there was a lion near them). Analogously, nearly nobody discovered how to *hunt* lions by cooperating with other troglodyte. In the case of diamonds (dinosaurs), most of the people knew that they were sometimes dangerous and sometimes profitable (as in the case of hearts). However, there was a quite curious difference. As the profit obtained when hunting a dinosaur was too large, most of the students tried to elaborate complex theories to try to predict when they could obtain profit from diamonds. Most of them did not obtain the correct rule, but rules similar to "diamonds give a profit when there are two diamonds together and there is another black figure near them, otherwise you obtain a failure". As the profit was quite high, they used to try to hunt the dinosaurs in spite of the risk of obtaining a failure.

IMPLEMENTING THE ARTIFICIAL LEARNING SYSTEM

Let us now consider how we have implemented our automatic learning system. In particular, some notions required to construct the intelligence algorithm of the agents will be defined. The concrete algorithm will be presented in Section 5.2.

Regarding the elements of the environment, they will be grouped into sorts, some of them representing *intelligent* troglodytes. One of the main characteristics of these thinkers is that they can remember past environment configurations, which is the basis for being able to apply an intelligence mechanism. Remember that cognitivism (in humans) focuses on memory, that is, the way in which information (rules) is stored in the memory and how it is retrieved to solve a new problem. Let us remark that the elements belonging to non-intelligent sorts will be treated in a generic way and they will not have any additional information. For instance, in the Troglodytes system, each troglodyte will be distinguished from the others. In particular, they will have attached information about their memories. On the contrary, each palm-tree will not be distinguished from the rest of palm-trees, and they will not have any additional information.

The main characteristics of the intelligent elements of an environment can be split into two different concepts: *memory* and *life*. In our model, the memory will be the basis of the intelligence of agents, since it will represent both the cases and the rules the agents create from them. Specifically, the memory of a thinker consists of two different types of memory depending on when it has been obtained. Then, we can distinguish between a *short term memory* (or, using the terminology of (Wang & Wang, 2002), a *sensory buffer memory*) and a *long term memory*. Actually, as we will see later, the first kind of memory contains only cases, while the second contains both cases and rules (represented in a unique generalized form by considering that cases are the *less general* kind of rule). On the other hand, the life of a thinker consists of an *amount of health*. A thinker is defined as a tuple of three elements: Its current location, its memory, and its health points. The health points will be defined as a counter. However, the memory of the intelligent elements is more complex to be defined. For that reason, we devote the following subsection to explain its basic structure. The interested reader is referred to (López, Rodríguez & Rubio, 2003) for more details.

Dealing with Memories

The *short term sequence* of a thinker contains the perceptions detected by the thinker in her last moves, that is, the short term memories. By using this memory, when an important event happens (i.e., either a success or a fail), past steps which yielded to such an event can be recorded in order to either repeat or avoid them in future situations. The memories of the sequence are sorted from the earliest to the latest. Each new move will introduce a new memory as first element and all of the other memories will be shifted one position.

The *long term set* of a thinker contains the memories which are related with past successes or fails, that is, the long term memories. They will have attached a score indicating the relevance of such memory, with a positive value if it was a success and a negative value if it was a fail. The memories belonging to this set will be used to relate present situations to past situations. If there is a match between the current situation and any memory of the long term set, the thinker will take into account the score attached to the memory, and so it may decide to perform the best move according to past experiences. Let us remark that this group of long term memories should not be sorted in a first-in-first-out structure, as *important* experiences should not be easily forgotten, even if they took place a long time ago. Thus, those memories used to remember the *weakest* stimuli (i.e. the stimuli whose successes or fails were less impressive) will be removed first than those related to the *strongest* stimuli. Besides, the stimulus strength of long term memories is periodically decreased as a consequence of the forgetfulness. Thus, these memories will have to be reinforced from time to time to remain in the memory. By doing so we try to avoid noise measurements of a particular experience, that may not be general enough, and that could be seen as a psychological *trauma* in the troglodyte due to a singular fact in her infancy. Besides, old memories that become obsolete should be replaced by newer and more significant experiences.

The two groups of memories described before are related as follows. When an important event happens, all the perceptions of the short period sequence are marked according to the nature of the event (success or fail). Then, they are stored in the long term set. Thus, when something interesting happens, a short term memory becomes a long term memory. In order to describe the mechanism, let us consider an example in the Troglodytes system. Let *t* be a troglodyte with a short term sequence of size 3. Let us suppose that *t* finds a palm-tree and tries to get its bananas. However, there is a lion that stays near the palm-tree, and the lion bites *t*. As a result, *t* will transfer her three last short term memories to her long term set, associating a negative score with all of them. This operation will improve the behavior of *t* in the future, because from now on she will avoid situations that are similar to those she keeps as bad memories.

Let us remark that by copying memories from the short term sequence into the long term set, we do not necessarily create a new memory in the long period set. In fact, during the transferring process the new incoming memories can be *unified* with some other memories that already belong to the long period set. This will be done when the incoming memory *matches* another previously existing memory in the long term set. A match will be possible if both memories represent simultaneously a success or a fail, and the difference between respective intervals is *low*. In this case, we will create a new single memory that will be more general and will embrace both memories (that is, we will create a *rule*).

Intelligence Algorithm

In this section we present the algorithm used to deal with the intelligence of the thinkers.The algorithm rules over all the acts of the thinkers, whose behavior will be determined by their recorded long term memories in each situation. We consider that the evolution of the environment is obtained by means of the evolution of the thinkers. The steps that each thinker of the environment has to perform in each turn are the following:

- The thinker considers the possibility of moving itself from its current location to any other, or even to stay in the same place, by taking into account the perceptions detected from each possible destination. To do that, the thinker considers the memory within the Memory in the Long Term (*MLT* from now on) which best matches with each possible destination. This is done by comparing the differences between the hypothetic perception at each destination and all of its memories in the *MLT*. If the smallest difference between a perception and any memory exceeds a given maximal difference *d*, the hypothetic destination is set as unmatched. Next, it chooses the movement which matches the highest scored memory. If no movement is matched, it chooses randomly among all the possibilities without taking into consideration the memories stored. Besides, if the matched perceptions correspond only with memories having negative scores, it will prefer to choose any unmatched movement.
  Once it arrives to its new location, all of the memories belonging to its Memory in the Short Term (*MST* from now on) are displaced one position, the last memory is destroyed, and the new perception detected from its new location enters as the first memory item.
- If an important event happens (i.e., either a success or a fail) as a result of its moving, let *sc* be the score associated with it. The value of *sc* will be positive if the event is a success, and it will be negative if it is a fail. After that, we perform the following steps:

All the memories belonging to its *MST* are marked with a success or a failure mark, depending on the nature of the event.

Let p be the position of each memory in the *MST*. For each memory in the *MST* we perform the following operations:

- We look for the memory in the *MLT* with a score signal according to the event nature (that is, either a positive one if *sc* is positive, or a negative one if *sc* is negative) which best matches our memory of the *MST* (that is, with the lowest difference).
- If the difference with that nearest memory is less than a given maximal difference d, then that memory in the *MLT* is unified with the treated memory. The score of the resulting memory increments the old value with , being m the maximal allowed size of the *MST*, and p the position of the memory in *MST*.
  However, if the difference exceeds d, then the memory of the *MST* enters itself as a new element of the *MLT*.
  All of the resulting memories of the *MLT* are treated again looking for new memories to unify. The inconsistencies are eliminated by using the described procedure. Then, the memories with the scores closer to 0 (i.e., the less important memories) are removed from the set if the maximal number of memories is exceeded.
- All the positive (resp. negative) scores of the remaining memories of the *MLT* are slightly reduced (resp. increased). Thus, memories have to be frequently refreshed to stay inside the set of relevant long period memories.

MEASURING AUTOMATIC LEARNING CURVES

We have implemented the Troglodytes system in C++, and we have conducted some experiments with them. In order to be able to extract statistics about the behavior of the troglodytes, we have analyzed how they evolve during 1000 steps, taking into account that in each step we analyze 2000 troglodytes. In the top part of Figure 3 we depict the evolution along time of the number of points obtained in mean by the troglodytes in each step. The bottom part of the same figure depicts the evolution of the mean health of the troglodytes.

As it can be seen, at the beginning troglodytes use to obtain bad results, because they behave randomly. As times goes by, they start obtaining more information about the rules, and then their behavior starts to be not so random. It is easy to see that their skills are continously improving, obtaining a learning curve (top part of Figure 3) close to a logarithmic function.

Another interesting point of the experimentation with our system is to observe how troglodytes learn to *cooperate* among them. In the case of hunting, some troglodytes learned that being near another troglodyte is good, because they act as helpers when hunting and therefore they can eat afterwards. Unfortunately, learning to cooperate is not always good, as trying to cooperate with a troglodyte that does not know how to hunt will not allow any of them to hunt. Let us remark that troglodytes take into account the number of troglodytes surrounding them, but our current definition of their behavior does not allow them to identify *specific* troglodytes. Thus, a troglodyte who learned to be a hunting helper could follow any troglodyte crossing her way, independently of whether such a troglodyte knows how to hunt or not.

CONCLUSION

Let us compare the results of the experiment done with humans and the results of the experiment done with the program. First, when analyzing only the conduct of the subjects, we find a quite relevant difference at the beginning of the test:

Figure 3. Automatic experiments

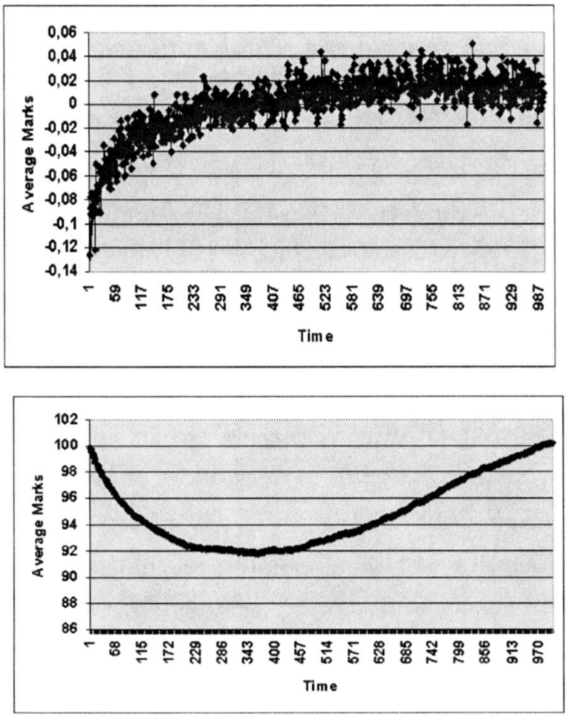

Troglodytes behavior is worse than human behavior at the beginning (i.e. around 200 steps) of the experiments. The reason is that the program has no information at all about the rules of the game, so the troglodytes behave in a completely random way. In the case of the students, although they did not know the rules of the game, they had previous knowledge about games (e.g. the know that empty places are usually safe, but not much profit can be obtained from being isolated). Thus, from a cognitivist point of view, they retrieved from their memory this information and assumed a not so random behavior from the first step. In fact, students used to show passive behaviors at the beginning, reducing their movements and trying not to be very close to many objects at the same time. In this sense, we could think that the program was far away from obtaining a good mark in the Turing test. However, if we remove the initial steps, the observable external behaviors of both humans and automatic troglodytes starts to be much closer. Hence, we are working on a modified version of our experiment providing a different initial behavior to the troglodytes.

From a constructivist point of view (not considered by Turing in his test), the differences seem bigger, although in fact they are not that big. When we analyze the rules inferred by the troglodytes, they are more complex, and they usually include irrelevant data. In the case of humans, the inferred rules are usually simpler. However, as we explained in Section 4, humans also build/assemble/deduce quite complex rules in order to be able to hunt dinosaurs. Let us remind that the profit (reinforcement) obtained when hunting a dinosaur was quite large. So, humans try to do their best to analyze that particular situation and to repeat that particular behavior. Let us remark that troglodytes also devote more effort to deal with those situations where the profit is bigger, as they are motivated by this reinforcement. Thus, the strategies are somehow similar in both cases.

ACKNOWLEDGMENT

First of all, we would like to thank the collaboration of the students who volunteer to be part of our experiment. In addition, we would also like to thank Natalia López for her valuable technical support during the development of this work.

REFERENCES

de la Encina, A., Hidalgo-Herrero, M., Rabanal, P., Rodríguez, I., & Rubio, F. (2008). Testing the behaviour of entities in a cognitive language. *International Journal of Cognitive Informatics and Natural Intelligence*, *2*(1), 29–43. doi:10.4018/jcini.2008010103

Flax, L. (2007). Cognitive modelling applied to aspects of schizophrenia and autonomic computing. *International Journal of Cognitive Informatics and Natural Intelligence*, *1*(2), 58–72. doi:10.4018/jcini.2007040104

Gagne, R. (1985). *The conditions of learning* (4th ed.). New York: Holt, Rinehart and Winston.

Hauser, L. (1997). Searle's chinese box: Debunking the chinese room argument. *Minds and Machines*, *7*, 199–226. doi:10.1023/A:1008255830248

Haygood, R., & Bourne, R. (1965). Attribute- and rule-learning aspects of conceptual behavior. *Psychological Review*, *72*(3), 175–195. doi:10.1037/h0021802

Kinsner, W. (2007). Is entropy suitable to characterize data and signals for cognitive informatics? *International Journal of Cognitive Informatics and Natural Intelligence*, *1*(2), 34–57. doi:10.4018/jcini.2007040103

López, N., Núnez, M., & Pelayo, F. L. (2007). A formal specification of the memorization process. *International Journal of Cognitive Informatics and Natural Intelligence*, *1*(4), 47–60. doi:10.4018/jcini.2007100104

López, N., Rodríguez, I., & Rubio, F. (2003). Defining meta-adaptable living agents. *Second IEEE International Conference on Cognitive Informatics* (pp. 161–170). IEEE-CS Press.

Moore, E. F. (1962). Machine models of self reproduction. *American Mathematical Society. Proceedings of Symposia in Applied Mathematics*, *14*, 17–33.

Piaget, J. (1973). *Introduction à l'Épistemologie genetique*. Paris: PUF.

Rajlich, V., & Xu, S. (2007). Constructivist learning during software development. *International Journal of Cognitive Informatics and Natural Intelligence*, *1*(3), 78–101. doi:10.4018/jcini.2007070106

Saygin, A., Cicekli, I., & Akman, V. (2000). Turing test: 50 years later. *MANDMS: Minds and Machines*, *10*, 463–518. doi:10.1023/A:1011288000451

Searle, J. (1980). Minds, brains and programs. *The Behavioral and Brain Sciences*, *3*, 417–424. doi:10.1017/S0140525X00005756

Searle, J. (1990). Is the brain's mind a computer program? *Scientific American*, *3*(262), 26–31. doi:10.1038/scientificamerican0190-26

Turing, A. (1950). Computing machinery and intelligence. *Mind*, *59*(236), 433–460. doi:10.1093/mind/LIX.236.433

Wang, Y. (2002). On cognitive informatics. *1st IEEE International Conference on Cognitive Informatics* (pp. 34–42). IEEE.

Wang, Y., & Wang, Y. (2002). Cognitive models of the brain. *First IEEE International Conference on Cognitive Informatics* (pp. 259–269). IEEE-CS Press.

ENDNOTE

1 Research partially supported by the Spanish MCYT projects TIN2006-15578-C02-01, the Junta de Castilla-La Mancha project PAC-06-0008-6995, and the Marie Curie project MRTN-CT-2003-505121/TAROT.

This work was previously published in International Journal of Cognitive Informantics and Natural Intelligence, Volume 3, Issue 3, edited by Yingxu Wang, pp. 12-26, copyright 2009 by IGI Publishing (an imprint of IGI Global)

Chapter 14
Classification of Breast Masses in Mammograms Using Radial Basis Functions and Simulated Annealing

Rafael do Espírito Santo
Universidade de São Paulo, Universidade Nove de Julho, & Instituto Israelita de Pesquisa e Ensino Albert Einstein, Brazil

Roseli de Deus Lopes
Universidade de São Paulo, Brazil

Rangaraj M. Rangayyan
University of Calgary, Canada

ABSTRACT

We present pattern classification methods based upon nonlinear and combinational optimization techniques, specifically, radial basis functions (RBF) and simulated annealing (SA), to classify masses in mammograms as malignant or benign. Combinational optimization is used to pre-estimate RBF parameters, namely, the centers and spread matrix. The classifier was trained and tested, using the leave-one-out procedure, with shape, texture, and edge-sharpness measures extracted from 57 regions of interest (20 related to malignant tumors and 37 related to benign masses) manually delineated on mammograms by a radiologist. The classifier's performance, with pre-estimation of the parameters, was evaluated in terms of the area A_z under the receiver operating characteristics curve. Values up to A_z = 0.9997 were obtained with RBF-SA with pre-estimation of the centers and spread matrix, which are better than the results obtained with pre-estimation of only the RBF centers, which were up to 0.9470. Overall, the results with the RBF-SA method were better than those provided by standard multilayer perceptron neural networks

DOI: 10.4018/978-1-60960-553-7.ch014

INTRODUCTION

Breast Cancer

Mammography is the most efficient method for early diagnosis of breast cancer. In Canada and the United States, mammography is used for early detection of breast cancer in asymptomatic women between the ages of 50 and 69 years, or women under the age of 40 years with high risk of breast cancer. In clinical diagnostic procedures, decisions are made based upon multiple mammographic features, which are used to classify a case as normal or abnormal. By studying the masses present in a mammogram, a radiologist can decide whether benign breast disease or cancer is present or not, and in the case of a positive finding, request a biopsy for a final diagnosis. However, studies have shown that the positive predictive value (PPV, given as the ratio of the number of breast cancers found to the total number of biopsies performed) of such procedures is only 15% to 30% (Kupinski, & Anastasio, 2009). Considering the traumatic nature and cost of biopsy, it is important to increase the PPV without reducing the sensitivity of mammography in breast cancer detection.

Several studies have shown the potential of computer-aided diagnosis (CAD) procedures in increasing the diagnostic accuracy by reducing the false-negative rate while increasing the PPV of diagnostic procedures based upon mammographic abnormalities (Rangayyan, Mudigonda, & Desautels, 2000; Hadjiiski, Sahiner, Chan, Petrick, & Helvie, 1999; Alto, Rangayyan, & Desautels, 2005; André, & Rangayyan, 2006; Mudigonda, Rangayyan, & Desautels, 2000; Mudigonda, Rangayyan, & Desautels, 2001; Sahiner, Chan, Petrick, Helvie, & Goodsitt, 1998; Sahiner, Chan, Petrick, Helvie, & Hadjiiski, 2001). Such procedures can decrease the risk of wrong diagnosis, help radiologists in the analysis of difficult cases, and help in deciding upon a recommendation of biopsy. CAD methods can be implemented to identify masses in a mammogram, to extract their features (such as measures of texture, edge-sharpness, and shape), and to classify them as benign or malignant (Rangayyan, Mudigonda, & Desautels, 2000; Hadjiiski, et al. 1999; Alto, et al. 2005; André, & Rangayyan, 2006; Mudigonda, et al. 2000; Mudigonda, Rangayyan, & Desautels, 2001; Sahiner, Chan, Petrick, Helvie, & Goodsitt, 1998; Sahiner, et al. 2001).

The aim of the present work is to investigate the problem of classifying mammographic masses with high accuracy, using measures of shape complexity, edge-sharpness, and texture (Rangayyan, et al. 2000; Hadjiiski, et al. 1999; Alto, et al. 2005; André, & Rangayyan, 2006; Mudigonda, et al. 2000; Mudigonda, et al. 2001; Sahiner, et al. 1998; Sahiner, et al. 2001). We propose a classifier using radial basis functions (RBF) (Haykin, 1999) and simulated annealing (SA) (Haykin, 1999), along with nonlinear and combinational optimization (Haykin, 1999; Broomhead, & Lowe, 1988). The results of the proposed approach are compared with the results obtained by André and Rangayyan (2006) and Alto et al. (2005), who used standard artificial neural networks (ANNs) and linear discriminant analysis (LDA), respectively, with the same dataset and features as in the present study.

Radial Basis Functions for Pattern Classification

The design of a pattern classifier can be viewed as an attempt to solve an optimization problem, known in statistics as stochastic approximation. In this approach, the learning process is the same as finding a surface in a multidimensional space that provides the best adaptation of the data used to train the classifier. On the other hand, the ability of generalization of the classifier is similar to using the multidimensional surface as above to interpolate the data to test the classifier. This motivates the use of RBFs (Haykin, 1999) to design ANNs that can separate classes, that is, perform pattern classification. In the context of ANNs (Haykin, 1999), the hidden layer provides a set of functions that constitutes the generators of bases in a multidimensional space. The bases permit the representation of the data, provided to the ANN as input, in the space of the hidden layer (Haykin, 1999).

Broomhead and Lowe (1988) explored the design of ANNs using RBFs. Other related works were reported by Powell (1985), Moody and Darken (1989), Renals (1989), and Poggio and Girosi (1990). The architecture of an ANN using RBFs has three distinct layers. The first layer is the layer where the data for training the network are connected. The second layer is a space that has a higher dimension as compared to the input layer. The third layer is the output of the network, and is the layer where the responses of the network are collected, regarding the activations made by the input data. The transformation between the input-layer space and the hidden-layer space is nonlinear, whereas the transformation between the hidden-layer space and the output-layer space is linear (Cover, 1965). The classification of nonlinearly separable populations appears to be linear when the classification is performed in a nonlinear multidimensional space (Haykin, 1999).

The use of RBFs as an ANN classifier provides some useful capabilities. The capability to separate samples that are not linearly separable is important, because most of the physical applications are based on nonlinear input information. However, a problem with any type of ANN lies in the capability of generalization from a training set to a test set in a real application. Generalization can be influenced by the following factors: the architecture of the network (André, & Rangayyan, 2006; Haykin, 1999); the complexity of the problem (Haykin, 1999; Lau, 1991); the size of the training set and how representative the set is of the general population of the data being considered (Lau, 1991); and the methods used to solve the optimization problem. André and Rangayyan (2006) studied the questions mentioned above, and suggested methods to apply and evaluate ANNs for the classification of breast masses. The present work is a continuation of the work of André and Rangayyan (2006), using the same dataset.

Linear and Nonlinear Methods of Optimization

In this work, we propose an ANN classifier based upon RBFs (do Espírito Santo, de Deus Lopes, & Rangayyan, 2005) and two techniques to solve the optimization problem: nonlinear optimization using the Levenberg-Marquart (LM) method and combinational optimization using SA (Press, Teukolsky, Vetterling, & Falnnerry, 1992). The proposed classifier, RBF-SA, has only one stage, but the training of the network is carried out in two phases. In the first phase, pre-optimization, the parameters of the network are estimated using SA. In the second phase, the final optimization, the parameters are estimated using the LM method. We explore the performance of the RBF-SA classifier with different combinations of the optimization techniques and training procedures.

THE PROPOSED RBF-SA METHOD

The proposed RBF-SA method is a classifier that discriminates samples – in the present study, features of mammographic masses and tumors – as malignant or benign. The structural model of the proposed classifier is an RBF network, which consists of a linear combination of multivariate Gaussian functions with center *ci* and standard deviation *σi*, defined as

$$F(\boldsymbol{x}) = \sum_{i=1}^{N} w_i \left(\exp\left(-\frac{1}{2\sigma_i^2} \left\| \boldsymbol{x} - \boldsymbol{c}_i \right\|^2\right) \right) \qquad (1)$$

Here, $F(\boldsymbol{x})$ is the network output for a set of features or input vector $\boldsymbol{x}$ (at the input layer); the exponential functions (RBFs) are the hidden-layer activation functions; each *wi* is the synaptic weight connection between $\boldsymbol{x}$ and the output layer via the

RBF (Haykin, 1999), and N is the dimension of the multidimensional space of the hidden layer (Haykin, 1999). The network training procedure consists of estimating the parameters c_i, σ_i, and all of the values of w_i. Several different strategies are available for training; the choice of a particular strategy depends on how the RBF centers are specified. Essentially, there are three possibilities (Haykin, S., 1999):

1. The centers are fixed and are selected in a random manner.
2. The centers are self-selected during a supervised training session.
3. The centers are selected in a supervised manner.

We propose a new approach to RBF network training, where two optimization techniques are used: nonlinear and combinational optimization (Haykin, 1999). In this approach, the network parameters are estimated in two phases, as follows: In the first phase, up to two parameters of the network (c_i and σ_i) are pre-optimized using a combinational optimization method, that is, the SA algorithm (Haykin, 1999; Broomhead, & Lowe, 1988; Kirkpatrick, Gelatt, & Vecchi, 1983). To implement the pre-optimization (pre-estimation) of c_i and σ_i, the following parameters are specified (Haykin, 1999;Broomhead, et al., 1988):

- *T*, temperature cooling schedule of the system (Broomhead, et al., 1988; Powell, 1985);
- *Factordec*, temperature decay factor (typically 0.01, 0.1, and 1.0 for slow, moderate, and fast decay, respectively);
- *Itry*, number of Metropolis-Monte-Carlo attempts (typically 50, 100, 150, or 200);
- *k*, search factor (typically 1.0, 0.1, or 0.09);
- *N*, dimension of the RBF space of the hidden layer (Haykin, 1999; Broomhead, et al., 1988).

In the second phase, all of the network parameters (c_i, σ_i, and w_i), including those pre-optimized in the first phase, are completely estimated employing the LM nonlinear optimization technique (Press, et al. 1992).

DATABASE OF IMAGES AND FEATURES OF BREAST MASSES

The database used in the present work consists of 57 regions of interest (ROIs) extracted from mammograms obtained from Screen Test: Alberta Program for the Early Detection of Breast Cancer (Alto, et al., 2005). The mammograms are of 20 women, with 22 breasts affected by cancer, exhibiting a total of 28 masses visible as 57 ROIs on 45 mammograms. The diagnosis of each case was proven by biopsy. Out of the 57 ROIs, 37 are related to benign masses and 20 to malignant tumors. The images were digitized to a resolution of 50 μm per pixel and quantized to 12 bits. However, texture analysis was performed with requantization to 8 bits per pixel. The ROIs with mass or tumor were manually identified by a radiologist experienced in screening mammography, who also drew the contour of the mass in each ROI; see Alto et al. (2005) for details.

The majority of benign masses display smooth and round shapes, whereas most malignant tumors possess rough shape with ill-defined contours (Rangayyan, et al., 2000; Alto, et al., 2005; Mudigonda, et al., 2000; Mudigonda, et al., 2001; Sahiner, et al., 2001). In addition, malignant tumors usually display heterogeneous density and texture, with a gradual transition from a central high-density region to the surrounding tissues. On the other hand, most benign masses have homogeneous density and texture, with a sharp or well-circumscribed transition from the mass to the surrounding area. Alto et al. (2005) and André and Rangayyan (2006) used combinations of shape, edge-sharpness, and texture features to represent the above-mentioned characteristics of masses and tumors and perform pattern classification. All of the features were computed using contours that were manually drawn by an expert radiologist specialized in mammography and breast cancer screening (Alto, et al., 2005). In the present study, we use a subset of the features used by Alto et al. (2005) and André and Rangayyan (2006), selected based upon their performance in pattern classification using several methods such as LDA, logistic regression, Mahalanobis distance, *k*-nearest neighbors, and precision of content-based retrieval (Alto, et al., 2005). The features used include the shape factor of fractional concavity (F_{cc}), the texture measure of sum entropy (F_8), and a measure of edge-sharpness known as acutance (A) (Rangayyan, et al., 2000; Alto, et al., 2005; Mudigonda, et al., 2000; Mudigonda, et al., 2001).

Fractional concavity is a measure of the portion of the indented length to the total contour length; it is computed by taking the sum of the lengths of the concave segments and dividing it by the total length of the contour (Rangayyan, et al., 2000). Benign masses have fewer, if any, concave segments than malignant tumors; hence, benign masses are expected to possess lower values of F_{cc} than malignant tumors.

The texture feature F_8 was computed using a ribbon of pixels around the margin of each mass (Alto et al., 2005; Mudigonda, et al., 2000; Mudigonda, et al., 2001). The ribbon of width of 8 mm was obtained by dilating the mass boundary after filtering and downsampling the mammograms to an effective resolution of 200 μm per pixel. A typical benign mass is expected to have homogeneous density, whereas a malignant tumor has a heterogeneous density distribution; this difference is expected to be captured by F_8, representing a measure of entropy, with higher values for malignant tumors than for benign masses.

Acutance is a measure of the sharpness or change in density across a mass margin (Mudigonda, et al., 2000). To compute A, perpendiculars (normals) to the contour are identified at every pixel on the contour, and differences are computed using pairs of pixels inside and outside the contour (mass region) for several distances, and averaged. Then, the normalized root-mean-squared value of the averaged differences is obtained around the entire contour. Benign masses are expected to possess a rapid transition for a dense inner region to the surrounding normal tissue, typically of lower density. On the other hand, malignant tumors have ill-defined borders with a slow transition of density. Therefore, malignant tumors are expected to have lower values of A than benign masses.

PATTERN CLASSIFICATION EXPERIMENTS AND RESULTS

Several classification experiments were conducted following two different training strategies with the proposed classifier:

1. The RBF network was trained only with the LM technique (Press, et al. 1992).
2. The RBF network was trained with SA as a pre-estimation stage and LM as a complementary estimation stage. We call this strategy RBF-SA (do Espírito Santo, de Deus Lopes, & Rangayyan, 2005), and compare its performance with pre-estimation of only the centers or the centers and the spread matrix of the RBFs.

In order to analyze the influence of the number of features used on the performance of RBF-SA, different combinations of the features F_{cc}, A, and F_8, labeled as feature sets S1 – S7, were used during the training phase; the feature sets studied are listed in Table 1.

The results shown in Table 2 correspond to the experiments conducted by training the classifiers with the sets S1 to S7 of the features of the 57 mammographic ROIs as described in the previous section, and using different optimization techniques. The training for each case was per-

Table 1. Sets of features used for training the RBF-SA network

Set	Features
S1	F_{cc}
S2	A
S3	F_8
S4	F_{cc} and A
S5	F_{cc} and F_8
S6	A and F_8
S7	F_{cc}, A, and F_8

formed until the best performance was obtained (by trial and error), using the leave-one-out procedure (Haykin, 1999). The testing session was conducted with the same set of features as used in the training phase. The validation of the performance of the classifier was carried out by using the receiver operating characteristics (ROC) curve, which represents the variation of the true-positive fraction (TPF) versus the false-positive fraction (FPF) in pattern classification. The area under the ROC curve (A_z) is used as a consolidated measure of classification accuracy or performance (Kupinski, et al., 1999; Metz, 1986; Woods & Bowyer, 1997).

Figure 1 shows the ROC curves obtained after training and testing the RBF-SA network with the feature set S4. The ROC curves were generated by using the ROCKIT package, and show the classifier's performance in two situations: with pre-estimation of only the centers, and with pre-estimation of both the centers and the spread matrix. During the training session, the value of the parameters T and k were set to 4000 and 0.1, respectively. The dimension of the hidden layer space (N) was 12 for training with S4. The parameter *Itry* (Metropolis-Monte-Carlo attempts) had a maximum value of 100 (with pre-estimation of the centers) and 50 (with pre-estimation of the spread matrix). The highest classification accuracy (A_z=0.9812), in this case, was obtained when

Table 2. Results of area A_z under the ROC curve in classifying breast masses as benign or malignant, using different optimization techniques and feature sets

				RBF-SA			
		ANN ([5])		Pre-estimation of the centers		Pre-estimation of the centers and spread	
Features used	LDA ([4])	A_z	Simulation time[1]	A_z[2]	Simulation time[3]	A_z[4]	Simulation time
F_{cc}	0.99	0.9967	12min 17s	0.9470	18h 20min 15s	0.9992	75h 52min 13s
A	0.74	0.7491	12min 25s	0.6753	19h 00min 15s	0.7902	79h 30min 16s
F_8	0.68	0.6864	12min 58s	0.6553	21h 00min 15s	0.7202	80h 30min 16s
F_{cc} and A	0.98	0.9872	14min 18s	0.9100	20h 00min 15s	0.9812	223h 17min 6s
F_{cc} and F_8	0.99	0.9898	16min 1s	0.9400	22h 00min15s	0.9888	193h 17min 6s
A and F_8	0.76	0.7650	13min 57s	0.8800	20h 00min 15s	0.9001	197h 10min 3s
F_{cc}, A, and F_8	0.99	0.99	22min	0.9600	21h 40min 15s	0.9997	272h 17min 6s

[1] ANN simulation time: The ANN code was written in MATLAB 7(R13) and run on a Pentium IV computer, with a speed of 1.8 GHz and RAM of 256 Mbytes.

[2] RBF-SA configuration: T = 4000; Factordec = 0.1; Itry = 100 (centers); k = 0.1; total cooling iteration = 50; and 3, 12 – 36, and 1 neurons in the input, hidden, and output layers, respectively.

[3] Simulation time: The RBF-SA code was written in MATLAB 7(R13) and run on a Pentium IV computer, with a speed of 1.8 GHz and RAM of 256 Mbytes.

[4] RBF-SA configuration: T = 4000; Factordec = 0.1; Itry = 100 and 50 (spread matrix); total cooling iteration = 300; 3, 12 – 36, and 1 neurons in the input, hidden, and output layers, respectively.

Figure 1. ROC curves for RBF-SA with the feature set S4 (F_{cc} and A) and the parameters listed in Table 2. The areas under the two curves are 0.91 (solid) and 0.9812 (dashed).

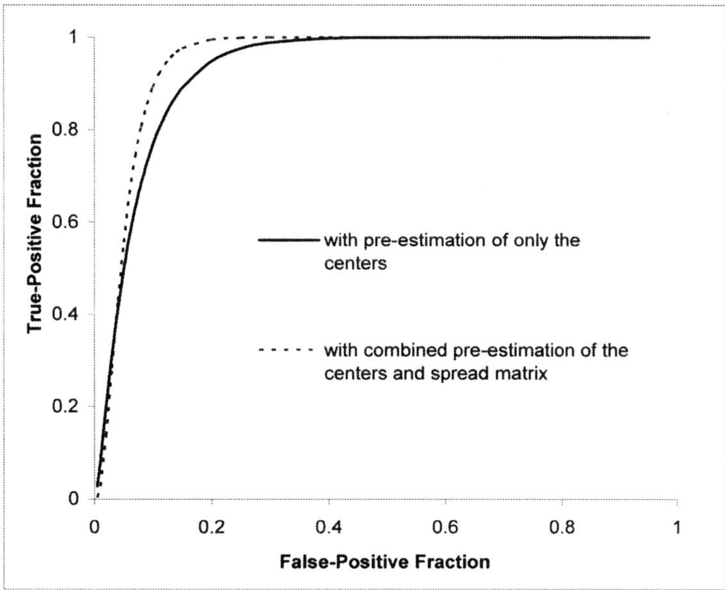

the classifier was trained with a combined pre-estimation of the centers and the spread matrix.

In the classification experiments conducted by Alto et al. (2005) with the same set of features used in our study and LDA as the classifier, the following values of A_z were obtained: 0.99 (F_{cc}); 0.74 (A); 0.68 (F_8); 0.98 (F_{cc} and A); 0.99 (F_{cc} and F_8); 0.76 (A and F_8); and 0.99 (F_{cc}, A, and F_8). The best value of $A_z = 0.9997$ (F_{cc}, A and F_8) that we have obtained in our simulation compares well with the corresponding result of Alto et al. (2005). Our results are also comparable to the values of A_z obtained in the experiments carried out by André and Rangayyan (2006) using standard ANNs for the stronger feature sets (2006). As can be seen in Table 2, the best performance of RBF-SA, provided in the seventh column, is comparable to or better than the performance of the standard ANN, shown in the third column. It is worth noting that the RBF-SA method has provided relatively large gains in A_z values with the comparatively weak features of A and F_8 that provided low accuracy with other classifiers (Alto, et al., 2005; André & Rangayyan, 2006).

DISCUSSION

Considering the values of A_z shown in Table 2, the performance of RBF-SA in classifying breast masses as benign or malignant using the features F_{cc}, A, and F_8 is improved when at least one RBF parameter is pre-estimated. In Table 2, an improvement in classifier performance is observed from the fifth column to the seventh column with all sets of features. The RBF results are better than those given by a standard ANN for three of the seven cases listed: for feature A 0.7902 *vs* 0.7491; for feature F_8 0.7202 *vs* 0.6864; and for features A and F_8, 0.9001 *vs* 0.7650. The increase in A_z for two of the remaining cases and the decrease in A_z for two other cases illustrated are small. Alto et al. (2005) observed that classifying masses and tumors with features from multiple perspectives (such as shape, edge-sharpness, and texture) is better than classifying them with only one type of features (such as shape); our experiments show the same tendency.

At the beginning of the training sessions, several combinations of initial cooling temperature

(T) and the number of Metropolis-Monte-Carlo attempts were used for each additional input feature. When changes in the parameters produced no further improvement in the performance of the RBF-SA, the training session was repeated by maintaining all of the RBF-SA parameters unchanged in order to investigate only the effect of sequential pre-estimation on the performance of the RBF-SA method. The result of such a scheme, shown in Table 2, reveals a gain in the accuracy, as the RBF-SA classifier was trained with pre-estimation of the centers followed by pre-estimation of the spread matrix (sequential pre-estimation). Pre-estimation of only the centers of the RBF does not guarantee good classification accuracy. As can be seen in Table 2, the values of A_z in the seventh column are better than the values in the fifth column.

Training a classifier under conditions as above tends to be highly time-consuming. For instance, to obtain good performance of RBF-SA with the feature F_{cc}, as shown in Table 2 (eighth column), a pre-estimation training time of 75 hours, 52 minutes, and 13 seconds (about three days) was needed, using T = 4,000, 50 cooling iterations, 100 Metropolis-Monte-Carlo attempts, and N=12. However, to obtain a similar performance with F_{cc}, A, and F_g, as shown in the last row of Table 2, 400 additional iterations were necessary; in this case, the pre-estimation phase took about 272 hours, 17 minutes, and 6 seconds (about 11 days).

The training parameters of the RBF-SA classifier are difficult to be determined in a systematic way. In our simulations, the parameters were established experimentally. The performance of the classifier depends upon the combination of the values of the parameters and the pre-estimation phases. The accuracy of the RBF-SA classifier in classifying the feature sets as benign or malignant was high only when the RBF-SA classifier was trained with pre-estimation of the center and spread matrix.

The RBF-SA is a classifier based on combinational optimization that consists of finding the best values of the centers, synaptic weights, and spread matrix of the network (the optimal network solution) in relation to the training cases provided at the input layer. The optimization problem is solved when an optimal hyperplane that separates the nonlinearly separable population is estimated in the high-dimensional space created by the hidden layer. In neural network theory, the objective of solving an optimization problem is to minimize a cost function defined as being the sum of all network energy levels. Generally, most of the practical applications require the number of the neurons in the network to be extremely high in order to estimate the globally optimal hyperplane. Yet, deterministic algorithms based on successive gradient computation, such as an ANN, when used to solve optimization problems, have difficulties in finding the global minimum. It is most probable that they find a local minimum, a point that still has a significant level of energy. There is no guarantee that methods based on gradient computation find the optimal solution (global minimum or global maximum). The probability of occurrence of local minima during the search for the optimal value is high. As a result, the training of a classifier implemented with such an approach has limitations, because the training procedure could get trapped at a local minimum instead of reaching the globally optimal point. The SA algorithm, however, always finds the global minimum as the solution of an optimization problem. This is because, contrary to the ANN, the RBF-SA procedure does not depend on gradient computation to solve optimization problems; rather, it translates optimization problems into a search for states of minimal temperature in physical systems (cooling process). Although the physical concept of temperature is not fully compatible with the principles of neural networks, it is possible for the SA algorithm to "escape" from local minima when the system operates at a non-zero temperature, provided that the maximum temperature of the physical system is high enough; the cooling process is time-consuming. On the other hand, with standard ANNs, although several strategies have been proposed for the design of efficient networks for pattern classification (André & Rangayyan, 2006), several parameter sets may have to be evaluated before arriving at an ad hoc selection.

To deal with the computational time requirement of the RBF-SA method, we propose to use high-performance computing and programming with cluster computers. Note that the heavy computational requirement is for the training phase only: this does not affect the practical application of a trained classifier.

The present work fits well in the current trend of the use of advanced computational methods such as genetic programming for feature selection (Nandi, Nandi, Rangayyan, & Scutt, 2006) and advanced optimization techniques such as SA for improved design of CAD systems (Sun, & Qian, 2002). Furthermore, our work addresses the current interest in modeling human expertise and implementing decision-making methods using multidisciplinary combinations of several scientific and engineering disciplines, including informatics, computing, software engineering, and cognitive sciences; the integration of the areas and applications mentioned above contributes to the emerging area of cognitive informatics (Azevedo, & Lajoie, 1998; Chiew, & Wang, 2003). Powerful computational techniques as above could improve the performance of CAD systems beyond the levels provided by conventional computer vision and pattern classification methods.

CONCLUSION

We have presented the results of an investigation on the use of RBF-SA to classify features of mammographic masses as benign or malignant, using combinations of shape, texture, and edge-sharpness measures, along with pre-estimation of the parameters involved. The RBF neural network is trained with supervised selection of the centers; the weights of the output layer can provide better generalization performance than standard multilayer neural networks. During the learning process, the SA algorithm is used to select the centers of the RBF and other training parameters with the minimal level of energy. In this manner, the learning process includes steps to find the optimal solution of an optimization problem. On the other hand, in the standard multilayer neural network implementation, there is no guarantee that the learning process will lead to an optimal solution.

The RBF-SA method has produced high classification accuracy even with relatively weak features of mammographic images of breast masses related to texture and edge-sharpness that provided low accuracy with other classification methods. High accuracies of up to 0.9997 in terms of the area under the ROC curve demonstrate the positive influence of pre-estimation of the parameters on the performance of RBF-SA, and indicate the potential use of this technique in decision making in CAD systems. The proposed methods could be used in CAD systems as a component of a framework for diagnostic decision based upon cognitive informatics.

REFERENCES

Alto, H., Rangayyan, R. M., & Desautels, J. E. L. (2005). Content-based retrieval and analysis of mammographic masses. *Journal of Electronic Imaging*, *14*(2), 1–17. doi:10.1117/1.1902996

André, T. C. S. S., & Rangayyan, R. M. (2006). Classification of tumors and masses in mammograms using neural networks with shape and texture features. *Journal of Electronic Imaging*, *15*(1), 1–10. doi:10.1117/1.2178271

Azevedo, R., & Lajoie, S. P. (1998). The cognitive basis for the design of a mammography interpretation tutor. *International Journal of Artificial Intelligence in Education*, *9*, 32–44.

Broomhead, D. S., & Lowe, D. (1988). Multivariable functional interpolation and adaptive networks. *Complex Systems*, *2*(3), 269–303.

Chiew, V., & Wang, Y. (2003). From cognitive psychology to cognitive informatics. In *Second IEEE International Conference on Cognitive Informatics, ICCI'03*, London, UK, (pp. 114-120).

Cover, T. M. (1965, June). Geometrical and statistical properties of systems of linear inequalities with applications in pattern recognition. *IEEE Transactions on Electronic Computers*, *EC-14*, 326–334. doi:10.1109/PGEC.1965.264137

do Espírito Santo, R., de Deus Lopes, R., & Rangayyan, R. M. (2005). Classification of mammographic masses using radial basis functions and simulated annealing with shape, acutance, and texture features. In *Proc. 3rd IASTED International Conference on Biomedical Engineering*, Innsbruck, Austria, (pp. 164-167).

Hadjiiski, L., Sahiner, B., Chan, H.-P., Petrick, N., & Helvie, M. (1999). Classification of malignant and benign masses based on hybrid ART2LDA approach. *IEEE Transactions on Medical Imaging*, *18*(12), 1178–1193. doi:10.1109/42.819327

Haykin, S. (1999). *Neural Networks: A Comprehensive Foundation*. Upper Saddle River, NJ: Prentice Hall.

Kirkpatrick, S., Gelatt, C. D., & Vecchi, M. P. (1983). Optimization by simulated annealing. *Science*, *220*(4598), 671–680. doi:10.1126/science.220.4598.671

Kupinski, M. A., & Anastasio, M. A. (1999). Multiobjective genetic optimization of diagnostic classifiers with implication for generating receiver operating characteristic curves. *IEEE Transactions on Medical Imaging*, *18*(8), 675–685. doi:10.1109/42.796281

Lau, C. G. Y. (Ed.). (1991). *Neural Networks: Theoretical Foundation and Analyses*. Piscataway, NJ: IEEE Press.

Metz, C. (1986, September). ROC methodology in radiologic imaging. *Investigative Radiology*, *21*, 720–733. doi:10.1097/00004424-198609000-00009

Moody, J., & Darken, C. J. (1989). Fast learning in networks of locally-tuned processing units. *Neural Computation*, *1*(2), 281–294. doi:10.1162/neco.1989.1.2.281

Mudigonda, N. R., Rangayyan, R. M., & Desautels, J. E. L. (2000). Gradient and texture analysis for the classification of mammographic masses. *IEEE Transactions on Medical Imaging*, *19*(10), 1032–1043. doi:10.1109/42.887618

Mudigonda, N. R., Rangayyan, R. M., & Desautels, J. E. L. (2001). Detection of breast masses in mammograms by density slicing and texture flow-field analysis. *IEEE Transactions on Medical Imaging*, *20*(12), 1215–1227. doi:10.1109/42.974917

Nandi, R. J., Nandi, A. K., Rangayyan, R. M., & Scutt, D. (2006). Classification of breast masses in mammograms using genetic programming and feature selection. *Medical & Biological Engineering & Computing*, *44*(8), 683–694. doi:10.1007/s11517-006-0077-6

Poggio, T., & Girosi, F. (1990). Regularization algorithms for learning that are equivalent to multilayer networks. *Science*, *247*(4945), 978–982. doi:10.1126/science.247.4945.978

Powell, M. J. D. (1985). Radial basis functions for multivariable interpolations: A review. *IMA Conference on Algorithms for the Approximations of Functions and Data.* RMCS, Shrivenham, UK, (pp. 143-167).

Press, H. P., Teukolsky, S. A., Vetterling, W. T., & Falnnerry, B. P. (1992). *Numerical Recipes in C: The art of Scientific Computing*. UK: Cambridge University Press.

Rangayyan, R. M., Mudigonda, N. R., & Desautels, J. E. L. (2000). Boundary modeling and shape analysis methods for classification of mammographic masses. *Medical & Biological Engineering & Computing*, *38*, 487–495. doi:10.1007/BF02345742

Renals, S. (1989). Radial basis functions network for speech pattern classification. *Electronics Letters*, *25*(7), 437–439. doi:10.1049/el:19890300

ROCKIT 0.9 B – Beta Version: www.radiology.uchicago.edu/krl/KRL_ROC/software_index.htm.

Sahiner, B. S., Chan, H.-P., Petrick, N., Helvie, M. A., & Goodsitt, M. M. (1998). Computerized characterization of masses on mammograms: The rubber band straightening transform and texture analysis. *Medical Physics*, *25*(4), 516–526. doi:10.1118/1.598228

Sahiner, B. S., Chan, H.-P., Petrick, N., Helvie, M. A., & Hadjiiski, L. M. (2001). Improvement of mammographic mass characterization using spiculation measures and morphological features. *Medical Physics*, *28*(7), 1455–1465. doi:10.1118/1.1381548

Sun, X., & Qian, W. (2002). System-oriented optimization of CAD for mass detection in digital mammography. In D.P. Chakraborty & E.A. Krupinski (Eds.), *Proc. SPIE Medical Imaging 2002: Image Perception, Observer Performance, and Technology Assessment, 4686*, 273-278. Bellingham, WA: SPIE.

Woods, K., & Bowyer, K. W. (1997). Generating ROC curves for artificial neural networks. *IEEE Transactions on Medical Imaging*, *16*(3), 329–337. doi:10.1109/42.585767

This work was previously published in International Journal of Cognitive Informantics and Natural Intelligence, Volume 3, Issue 3, edited by Yingxu Wang, pp. 27-38, copyright 2009 by IGI Publishing (an imprint of IGI Global)

Chapter 15
Advances in the Quotient Space Theory and its Applications

Liquan Zhao
Nanjing University of Finance and Economics and Anhui University, China

Ling Zhang
Anhui University, China

ABSTRACT

Quotient space theory (QST), a new granule computing tool dealing with imprecise, incomplete and uncertain knowledge, uses a triplet, including the universe, its structure and attributes, to describe a problem space or simply a space. As one of important theories of granular computing (GrC), QST is very helpful to the study of cognitive informatics (CI). This article summarizes the quotient space's model and its main principle. Then some basic operations on quotient space are introduced, and the significant properties of the fuzzy quotient space family are elaborated. Finally the main applications of quotient space theory are discussed.

INTRODUCTION

Cognitive Informatics (CI) is a transdisciplinary enquiry of the internal information processing mechanisms and processes of the brain and natural intelligence that draws together many fields, including modern informatics, computation, software engineering, artificial intelligence, cybernetics, neuropsychology, medical science, etc. (Wang, 2003a, b; Yao, 2006). Granular computing (GrC) which imitates the manner of human thinking and covers all the research of the theories, methodologies, technologies, and tools about granules is the foundation of artificial intelligence. It has been rapidly developed by the practical needs for problem solving (Zadeh, 1998; Yao & Zhong, 1999; Yao, 2000, 2005; Lin et al., 2002; Wang et al., 2003). It can be seen that GrC may offer a conceptual framework for study of CI (Lin, 1997; Zadeh, 1997; Zhang & Zhang, 2003; Yao, 2006; Zhao & Zhang, 2008).

DOI: 10.4018/978-1-60960-553-7.ch015

Quotient space theory (QST), a new granule computing tool to deal with imprecise, incomplete and uncertain knowledge, was introduced by Ling Zhang and Bo Zhang in 1989 (Zhang & Zhang, 1989a, b, 1990a, b). It aims at studying the relationship of transform and that of dependency among the worlds with different grain-size. It combines different granularities with the concept of mathematical quotient set, and the problem spaces with different grain sizes can be represented and analyzed hierarchically by a set of quotient spaces. While the essence of human cognition is that human can observe and analyze a problem at different grain sizes but also translate from one granule world to the others with no difficulty, back and forth freely. So the study of QST is naturally helpful to the development of CI.

Compared with other granular computing theories, QST has more powerful abilities of representation and absorption. The reason is that structure is introduced in its model. It can not only describe the elements in the universe and different structure relationship among the elements, but also define a lot of different attribute functions and operations, which makes it translating from one granule world to the others easily. It can absorb the methods of most theories which are relatively mature at the present time, such as decision tree (Zhang et al., 2004c; Zhang & Zhang, 1992), fuzzy set (Zhang & Zhang, 2003, 2005b, c), rough set (Zhao & Zhang, 2008; Zhang et al., 2004b; Zhang & Zhang, 2003), analysissitus (Zhang & Zhang, 1992), evidence theory (Zhang & Zhang, 1992), probability theory (Zhang & Zhang, 1992), wavelet analysis (Zhang & Zhang, 2005a), etc.

This article summarizes the quotient space's model and its main principles including false-preserving, true-preserving, weak false-preserving, weak true-preserving and quotient approaching. Then we introduce its basic operations on quotient space such as projection, combination, quotient operation, quotient restriction and quotient approaching. Besides, the significant properties of the fuzzy quotient space are elaborated. Finally we discuss its main applications in the areas like machine learning, biological sequence alignment, fuzzy control and communication countermeasure reconnaissance.

THE MODEL OF QUOTIENT SPACE THEORY AND ITS MAIN FEATURES

The quotient space theory combines the different granularities with the concept of mathematical quotient set and represents a problem by a triplet (X, f, T). where X is the set of our discussing object, namely the universe; $f(\bullet)$ is the attribute function of universe X, and it may be multidimensional, $f = (f_1, \ldots, f_n)$, $f_i: X \rightarrow Y_i$, which can be divided into two classes, the condition attribute and the decision attribute, denoted by the set C and D respectively, then $f = C$ Y D; T is the structure of universe X, namely the interrelations of elements.

When we view the same universe X from a coarser grain size, that is, when we give an equivalence relation R on X, we can get a corresponding quotient set denoted by $[X]$, where $[X] = \{[x] \mid [x] = \{y \mid xRy\}, x \in X\}\}$. Then viewing $[X]$ as a new universe, we must have the corresponding coarse-grained space $([X], [f], [T])$ called a quotient space of (X, f, T). Similarly we can construct a quotient space of (X, f, T) by taking a coarser grain size on T or f.

The Construction of Quotient Space

The Construction of Quotient Structure [T]

Definition 2.1. A problem space ([X], [f], [T]) is called a semi-order space if there exists a relation "<" among part of elements on X and satisfies: (1) if x < y and y < x, then x = y; (2) if x < y and y < z, then x < z. If the condition (1) does not hold, the relation < is called a pseudo-semi order relation.

Definition 2.2. Given a semi-order space (X, T) and an equivalence relation R on X. If the quotient space ([X], [T]) corresponding to R, is also a semi-order space, then we say that R and T are compatible, or R is compatible, for short.

When T is a topological structure on X, $[T] = \{u \mid p^{-1}(u) \in T, u \subset [X]\}$ $p: X \to [X]$. When T is a semi order structure on X, to be order-preserving we will construct it by the following algorithms.

Algorithm 2.3.

Step 1. Find the basis B, where $B = \{u(x) \mid x \in X\}$, $u(x) = \{y \mid x < y, y \in X\}$.

Step 2. Find the right-order topology $[T_R]$, where $[T_R] = \{u \mid p^{-1}(u) \in T_R, u \subset [X]\}$.

Step 3. Define a relation $<$ on $[T_R]$: for $x, y \in X$, $x < y \Leftrightarrow \forall u(x), y \in u(x)$,where $u(x)$denotes the open neighborhood of x.

Step 4. If $[T_R]$ is a semi-order or pseudo-semi order, end.

Generally it is a pseudo-semi order on the relation $<$, denoted by $[T_R]_s$, but we can modify R by mergence or decomposition algorithm, and make it compatible.

Algorithm 2.4. Mergence algorithm

Let $[X] = \{a_i, i=1,...,n\}$

Step 1. Find, $i = 1,..., n$. where$= \{a_l \mid a_l < a_i, a_i < a_l\}$.

Step 2. Find the equivalence relation corresponding to the partition$\{, i = 1,..., n\}$.

Algorithm 2.5. Decomposition algorithm

Step 1. Let $k = 0, X_0 = X, A = \varnothing$.

Step 2. If $[X_k] = \varnothing$, stop. The equivalence relation corresponding to the partition A.

Step 3. Choose $a \in [X_k]$, find $C_X(a)$, where $C_X(a)$ is the semi-order closure of a on X_k, and $(a) = \{y \mid x_1, x_2 \in a, x_1 < y < x_2\}$.

Step 4. If $(a) = a$, let $A = A$ Y $\{a\}$, $X_{k+1} = a$ Y $\{x \mid x \in X_k /a\}$, then $[X_k] \leftarrow [X_k] /a$, $X_k \leftarrow X_{k+1}$ go to Step 2.

Step 5. Find a_k, where $a_k = \partial_B(a) = \{x|x \in a, \forall y \in B, x < y \to y \in a\}$ is the boundary of a on $B((a))$.

Step 6. Let $A = A$ Y $\{a_k\}$, $X_{k+1} = a_k$ Y $\{x \mid x \in X_k /a_k\}$, $a^k = a/a_k$.

Step 7. If $a^k = \varnothing$, then $[X_k] \leftarrow [X_k] /a$, $X_k \leftarrow X_{k+1}$ go to Step 2.

Step 8. $a \leftarrow a^k$, $X_k \leftarrow X_{k+1}$, go to Step 5.

Proposition 2.6. If there exists such a quotient semi-order, the quotient semi-order can be found by above algorithms and it is an optimal one. If there does not exist, then we cannot find it no matter what approaches we use (Zhang & Zhang, 1992).

The Construction Methods of Quotient Attribute [f]

When X is unstructured or approximately unstructured, it is a rough set. Quotient attribute $[f]$ can be got by the methods such as statistical method, closure method, combination method and so on.

1. Statistical method: $[f](a)$ can be defined as any statistic of $f(a) = \{f(x) \mid x \in a\} (a \in [X])$.
2. Closure method: $[f](a)$ can be defined as some point in $C(f(a))$, the convex closure of $f(a)$. For example, the average of $f(x)$, sup $f(x)$, inf $f(x)$ or the center of $C(f(a))$, with $x \in a$.
3. Combination method: $[f](a)$ can be defined as or.

When X is structured, a variety of $[f]$ can be defined. (see more details in (Zhang & Zhang, 1992))

Its Main Features

Definition 2.7. Assume **R** is the whole equivalent relations on X, $R_1, R_2 \in \mathbf{R}$. If when x R_1 y we have x R_2 y, then R_1 is called finer than R_2, denoted by $R_2 < R_1$(Zhang & Zhang, 1992).

Proposition 2.8. **R** is a complete semi-order lattice under the previous relation <.

In a coarser grain-size space, some information is lost, thus we can simple the problem when we discuss the problem in this space, but the most important thing is to solve the problem. Generally we have some features as follows:

Proposition 2.9. (Homomorphism or No-solution preserving or False-preserving principle) If a problem has no solution in its corresponding coarse-grained space, then it must have no solution in its original space.

Proposition 2.10. (True-preserving principle I) If a problem has a solution in ([X],[f],[T]), $\forall x \in X$, $p^{-1}([x])$ is a connected set in X, then it must have a solution in (X, f, T), where p: X → [X] is a natural projection.

Proposition 2.11. (True-preserving principle II) If a problem has a solution in two semi-order quotient spaces (X_1, f_1, T_1) and (X_2, f_2, T_2), then it must have a solution in their combination space (X_3, f_3, T_3).

By statistic theory we can generalize the false-preserving principle (true-preserving principle) to the case of probability.

Proposition 2.12. (Weak false-preserving principle) Let it be false when its degree of belief is less than a($0 < a < 1$). If the conclusion deduced from its corresponding coarse-grained space is false, then the conclusion deduced from its original space must be false.

Proposition 2.13. (Weak true-preserving principle) If the probability of a problem with a solution is a in its quotient space, then it is more than a($0 < a < 1$) in its finer quotient space.

The two principles are very important in the reasoning process of quotient space model. By the false-preserving principle, we know if we want to decide that a problem has no solution, we can judge it in its corresponding coarse-grained space, the size of which is smaller, so the mount of calculation is smaller.

By the true-preserving principle, we can also reduce the computational complexity of problem solving. Let the size of two semi-order quotient spaces be s_1 and s_2 respectively. In general the maximum size of their combination quotient space can be $s_1 s_2$. Thus we can transform the problem the size of which is $s_1 s_2$ into two problems with the size of s_1 and s_2 respectively, namely the computational complexity of problem solving is reduced from multiplication to addition.

Proposition 2.14. (Quotient approximation principle) If the series {(X_i, T_i)} of quotient spaces converges to (X, T) with respect to their grain-sizes, then its attribute f_i converges to f, where f and f_i are the performance of the system (X, T) and (X_i, T_i) respectively.(see the definition of converge in (Zhang & Zhang, 2005a))

BASIC OPERATIONS PROJECTION OF QUOTIENT SPACE

Given an equivalent relation R on X and its corresponding quotient set denoted by X_1, the projection is to obtain the inference structure of X_1 through the known inference structure of X. In general if an equivalent relation R is given, the natural projection p and quotient topology T_1 are uniquely defined. As long as the global information method denoting attribute a extracted from the local information of a set is determined, $p: f \rightarrow f_1$ is also unique. When T is a semi order and at the same time R_1 and T_1 are incompatible, we should change R_1 to R′ which is compatible with T_1. Then let the quotient space corresponding to R′ be X_2. We can replace X_1 by X_2 to carry on projection, reasoning and analysis.

Combination of Quotient Space

Assume that the original problem space (X, f, T) is unknown in advance. The combination problem is how to get the new states and properties of a finer quotient space (X_3, f_3, T_3) based on the known states and properties of two quotient spaces (X_1, f_1, T_1) and (X_2, f_2, T_2) of (X, f, T), which accords with human's cognition process, i.e., one usually learns things starting from local fragments and gradually integrating them to form a global picture. All constraint or satisfaction problems, all reasoning process and others can be regarded as the process of combination. In (Zhang & Zhang, 1992) they successfully explain the CT approach of axon tomography by the combination model of quotient space, which is a special case of the combination model. They also successfully deduce D-S composition law by the methods of least square and maximum entropy.

The Combination of Domains

Assume that R_1 and R_2 are two equivalence relations on X with respect to X_1 and X_2. Since all equivalence relations on X form a semi-order lattice under the relation <, the combination of R_1 and R_2 can be defined as their least upper bound among the lattice. Then the partition X_3 corresponding to R_3 is $\{a_i \text{ I } b_j | a_i \in X_1, b_j \in X_2\}$, where $X_1 = \{a_i\}$, $X_2 = \{b_j\}$. Obviously X_3 is the finest space got from X_1 and X_2, and it is also the coarsest one, that is the least upper bound, among the spaces that satisfy the combination principle. In some sense it is optimal, so the combination principle is reasonable.

The Combination of Structures

When T_1 and T_2 are topological structures, their combination can be defined as the least upper bound of T_1 and T_2 in the semi-order lattice formed from all topologies on X, i.e., T_3 is a topology formed from the basis $B = \{u_i \text{ I } v_j | u_i \in T_1, v_j \in T_2\}$, with $T_1 = \{u_i\}$ and $T_2 = \{v_j\}$.

If T_1 and T_2 are two semi-order structures, their combination is based on the following algorithms.

Algorithm 3.1.

Step 1. Find the least upper bound of X_1, X_2 denoted by X_3.

Step 2. Find the right-order topologies corresponding to T_1 and T_2, denoted by T_{1R} and T_{2R}.

Step 3. Find the least upper bound of T_{1R} and T_{2R}, denoted by T_{3R}.

Step 4. Find the semi order from T_{3R} denoted by T_3.

Algorithm 3.2.

Step 1. Find the least upper bound of X_1, X_2 denoted by X_3.

Step 2. Find T_3' from X_3. For all $x_1, x_2 \in X_3$, if $x_1 = a_1 \cap b_1$, $x_2 = a_2 \cap b_2$, with $a_i \in X_1$, $b_i \in X_2$ $(i = 1,2)$ then the relation < is defined as $x_1 < x_2 \Leftrightarrow a_1 < a_2$ and $b_1 < b_2$, especially when $b_1 = b_2$, $x_1 < x_2 \Leftrightarrow a_1 < a_2$.

Proposition 3.3. T_3 and T_3' obtained by the above algorithms are identical, and T_3 is the finest semi-order on X_3, where p_i: $(X_3, T_3) \rightarrow (X_i, T_i)$, i = 1,2 is an order-preserving map.

The Combination of Attributes

Given two quotient spaces (X_1, f_1, T_1) and (X_2, f_2, T_2), their combination satisfies the two following conditions.

1. $pf_3 = f_i$, $i = 1, 2$, where p_i:$(X_3, T_3, f_3) \rightarrow (X_i, T_i, f_i)$ is a natural projection;
2. $D(f_3, f_1, f_2) = D(f, f_1, f_2)$ or $D(f, f_1, f_2)$, where $D(f, f_1, f_2)$ is a given optimal judging criteria, f ranges over all attributes functions on X_3 that satisfy condition (1).

Its optimal judging criteria are relevant to the problem itself, and can be determined by its additional information. Its optimal criteria function are different, that is, $pf = f_i$ are different, while the methods obtained its whole information from local information of a problem are different. Thus we cannot get general criteria. We assume that the information is inerrable in the above discussion, but in the actual problem it usually has some error, see (Zhang & Zhang, 1992) for more details.

Quotient Operation

An operation also indirectly presents certain relationship among the elements of the domain. Given an operation N on the known domain X, our concern is how to get the quotient operation N_1 on its corresponding quotient space X_1, and make the projection p from (X, N) to (X_1, N_1) a homomorphism-preserving map. In general, that quotient operation does not exist, but there exists a unique least upper bound and a unique greatest lower bound quotient operation, and their corresponding quotient spaces are the finest and the coarsest respectively, which may be not very ideal, and can be improved by step-by-step subdivision or cut and try method. If we want to obtain the combination of two quotient operations N_1 and N_2, we can get it by the preceding methods on X_3 which is the least upper bound of their corresponding quotient spaces X_1 and X_2.

Quotient Constraint

While carrying on analysis, inference and diagnosis to a system, we are often faced with various constraints. It is necessary to know how the constraints are transformed in different grain spaces when constructing different grain space models.

Definition 3.4. Assume that C is a constraint of X and Y, X_1 and Y_1 are X and Y's quotient space respectively. () is called a quotient constraint of X_1 and Y_1, if =, where () is a inner(outer) quotient constraint, with = $\{(a, b)| \forall x \in a, y \in b, (x, y) \in C, (a, b) \in X_1 \times Y_1\}$, = $\{(a, b)| \exists x \in a, y \in b, (x, y) \in C, (a, b) \in X_1 \times Y_1\}$, = p(C), p: $(X, Y) \rightarrow (X_1, Y_1)$.

To increase the speed of problem solving, sometimes we can reduce properly, and choose $C^*(\subset C \subset)$ as a constraint of X_1 and Y_1. It, however, cannot generally satisfy homomorphism principle, and we can improve it by different back trace techniques. If there are more than one constraints of X and Y, we can choose certain of their combinations as a constraint C. If we are to obtain the combination of two quotient operations C_1 and C_2, we can choose $C^*(_3 \subset C \subset _3)$ as a constraint of X_3 and Y_3 by the preceding methods, where $_3 = (C_1) \cap (C_2)$, $_3 = (C_1)$ Y (C_2), p_i: $(X_3, Y_3) \rightarrow (X_i, Y_i)(i = 1, 2)$.

Quotient Approximation

If the performance of a system (X, T) is described by an attribute function f, the quotient function is $[f]$ in its corresponding quotient space $[X]$, then the analysis of its performance is the analysis of f, and the study of quotient approximation is that of the quotient function approximation, where the quotient function $[f](a)$ on $[X]$ is defined as the convex closure of f on X, i.e., $[f](a) \in C(f(x), x \in a)(a \in [X])$.

Definition 3.5. Let $R_1,...,R_n,...$be a series of equivalence relations on X. If $d(R_i) \rightarrow 0$, then the series of their corresponding quotient spaces$[X]_1,..., [X]_n,...,$ converges to X with respect to its grain-size, where $d(R_i) = \{d(a)\}$is called the fineness (grain-size) of R_i, d(a) is the diameter of a.

Definition 3.3 Given a system (X, T), its performance f is called the quotient approachable, if there exits a series of quotient spaces $[X]_1,...,$ $[X]_n,...,$ when it converges to X with respect to their grain-sizes, then the performance $[f]_i$ also converges to f.

Definition 3.6. Assume $f:(X, d) \rightarrow ([X], d_1)$, for any series of quotient spaces $[X]_1,..., [X]_n,...,$ let the series of their corresponding quotient functions be $[f]_1,..., [f]_n,...,$if $[X]_i$ converges with respect to X, then also converges to f on a metric space $(M(x), d_M)$ composed by all bounded functions on X, where (x) = $[\, f\,]_i\, (a)(\forall x \in X, x \in a, a \in [X]_i)$, $d_M(f_i, f_j) = \sup\{d_1(f_i(x), f_j(x))| x \in X\}$.

Proposition 3.7. Assume that (X, d) is a metric space and $f:X \rightarrow R^n$ is a measurable function. The necessary and sufficient condition for the quotient space approachability of f is that f is bounded on X; the necessary and sufficient condition for quotient space absolutely approachability of f is that f is consistently continuous on X.

When we discuss the above quotient space approximation, we partition X into subsets with different grain-size and the subsets can overlap each other. In this case, we call the quotient spaces pseudo-quotient spaces. The above conclusions we got still hold for a series of pseudo-quotient spaces.

Proposition 3.8. Let $\{[f]_i\}$ be a series of quotient function approximations to function f on a metric space (X, d), then we have:

1. *The i-th quotient function $[f]_i(\{f_{ik}\})$ is the f(x)'s coefficient of the general Haar wavelet expansion with respect to the scaling basis functions $\{\varphi_{im}, m = 1, 2,..., 2^i\}$.*
2. The i-th quotient increment function2$\{d_{im}\}$ is the f(x)'s coefficient of the general Haar wavelet expansion with respect to the basis functions (see the definitions of $\{f_{ij}\}$ and $\{d_{ij}\}$ in (Zhang & Zhang, 2005a)).

The proposition connects the quotient approximation of function with multiresolution analysis, which provides the quotient approximation with powerful mathematical support.

The wavelet analysis as viewed from functional perspective is to find a proper set of basis functions (wavelets) in a function space for a given function so that the function can be expanded based on the basis and then be analyzed. It owns better versatility, but generally it's rather difficult to construct a proper set of basis functions according to the characteristics of a concrete research object.

The quotient space approximation is to choose a proper partition "on line" for a given function and use a proper quotient function to approach. This is an ad hoc approach and has some flexibility. It is relatively easier to present a method of constructing a proper quotient function according to the characteristics of a concrete research object,

FUZZY QUOTIENT SPACE

Definition 4.1. Let be a family of all fuzzy sets on X × X, and ∈. is called a fuzzy equivalence relation on X, if it satisfies:

1. $\forall x, (x, x) = 1$;
2. $\forall x, y, (x, y) = (y, x)$;
3. $\forall x, y, z \in X, (x, z) \geq (\min((x, y), (y, z)))$.

The definition is reasonable, for a set which satisfies some certain conditions on its product space is an equivalence relation on X, then a fuzzy set which satisfies some certain conditions naturally corresponds to a fuzzy equivalence relation. It has the following characteristics:

1. If = 0 (or 1), then the definition defined above is a crisp equivalence relation on X;
2. If $R_\lambda = \{(x, y)|(x, y) \geq \lambda\}(0 \leq \lambda \leq 1)$, then R_λ, a cut relation of, is also a crisp equivalence relation;
3. If $\forall a,b \in [X], d(a, b) = 1 - (x, y)(\forall x \in a, y \in b)$, then $d(\cdot, \cdot)$ is a distance function on $[X]$, and $([X], d)$ is the quotient structure space corresponding to ;
4. If $[X]$ is defined as $X(\lambda) = \{[x] = \{y \mid (x, y) \geq \lambda| x \in X\}$, then $(X(\lambda), d_\lambda)$ is the quotient structure space corresponding to, where $d_\lambda([x], [y]) = 1\ (x, y), x \in [x], y \in [y]$, for $(x, y) \geq \lambda$, = 1 and for $(x, y) < \lambda$, = $(x, y)/\lambda$.

Definition 4.2. Assume that (X, d), d(a, b)≤1 for ∀ a, b ∈ [X], is a metric space. For ∀ x, y, z ∈ (X, d), if there does not exist any number within the array {d(x, y), d(y, z),d(x, y)} such that it is great than other two numbers, then d is called a normalized equicrural distance.

Proposition 4.3. The following statements are equivalent, i.e.,

1. Given a fuzzy equivalence relation (x, y) on X.

2. Given a normalized equicrural distance d(·, ·) on some quotient set of X.
3. *Given a hierarchical structure {X(λ), 0 ≤ λ ≤ 1}on X.*

The third is the most essential, for a hierarchical structure presents a knowledge with certain structure of granularity, which can be obtained by the following two methods: one is to obtain a binary function *f*(*x*, *y*) on *X* from the problem space (*X*, *f*, *T*) and construct a hierarchical structure on *X*; the other is to obtain a unitary function *f*(*x*) on *X* from the problem space (*X*, *f*, *T*) by contour line approach and construct a hierarchical structure on *X*. Then the computing on a fuzzy granularity space can be transformed into that on a quotient structure space ([*X*], *d*). Thus we can carry on studying by the quotient space theory of precise granularity.

Definition 4.4. Given a fuzzy equivalence relation (x, y) on X and A ⊂ X. Let be a corresponding fuzzy set of A. If μ(x)((x)) = $\sup_y$ {(x, y)|y ∈ A}, then μ(x)((x)) is called a structural definition of membership functions.

Thus we can generalize a crisp set *A* to a fuzzy set by a fuzzy equivalence relation. The space constructed by these fuzzy sets is called their corresponding fuzzy quotient space, and different fuzzy equivalence relations can correspond to the same hierarchical structure.

Definition 4.5. If fuzzy equivalence relations $_1$ and $_2$ have the same corresponding hierarchical structure, then $_1$ and $_2$ are called isomorphic; if fuzzy subsets and belong to a corresponding totally ordered space, then and are called isomorphic, where relation < is defined as [x] < [y] ⇔ (x) ≤ (y), x ∈ [x], y ∈ [y].

Proposition 4.6. Given a set family **A** = {A_1,...,A_n}. From $_1$ and $_2$, the families = {$_1$,..., $_n$} and = {$_1$,..., $_n$} of fuzzy subsets can be defined. After performing a finite number of set operations (complement, intersection, union, etc.) over and, we have new families = {$_1$,..., $_m$} and = {$_1$,..., $_m$} of fuzzy subsets. If $_1$ and $_2$ are isomorphic, then and are also isomorphic.

Thus although a membership function can be defined differently, we can get the same or similar structural explanation by fuzzy inference, which suggests that the fuzzy inference has great robustness.

Assume that the membership function of a fuzzy set on *X* is defined as μ:*X* → [0,1]. If μ is regarded as a attribute function on *X*, then an operation on fuzzy quotient sets corresponds to a projection, combination or decomposition operation on attribute functions. In order to lower the representation cost of *fuzzy*, we can transform a membership function on *X* into a membership function [μ]:[*X*]→[0,1] on [*X*], which satisfied the weak true-preserving and weak false-preserving principle. Given a fuzzy set and its membership function, we can get many different membership functions on its quotient space. Thus we can get some useful characteristics, which improve the validity of fuzzy inference.

APPLICATIONS OF QUOTIENT SPACE THEORY

Quotient space theory is a very practical subject. It has born abundant fruits in many fields like image processing (Liu et at., 2004, 2005; Liu & Huang 2005), pattern recognition (Wang et al., 2003; Wu et al., 2000; Wu, 2000), data mining (Xu & Zhang, 2005; Mao, 2004; Bu, 2002), machine learning (Zhang & Zhang, 1999; Mao et al., 2004a; Wu et al., 2005), biological sequence alignment (Mao et al., 2004b), fuzzy control (Zhang et al., 2004a), communication countermeasure reconnaissance (Wang, 2004), etc.

Uncertain Reasoning

Through the methods of hierarchy and combination of quotient space, the reasoning problem of uncertain information can be represented and solved. A distinct difference between the quotient space theory and other theories used to deal with imprecise, incomplete and uncertain knowledge is that it's unnecessary to provide any prior information, thus making it relatively objective to represent and deal with uncertainty of the problem.

Machine Learning

We can first use the quotient space theory to analyze a problem from different grain size spaces, different hierarchical structures, making the research objects conveniently transform their fineness of grain size according to our needs. And then on the suitable grain-size space we can obtain the rules among the objects (data) using methods concerning machine learning such as covering algorithm, SVM algorithm, generic algorithm and so on.

Data Mining

In the quotient space theory we can use constructional machine learning methods to carry on data mining, which is especially useful in dealing with large datasets and very large datasets. The granularities of structure of universe are introduced, which, compared with rough set theory, makes the discretization processing on data mining of data warehouse unnecessary, and the computational complexity is greatly reduced. What's more, it makes the mined rules more accurate than the rules presented in single data table by combinational methods.

Biological Sequence Alignment

Biological sequence alignment is in essence a hierarchical structure, namely an ordered chain of quotient space. The quotient space methods are different from others methods such as recurring, heuristic, tree alignment, BLAST, FAST and so on, which are all based on a specific scoring system. For different scoring systems, different measurement representation and different scoring law, our concern is to get its hierarchical structure rather than its representation form and scoring state, As long as the structure is the same, its corresponding optimal alignment is also the same.

Fuzzy Control

By the quotient space theory, we can solve the problem of exponential explosion of fuzzy control rules. Furthermore, through continuous changing of control granularity we can roughly adjust the parameters of control system on coarser granularities while make delicate adjustment on finer granularities. In this way, we can improve the fuzzy control system in terms of the control indexes of precision and speed so that the system can achieve ideal performance of both stable state and transient state.

Communication Countermeasure Reconnaissance

Communication countermeasure reconnaissance refers to the use of communication countermeasure reconnaissance equipment to search, intercept and capture the enemy's radio communication signals, and carry on measurement, analysis, recognition, goniometry and orientation on the signals so as to obtain technical parameters like signal frequency, electrical level, modulation system and information like communication mode and characteristics, the structure and attribute of communication network. By the quotient space theory we can search communication signals at different granularity's clustering, analyze and deal with the signals at different granularities and then recognize them so as to improve reconnaissance power.

CONCLUSION

The quotient space theory aims at studying the relationship among different grain-size quotient spaces, the combination, mergence, decomposition of quotient space, carrying on problem solving, reasoning, analysis at different grain-size quotient spaces and get the solution in the original problem space. It is an important branch of granule computing. With its great representation power and absorbency, it is undoubtedly the most challenging theory for the development of contemporary artificial intelligence as compared with the fuzzy theory, rough theory and others.

ACKNOWLEDGMENT

The work is supported by Basic Research Program of Education Office of Jiangsu Provice, Grant No. 07KJA52004, and Jiangsu Province Advanced School Project of Natural Science, Grant No. 07KJD520069. I wish to thank Professor Yingxu Wang for introducing me to the interesting field of CI.

REFERENCES

Bu, D. B., Bai, S., & Li, G. J. (2002). Principle of Granularity in Clustering and Classification. [in Chinese]. *Chinese Journal of Computers*, *25*(8), 810–816.

Lin, T. Y. (1997). Granular computing, announcement of the BISC Special Interest Group on Granular Computing.

Lin, T. Y., Yao, Y. Y., & Zadeh, L. A. (2002). *Data mining, Rough Sets and Granular Computing*. Heidelberg: Physica-Verlag.

Liu, R. J., & Huang, X. W. (2005). The granular theorem of quotient space in image segmentation. [in Chinese]. *Chinese Journal of Computers*, *28*(10), 37–40.

Liu, R. J., Huang, X. W., & Meng, J. (2005). Texture Image Segmentation Based on Quotient Space Granularity Synthesis. *Asian Journal of Information Technology*, *4*(3), 61–67.

Liu, R.J., Huang, X.W., Meng, J., & Zhong, X.R. (2004). Texture image segmentation based on quotient space. *Computer applications, 14*(7), 37-40(in Chinese).

Mao, J. J., Wu, T., Zheng, T. T., & Zhang, L. (2005). Algorithm of hierarchical competitive covering networks based on quotient space. [in Chinese]. *Microcomputer Development*, *14*(4), 37–39.

Mao, J. J., Zhang, L., & Xu, Y. S. (2004a). Fuzzy clustering analysis based on quotient space and information granularity. [in Chinese]. *Operations Research and Management Science*, *13*(4), 25–29.

Mao, J. J., Zheng, T. T., & Zhang, L. (2004b). Biological sequence alignments based on quotient space. [in Chinese]. *Computer Engineering And Applications*, *34*(14), 15–17.

Wang, G., Liu, Q., Yao, Y. Y., & Skowron, A. (2003). *Rough sets, Fuzzy sets, Data mining, and Granular Computing*. Berlin: Springer. doi:10.1007/3-540-39205-X

Wang, L. W. (2004). *Applications of quotient space and the constructive learning method in the communication countermeasure reconnaissance* [Ph. D dissertation]. Anhui University, Hefei, China(in Chinese).

Wang, L. W., Zhang, L., & Zhang, M. (2003). A method of pattern classification based on RS and NCA. In *Proceedings of International Conference on Machine Learning and Cybernetics* (pp. 3090-3094), Xi'an, China.

Wang, Y. (2003a). Cognitive Informatics: a new transdisciplinary research field,. *Brain and Mind: A Transdisciplinary Journal of Neuroscience and Neurophilosophy, 4*(2), 115-127.

Wang, Y. (2003b). On Cognitive Informatics. *Brain and Mind: A Transdisciplinary Journal of Neuroscience and Neurophilosophy, 4*(2), 151-167.

Wu, M. (2000). *The research on design of the classifier for large scale pattern recognition* [Ph. D dissertation]. Tsinghua University, Beijing, China (in Chinese).

Wu, M., Zhang, B., & Zhang, L. (2000). A neural network based classifier for handwritten Chinese character recognition. In *Proceedings of the 15th International Conference on Pattern Recognition* (pp. 561-568), Barcelona.

Wu, T., Zhang, L., & Zhang, Y. (2005). Kernel Covering Algorithm for Machine Learning. [in Chinese]. *Chinese Journal of Computers, 28*(8), 1295–1300.

Xu, F., & Zhang, L. (2005). An analysis of uneven granules clustering based on quotient space. [in Chinese]. *Computer Engineering, 31*(3), 26–28.

Yao, Y. Y. (2000). Granular computing: basic issues and possible solutions. In *Proceedings of the 5th Joint Conference on Information Sciences* (pp. 186-189), Atlantic City, New Jersey, USA.

Yao, Y. Y. (2005). Perspectives of granular computing. In *Proceedings of IEEE International Conference on Granular Computing* (pp. 85-90), Beijing, China.

Yao, Y. Y. (2006). Granular Computing and Cognitive Informatics. In *Proceedings of 5th IEEE International Conference on Cognitive Informatics,* (pp.17-18). Beijing, China. IEEE CS Press.

Yao, Y. Y., & Zhong, N. (1999). Potential applications of granular computing in knowledge discovery and data mining. In *Proceedings of World Multi-conference on Systems* (pp. 573–580). Cybernetics and Informatics.

Zadeh, L. A. (1997). Towards a theory of fuzzy information granulation and its centrality in human reasoning and fuzzy logic. *Fuzzy Sets and Systems, 19*(1), 111–127. doi:10.1016/S0165-0114(97)00077-8

Zadeh, L. A. (1998). Some reflections on soft computing, granular computing and their roles in the conception, design and utilization of information/intelligent systems. *Soft Computing, 2*(1), 23–25. doi:10.1007/s005000050030

Zhang, C. J., Li, Y., & Zhang, L. (2004a). Realizing the high-precision fuzzy control based on the theory of quotient space methods of granular computing. [in Chinese]. *Computer Engineering And Applications, 40*(11), 37–39.

Zhang, L., & Zhang, B. (1989a). Quotient space model (I) of qualitative reasoning. [in Chinese]. *Anqing Normal College Journal, 7*(1), 1–8.

Zhang, L., & Zhang, B. (1989b). Mathematic model of quotient space of problem description. [in Chinese]. *Chizhou College Journal, 8*(1), 15–20.

Zhang, L., & Zhang, B. (1990a). Quotient space model (II) of qualitative reasoning. [in Chinese]. *Anqing Normal College Journal, 8*(1), 15–20.

Zhang, L., & Zhang, B. (1990b). Computational complexity of problem solving of quotient space model. [in Chinese]. *Anqing Normal College Journal, 8*(2), 1–7.

Zhang, L., & Zhang, B. (1992). *Theory of Problem Solving and Its Applications. North-Holland.* Elsevier Science Publishers.

Zhang, L., & Zhang, B. (1999). A geometrical representation of Mc-Culloch-Pitts neural model and its applications. *IEEE Transactions on Neural Networks, 10*(4), 925–929. doi:10.1109/72.774263

Zhang, L., & Zhang, B. (2003). Theory of fuzzy quotient space (methods of fuzzy granular computing). [in Chinese]. *Journal of Software, 14*(4), 770–776.

Zhang, L., & Zhang, B. (2004). The quotient space theory of problem solving. *Fundamenta Informaticae, 59*(2,3), 278-298.

Zhang, L., & Zhang, B. (2005a). A quotient space approximation model of multiresolution signal analysis. *Journal of Computer Science & Technology, 20*(1), 92–108.

Zhang, L., & Zhang, B. (2005b). Fuzzy reasoning model under quotient space structure. *Information Sciences, 173*(4), 353–364. doi:10.1016/j.ins.2005.03.005

Zhang, L., & Zhang, B. (2005c). The structure analysis of fuzzy sets. *International Journal of Approximate Reasoning, 40*(1-2), 92–108. doi:10.1016/j.ijar.2004.11.003

Zhang, Y. P. (2002). A repeated cover algorithm of achieving characteristic rule. [in Chinese]. *Journal of Anhui University, 26*(2), 9–13.

Zhang, Y. P., Zhang, L., & Wu, T. (2003). A constructive self-adjusting and probabilistic decision-making classifier. [in Chinese]. *Microcomputer Development, 13*(7), 85–87.

Zhang, Y. P., Zhang, L., & Wu, T. (2004b). The representation of different granular worlds: a quotient space. [in Chinese]. *Chinese Journal of Computers, 27*(3), 328–333.

Zhang, Y. P., Zhang, L., & Wu, T. (2005). A multiside increase by degrees algorithm at machine learning. [in Chinese]. *ACTA Electronica Sinica, 33*(2), 327–331.

Zhang, Y. P., Zhang, L., & Xia, Y. (2004c). To compare the theory of quotient space with rough set. [in Chinese]. *Microcomputer Development, 14*(10), 21–24.

Zhao, L., & Zhang, L. (2006). Research in Quotient Space Theory Based on Structure. In *Proceedings of 5th IEEE International Conference on Cognitive Informatics,* (pp. 309-313). Beijing, China. IEEE CS Press.

This work was previously published in International Journal of Cognitive Informantics and Natural Intelligence, Volume 3, Issue 3, edited by Yingxu Wang, pp. 39-50, copyright 2009 by IGI Publishing (an imprint of IGI Global)

Chapter 16
Important Attributes Selection Based on Rough Set for Speech Emotion Recognition

Jian Zhou
Anhui University, China, & Chongqing University of Posts and Telecommunications, China

Guoyin Wang
Chongqing University of Posts and Telecommunications, China

Yong Yang
Chongqing University of Posts and Telecommunications, China

ABSTRACT

Speech emotion recognition is becoming more and more important in such computer application fields as health care, children education, etc. In order to improve the prediction performance or providing faster and more cost-effective recognition system, an attribute selection is often carried out beforehand to select the important attributes from the input attribute sets. However, it is time-consuming for traditional feature selection method used in speech emotion recognition to determine an optimum or suboptimum feature subset. Rough set theory offers an alternative, formal and methodology that can be employed to reduce the dimensionality of data. The purpose of this study is to investigate the effectiveness of Rough Set Theory in identifying important features in speech emotion recognition system. The experiments on CLDC emotion speech database clearly show this approach can reduce the calculation cost while retaining a suitable high recognition rate.

INTRODUCTION

As one of the main cognitive processes at the perception layer of the *Layered Reference Model of the Brain*(LRMB), emotion is a personal feeling derived from one's current internal status, mood, circumstances, historical context and external stimuli(Wang,2005; Wang,2007; Griffith & Greitzer,2007). It is one of the most important challenges in speech technologies to recognize the speaker's emotional states and give a suitable feedback. The objective of speech emotion

DOI: 10.4018/978-1-60960-553-7.ch016

recognition is to determine the emotional state of the speaker out of the speech samples(Cowie, Douglas-Cowie, Tsapatsoulis, Votsis, Kollias, Fellenz, et al.,2001). Speech emotion recognition is becoming more and more important in such computer application fields as health care, children education, etc.

Undoubtedly, the classifier is very important for emotion recognition, however, the performance (runtime cost or classification accuracy) can also be significantly improved via removing irrelevant and redundant features (Jain & Chandrasekaran, 1983; Hall, 1999). There are many algorithms have been proposed to select good feature subset for speech emotion recognition system in the last years (Ververidis., Kotropou & Pitas, 2004; Razak & Komiya, 2005; Dellaert, Polzin & Waibel, 1996; Oudeyer, 2003; Kwon & Chan, 2003; Wang & Guan, 2005), however, it is time-consuming for these traditional feature selection method used in speech emotion recognition to determine an optimum or suboptimum feature subset.

Rough Set Theory (RST) proposed by Pawlak in 1980s offers an alternative, formal and methodology that can be employed to reduce the irrelevant and redundant features of dataset (Pawlak,1984; Orlowska,1997; Peter & Skowron,2000; Yao,2006). Attribute reduction algorithms of RST can select the most information rich attributes in a dataset without transforming the data while attempting to minimize information loss during the selection process. Unlike statistical correlation-reducing approaches, it requires no prior assumption and retains the semantics of the original data. Relying on simple set operations makes it suitable as a preprocessor for techniques that are much more complex.

In this paper, a feature selection method based on rough set theory is proposed for speech emotion recognition. Figure 1 illustrates the block diagram of the process flow in an emotion recognition system based on the proposed attribute selection method. A comparison of recognition accuracy with and without the feature selection process step is done on CLDC emotion speech database An accuracy of 74.75% with only 13 features is got. The experiments clearly show this approach can keep high recognition rate and reduce the calculation cost.

BASIC CONCEPTS OF ROUGH SET THEORY

In rough set theory, data is stored in a table, which may be called decision table. Rows of the decision table stand for objects, and columns show attributes which are divided into two disjoint groups called condition and decision attributes respectively. Each row of a decision table induces a decision rule. Decision rules are closely connected with approximations. Roughly speaking, certain decision rules describe lower approximation of decisions in terms of conditions, whereas uncertain decision rules refer to the boundary region of decisions. For the convenience of discussion, some basic concepts of rough set are introduced as follows.

Definition 1. An information system is defined as a quadruple tuple such that $S=(U,R,V,f)$, where U is a finite set of objects and $R=C\cup D$ is a finite set of attributes, C is the condition attribute set and $D=\{d\}$ is the decision attribute set. With every attribute $a\in R$, set of its values V_a is associated. Each attribute a determines function $f_a: U\rightarrow V_a$.

Definition 2. For a subset of attributes $B\subseteq A$, the indiscernibility relation is defined by $Ind(B)=\{(x,y)\in U\times U: a(x)=a(y), \forall a\in B\}$.

Definition 3. The lower approximation $B_{\_}(X)$ and the upper approximation $B^{-}(X)$ of a set of objects $X\subseteq U$ with reference to a set of attributes $B\subseteq A$ may be defined in terms of the classes in the equivalence relation, as follows:

$$B_{\_}(X)=\cup\{E\in U/Ind(B)\big|E\subseteq X\}$$

Figure 1. Diagram of speech emotion recognition system using RST as its feature selection processor

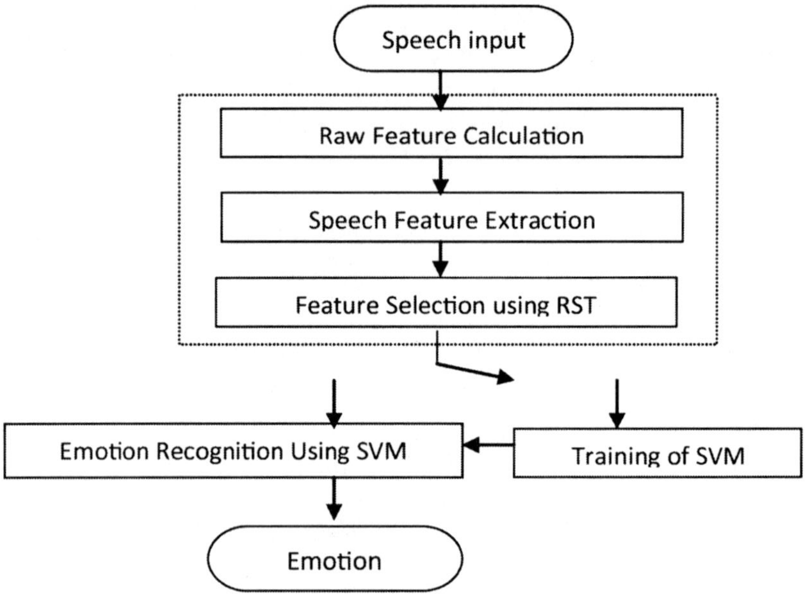

$$B^{-}(X) = \cup\{E \in U / Ind(B) \big| E \cap X \neq \Phi\}$$

Definition 4. $POS_P(Q) = \cup_{x \in U / Ind(Q)} P_{-}(X)$ is the P positive region of Q, where P and Q are both attribute sets of an information system.

Definition 5. A reduction of P in an information system is a set of attributes $S \subseteq P$ such that all attributes $a \in P - S$ are dispensable, all attributes $a \in S$ are indispensable and $POS_S(Q) = POS_P(Q)$. We use the term $RED_Q(P)$ to denote the family of reductions of P. $CORE_Q(P) = \cap RED_Q(P)$ is called as the Q-core of attribute set P.

Attribute reduction is one of the most important contributions of rough set theory. A reduct is a subset of attributes that is jointly sufficient and individually necessary for preserving the same information under consideration as provided by the entire set of attributes. After a process of discretization, it is possible to find a subset (namely, a reduct) of the original attributes that are the most informative; all other attributes can be removed from the dataset with minimal information loss. From the dimensionality reduction perspective, features containing sufficient information are those that are most predictive of the class attribute.

Generally speaking, there are two main approaches to finding rough set reducts: those that consider the degree of dependency and those that are concerned with the discernibility matrix (Jensen & Shen, 2007). Although the latter approaches may be guaranteed to discover all minimal subsets, it is a costly operation rendering the method impractical for even medium-sized datasets. There are three search control strategies in the former approaches: the addition strategy, the deletion strategy, and addition-deletion strategy. The addition strategy starts with the empty set and consecutively adds one attribute at a time until we obtain a reduct, or a superset of a reduct. The deletion strategy starts with the full set and consecutively delete one attribute at a time until a reduct is obtained finally. The addition-deletion strategy re-apply the deletion strategy on the superset of a reduct produced by the addition strategy. Given specific control strategy, one is able to derive most of the existing algorithms by adopting different heuristics or fitness functions such as information entropy, etc. for attribute selection(Yao,2006).

In this paper the CEBARKNC algorithm which belongs to the former approach using the deletion strategy (Yu, Wang, Yang & Wu,2002) is used to select speech features. This approach is based on the entropy heuristic employed by machine learning techniques such as C4.5 (Quinlan, 1993). It is concerned with examining a data set and determining those attributes that provide the most gain in information. The entropy of attribute P (which can take values $p_1, p_2, \ldots, p_m$) with respect to the conclusion D (of possible values $d_1, d_2, \ldots, d_n$) is defined as follow:

$$H(D \mid P) = -\sum_{i=1}^{n} p(p_i) \sum_{j=1}^{m} p(d_j \mid p_i) \log(p(d_j \mid p_i))$$

where $p(d_j \mid p_i) = \dfrac{|d_j \cap p_i|}{|p_i|}$

To illustrate the fundamental ideas and operations of the entropy-based reduction, an example dataset shown in Table 1 will be used. An entropy based reduction algorithm CEBARKNC is described as follows:

Input: a decision table with condition attributes set C and decision attributes set D.
Output: a reduction set B of the condition set C.
Step1: Calculate $H(D \mid C)$;
Step2: For every $a_i \in C$, calculate $H(D \mid \{a_i\})$; sort a_i's based on $H(D \mid \{a_i\})$ in descending order;
Step3: Let $B = C$;
Step4: For i=1 to |C| do
(1) Calculate $H(D / B - \{a_i\})$;
(2) if $H(D \mid C) = H(D \mid B - \{a_i\})$, then $B := B - \{a_i\}$.

Table 1. An example dataset

	Condition attribute (C)				Decision attribute
U	**a**	**b**	**c**	**d**	**e**
1	1	1	1	1	1
2	1	1	1	2	1
3	2	1	1	1	2
4	3	2	1	1	2
5	3	3	2	1	2
6	3	3	2	2	1
7	2	3	2	2	2
8	1	2	1	1	1
9	1	3	2	1	2
10	3	2	2	1	2
11	1	2	2	2	2
12	2	2	1	2	2
13	2	1	2	1	2
14	3	2	1	2	1

Returning to the example dataset, the algorithm first calculates $H(D \mid C)$ and the condition entropy of each individual attribute is recorded in Table 2.

Each attribute will be sorted in descending order of condition entropy, the new sorted set is {b, d, c, a}. The algorithm removes attribute b from C first and calculate the condition entropy $H(D \mid \{B - b\})$, the attribute b can be really removed from the attribute set because $H(D \mid \{B - b\}) = H(D \mid C)$. The set B is updated as {a,c,d}. when processing the attribute d, we found that d can not be deleted from set B because $H(D \mid \{B - d\}) \neq H(D \mid C)$,other two attributes are processed in iteration in the same way thereafter. The result of each step is recorded in Table 3.

At last, the subset {a, c, d} is chosen as the results. The dataset can now be reduced to these features only. The returned subset is called a rough set reduction.

Table 2.

Subset	Entropy
{a,b,c,d}	0
{*a*}	0.69
{*b*}	0.911
{*c*}	0.78
{*d*}	0.89

Table 3.

Step	Subset	Condition Entropy	Attribute reduction
1	{a, b, c, d} −{b}	0	remove b
2	{a, c, d}-{d}	0.34	Can not remove d
3	{a, c, d}-{c}	0.5	Can not remove c
4	{a, c, d}-{a}	0.65	Can not remove a

FEATURE SELECTION OF EMOTION RECOGNITION

In order to get raw features, an utterance is firstly input into the speech preprocessor which includes A/D transform, noise scatter and pre-emphasis which is conducted using the following function: $H(z) = 1 - \mu z^{-1}$ where $\mu = 0.97$ .Each utterance is divided into a number of frames with the same length. There is a fold on adjacent frames in order to smooth the features. The window size determines the number of samples used to generate a frame. Since the sampling rate of the utterances used in this paper is 16khz, the window duration is 32 ms(512 points) and the shared data in the windows overlap from frame to frame is 8 ms(128 points). This means that a feature vector is calculated from the speech data with 32 ms long. To estimate the energy contour as shown in Figure 2, a simple short-term energy function $M_n = \sum_{m=0}^{N-1} \left| x_n(m) \right|$ is used, where M_n is the energy of frame n and N=512.

The pitch contour as shown in Figure 2 is derived by estimating the pitch from energy peaks of the short-term autocorrelation function as follows:

$$R_n(k) = \sum_{m=0}^{N-1-k} x_n(m) x_n(m+k)$$

To estimate the first 3 formant contours as shown in Figure 3, we use the method proposed by Cheng (Cheng, 2003)who uses LPC to cal-

Figure 2. Energy and Pitch Contour

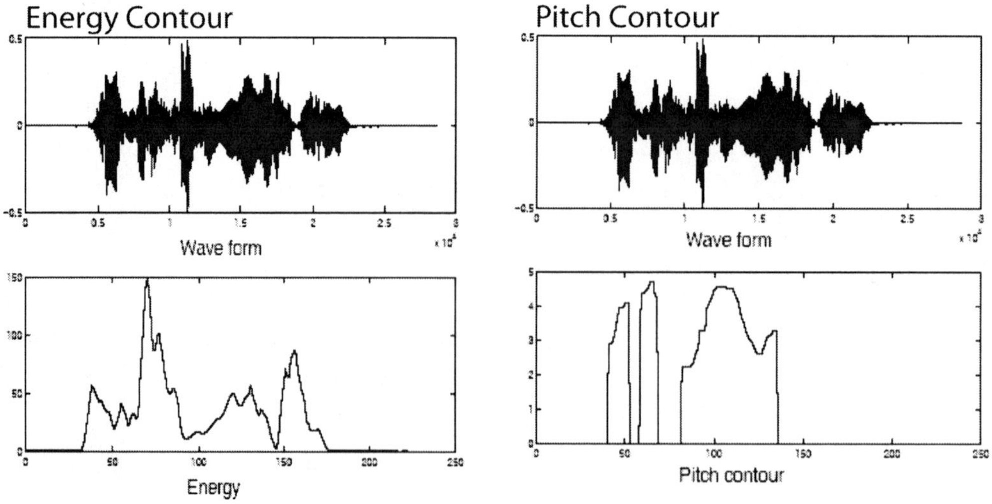

culate prediction coefficients, then the power stability of the track is estimated, and finally peak picking is done to get the frequency of formant. The 37 secondary statistical features (Polzin & Waibel,2000; Razak & Komiya, 2005; Dellaert & Polzin,1996; Oudeyer,2003; Kwon & Chan,2003) presented in section 2 are derived from the four raw features. The 37 features include energy related features,pitch related features,formant related features and speech rate related feature. Table 2 is a detailed list about the 37 features.

In Table 4, the features 1~12 are statistical properties of the energy contour, the features 13~24 are statistical properties of the pitch contour, the features 25~36 are statistical properties of the three formants contour, the feature 37 is the speech rate, which is equal to the number of words voiced per second.

The emotion speech database used in this paper comes from Chinese Linguistic Data Consortium(CLDC). There are two male speakers and two female speakers. 50 different sentence materials are used. Each speaker utters each sentence with 6 different emotions: happiness, anger, surprise, sadness, fear, and normal respectively. Accordingly, 1200 utterances are got with 200 utterances for each emotion state. All these utterances are digitized (16khz, 16bits).

In order to evalute the efficiency of this feature reduction method,a classification based on the total 37 features is carried out. A SVM((Burges, 1998))classifier is constructed with RBF kernel and trained with 600 speech data random selected. The total 1200 speech data is used to test and an average accuracy of 77.91% is got. Table 5 shows the classification result based on the total 37 features.

An attribute reduction via CEBARKNC is done thereafter, the following 13 features are selected.

- Energy features: 8, 10, 12
- Pitch features: 15, 16, 18, 19, 23
- Formant features:27, 30, 31, 34

In order to evaluate the efficiency of these 13 features, a SVM classifier is reconstructed with the same conditions as before. The experiment result is shown in Table 6.

From Table 5 and Table 6 we can find some interesting results. At first, both the 37 features and the 13 features selected through attribution reduction have good performance with high rec-

Figure 3. The first formant contour

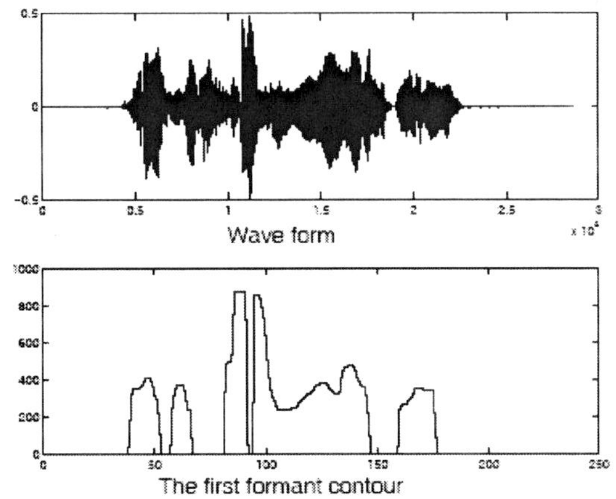

Table 4. Speech features used for emotion recognition

No.	Feature Description	No.	Feature Description
1	Maximum value of energy	20	Mean duration of rising slopes (pitch)
2	Mean value of energy	21	Maximum value of falling slopes (pitch)
3	Median value of energy	22	Mean value of falling slopes (pitch)
4	Variance of energy	23	Maximum duration of falling slopes(pitch)
5	Maximum value of rising slopes (energy)	24	Mean duration of falling slopes (pitch)
6	Mean value of rising slopes(energy)	25	Maximum value of the first formant
7	Maximum duration of rising slopes(energy)	26	Mean value of the first formant
8	Mean duration of rising slopes(energy)	27	Median value of the first formant
9	Maximum value of falling slopes(energy)	28	Variance of the first formant
10	Mean value of falling slopes(energy)	29	Maximum value of the second formant
11	Maximum duration of falling slopes(energy)	30	Mean value of the second formant
12	Mean duration of falling slopes(energy)	31	Median value of the second formant
13	Maximum value of pitch contour	32	Variance of the second formant
14	Mean value of pitch contour	33	Maximum value of the third formant
15	Median value of pitch contour	34	Mean value of the third formant
16	Variance of pitch contour	35	Median value of the third formant
17	Maximum value of rising slopes (pitch)	36	Variance of the third formant
18	Mean value of rising slopes(pitch)	37	Speech rate
19	Maximum duration of rising slopes(pitch)		

Table 5. Result of speech emotion recognition based on SVM

	Normal	Surprise	Anger	Sadness	Happiness	Fear
Normal	**181**	5	7	11	18	4
Surprise	2	**156**	5	3	22	12
Anger	6	12	**172**	1	15	1
Sadness	3	5	4	**153**	8	38
Happiness	6	16	11	3	**132**	4
Fear	2	6	1	29	5	**141**
Total	**200**	**200**	**200**	**200**	**200**	**200**
Accuracy	90.50%	78.00%	86.00%	76.50%	66.00%	70.50%
Total accuracy	77.91%					

ognition rate of 77.91% and 74.75% respectively. This result confirms that the four kinds of features are all important for emotion recognition. The second but the most remarkable conclusion about these results is that we can get an accuracy of 74.75% with only 13 features comparing with the accuracy of 77.91% using the total 37 features. The amount of the features selected through attribution reduction has been decreased 64.86%, but the accuracy is just reduced by 3.16%. Compared with the recognition system using 37 features, the performance of the classification using 13 features is increased greatly since the computation cost is reduced highly. This indicates that

Table 6. Result of speech emotion recognition based on Rough Set and SVM

	Normal	Surprise	Anger	Sadness	Happiness	Fear
Normal	**161**	7	8	10	18	7
Surprise	6	**152**	7	3	24	11
Anger	8	13	**170**	5	17	5
Sadness	5	2	2	**144**	9	30
Happiness	12	15	9	5	**128**	5
Fear	8	11	4	33	4	**142**
Total	**200**	**200**	**200**	**200**	**200**	**200**
Accuracy	80.50%	76.00%	85.00%	72.00%	64.00%	71.00%
Total accuracy	74.75%					

rough set could be used to select efficient subset of features which are able to classify the utterances with a high accuracy, and reduce the computation complexity.

CONCLUSION

A novel application of rough set theory is proposed in this paper. Other than traditional feature selection methods proposed in most literatures about speech emotion recognition, a feature selection method of emotional speech based on rough set is proposed in this paper. In order to evaluate the efficiency of the proposed method, SVM is taken as the classifier to recognize six emotion statuses with the total 37 features. 13 features are selected and correct recognition rate of 74.75% is resulted. These 13 features consist with empirical study and theory study in emotion research area. Since there is a process of discretization before feature reduction, information may be lost or changed in the process, the result of attribute reduction would be affected and the classification accuracy is sacrificed in some extent. However, the selecting process is easier understandable than other feature selection algorithms and the amount of the features selected through the proposed method has been decreased 64.86% while the accuracy is just reduced by 3.16%. The other advantage of this feature selection method is that there is no need to train the classifier when doing feature selection. So it is proved to be effective since its computation cost is reduced greatly. Because of its independency to the classifier, the feature selection method is universally used to other classifiers, and the feature selection step can be offline done which can make the speech emotion system more effective and fast. In the future, bi-module emotion recognition system through both speech and facial features will be studied. Furthermore, the features that are unique for speech emotion recognition will be studied.

ACKNOWLEDGEMENT

This work is supported by the National Natural Science Foundation of P. R. China (No. 60773113), Natural Science Foundation Project of CQ CSTC(No. 2008BA2041)and Natural Science Foundation Project of CQ CSTC(No. 2007BB2445).

REFERENCES

Burges, C. J. C. (1998). A Tutorial on Support Vector Machines for Pattern Recognition. *Data Mining and Knowledge*, *2*(2), 121–167. doi:10.1023/A:1009715923555

Cheng, X. M. (2003). The Method Analysis of Formant Parameters Picked-up in Sensibility Speech Communication. *Journal of Huzhou Teachers College*, *25*(6), 76–80.

Cowie, R., Douglas-Cowie, E., Tsapatsoulis, N., Votsis, G., Kollias, S., & Fellenz, W. (2001). Emotion recognition in human-computer interaction. *IEEE Signal Processing Magazine*, *18*(1), 32–80. doi:10.1109/79.911197

Dellaert, F., Polzin, T., & Waibel, A. (1996). Recognizing Emotion in Speech. In . *Proceedings of the ICSLP*, *96*, 1970–1973.

Griffith, D., & Greitzer, F. (2007). Neo-Symbiosis: The Next Stage in the Evolution of Human Information Interaction. *International Journal of Cognitive Informatics and Natural Intelligence*, *1*(1), 39–52. doi:10.4018/jcini.2007010103

Hall, M. A. (1999). *Correlation based Feature Selection for Machine Learning.* Doctoral dissertation, Department of Computer Science, The University of Waikato, Hamilton, New Zealand.

Jain, A. K., & Chandrasekaran, B. (1983). Dimensionality and sample size considerations. In P.R. Krishnaiah & L.N. Kanal (Eds.), *Pattern Recognition Practice, 2*(39), 835-855.

Jensen, R., & Shen, Q. (2007). Rough set based feature selection: A review . In *Rough Computing*. Theories, Technologies and Applications. doi:10.4018/9781599045528.ch003

Kwon, O. W., Chan, K., Hao, J., & Lee, T. W. (2003). Emotion Recognition by Speech Signals. *EuroSpeech*, *2003*, 125–128.

Orlowska, E. (1997). *Incomplete Information: Rough Set Analysis*. Springer Verlag.

Oudeyer, P. Y. (2003). The Production and Recognition of Emotions in Speech: features and algorithms. *International Journal of Human-Computer Studies*, *59*(1), 157–183. doi:10.1016/S1071-5819(02)00141-6

Pawlak, Z. (1984). Rough Classification. *International Journal of Man-Machine Studies*, *20*(5), 469–483. doi:10.1016/S0020-7373(84)80022-X

Pawlak, Z. (1991). *Rough Sets: Theoretical Aspects of Reasoning about Data.* Dordrecht: Kluwer Academic Publishing.

Peter, J., & Skowron, A. (2002). A rough set approach to knowledge discovery. *International Journal of Intelligent Systems*, *17*, 109–112. doi:10.1002/int.10010

Polzin, T. S., & Waibel, A. (2000). Emotion-Sensitive Human Computer Interfaces. In *Proceedings of the ISCA Workshop on Speech and Emotion,* (pp. 201-206).

Quinlan, J. R. (1993). *C4.5: Programs for Machine Learning. The Morgan Kaufmann Series in Machine Learning*. San Mateo, CA: Morgan Kaufmann Publishers.

Razak, A. A., Komiya, R., & Abidin, M. I. Z. (2005). Comparison Between Fuzzy and NN Method for Speech Emotion Recognition. *Proceedings of the ICITA'05*, Sydney, (pp. 297-302).

Ververidis, D. C. Kotropou, & Pitas, I. (2004). Automatic Emotional Speech Classification. *Proceedings of the ICASSP2004*, Canada, (pp. 593-596).

Wang, G. Y., Zheng, Z., & Zhang, Y. (2002). RIDAS-A Rough Set Based Intelligent Data Analysis System. *Proceedings of, ICMLC2002*, 646–649.

Wang, Y. (2005). On the cognitive processes of human perception. *Proceedings of ICCI'05*, (pp. 203-210).

Wang, Y. (2007). On the Cognitive Processes of Human Perception with Emotions, Motivations, and Attitudes. *International Journal of Cognitive Informatics and Natural Intelligence*, *1*(4), 1–13. doi:10.4018/jcini.2007100101

Wang, Y. J., & Guan, L. (2005). Recognition human emotion from audiovisual information. *Proceedings of the ICASSP*, *05*, 1125–1128.

Yao, Y. Y., Zhao, Y., & Wang, J. (2006). On reduct construction algorithms, Rough Sets and Knowledge Technology. *Proceedings of RSKT*, *2006*, 297–304.

Yu, H., Wang, G. Y., Yang, D. C., & Wu, Z. F. (2002). Knowledge Reduction Algorithms Based on Rough Set and Conditional Information Entropy. *Proceedings of the Society for Photo-Instrumentation Engineers*, *4730*, 422–431.

This work was previously published in International Journal of Cognitive Informantics and Natural Intelligence, Volume 3, Issue 3, edited by Yingxu Wang, pp. 51-60, copyright 2009 by IGI Publishing (an imprint of IGI Global)

Chapter 17
A User-Driven Ontology Guided Image Retrieval Model

Lisa Fan
University of Regina, Canada

Botang Li
University of Regina, Canada

ABSTRACT

The demand for image retrieval and browsing online is growing dramatically. There are hundreds of millions of images available on the current World Wide Web. For multimedia documents, the typical keyword-based retrieval methods assume that the user has a specific goal in mind by using accurate query keywords in searching a set of images. Whereas the users may face with a repository of images whose domain is less known and content is semantically complicated, or the users may only generally know what they search for. In these cases it is difficult to decide what exact keywords to use for the query. In this article, we propose a user-centered image retrieval method that is based on the current Web, keyword-based annotation structure, and combining Ontology guided knowledge representation and probabilistic ranking. A prototype of web application for image retrieval using the proposed approach has been implemented. The model provides a recommendation subsystem to support and assist the user modifying the queries and reduces the user's cognitive load with the searching space. Experimental results show that the image retrieval recall and precision rates increased and therefore demonstrates the effectiveness of the model.

INTRODUCTION

Image retrieval is a human centered task. Images are created by people and are ultimately retrieved and used by people for human related activities. Since the past decade, online image retrieval has been becoming one of the most popular topics on the Internet. The number of images available in online repositories is growing dramatically. For example, Flickr.com is hosting more than

DOI: 10.4018/978-1-60960-553-7.ch017

50 million member-submitted images on their Web site (Terdiman, D.). And the giant search engine company Google claimed that they had indexed more than 880 millions images since 2004 (www.google.com). The typical method of image retrieval using mostly by the industry is to create a keyword-based query interface above the multimedia database (Agosti, M., & Smeaton, A., 1996). There are two major problems in keyword-based image retrieval. The first one is the retrieval quality problem from the search result. The keyword annotation of image documents has low capability to analyze semantic relations among keywords, such as synonym, homonym and antonym. Taking the topics of images as an example, it is nearly impossible to include all the synonyms of the topic keywords in the annotation for every image. The reality is that if the images are annotated with keywords having same meanings with users input but in different terms, those images are not able to be retrieved by the keyword-based retrieval system. The second problem is that keyword-based search method always assumes that users have the exact searching goal in their mind (Hyvonen, E., Saarela, S., & Viljanen, K., 2003). However, in the real world application, the case is that most of them only hold a general interest to explore the images, and have a vague knowledge about the domain topic. They may not know what specific query keywords should use. As a result, a recommendation or a support subsystem helping users to modify their queries is needed.

Semantic Web technologies have been expected to improve the quality of information retrieval on the Web (Berners-Lee, T., Hendler, J., & Lassila, O., 2001; Berners-Lee, T.). In this article, we proposed a hybrid image retrieval model that uses a Web Ontology-based reasoning component and combining Bayesian Network model to improve the quality of image retrieval. Our proposed method returns more query keywords as recommendations that are semantically related to the user input keywords so that it can assist the users to explore more relevant images.

The rest of the article is organized as follows. In section 2, we reviewed the related research works in the area of image retrieval. In section 3, we present the proposed Ontology-guided model; precisely including the rules for Ontology reasoning and how to define the Bayesian Network model for ranking. In section 4, an image retrieval application for evaluating our model is presented in this implementation section. Section 5 is the conclusion and the future works.

Keyword-Based and Content-Based Image Retrieval

Traditionally, there are two main research approaches in the area of image retrieval. One is keyword-based image retrieval. This approach is to create a set of keywords as metadata to describe the images and then associate it to the image document. As a result, it is also called keyword annotation. Based on the keyword annotations, the system can apply keyword-based information retrieval techniques to search the images (Long, F., Zhang, H., Feng, D.D., 2003). Searchers try to analyze the text around image to improve the web image retrieval. However, huge amount of the images on Web, such as personal uploaded photo gallery, still lack of adequate text description.

The other approach is content-based image retrieval that mainly focuses on studying and analyzing the visual elements of the images. Through this approach, one is able to query the target images using criterions, such as color, shape and texture. A number of efficient content-based image retrieval systems have been presented in the last few years. For example, a database perspective of image annotation and retrieval has been studied by G. Carneiro and N. Vasconcelos (2005); a statistical approach of automatic linguistic indexing model was presented by J. Li and J. Z. Wang (2003); a machine learning approach is applied to study ancient art was presented by J.

Li and J. Wang (2004). However, in the scenarios of the online Web image retrieval, content-based approach is still hard to meet the requirement of immediate response of retrieval result to users. Features input, such as color, shape and texture, is still not suitable and realistic for most of the online users. Furthermore, it is difficult for the systems to deal with the features such as human emotions and perceptions of the images. In our study, we realized that both keyword-based and content-based approach is hard to solve the semantic problem. A solution is to incorporate the Ontology technology into image retrieval.

Ontology-Based Image Retrieval

With the advent of Semantic Web technology, information retrieval is able to widely be benefit from this ambitious technology which is being expected as the next generation of Internet. When searching information on the Web, it is very common that there are numerous different terms representing the same meaning of certain online resources. A solution to this problem is to provide a third part component of information collection, Ontology (Davies, J., Fensel, D., & Harmelen, F.V., 2003). Ontology is playing a key role as the core element of knowledge representation on the Semantic Web for machine understanding. Basically, Ontology consists of vocabularies and their relations to describe the existing things around us in the world. Some effort has been made for image retrieval using Semantic Web techniques, for example the case study of a view-based image search method using semantic annotation and retrieval from E. Hyvonen, S. Saarela and K. Viljanen (2003). Nevertheless, applying semantic annotation to the existing image resources will encounter the same difficulty with realizing the whole Semantic Web vision. Benjamins has pointed out that too little semantic content are available on the current Web, and shifting the current keyword annotation of the current Web content onto semantic annotation is also a big challenge (Kant, S., & Mamas, E., 2005). In order to discover an approach widely available for the existing content, our study is focusing on utilizing Ontology techniques and improving the reusability of the keyword annotation.

A successful image retrieval method should have a recommendation system to assist users to find what they actually need. To compute the relevance of the recommendations, however, the logic-based Ontology technology lacks of the capability to support plausible reasoning (Costa, P.C.G., Laskey, K.B., & Laskey, K.J., 2005; Benjamins, V. R., Centreras, J., Corcho, O., & Gomez-Perez, A., 2002). Because the propositions are either true or false, the kind of classical tradition of monotonic deductive reasoning can only answer the "Yes" or "No" questions. For instance, if a user makes a query with the keyword "Car", but the system only contains the index of the documents with the keywords "Vehicle" and "Honda Civic". By looking up into the Ontology, the machine is able to understand that the concept "Car" is a subclass of the concept "Vehicle", and "Honda Civic" is an instance of the class "Car". Which is closer to the user's query? From the pure logic-based approach, it can hardly answer the question without the totally matched result. Therefore, a method combining Ontology and

Bayesian Network technology to support uncertain reasoning for the retrieval system is proposed.

Bayesian Network

A Bayesian belief network is graphical representation of a set of random variables and their dependencies. It provides an effective knowledge representation which imitates knowledge structures used by the human mind.

Bayesian networks (BN) proposed by Judea Pearl in 1988 has become a very popular topic of uncertain reasoning in the artificial intelligent community (Pearl, J., 1988). There is an increasing number of successfully applying BN into a range of domains, such as medical diagnosis program, knowledge expert support system, classification and information retrieval etc. Based on the probability theory, Bayesian networks consist of the key concept, jpd (joint probability distribution) and a set of CPTs (Conditional Probability Table). In order to answer a query, typically, a DAG (Directed Acyclic Graph) is needed to be constructed. According to the DAG, a set of corresponding CPTs can be obtained. The jpd and CPTs can be represented as the following equation.

$$P(U) = P_1(a_1|c_1) \times P_2(a_2|c_2) \ldots\ldots \times P_i(a_i|c_i)$$

$U = \{a_1, a_2, \ldots, a_i\}$ denotes a finite set of discrete variables, and c_i denotes the fixed configuration of the parent p_i. According to the definition of the JPD (www.google.com), if we sum up all the variables in $P(U)$, the result will be 1.0, which can be also written in an equation as follows.

$$\sum_{a_1 \cdots a_i} P(U) = 1.0$$

Another important notion in Bayesian Network is CI (Conditional Independency). For avoiding the big table jpd acquisition problem, Bayesian Network utilizes the CI to break the big jpd table $P(U)$ into smaller tables consisting of $P_1(a_1|c_1)$, $P_2(a_2|c_2)$, ... $P_i(a_i|c_i)$.

A CI in a CPT $P(A|B)$ can be written in the following equation.

$$P(A|B) = \frac{P(A,B)}{P(B)}$$

The conditional probability $P(A|B)$ indicates the dependency and causal relation between variable A and variable B. Moreover, it also implies that node B is parent of node A in the DAG. Answering a query $P(A = v_i)$ from Bayesian networks means reasoning the probability of node A when given the value of v_i. By utilizing this theory, we can build up a BN model for computing the neighboring information of the Ontology in image retrieval. Inspired by the BN model for information retrieval from B. Ribeiro-Neto, I. Silva and R. Muntz (2000), we built up a BN model for computing the relevance ranking of the neighbours of the input concept in the Ontology graph.

Bayesian networks are used to model the causal relationships that exist in the context of image semantics. This will allow a system to associate query keywords with other semantic concepts to some degree of belief. Hence, image queries will be more intelligently handled and will yield better results, a direct result of understanding the dependencies and relationships between different semantic concepts. The causal relationships can be patterned using the provided web ontology, which already represents a set of concepts and their interrelations.

In the latest URSW05 workshop, some researchers prefer adding probabilistic extension into Ontology to enable the uncertain reasoning capability within Ontology, for instance the extension to OWL for representing particular Bayesian Network model (Ding, Z., & Peng, Y., 2004). Web Ontology is designed for representing and sharing knowledge in a distributed environment through Internet but uncertain reasoning models are usually application-independent. As a result, our proposed method prefers running the BN inference after the reasoning from Ontology. In this model, the Ontology is served as the source providing extend metadata for the BN Computation.

THE PROPOSED ONTOLOGY GUIDED IMAGE RETRIEVAL MODEL

In general, the process of our approach for image retrieval can be divided into three phases. Phase one is that once the system acquires the input keyword k from the user, it will try to look for the target images, which are annotated with the same keywords. If the target images record is found from the database, it will return the hits of those images. Otherwise, it requests to proceed for the second retrieval step, sending the keywords set to the Ontology reasoner.

In phase two, the Ontology reasoner is trying to find all the neighboring information of the input keyword. The keyword k is retrieved inside the Ontology by the Ontology reasoner which relies on rule-based inference technique. The reasoner will look up the conceptual graph to find the matched node which represents the concept k. If the matched node is not found, the reasoner will return an empty set. If the matched node denoted as k' is found, then k' will be marked.

k' is taken as an input parameter in the Ontology reasoner again. At the same time, we set the following rules for the reasoner.

If k' is a class entity in the graph, then the reasoner will return all of its properties $P = \{p_1, p_2,...p_i\}$, values of those properties $V = \{v_1, v_2,... v_i\}$, instance $I = \{i_1, i_2,...i_i\}$, subclasses $SB = \{sb_1, sb_2,...sb_i\}$, superclasses $SP = \{sp_1, sp_2,...sp_i\}$ and equivalence classes $E = \{e_1, e_2,...e_i\}$.

If k' is a property entity in the graph, then the reasoner will return all of its related classes $C = \{c_1, c_2,...c_i\}$, and the values of this property in its property domain $V = \{v_1, v_2,...v_i\}$.

If k' is an instance of specific classes in the graph, then the reasoner will return all of those classes $Cl = \{cl_1, cl_2,...cl_i\}$.

Collecting all of the above nodes and denoting the set as N, the reasoner send N back to the retrieval system. To avoid extra computation time for the ranking, the retrieval system filters out some of the elements in the set N, which are not included in the database, and then we have a smaller set N'.

The third step, as shown on Figure 1, after the nodes selection process, the nodes set N' will be passed to the BN model for relevance computation. In order to construct the BN model, we need to define what the role of the BN computation is playing in our hybrid image retrieval model. When the Ontology reasoner returns a set of corresponding nodes set N' to the query keyword k,

Figure 1. The retrieval procedure of our hybrid method

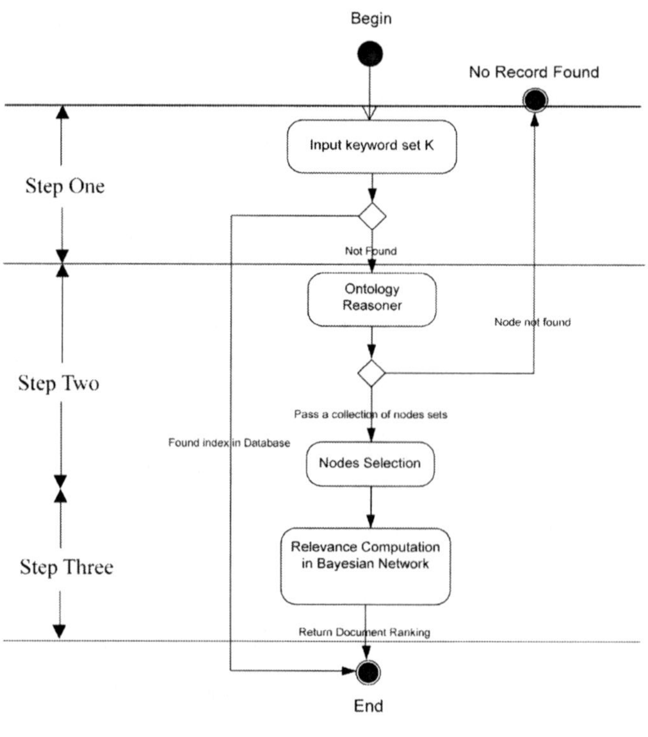

those nodes are all the neighbors of the original query keyword. However, sent from the Ontology reasoner which is based on the rule-based inference engine, the returning nodes have no ranking order for relevance. Thus, the BN model is used to computing the relevance from each returning node to the query keyword *k* for providing ranking of images collection.

A Bayesian Network is a compression of the joint probability distribution (jpd) of a set of random variables (U). Let $P_i(U)$ be joint probability distribution (jpd) of the all the relevance factors of the node N_i corresponding to the concept *k'* corresponding to the input keyword. Because $P_i(U)$ is a jpd, we can have the notion, $0 \leq P_i(U) \leq 1.0$. It is possible that there are many variables for the problem domain, so that the whole jpd table $P_i(U)$ is too large to be obtained. (Figure 2)

Therefore, we need to find out the conditional independencies (CIs) among the variables of this jpd. To understand the meaning of the relevance between the return node and the input keyword, we can also call the relevance between node and *k'* semantic distance, because Ontology defines the semantics of the vocabularies. If the semantic distance is shorter, it means the node is more relevant to the concep t*k'*. Therefore, if we can find a marginal $P_i(X)$ of $P_i(U)$, $X \subseteq U$, representing the probability of how relevant between the returning node and k_i. If $P_i(X) = 1.0$, we can claim that is exactly what user queried. Otherwise, if $P_i(X)=0.0$, it means that is irrelevant to the user's query.

In other words, if the value of the probability of $P_i(X)$ is closer to 1.0, it indicates the semantic distance of node is shorter than the one whose value is closer to 0.0.

Because $P_i(U)$ represents how relevant it is between the returning node and the concept node *k'* to the original query keyword, we can define the variables as follows:

- *sb*: is a subclass of node *k'*;
- *sp*: is a superclass of node *k'*;
- *i*: is an instance of node *k'*;
- *e*: is an equivalence class of node *k'*;
- *p*: is one of the properties of node *k'*;
- *Pr*: *k'* is a property entity in the Ontology;
- *In*: *k'* is a instance entity in the Ontology;
- *Cl*: *k'* is a class entity in the Ontology;
- *v*: is a value of *k'*, if *k'* is a property entity in the Ontology;
- *c*: is a class of *k'*, if *k'* is a instance entity or a property entity.

Figure 2. The neighbouring information of the concept k' in the hierarchical structure graph of the Ontology

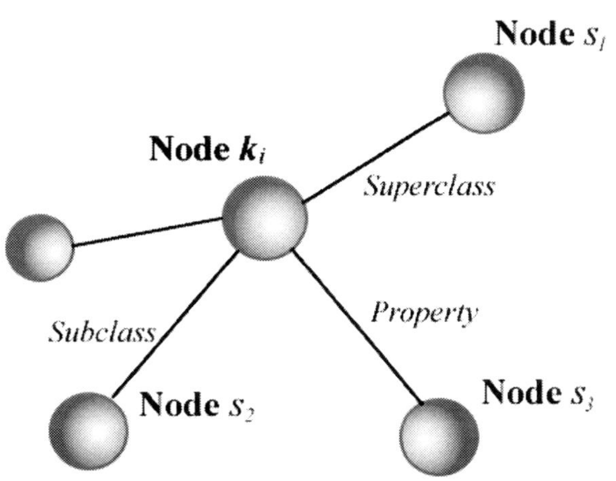

All the above variables are binary, having a domain $D = \{Yes, No\}$. Because the probability $P(n)$ is that we expect to compute, therefore, in the next step, we need to define the CIs for this BN model.

Assume that we are the knowledge experts in ontologies constructions. The CIs of the nodes extracted from ontologies can be defined. Once we have *k'*, we will know if it is a class, property, or instance entity in the Ontology. As a result, we can acquire $P(Cl|\mathrm{Pr}, In)$. If *k'* is a property entity in the Ontology, then we have $P(v, c|\mathrm{Pr})$. If *k'* is an instance entity, we have $P(c|In)$.

If a node is a subclass of node k_i, it indicates that it is impossible to be a superclass of the node k_i. Then we can write it as $P(sb|sp)$. If a node is an equivalence class of node k_i, it indicates that it is not either a subclass or superclass of the node k_i. Then we have $P(sb, sp|e)$. When the original concept node k_i takes different value, for example, as a class entity or a property entity, the return nodes set will be different.

Consequently, we can obtain another conditional probability table (CPT), $P(i, r, e, p|k)$. All the CPTs can be listed as follows.

$P(Cl|\mathrm{Pr}, In)$
$P(v, c|\mathrm{Pr})$
$P(c|In)$
$P(n|sb, sp, i, r, e, p)$
$P(i, r, e, p|Cl)$
$P(sb|sp)$
$P(sb, sp|e)$

According to the above CPTs, we can draw the DAG (directed acyclic graph) as shown in Figure 3.

According to the DAG, we can simplify the jpd into Table 1

$$P(U) = P(\mathrm{Pr}) \times P(In) \times P(v|\mathrm{Pr}) \times P(c|\mathrm{Pr}, In) \times P(Cl|\mathrm{Pr}, In) \times P(i, r, e, p|Cl) \times P(sb, sp|e) \times P(sb|sp) \quad (1)$$

Once we have the whole $P(U)$ equation (1), the values of CPT tables are ready to be assigned for the initialization work. Once the Ontology reasoner returns the set *N'* of nodes, we can obtain the values of the variables of *k, e, sb, sp, i, r,* and *p*. Post those values as evidences to the BN network, propagating by the HUGIN (Jensen, F.V., Lauritzen, S.L., & Olesen, K.G., 1990) or Shafer-

Figure 3. The DAG of our relevance model

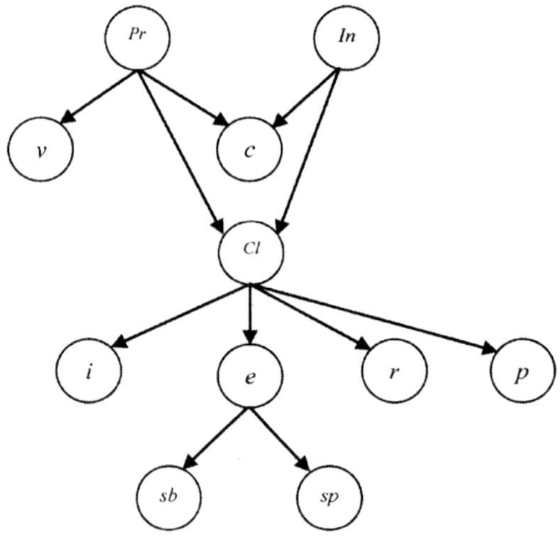

Table 1 .The comparison of two methods

Approach	Description	Providing Semantics Recommendation	Numbers of Images in Search Result
Ontology & BN based Retrieval	Using Ontology for semantic reasoning, and BN for ranking computation	Yes	39
Keyword-based Retrieval	Find matched keyword in the database	No	18

Shenoy (1990) method. Then the probability $P(U)$ can be queried.

Using this method for computing all the returning nodes, the ranking factors are obtained for each node.

With these factors, the system can retrieve the related images documents from the database. After sorting the images by the relevance, eventually, the system can have a descending order of the images list of the retrieval result.

The set of returned concepts with the relevance can be served as query keywords of recommendations to the users on top of the search result, because those keywords are the neighbors of the input query keyword, and are semantically related to users query. This recommendation system can help users address the problem that they have an unclear search goal in mind. A simple and long ranked list of retrieval result can not satisfy the need of the users. Lots of researchers have proposed to provide a better organized search result. For example, online search result clustering. For the reason that the amount of the online images growing amazingly every day, the search result may return hundreds or thousands of images to the users, since we use our approach to increase the recall rate. Thus, our recommendations with keywords assisting users to find the desired images can be viewed as a search result categorization consisting of clusters of the meaningful keywords.

IMPLEMENTATION AND DISCUSSION RESULTS

We have implemented our prototype hybrid retrieval model by combining Ontology and BN techniques into a Web application. The experiment demonstrates some promising results. We currently use the images relates to automobile only. In our experiment, a query with the keyword "car" is input into the search system. The images annotated with keywords, such as "automobile" or "honda", which are the equivalent to "car" or a brand name of "car", are shown in Figure 4.

Figure 4. A screenshot of the search result from a query "car"

In this implementation application, all the images are stored in a local web site and indexed with keywords in a local database. We assume that all the images are annotated well with keywords representing the topics of those images.

The reason that we set this assumption is that one of the purposes in this research project is to reuse the existing resource of annotation on the Web without any modification or attachment to it.

The Ontology reasoning component here serves as a Web service. The reason we wrap the Ontology reasoner logic into a SOAP XML Web service is because we have considered that the Ontology can be possibly selected from a remote online Ontology provider, and SOAP XML Web Services have been considered as an efficient choice for remote information access crossing platform and organizational boundaries. In the experiment from our application, we finished our programming with help of the existing research and tools of Semantic Web and Bayesian Networks. For instance, we implemented a demo Ontology in Ontology Web Language (OWL) using a Java Ontology APIs, Jena, from HP research lab (McGuinness, D.L., & Harmelen, F.V., 2004; McBride, B.). Our Bayesian Network computation model was constructed and inferred by the MSBNx (Microsoft Bayesian Network Tookit) from Microsoft research (Microsoft).

There are 210 images in our local image library, and 18 different keywords annotating the images, stored in database. There is a database indexing for all the image resources running in our system as well. In this experiment, all the images are stored in the local Web site. First of all, this application provides both Ontology & BN based search and Keyword-based search methods. Secondly, each related keyword returned from the Ontology reasoner was ranked by the semantic distance between the original input keyword node and the returned concept node from Ontology.

The shorter semantic distance between two nodes the higher position the image will be put on the search result ranking. The Bayesian Network model plays a critical key role in the semantic distance computation. Figure 5 shows the detail information from the search result. In this example, an image of a vehicle is annotated with the only keyword "vehicle". If using the classical keyword match search, this image is not able to be retrieved. In our experiment, the query keyword "car" is input into our system. After the Ontology reasoning, keyword "vehicle" is returned as the superclass of concept "car" in the conceptual graph. "vehicle" is passed into the BN compution as an evidence in the BN belief network model we defined earlier. Then after the network propagation, a relevance probability of "vechicle", also call semantic distance value, is sent back to the main process.

The comparing result of using different retrieval methods from one simple query, "car", is listed in Table 1. According to the experiment data from the table, there are only 15 images which are directly annotated with the keyword, "car", and 39 images with other related keywords, such as "vehicle", "honda civic", and "automobile". The recommendation system works as planned and the recommendations from the search result contain the recommended images resource for users to navigate the search result. Take an ex-

Figure 5. A detailed view of one return image item

ample from the search in Figure 4, those recommended keywords, such as "automobile", "honda" and "toyota", are the neighbor nodes of the input keyword, "car", generated from the Ontology reasoning.

The recall rates of our search approach are mainly affected by the Ontology reasoning, because, if the Ontology does not contain the concept keyword which appears in the image annotation and represents the topic of the images, those images will not be retrieved. In others words, under this situation, the recall rate are lowered. Another factor to affect the recall rate of our proposed approach is the image annotation. However, at the beginning of our implementation, we assume that all the images are well annotated with keywords representing topics. Therefore, our experiment result is not affected by this factor. With the assumption in our experiment, it is indicates that all the retrieved records from the search result are relevant to the user's query, since the keywords of those retrieved records are semantically relevant to the query keyword and ranked with relevance rate.

To sum up, the experiment data has shown that the Ontology & BN based method returns far more related records from the images library. Furthermore, the search result of our method not only covers the one from the keyword-based search, but also provides semantic recommendation for the end users to navigate around the result set of images, to solve the problem that users do not have the exact search goal in their mind.

CONCLUSION AND FUTURE WORK

In this article, we have presented an approach of the user-driven Ontology guided image retrieval. It combines the certain reasoning techniques based on logic inside Ontology and the uncertain reasoning technique based on Bayesian Network to provide users the enhanced image retrieval on the web. More significantly, our approach is for easily plugging in an external Ontology in the distributed environment and assist user searching for a set of images effectively. In addition, to obtain a faster real-time search result, the Ontology query and BN computation should be run on the off-line mode, and the results should be stored into the indexing record.

Although our retrieval prototype has shown that it can retrieve more images than the keyword-based method, there are still some problems that we found from our experiment. The returned concepts from the Ontology reasoner are ranked relying on their basic relations inside Ontology with the original input keyword. For example, a brand name, "Honda", and model name, "Civic", both are the values of the properties of the concept, "car". Which one is closer to the user query? Our model still cannot answer this kind of precise question in semantics. This will be studied in our future research work. We are also going to apply a large scale remote Ontology into our system. And searching a larger scale image library will be considered. User profile oriented recommendations should take into consideration in our future works since users have different search experiences, search strategies and knowledge about the domain topics. Personalized user recommendations would serve better to the user's individual goals.

REFERENCES

Agosti, M., & Smeaton, A. (1996). *Information retrieval and hypertext.* New York: Kluwer.

Benjamins, V. R., Centreras, J., Corcho, O., & Gomez-Perez, A. (2002). Six Challenges for the Semantic Web. *ISWC2002.*

Berners-Lee, T. (n.d.). *Semantic Web Road Map.* World Wide Web consortium, http://www.w3.org/DesignIssues/Semantic.html.

Berners-Lee, T., Hendler, J., & Lassila, O. (2001). *The Semantic Web.* Scientific American.

Carneiro, G., & Vasconcelos, N. (2005). A Database Centric View of Semantic Image Annotation and Retrieval. *Proceedings of ACM Conference on Research and Development in Information Retrieval.*

Costa, P.C.G., Laskey, K.B., & Laskey, K.J. (2005). PR-OWL: A Bayesian Ontology Language for the Semantic Web. *URSW`05.*

Davies, J., Fensel, D., & Harmelen, F. V. (2003). *Towards the Semantic Web -- Ontology-Driven Knowledge Management.* Wilq.

Ding, Z., & Peng, Y. (2004). A Probabilistic Extension to Ontology Language OWL. *Proceedings of the 37th Hawaii International Conference on System Sciences.*

Google. http://www.google.com/press/pressrel/6billion.html.

Google Achieves Search Milestone with Immediate Access To More Than 6 Billion Items.

Hyvonen, E., Saarela, S., & Viljanen, K. (2003). Intelligent Image Retrieval and Browsing Using Semantic Web Techniques – A Case Study. *The International SEPIA Conference*

Jensen, F.V., Lauritzen, S.L., & Olesen, K.G. (1990). *Bayesian Updating in Causal Probabilistic Network by Local Computation.*

Kant, S., & Mamas, E. (2005). Statistical Reasoning – A Foundation for Semantic Web Reasoning. *URSW`05.*

Li, J., & Wang, J. (2003). Automatic Linguistic Indexing of Pictures by a Statistical Modeling Approach. *IEEE Trans. on Pattern Analysis and Machine Intelligence.*

Li, J., & Wang, J. (2004). Studying digital imagery of ancient paintings by mixtures of stochastic models. *IEEE Transactions on Image Processing.*

Long, F., Zhang, H., & Feng, D. D. (2003). *Fundamentals of Content-based image retrieval. Multimedia Information Retrieval and Management- Technological Fundamentals and Applications.* Springer.

McBride, B. (n.d.). *Jena: Implementing the RDF Model and Syntax Specification,* http://www.hpl.hp.com/personal/bwm/papers/20001221-paper.

McGuinness, D. L., & Harmelen, F. V. (2004). OWL Web Ontology Language Overview. *W3C,* http://www.w3.org/TR/owl-features/.

Microsoft Bayesian Network Toolkit. http://research.microsoft.com/adapt/MSBNx/.

Pearl, J. (1988). *Probabilistic Reasoning in Intelligent Systems: Networks of Plausible Inference,* (pp. 1-20).

Ribeiro-Neto, B., Silva, I., & Muntz, R. (2000). *Bayesian Network Models for Information Retrieval. Soft Computing in Information Retrieval: Techniques and Applications* (pp. 259–291). Springer.

RIEMANN (Research on Intelligent Media Annotation), Pennsylvania State University, http://wang.ist.psu.edu/IMAGE/.

Shafer, G.R., & Shenoy, P.P. (1990). *Probability Propagation Annals of Mathematics and Artificial Intelligence.*

Terdiman, D. (n.d.). 'Tagging' gives Web a human meaning. *CNET News.com.* http://news.com.com/Tagging+gives+Web+a+human+meaning/2009-1025_3- 5944502.html.

Veltkamp, R., & Tanase, M. (2000). Content-based Image Retrieval Systems: A Survey. Technical Report UU-CS-2000-34, Utrecht University.

This work was previously published in International Journal of Cognitive Informantics and Natural Intelligence, Volume 3, Issue 3, edited by Yingxu Wang, pp. 61-72, copyright 2009 by IGI Publishing (an imprint of IGI Global)

Section 4

Chapter 18
On Cognitive Foundations of Creativity and the Cognitive Process of Creation

Yingxu Wang
University of Calgary, Canada

ABSTRACT

Creativity is a gifted ability of human beings in thinking, inference, problem solving, and product development. A creation is a new and unusual relation between two or more objects that generates a novel and meaningful concept, solution, method, explanation, or product. This article formally investigates into the cognitive process of creation and creativity as one of the most fantastic life functions. The cognitive foundations of creativity are explored in order to explain the space of creativity, the approaches to creativity, the relationship between creation and problem solving, and the common attributes of inventors. A set of mathematical models of creation and creativity is established on the basis of the tree structures and properties of human knowledge known as concept trees. The measurement of creativity is quantitatively analyzed, followed by the formal elaboration of the cognitive process of creation as a part of the Layered Reference Model of the Brain (LRMB).

INTRODUCTION

Creativity is a gifted ability of human beings in thinking, inference, problem solving, and product development (Beveridge, 1975; Csikszentmihalyi, 1996; Holland, 1986; Matlin, 1998; Smith, 1995; Sternberg & Lubart, 1995; Wang et al., 2006; Wilson & Keil, 1999). Human creativity may be classified into three categories known as the abstract, concrete, and art creativities. A scientific (abstract) creation is usually characterized by a free and unlimited creative environment where the

DOI: 10.4018/978-1-60960-553-7.ch018

goals and paths for such a creation is totally free and unlimited; while an engineering (concrete) creation is characterized by a limited creative environment where a creative problem solving is constructed by a certain set of goals, paths, and available conditions. The third form of creation is the art (empirical) creation that generates a novel artifact that attracts human sensorial attention and perceptual satisfactory.

Creativity has been perceived diversely and controversially in psychology, intelligence science, and cognitive science (Csikszentmihalyi, 1996; Guiford, 1967; Leahey, 1997; Mednich & Mednich, 1967; Matlin, 1998; Sternberg & Lubart, 1995; Wallas, 1926; Wang et al., 2009a, 2009b). Creativity may be treated as a form of art that generates unexpected results by unexpected paths and means. It may also be modeled as a scientific phenomenon that generates unexpected results by purposeful pursuits. In 1998, Matlin perceived that creativity is a special case of problem solving (Matlin, 1998). From this perspective, he defined creativity as a process to find a solution that is both novel and useful. However, problem solving often deals with issues for a certain goal with unknown paths. Therefore, creation is much more divergent that deals with issues of both unknown goals and unknown paths for a problem under study.

The nature of creations is a new and unusual relation between two or more objects that generates a novel and meaningful concept, solution, method, explanation, or product. This article investigates into the cognitive process of creation and creativity as a higher-layer life function. Cognitive foundations of creativity are explored on such as the space of creativity, the approaches to creativity, the relationships of creation and problem solving, and the attributes of creative researchers. A set of mathematical models of creation and creativity is developed by studying the tree structures and properties of human knowledge known as concept trees. On the basis of the concept tree, the measurement of creativity is quantitatively analyzed. The cognitive process of creation is rigorously elaborated with Real-Time Process Algebra (RTPA) (Wang, 2002a, 2007a, 2008a, 2008e), which provides a formal explanation of human creativity.

COGNITIVE FOUNDATIONS OF CREATIVITY[2]

Human creativity as a gifted ability is an intelligent driving force that brings something into existence.

Definition 1. *Creativity is the intellectual ability to make creations, inventions, and discoveries that brings novel relations and entities or unexpected solutions into existence.*

Definition 2. *A creation is a higher cognitive process of the brain at the higher cognitive layer that discovers a new relation between objects, attributes, concepts, phenomena, and events, which is original, proven true, and useful.*

Wallas identified five stages in a creative process (Wallas, 1926) as follows: (1) *preparation*, (2) *incubation*, (3) *insight*, (4) *evaluation*, and (5) *elaboration*. Csikszentmihalyi pointed out that creativity can best be understood as a confluence of three factors: a *domain* that consists of a set of rules and practices; an *individual* who makes a novel variation in the contents of the domain; and a *field* that consists of experts who act as gatekeepers to the domain, and decide which novel variation is worth adding to it (Csikszentmihalyi, 1996).

Various creativities and creation processes may be identified such as free/constrained creativity, analytic/synthetic creativity, inference-based creativity, problem-solving-based creativity, and scientific/ technological/art creativity. The entire set of creativities can be classified into three categories according to their creation spaces, approaches, and problem domains as summarized in Table 1.

Table 1. Taxonomy of creativity and creation

No.	Category	Type of creation	Description	Reference
1	Creation space	Free	A creation process with an unlimited creation space S_c, which is determined by unconstrained sets of alternatives N_a, paths N_p, and goals N_g.	Def. 4
2		Constrained	A creation process with a limited creation space S'_c where one or more conditions such as the goals N'_g, paths N'_p, or alternatives N'_a, are limited.	Def. 5
3	Approach	Analytic	A top-down creation process that discovers a novel solution to a given problem by deducing it to the subproblem level where new or existing solutions may be found.	Def. 7
4		Synthetic	A bottom-up creation process that discovers a novel solution to a given problem by inducting it to a superproblem where new or existing solutions may be found.	Def. 8
5		Inference-based	An abstract creativity based on the deductive, inductive, abductive, and analogy inference methodologies.	Def. 9
6		Problem-solving-based	A novel solution for a given problem by creative goals and/or creative paths.	Def. 15
7	Domain	Scientific (abstract)	A free and unlimited creative environment where the goals and paths for such a creation is totally free and unlimited.	Section 1
8		Technological (concrete)	A limited creative environment where a creative problem solving is constructed by a certain set of goals, paths, and available conditions.	Section 1
9		Art (empirical)	A free and unlimited creative environment where a novel artifact is generated that attracts human sensorial attention and perceptual satisfactory	Section 1

Definition 3. *A creation space Θ is a Cartesian product of a nonempty set of baseline alternatives A, a nonempty set of paths P, and a nonempty set of goals G, i.e.:*

$$\Theta \triangleq A \times P \times G$$

where × represents a Cartesian product.

The Space of Creativity

On the basis of the creation space, the nature of free and constrained creativities can be explained.

Definition 4. *A free creativity is a creation process with an unlimited creation space S_c, $S_c \subseteq \Theta$, which is determined by unconstrained sets of alternatives N_a, paths N_p, and goals N_g, i.e.:*

$$S_c \triangleq N_a \bullet N_p \bullet N_g$$
$$= \#A \bullet \#P \bullet \#G$$

where # is the cardinal calculus that counts the number of elements in a given set.

Equation 2 indicates that the creative space of a free creation may very easily turn to be infinitive, because N_a, N_p, and N_g can be extremely large. Therefore, the cost or difficulty of creation is often extremely high. That is, only mechanical and exhaustive search is insufficient in most cases for potential creations and discoveries, if it is not directed by heuristic and intelligent vision. In other words, creations and discoveries are usually achieved only by chance of purposeful endeavors of prepared minds, where an appreciation of highly unexpected result is always prepared. This is also inline with the empirical finding of Pasteur as he stated that "Creation always favorites prepared minds (Beveridge, 1975)."

Definition 5. *A constrained creativity is a creation process with a limited creation space S'_c, $S'_c \subseteq S_c \subseteq \Theta$, where one or more conditions such* *as the goals N'_g, paths N'_p, or alternatives N'_a, are limited, i.e.:*

$$S'_c \triangleq N'_a \bullet N'_p \bullet N'_g$$
$$= \#A' \bullet \#P' \bullet \#G',\ A' \subset A \wedge P' \subset P \wedge G' \subset G$$

Usually, a scientific and art creation is characterized as a free creation process, while an engineering creation is featured as a constrained creation process.

Approaches to Creativity

A variety of typical and sometimes controversial approaches to creation have been identified in literature, such as divergent production (Guiford, 1967), remote association test (Mednich & Mednich, 1967), analysis/synthesis (Wang et al., 2006), and inferences (Wang, 2007c).

Wallas (1926), Beverage (1957), and Smith (1995) pointed out an important phenomenon in human creativity known as *incubation*.

Definition 6. *Incubation is a mental phenomenon that a breakthrough in problem solving may not be achieved in a continuous intensive thinking and inference until an interrupt or interleave action is conducting, usually in a relax environment and atmosphere.*

The cognitive mechanism of incubation can be explained by the subconscious processes of the brain related to thinking and inference, such as perception, imagination, and unintentional search, which are involved in complex thinking and long chains of inferences. Whenever there is an impasse, incubation may often lead to a creation under the effect of active subconscious processes. Incubation has been observed to play an active role in the creation process.

The approaches to creativity can be categorized into the analytic, synthetic, and inference approaches as described below.

Definition 7. *An analytic creativity is a top-down creation process that discovers a novel solution to a given problem by deducing it to the subproblem level where new or existing solutions may be found.*

Definition 8. *A synthetic creativity is a bottom-up creation process that discovers a novel solution to a given problem by inducting it to a superproblem where new or existing solutions may be found.*

Definition 9. *An inference creativity is an abstract creativity based on the deductive, inductive, abductive, and analogy inference methodologies.*

The inference methodologies as a fundamental approach to creativity have been formally studies in (Wang, 2007c).

Creation vs. Problem Solving

As creativity is a novel or unexpected solution to a given problem, a creation may be perceived as a special novel solution where the problem, goal, or path is usually unknown. Therefore, the study of the generic theory of creativity can be reduced to the theory of problem solving (Wang & Chiew, 2009). The theoretical framework of problem solving can be modeled as follows.

Definition 10. *A problem solving is a cognitive process of the brain that searches or infers a solution for a given problem in the form of a set of paths to reach a set of given goals.*

Definition 11. *Assuming the layout of a problem solving is a function $f: X \to \ldots \to Y$, the problem ρ is the domain of f, X, in general, and a specific instance x, $x \in X$, in particular, i.e.:*

$$\rho \triangleq (X \mid f: X \to \ldots \to Y) \quad (4)$$

Figure 1. A solution in creation trace T and problem solving

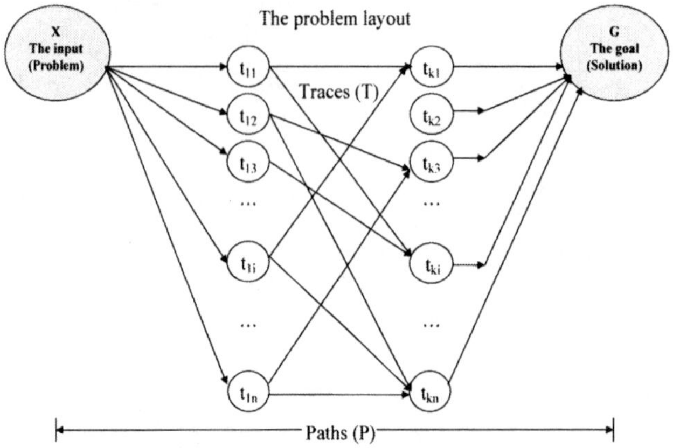

Equation 4 denotes that, in problem solving, a problem ρ is the fix point of a function in general, and the input of the function in particular. The former is the broad sense of a problem, and the latter is the narrow sense of a problem.

Problem solving is a process that seeks the generic function for a layout of problem, and determines its domain and codomain. Then, a solution in problem solving can be perceived as a concrete instance of a given function for the layout of the problem.

According to Definition 11, there are two categories of problems in problem solving: (a) The *convergent* problem where the goal of problem solving is given but the path of problem solving is unknown; and (b) The *divergent* problem where the goal of problem solving is unknown and the path of problem solving are either known or unknown.

Definition 12. *A goal G in problem solving is a terminal result Y of satisfactory in the creation space S_c of the problem ρ, which deduce X to Y by a sequence of inference in finite steps, i.e.:*

$$G \triangleq (Y \mid X \to ... \to Y), \;\; G \in \Theta$$

Definition 13. *A path P in problem solving is a 3-tuple with a nonempty finite set of problem inputs X, a nonempty finite set of traces T, and a nonempty finite set of goals G, i.e.:*

$$\begin{aligned} P &\triangleq (X, T, G) \\ &= X \times T \times G \end{aligned}$$

where the traces T is a set of internal nodes or possible subpaths that leads to the solution.

Definition 14. *A solution to a given problem ρ is a selected relation or function, S, which is an instance of the solution paths in P, i.e.:*

$$S \triangleq (X, T, G) \subseteq P, \;\; X, T, G \neq \varnothing$$

The solutions S and paths P in problem solving as modeled in Definition 14 can be illustrated in Figure 1.

Theorem 1. *The polymorphic solutions state that the solution space SS, $SS \subseteq \Theta$, of a given problem ρ is a product of the numbers of problem inputs N_x, traces N_t, and goals N_g, i.e.:*

$$\begin{aligned} SS &\triangleq N_x \bullet N_t \bullet N_g \\ &= \#X \bullet \#T \bullet \#G \end{aligned} \tag{8}$$

The polymorphic characteristic of the solution space contributes greatly to the complexity of problem solving and creations. It is noteworthy that the path $p(x,t,g) \in P$ in Definition 13 can be a simple or a complex function. A complex function that mapping a given problem into a solution goal may be very complicated depending on the nature of the problem.

According to Definition 13, in case $\#X = 0$, $\#G = 0$, or $\#T = 0$, there is no solution for the given problem. For a convergent problem, i.e. $\#G = 1$, the number of possible solutions $SS = \#X \bullet \#T$.

Definition 15. *A creation C is a novel and unexpected solution S_0, which is a subset of the entire set of SS that meets the criteria of novelty, originality, and utility, or the originality of the creation is true, i.e. $O = 1$, i.e.:*

$$C \triangleq (S_0 \mid S_0 \subseteq P \subseteq SS \wedge O = 1)$$
$$= X_0 \times T_0 \times G_0,\ X_0 \subseteq X \wedge T_0 \subseteq T \wedge G_0 \subseteq G \wedge O = 1$$

It is noteworthy that, although a creation C is a subset of the entire solutions S for a given problem, it is always the unknown and novel subset, which extends the entire solution set.

According to Definition 15, a creation is a search for the unknown goals, unknown paths, or both under a given problem or a set of coherent problems. Therefore, creations can be classified into the categories of goal-driven, method-driven, and problem-driven. Among them, the problem-driven creation is a fully open process because both goals and paths are unknown for the given problem.

Attributes of Inventors

A number of typical attributes sharing by inventors have been studied. In his book on *The Art of Scientific Investigation*, Beveridge (1957), a professor at Cambridge University, thought that the research scientists are fortunate in that in their work they can find something to give meaning and satisfaction to life. Beveridge identified a set of attributes required for researchers and inventors, such as enterprise, curiosity, initiative, readiness to overcome difficulties, perseverance, a spirit of adventure, a dissatisfaction with well-known territory and prevailing ideas, and an eagerness to try his own judgment, intelligence, imagination, internal drive, willingness to work hard, perseverance and tenacity of purpose (Beveridge, 1957). In the inventive theory of creation in psychology, Sternberg and Lubart's (1995) elicited the following set of attributes of inventors in psychology, such as intelligence, knowledge, motivation, appreciation, thinking style, and personality. Contrasting the two sets of attributes identifies above, it is interesting to note that the former would have understood scientific creation and invention deeper and with much insight than that of the psychological observations.

Beveridge believed that an insatiable curiosity and love of science are the two most essential attributes of scientists. He pointed out that a good maxim for researchers is look out for the unexpected. He described that creators are those whose imagination are fired by the prospect of finding out something never before found by man, and only for those will succeed who have a genuine interest and enthusiasm for discovery (Beveridge, 1957). Another crucial attribute is perseverance or persistence as Pasteur wrote: "Let me tell you the secret that has led me to my goal. My only strength lies in my tenacity (Dubos, 1950)." Pasteur has also revealed that "In the field of observation, chance favors only the prepared mind."

It is noteworthy that the above investigations into research and researchers have overlooked a more significant attribute for creativity and discovery ability, i.e., *mathematical skills* or the *abstract inference capability*, because mathematics plays the ultimate role of meta-methodology in science and engineering creativities. Actually, mathematical skills and abstraction capability are the most important foundation for scientific creation and invention, which enables a scientist to inductively generalize a hypothesis into the maximum scope, usually the infinitive or the universal domain based on limited sample empirical studies and/or mathematical/logical inferences. It is noteworthy that mathematics is the generic foundation of all science and engineering disciplines, as well as all scientific methodologies. To a certain extent, the maturity of a discipline is characterized by the maturity of its mathematical means (Bender, 2000; Zadeh, 1965, 1973; Wang, 2007a, 2008a, 2008b). One of the major purposes of cognitive informatics is to develop and introduce suitable mathematical means into the enquiry of natural intelligence, computational intelligence, cognitive science, and software science. The studies on denotational mathematics (Wang, 2008a, 2008b), such as system algebra (Wang, 2008d), concept algebra (Wang, 2008c), RTPA (Wang, 2002b, 2007a, 2008a, 2008e), and Visual Semantic Algebra (VSA) (Wang, 2009b) are fundamental endeavors towards the formalization of the entities that are conventionally hard-to-be-formalized.

According to cognitive informatics (Wang, 2002a, 2003, 2007b, 2009a; Wang & Wang, 2006; Wang et al., 2006, 2009a, 2009b), significant cognitive attributes related to creativity are those of knowledge organizational efficiency, searching efficiency, abstract ability, appreciation of new relations, curiosity, induction, and categorization, because those identified in the list are fundamental cognitive mechanisms and processes of the brain at the layers of meta-cognition and meta-inference according to the Layered Reference Model of the Brain (LRMB) (Wang et al., 2006), which are frequently used in supporting higher layer cognitive processes.

MATHEMATICAL MODELS OF CREATION AND CREATIVITY

On the basis of the discussions on the cognitive foundations of creativity, a more rigorous treatment of it can be developed in this section on the mathematical models of creation and creativity. The tree structure of human knowledge in term of concept trees and their properties are introduced. Then, a measurement model of creativity is quantitatively established.

The Tree Structure of Human Knowledge

It has been empirically observed that the tree-like architecture is a universal hierarchical prototype of systems across disciplines of not only science and engineering, but also sociology and living systems. The underlying reasons that force systems to take hierarchical tree structures are: a) The complexity of an unstructured system can easily grow out of control; b) The efficiency of an unstructured system can be very low; and c) The gain of system by coordination may diminish when the overhead for doing so is too high in unstructured systems.

An ideal structural form for modeling the knowledge system and creative space of humans is known as the complete tree (Wang, 2007a).

Definition 16. *A complete n-nary tree $T_c(n, N)$ is a normalized tree with N nodes in which each node of T_c can have at most n children, each level k of T_c from top-down can have at most n^k nodes, and all levels have allocated the maximum number of possible nodes, except only those on the rightmost subtrees and leaves.*

It is noteworthy in Definition 16, a tree said to be *complete* means that all levels of the tree have been allocated the maximum number of possible nodes, except those at the leave level and the rightmost subtress. The advantage of complete

Figure 2. Growth of complete binary trees

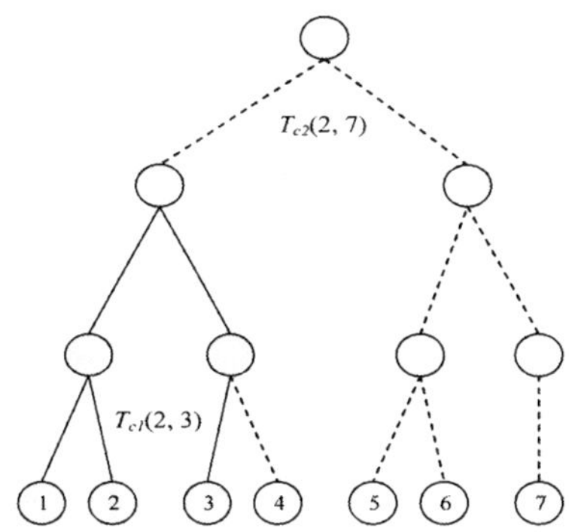

trees is that the configuration of any complete n-nary tree $T_c(n, N)$ is uniquely determined by only two attributes: the unified *fan-out n* and the number of leave nodes N at the bottom level. For instance, the growth of a system from complete tree $T_{c1}(n_1, N_1) = T_{c1}(2, 3)$ to $T_{c2}(n_2, N_2) = T_{c1}(2, 7)$ is illustrated in Figure 2.

Theorem 2. *The generic topology of normalized systems states that systems tend to be normalized into a hierarchical structure in the form of a complete n-nary tree.*

Systems are forced to be with tree-like structures in order to maintain equilibrium, evolvability, and optimal predictability. The advantages of tree structures of systems can be formally described in the following corollary.

Corollary 1. *Advantages of the normalized tree architecture of systems are as follows:*

a. Equilibrium: Looking down from any node at a level of the system tree, except at the leave level, the structural property of fan-out or the number of coordinated components are the same and evenly distributed.

b. Evolvablility: A normalized system does not change the existing structure for future growth needs.

c. Optimal predictability: There is an optimal approach to create a unique system structure $T_c(n, N)$ determined by the attributes of the unified fan-out n and the number of leave nodes N at the bottom level.

Properties of the Concept Tree of Knowledge Space

Based on the model of the complete tree, the topology of the knowledge space for creation can be denoted as a concept tree with each node of the n-nary complete tree as a concept.

Definition 17. *A concept tree, CT(n, N), is an n-nary complete tree in which all leave nodes N represent a meta-concept, and all remainder nodes beyond the leave level represent superconcepts.*

For instance, a ternary CT, $CT(n, N) = CT(3, 24)$, is shown in Figure 3. Since the CT is a complete tree, when the leaves (components) do not reach the maximum possible numbers, the right most leaves and subtrees of the CT will be left open.

A set of useful topological properties of CT is identified as summarized in the following corollary (Wang, 2007a).

Corollary 2. *An n-nary concept tree CT(n, N) with the total number of leaves nodes N possesses the following properties:*

(a) The maximum number of fan-out of any node $\overline{n}_{fo}$:

$$\overline{n}_{fo} = n \tag{10}$$

(b) The maximum number of nodes at a given level k, n_k:

Figure 3. A ternary concept tree CT(3, 24)

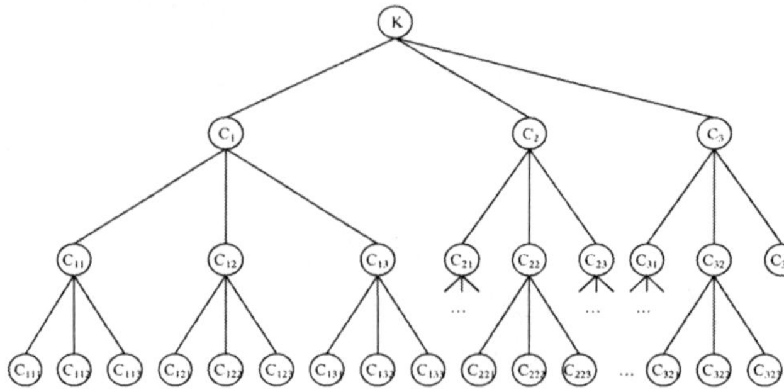

$$n_k = n^k \tag{11}$$

(c) The depth of the CT, d:

$$d = \left\lceil \frac{\log N}{\log n} \right\rceil$$

(d) The maximum number of nodes in the CT, N_{CT}:

$$N_{CT} = \sum_{k=0}^{d} n^k$$

(e) The maximum number of meta-concepts (on all leaves) in the CT, N_{max}:

$N_{max} = n^d$ *(f) The maximum number of subtrees (nodes except all leaves) in the CT, N_m:*

$$N_m = N_{CT} - N_{max} - 1 = \sum_{k=1}^{d-1} n^k \tag{15}$$

CT can be used to model and analyze the knowledge space of creativity. It also shows that a well organized knowledge tree in the brain is helpful for creation, because it can greatly reduce the cost for search.

Measurement of Creativity

On the basis of *CT*, a creation is modeled by the relational distances between two or more concepts in the concept tree.

Definition 18. *The relational distance of a creation, δ, is a sum of the distances δ_1 and δ_2 of a pair of concepts or objects c_1 and c_2 to their most closed parent node c_p in a given concept tree CT, i.e.:*

$$\delta(c_1, c_2) \triangleq \delta_1 + \delta_2 = |c_1 \leftrightarrow c_p| + |c_2 \leftrightarrow c_p|$$

where $\delta_i = |c_i \leftrightarrow c_p|$ denotes the distance between a concept i and its most closed parent concept p shared with another given concept.

According to Definition 18, the minimum creation distance $\delta_{min}(c_1, c_2) = 2$ when any pair of concepts at the same level of the *CT* under the same parent node.

Definition 18 can be extended to a more general case where multiple concepts are involved in a given *CT* as follows.

Definition 19. *The general relational distance of a creation, δ, is a sum of n, n > 1, subdistances δ_i, $1 \le i \le n$, between all individual concepts c_i and the most closed parent node c_p in the given knowledge space modeled by a CT, i.e.:*

$$\delta \triangleq \sum_{i=1}^{n} \delta_i = \sum_{i=1}^{n} |c_i \leftrightarrow c_p|$$

Example 1. *Given a knowledge space modeled by a CT as shown in Figure 3, any potential pairwise or multiple creation distances can be determined according to Definition 19 as follows:*

$$\begin{aligned}&\delta(c_{111}, c_{113}) = |c_{111} \leftrightarrow c_{11}| + |c_{113} \leftrightarrow c_{11}| = 1 + 1 = 2\\&\delta(c_{121}, c_{323}) = 3 + 3 = 6\\&\delta(c_{111}, c_{113}, c_{121}, c_{323}) = 3 + (3-2) + (3-1) + 3 = 9\end{aligned} \tag{18}$$

It is noteworthy that the creativity of a creation is proportional not only to its relational distance, but also to its originality and usefulness.

Definition 20. *Assume O = {0, 1} is a Boolean evaluation for the false or true originality of a creation, M the total number of nodes at level k in the d level creation space for a given concept tree CT. Then, the extent of creativity C is a product of the creation distance δ, the size of the creation space M, and its originality O, i.e.:*

$$\begin{aligned}C &\triangleq (\delta \bullet M) \bullet O\\&= \delta O \bullet \sum_{i=0}^{d-k} n^i\end{aligned}$$

where n is the fan-out of the given CT.

Example 2. *Based on the three solutions as given in Example 1, assume their originalities $O_1 = O_2 = O_3 = 1$, then the creativities of the three solutions can be quantitatively evaluated as follows:*

$$\begin{aligned}C_1 &= \delta_1 O_1 \bullet \sum_{i=0}^{d-k_1} n^i = 2 \bullet 1 \bullet \sum_{i=0}^{3-2} n^i = 2 \bullet (1+3) = 8\\C_2 &= \delta_2 O_2 \bullet \sum_{i=0}^{3-1} n^i = 6 \bullet (1+3+9) = 78\\C_3 &= \delta_3 O_3 \bullet \sum_{i=0}^{3-0} n^i = 9 \bullet (1+3+9+27) = 360\end{aligned}$$

Corollary 3. *The creativity of a creation is proportional to the product of the creative distance and the size of the creation space, subject to a satisfactory originality.*

THE COGNITIVE PROCESS OF CREATION[3]

With the cognitive and mathematical models of creation and creativity developed in previous sections, the process model of creation can be formally described in this section.

The Conceptual Model of the Creation Process

On the basis of Definitions 13, 14, and 15, a search-based creation process is modeled as shown in Figure 4, where an informal process of creation is divided into the following six steps: i) To define the problem; ii) To search the solution goals and paths; iii) To generate candidate solutions; iv) To identify and evaluate novel solutions; v) To represent creative solutions; and vi) To memorize creative relations.

It is noteworthy in Figure 4 that a number of lower layer cognitive processes, as represented by double-ended boxes, are adopted to carry out the creation process. These supporting processes for the creation process are those of ObjectIdentificationST, ConceptEstablishmentST, SearchST, QuatificationST, and MemorizationST according to LRMB (Wang et al., 2006).

The Formal Model of the Creation Process

On the basis of the conceptual model as given in Figure 4, a rigorous process model of creation can be formally described, as shown in Figure 5, using RTPA (Wang, 2002b, 2007a, 2008a, 2008e). The RTPA model formally explains the cognitive process of creation in the following six steps:

Figure 4. The cognitive process of creation

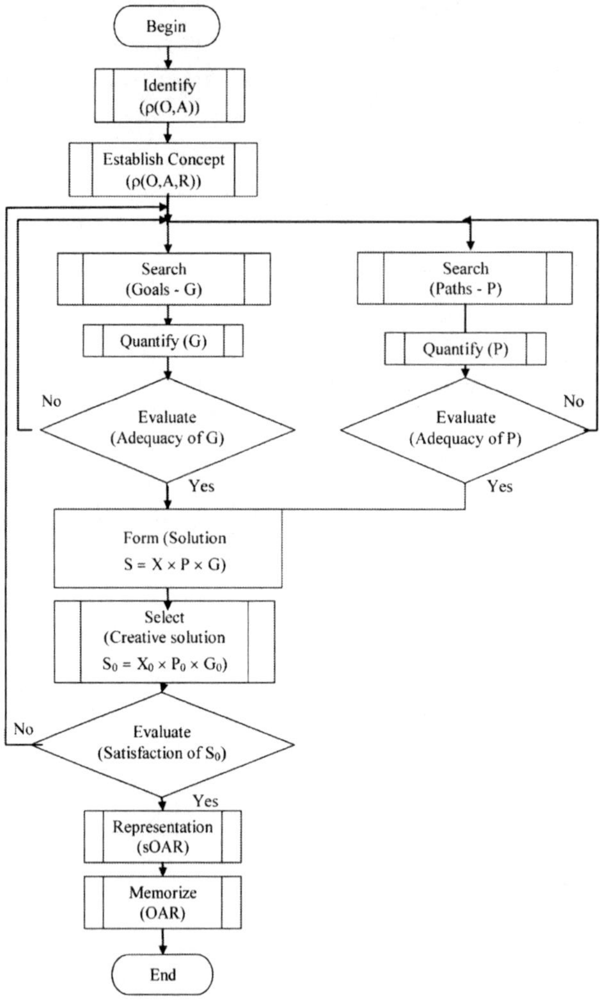

1. *To define the problem*: This step describes the problem ρS by identifying the related objects OSET and attributes ASET. Then, a problem concept ρST in the form of a sub-OAR model ρ(OSET, ASET, RSET)ST is established.
2. *To search the solution goals and paths*: In this step, the brain performs a parallel search for possible goals GSET and paths PSET of a set of potential solutions. External memory and resources may be searched if there is no available or sufficient GSET or PSET in the internal knowledge of the problem solver.
3. *To generate candidate solutions*: This step forms a set of possible solutions according to Equation 7, which is a Cartesian product of the searching results produced in Step (ii), i.e., SST = $X \times T \times G, S \subseteq P$.
4. *To identify and evaluate novel solutions*: This step evaluates each potential solution in SST as obtained in Step (iii) in order to find novel and creative solutions. Recursive searching actions may be executed if SST cannot satisfy the originality and utility criteria for a creation.

Figure 5. Formal description of the creation process in RTPA

```
The Cognitive Process of Creation

Creation_Process (I:: ρS; O:: S_0ST, OAR'ST) ST ≙
{ // I. Define the problem
   ↣ Identify (OSET ~ ρS)        // Representation of the problem
   ↣ Identify (ASET ~ ρS)
   ↣ ConceptEstablishment (ρST)

  // II. Search solution goals and paths
   → (             T
                   R                  ( ↣ Search (GSET)
       Satisfication of G = F
                                        ↣ Quantify (GSET)
                                        ↣ Evaluate (GSET)
                                      )
                   T
      ||           R                  ( ↣ Search (PSET)
       Satisfication of P = F
                                        ↣ Quantify (PSET)
                                        ↣ Evaluate (PSET)
                                      )
     )

  // III. Generate candidate solutions
   → XST := ρST
   → SST := (XST × PSET × GSET)ST

  // IV. Identify and evaluate novel solutions
      #SSET
   →  R  ( S_i(ρST, T_iSET, G_iSET)ST ∈ SST
      iN=1
             ↣ Evaluate (S_iST)
             → ( ◆ Sat(S_iST) ≥ k  // k: a satisfaction threshold
                   ↷ LB: V         // Jump to Label : Step V
               | ◆ ~
                   → ( ◆ GiveUpBL = F
                         ↷ LB: II
                     | ◆ ~
                         → ?
                     )
               )
         )

  // V. Represent creative solutions
   → S_0ST = (X_iST, T_iSET, G_iSET)ST        // The creative solution
   → OSET := ρST.OSET
   → ASET := ρST.ASET
   → RSET := S_0ST.T_i(X_iST, G_iSET)SET
   → sOARST := (OSET, ASET, RSET)ST          // Form a sub OAR

  // VI. Memorization
   → OAR'ST := OARST ⊎ sOARST
   ↣ Memorization (OAR'ST)
}
```

5. *To represent creative solutions*: This step creates a new sub-OARST to represent the creative solution(s) S_0ST, $S_0ST \subseteq SST$, obtained in Step (iv).
6. *To memorize creative relations*: This step incorporates and memorizes the solution(s) in the form of sub-OARST into the entire OARST model in the long-term memory of the brain, where â denotes a concept composition in long-term memory.

The cognitive process of creation developed in this section not only reveals the mechanism of basic human creation and invention process, but also indicates the approach to implement machine intelligence on creation and creative knowledge processing.

CONCLUSION

This article has presented the cognitive process of creation and creativity as a higher-level life function according to the Layered Reference Model of the Brain (LRMB). The cognitive foundations of creativity, such as the space of creativity, the approaches to creativity, the relationships of creation with problem solving, and the attributes of inventors, have been explored. A set of mathematical models of creation and creativity has been developed based on the hierarchical structures and properties of human knowledge known as concept trees. The measurement of creativity has been quantitatively analyzed. The cognitive process of creation has been described with Real-Time Process Algebra (RTPA), which provides a formal explanation of human creativity.

A creation has been defined as a novel and unexpected solution, which is a subset of the entire set of the creation space that meet the criteria of *novelty, originality,* and *utility*. The extent of creativity has been modeled as proportional to the product of the creative distance and the size of the creation space, subject to a satisfactory originality.

Various creativities and creation processes have been identified such as free/constrained creativity, analytic/synthetic creativity, inference-based creativity, problem-solving-based creativity, and scientific/ technological/art creativity. The entire set of creativities has been classified into three categories according to their creation spaces, approaches, and problem domains.

ACKNOWLEDGMENT

This work is partially sponsored by the Natural Sciences and Engineering Research Council of Canada (NSERC). The author would like to thank the anonymous reviewers for their valuable suggestions and comments on this work.

REFERENCES

Bender, E. A. (2000). *Mathematical Methods in Artificial Intelligence*. Los Alamitos, CA: IEEE CS Press.

Beveridge, W. I. (1957). *The Art of Scientific Investigation*. UK: Random House Trade Paperbacks.

Csikszentmihalyi, M. (1996). *Creativity: Flow and the Psychology of Discovery and Invention*. New York: HarperCollins.

Dubos, R. J. (1950). *Louis Pasteur: Freelance of Science*. Boston: Little, Brown & Co.

Guiford, J. P. (1967). *The Nature of Human Intelligence*. NY: McGraw-Hill.

Holland, J. H., Holyoak, K. J., Nisbett, R. E., & Thagard, P. R. (1986). *Induction: Processes of Inference, Learning, and Discovery*. Cambridge, MA: MIT Press/Bradford Books.

Leahey, T. H. (1997). *A History of Psychology: Main Currents in Psychological Thought* (4th ed.). Upper Saddle River, NJ: Prentice- Hall Inc.

Matlin, M. W. (1998). *Cognition* (4th ed.). Orlando, FL: Harcourt Brace College Publishers.

Mednich, S. A., & Mednich, M. T. (1967). *Examiner's Manual, Remote Associates Test*. Boston: Houghton Mifflin.

Smith, S. M. (1995). Fixation, Incubation, and Insight in Memory and Creative Thinking. In S.M. Smith, T.B. Ward, & R.A. Finke (Eds.), *The Creative Cognition Approach*. Cambridge, MA: MIT Press.

Sternberg, R. J., & Lubart, T. I. (1995). *Defying the Crowd: Cultivating Creativity in a Culture of Conformity*. NY: Free Press.

Wallas, G. (1926). *The Art of Thought*. New York: Harcourt-Brace.

Wang, Y., & Chiew, V. (2009). On the Cognitive Process of Human Problem Solving. *Cognitive Systems Research: An International Journal, 9*(4). UK: Elsevier.

Wang, Y., & Wang, Y. (2006). Cognitive Informatics Models of the Brain. [C]. *IEEE Transactions on Systems, Man, and Cybernetics, 36*(2), 203–207. doi:10.1109/TSMCC.2006.871151

Wang, Y. (2002a). Keynote: On Cognitive Informatics. *Proceedings 1st IEEE International Conference on Cognitive Informatics (ICCI'02)* (pp. 34-42). Calgary, Canada: IEEE CS Press.

Wang, Y. (2002b). The Real-Time Process Algebra (RTPA). [Springer.]. *Annals of Software Engineering: An International Journal, 14*, 235–274. doi:10.1023/A:1020561826073

Wang, Y. (2003). On Cognitive Informatics. *Brain and Mind: A Transdisciplinary Journal of Neuroscience and Neurophilosophy, 4*(3), 151-167.

Wang, Y. (2007a). *Software Engineering Foundations: A Software Science Perspective*. CRC Book Series in Software Engineering (Vol. II). New York: Auerbach Publications.

Wang, Y. (2007b). The Theoretical Framework of Cognitive Informatics. *International Journal of Cognitive Informatics and Natural Intelligence, 1*(1), 1–27.

Wang, Y. (2007c). The Cognitive Process of Formal Inferences. *International Journal of Cognitive Informatics and Natural Intelligence, 1*(4), 75–86.

Wang, Y. (2008a). On Contemporary Denotational Mathematics for Computational Intelligence. *Transactions of Computational Science*, *2*, 6–29. doi:10.1007/978-3-540-87563-5_2

Wang, Y. (2008b). Keynote: Abstract Intelligence and Its Denotational Foundations. *Proceedings 7th International Conference on Cognitive Informatics (ICCI'08).* CA, USA: Stanford University.

Wang, Y. (2008c). On Concept Algebra: A Denotational Mathematical Structure for Knowledge and Software Modeling. *International Journal of Cognitive Informatics and Natural Intelligence*, *2*(2), 1–19.

Wang, Y. (2008d). On System Algebra: A Denotational Mathematical Structure for Abstract System modeling. *International Journal of Cognitive Informatics and Natural Intelligence*, *2*(2), 20–42.

Wang, Y. (2008e). RTPA: A Denotational Mathematics for Manipulating Intelligent and Computational Behaviors. *International Journal of Cognitive Informatics and Natural Intelligence*, *2*(2), 44–62.

Wang, Y. (2009a). On Abstract Intelligence: Toward a Unified Theory of Natural, Artificial, Machinable, and Computational Intelligence. *International Journal of Software Science and Computational Intelligence*, *1*(1), 1–17.

Wang, Y. (2009b). On Visual Semantic Algebra (VSA): A Denotational Mathematical Structure for Modeling and Manipulating Visual Objects and Patterns. *International Journal of Software Science and Computational Intelligence*, *1*(4).

Wang, Y., Kinsner, W., & Zhang, D. (2009a). Contemporary Cybernetics and its Faces of Cognitive Informatics and Computational Intelligence. *IEEE Trans. on System, Man, and Cybernetics (B)*, *39*(4), 823–833. doi:10.1109/TSMCB.2009.2013721

Wang, Y., Kinsner, W., Anderson, J. A., Zhang, D., Yao, Y., Sheu, P., et al. (2009b). A Doctrine of Cognitive Informatics. *Fundamenta Informaticae*, *90*(3), 203–228.

Wang, Y., Wang, Y., Patel, S., & Patel, D. (2006). A Layered Reference Model of the Brain (LRMB). *IEEE Transactions on Systems, Man and Cybernetics. Part C, Applications and Reviews*, *36*(2), 124–133. doi:10.1109/TSMCC.2006.871126

Wilson, R. A., & Keil, F. C. (Eds.). (1999). *The MIT Encyclopedia of the Cognitive Sciences*. Cambridge, Mass: The MIT Press

Zadeh, L. A. (1965). Fuzzy Sets and Systems. In J. Fox (Ed.), *Systems Theory* (pp. 29-37). Brooklyn, NY: Polytechnic Press.

Zadeh, L. A. (1973). Outline of a New Approach to Analysis of Complex Systems. *IEEE Transactions on Systems, Man, and Cybernetics*, *1*(1), 28–44.

This work was previously published in International Journal of Cognitive Informantics and Natural Intelligence, Volume 3, Issue 4, edited by Yingxu Wang, pp. 1-18, copyright 2009 by IGI Publishing (an imprint of IGI Global)

Chapter 19
Modified Gabor Wavelets for Image Decomposition and Perfect Reconstruction

Reza Fazel-Rezai
University of North Dakota, USA

Witold Kinsner
University of Manitoba, Canada

ABSTRACT

This article presents a scheme for image decomposition and perfect reconstruction based on Gabor wavelets. Gabor functions have been used extensively in areas related to the human visual system due to their localization in space and bandlimited properties. However, since the standard two-sided Gabor functions are not orthogonal and lead to nearly singular Gabor matrices, they have been used in the decomposition, feature extraction, and tracking of images rather than in image reconstruction. In an attempt to reduce the singularity of the Gabor matrix and produce reliable image reconstruction, in this article, the authors used single-sided Gabor functions. Their experiments revealed that the modified Gabor functions can accomplish perfect reconstruction.

INTRODUCTION

Since its first formulation in 1984 (Grossmann, 1984), the wavelet transform has become a common tool in signal processing, in that, it describes a signal at different levels of detail in a compact and readily interpretable form (Daubechies, 1992).

DOI: 10.4018/978-1-60960-553-7.ch019

Wavelet theory provides a unified framework for a number of techniques which had been developed independently for various signal processing applications (Rioul, 1991). Different views of signal theory include multiresolution signal processing as used in computer vision, subband coding as developed for speech and image compression, and wavelet series expansion as developed in applied mathematics (Mallat, 1999).

In general, wavelets can be categorized in two types: real-valued and complex-valued. There are some benefits in using complex-valued wavelets. Gabor wavelet is one of the most widely used complex wavelets. More than a half of century ago, Gabor developed a system for reducing the bandwidth required to transmit signals (Gabor, 1946; Gabor, 1947). Since then, the Gabor function has been used in different areas of research such as image texture analysis (Porat, 1989; du Buf, 1991), image segmentation (Billings, 1976; Bochum, 1999), motion estimation (Magarey, 1998), image analysis (Daugman, 1988), signal processing (Qiu, 1997; Bastiaans, 1981), and face authentication (Duc, 1999). It should be noted that most of those areas rely on analysis and feature extraction, and not reconstruction.

In 1977, Cowan proposed that since visual mechanisms are indeed effectively bandlimited and localized in space, Gabor functions are suitable for their representation (Cowan, 1973). Other studies (Marcelja, 1980; Kulikowski, 1982; Pollen, 1985; Jones, 1987) assert that Gabor functions also well represent the characteristics of simple cortical cells, and present a viable model for such cells. Investigating "what does eye see best", Watson et al. have demonstrated that the pattern of two-dimensional Gabor functions is optimal (Watson, 1983). The main difficulty with the Gabor functions is that they are not orthogonal. Therefore, they do not have a perfect reconstruction condition and no straightforward technique is available to extract the coefficients. However, there has been some attempt for reconstruct images with an acceptable accuracy. For example, Wundrich (Wundrich *et. al,* 2002) developed an iterative method to reconstruct images from the magnitude of the Gabor wavelet. It is shown that the image can be detected after 1300 iteration. In this article, we present an analytical approach to overcome the difficulty with Gabor functions, and demonstrate their usefulness in the decomposition and reconstruction of still images. We show that an image decomposed using modified Gabor wavelets can be reconstructed perfectly.

The Gabor Decomposition

Notation

A one-dimensional (1D) signal $\boldsymbol{u} = [u_1, u_2, \ldots, u_N]^T$ is considered as an N-element complex column vector. Such signals are also considered as N-periodic sequences over integers Z. If the kth coordinate of $\boldsymbol{u}$ is expressed as either $\boldsymbol{u}_k$ or $\boldsymbol{u(k)}$, we have

$$u(i + kN) = u(i) \tag{1}$$

The norm of $\boldsymbol{u}$ is defined as the Euclidean norm

$$\left\|\mathbf{u}\right\| = \left(\sum_{i=1}^{N} \left|u_i\right|^2\right)^{1/2} \tag{2}$$

in order to relate to the energy of the signal.

The inner product of two signals $\boldsymbol{u}$ and $\boldsymbol{v}$ is defined as

$$< \mathbf{u}, \mathbf{v} > \; = \sum_{i=1}^{N} u_i v_i^* \tag{3}$$

where v_i^* denotes conjugate of the v_i.

The Matrix Form of the Gabor Transform

For a given 1D signal $\boldsymbol{s}$, a Gabor expansion (also known as the Gabor wavelet series) is expressed in terms of a Gabor function (mother wavelet) through time and frequency, that is

$$s(k) = \sum_{i=1}^{N} g_i(k) c(i) \quad \text{for} \quad k = 1, 2, \ldots, M; \quad N \le M \tag{4}$$

where g_i is the Gabor elementary function as discussed in Sec. 2.4 and *c(i)* is the coefficient. In a matrix notation, the above expansion can be written as

$$s = Gc \tag{5}$$

where the signal vector (***s***), the Gabor matrix (***G***) and the coefficient vector (***c***) are given by

$$s = \begin{bmatrix} s(1) \\ s(2) \\ \vdots \\ s(M) \end{bmatrix} \tag{6}$$

$$\mathbf{G} = \begin{bmatrix} g_1(1) & g_2(1) & \cdots & g_N(1) \\ g_1(2) & g_2(2) & \cdots & g_N(2) \\ \vdots & \vdots & \ddots & \vdots \\ g_1(M) & g_2(M) & \cdots & g_N(M) \end{bmatrix} \tag{7}$$

and

$$c = \begin{bmatrix} c(1) \\ c(2) \\ \vdots \\ c(N) \end{bmatrix} \tag{8}$$

In a two-dimensional (2D) case, the expansion of a 2D signal (for example, an image) into a set of 2D elementary functions is given by

$$S(k_1, k_2) = \sum_{i=1}^{N_1} \sum_{j=1}^{N_2} g_{ij}(k_1, k_2) C(i, j) \quad \text{for} \begin{cases} \mathrm{k}_1 = 1, \cdots, M_1; & N_1 \le M_1 \\ \mathrm{k}_2 = 1, \cdots, M_2; & N_2 \le M_2 \end{cases} \tag{9}$$

In a matrix notation, if the 2D elementary functions are separable, the 2D expansion can be written as

$$\mathbf{S} = \mathbf{G}_1 \mathbf{C} \mathbf{G}_2^{\ T} \tag{10}$$

where ***s*** is the 2D data matrix, ***c*** is the 2D expansion coefficient matrix

$$\mathbf{S} = \begin{bmatrix} S(1,1) & S(2,1) & \cdots & S(1, M_2) \\ S(1,2) & S(2,2) & \cdots & S(2, M_2) \\ \vdots & \vdots & \ddots & \vdots \\ S(M_1, 1) & S(M_1, 2) & \cdots & S(M_1, M_2) \end{bmatrix} \tag{11}$$

$$\mathbf{C} = \begin{bmatrix} C(1,1) & C(2,1) & \cdots & C(1, N_2) \\ C(1,2) & C(2,2) & \cdots & C(2, N_2) \\ \vdots & \vdots & \ddots & \vdots \\ C(N_1, 1) & C(N_1, 2) & \cdots & C(N_1, N_2) \end{bmatrix} \tag{12}$$

and G_i is the same Gabor matrix as defined in Equation (7).

Rotation and Modulation Operators

For any signal ***u***, the signal ***v*** is said to be the cyclic rotation ***u*** with rotation number *r* if the following holds

$$\mathbf{v} = \Re_r(\mathbf{u}) = \begin{bmatrix} u_{N-r+1} \\ u_{N-r+2} \\ \vdots \\ u_N \\ u_1 \\ u_2 \\ \vdots \\ u_{N-r} \end{bmatrix} \tag{13}$$

where R_r is the rotation operator with rotation number *r*.

The modulation (or frequency translation) of a signal ***u*** is signal ***w*** defined as

$$\mathbf{w} = \Im_k(\mathbf{u}) = \begin{bmatrix} u_1 \\ u_2 e^{-\frac{2\pi i}{N}k} \\ u_3 e^{-\frac{4\pi i}{N}k} \\ u_4 e^{-\frac{6\pi i}{N}k} \\ \vdots \\ u_N e^{-\frac{2(n-1)\pi i}{N}k} \end{bmatrix} \quad (14)$$

where T_k is the modulation operator with frequency *k*.

It can be shown that for every signal ***u*** and any value for *l*

$$F(\Re_l(\mathbf{u})) = \Im_l(F(\mathbf{u})) \quad (15)$$

and

$$F(\Im_{-l}(\mathbf{u})) = \Re_l(F(\mathbf{u})) \quad (16)$$

where *F* denotes the Fourier transform operator defined as

$$F(\mathbf{u}) = \frac{1}{N}\sum_{n=0}^{N} u(n) e^{-\frac{2\pi i}{N}n} \quad (17)$$

Based on these two operators, the Gabor matrix can be rewritten into a more convenient form as described in the next section.

The Gabor Matrix

Consider the Gabor mother function, ***g***, as defined in the following form

$$g_\sigma(n) = e^{-\pi(\frac{n}{\sigma})^2} \quad (18)$$

where the parameter σ determines the scale of the Gaussian in time domain. The Gabor matrix ***G*** with decomposition levels *L* is then defined as

$$G(L) = \begin{bmatrix} \mathbf{g}_\sigma \\ \Re_L(\mathbf{g}_\sigma) \\ \Re_{2L}(\mathbf{g}_\sigma) \\ \vdots \\ \Re_{N-1}(\mathbf{g}_\sigma) \\ \Im_1(\mathbf{g}_\sigma) \\ \Im_1(\Re_L(\mathbf{g}_\sigma)) \\ \Im_1(\Re_{2L}(\mathbf{g}_\sigma)) \\ \vdots \\ \Im_1(\Re_{N-1}(\mathbf{g}_\sigma)) \\ \vdots \\ \vdots \\ \Im_{L-1}(\mathbf{g}_\sigma) \\ \Im_{L-1}(\Re_L(\mathbf{g}_\sigma)) \\ \Im_{L-1}(\Re_{2L}(\mathbf{g}_\sigma)) \\ \vdots \\ \Im_{L-1}(\Re_{N-1}(\mathbf{g}_\sigma)) \end{bmatrix} \quad (19)$$

where *N* is the size of the vector ***s*** to be expanded using the Gabor matrix. It should be noted that the size of matrix ***G*** is *M*×*N*, where the size of Gabor mother function $\boldsymbol{g}_\sigma$ is *M*. If *N*<*M*, the relation in Equation (5) cannot be satisfied. Therefore, a criterion should be defined to find the best solution for the relation in Equation (5). The most commonly used approach is the least-squares error criterion. Therefore, a 1D solution to Equation (5) is the vector ***c***=***c*** that minimizes the objective function $(\boldsymbol{Gc}\text{-}\boldsymbol{s})^T(\boldsymbol{Gc}\text{-}\boldsymbol{s})$, and which can be found as

$$\hat{\mathbf{c}} = (\mathbf{G}^T\mathbf{G})^{-1}\mathbf{G}^T \quad (20)$$

Singularity

The *N*×*N* matrix $\boldsymbol{G}^T\boldsymbol{G}$ cannot always be inverted as it can be singular or nearly singular. To avoid the problem of matrix inversion, the singular value decomposition (SVD) can be used. The SVD states that any *M*×*N* matrix ***A*** can be decomposed as the product of an *M*×*N* column-orthogonal matrix ***U***, as $N \times N$ diagonal matrix $\boldsymbol{W}$ with positive or zero elements, and the transpose of an $N \times N$ matrix $\boldsymbol{V}$ which can be written as

$$\mathbf{A} = \mathbf{U}\mathbf{W}\mathbf{V}^T \tag{21}$$

with $\boldsymbol{U}\boldsymbol{U}^T = \boldsymbol{V}^T\boldsymbol{V} = \boldsymbol{I}$. Then the inverse of $\boldsymbol{A}$ is defined as

$$\mathbf{A}^{-1} = \mathbf{V}[diag(\frac{1}{w_i})]\mathbf{U}^T \tag{22}$$

In the 2D case, the solution to Equation (10) is given by

$$\hat{\mathbf{C}} = (\mathbf{G}_1^{\ T}\mathbf{G}_1)^{-1}\mathbf{G}_1^{\ T} \cdot \mathbf{S} \cdot \mathbf{G}_2(\mathbf{G}_2^{\ T}\mathbf{G}_2)^{-1} \tag{23}$$

The SVD can be also applied to this case, and the solution is given by

$$\mathbf{A}^{-1} = \mathbf{V}_1[diag(\frac{1}{w_{i1}})]\mathbf{U}_1^{\ T} \cdot \mathbf{S} \cdot \mathbf{U}_2[diag(\frac{1}{w_{i2}})]\mathbf{V}_2^{\ T} \tag{24}$$

If the matrix $\boldsymbol{A}$ is either singular or nearly singular, the diagonal matrix $\boldsymbol{W}$ will contain zero or very small values on its diagonal. Therefore, to invert the matrix, the very large associated elements should be replaced with zeros. However, this approximation affects the reconstructed results, and produce low value of the peak signal to noise ratio (PSNR) for the reconstructed images. In the next section, we introduce an approach to overcome this problem.

The Single-Sided Gabor Reconstruction

A typical Gabor function and its corresponding modulated signal are shown in Figure 1.a and Figure1.b. The Gabor matrix $\boldsymbol{G}$ developed from

Figure 1. Two-sided Gabor functions: (a) mother function. (b) modulated function and single-sided Gabor function: (c) mother function. (d) modulated function

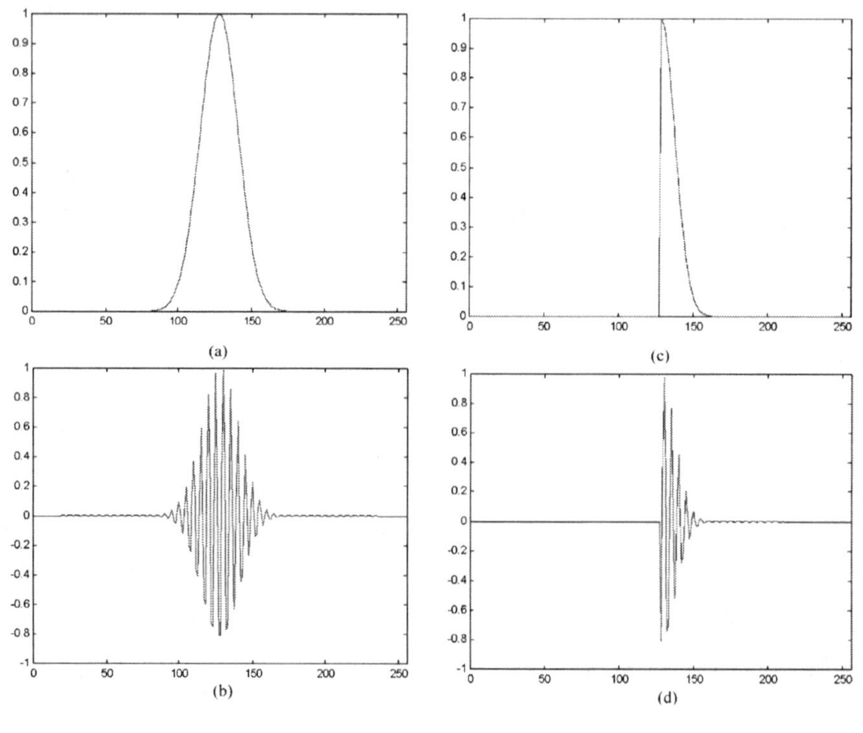

such a two-sided mother Gabor function is very close to singular (as described in the previous section). The main reason is that in this matrix (shown in Equation 16), there are two rows that are similar. Each row is created by each side of the Gabor function. To overcome this problem, a single-sided Gabor function (shown in Figure 1.c and Figure 1.d) was used to force the Gabor matrix to be no longer singular. Therefore, the single-sided Gabor function can be used for the decomposition and reconstruction of images to overcome the singularity problem exists in the two-sided Gabor wavelet decomposition and reconstruction.

Experimental Results

The single-sided Gabor decomposition and reconstruction algorithms for images have been implemented in MATLAB. Two test images have been used, including Lena and a model plane (Figure 2). The second test image is a computer generated image and was chosen because it has straight-line sharp edges which are not vertical and diagonal.

The Gabor matrix ***G*** (Equation 19) has real and complex components. Both depend on the rotation number (L). The real and imaginary forms of the Gabor matrix ***G*** for gray-level images with a two-level decomposition (L=**2**), four-level decomposition (L=**4**) and eight-level decomposition (L=**8**) are shown in Figure 3. Note that the values have been scaled between 0 to 255. It can been seen from these figures that the number of white bands corresponds to the rotation number (L). It can be shown that this number determines the number of subimages in decomposition.

Considering that the Gabor matrix (***G***) multiplies to the image matrix from the left and right (Equation 10), one would expect to have a result with four subimages, each similar to the original image. The results of applying the single-sided Gabor decomposition and reconstruction technique to the images are presented next.

Figure 2. Test images. (a) Lena as a test image #1. (b) Model airplane as a test image #2

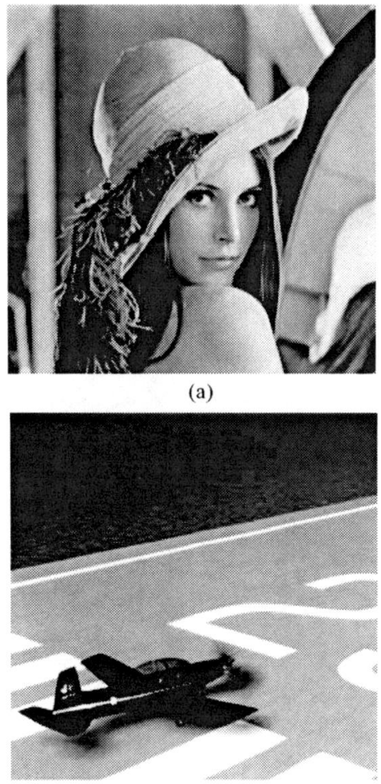

(a)

(b)

In the case of L=**2**, the real and imaginary parts of the Gabor decomposition to the two test images are shown in Figure 4. As expected, each of real and imaginary images has four subimages. The corresponding reconstructed images using Equation (23) are shown Figure 4. A very high PSNR for the reconstructed images was achieved, with 313.01 dB and 311.83 dB for the test image #1 and test image #2, respectively. The PSNR was calculated as

$$PSNR = 10\log\frac{255^2}{\sigma_d^{\ 2}} \tag{25}$$

where σ_d^2 is the mean square error defined using the following equation

Figure 3. The real and imaginary structure of the Gabor matrix with L=2, L=4, and L=8

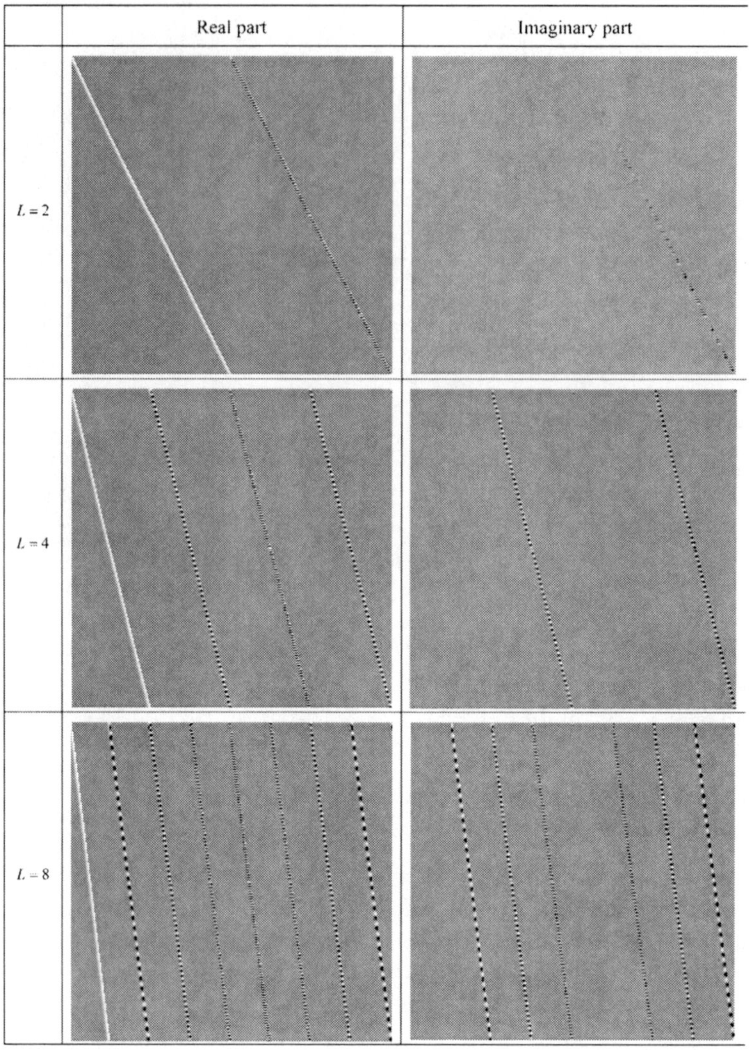

$$\sigma_d^2 = \frac{1}{N}\sum_{n=1}^{N}(\mathbf{S}(n) - \hat{\mathbf{S}}(n)) \tag{26}$$

It should be noted that the PSNR around 310 dB corresponds to the mean square error (σ_d^2) of 10^{-9}. This error results from numerical calculation because of finite precision of machine computation. Therefore, this reconstruction is as perfect as allowed by the computational resolution. The corresponding residual errors shown in Figure 5 also indicate a very good reconstruction in that there are no major visible patterns that would correspond to the original images.

The remaining decomposition and reconstruction results are shown in Figure 6 and Figure 7 (for $L = 4$) and Figure 8 and Figure 9 (for $L = 8$). Again, residual errors appear to be randomly distributed, without displaying contours of the objects in the original images. This confirms the suitability of the Gabor transform for decomposition and reconstruction of images for the human visual system (HVS).

Figure 4. Gabor decomposition of the test images with L = 2

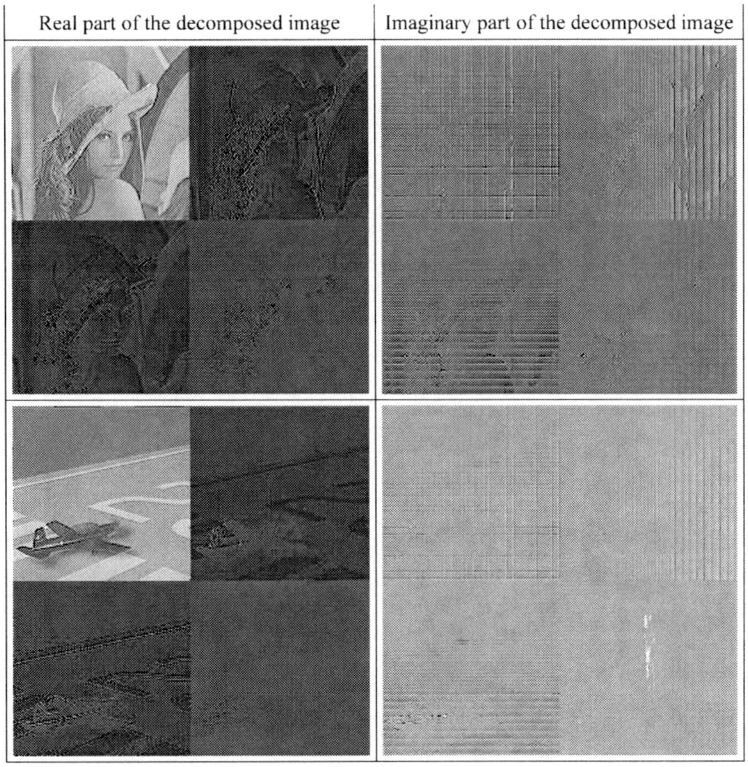

Figure 5. Reconstruction of the test images with L = 2

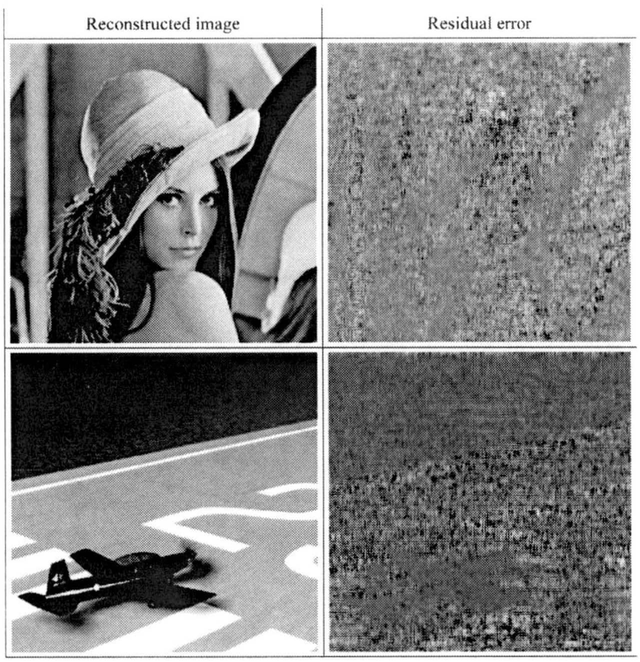

Figure 6. Gabor decomposition of the test images with L=4

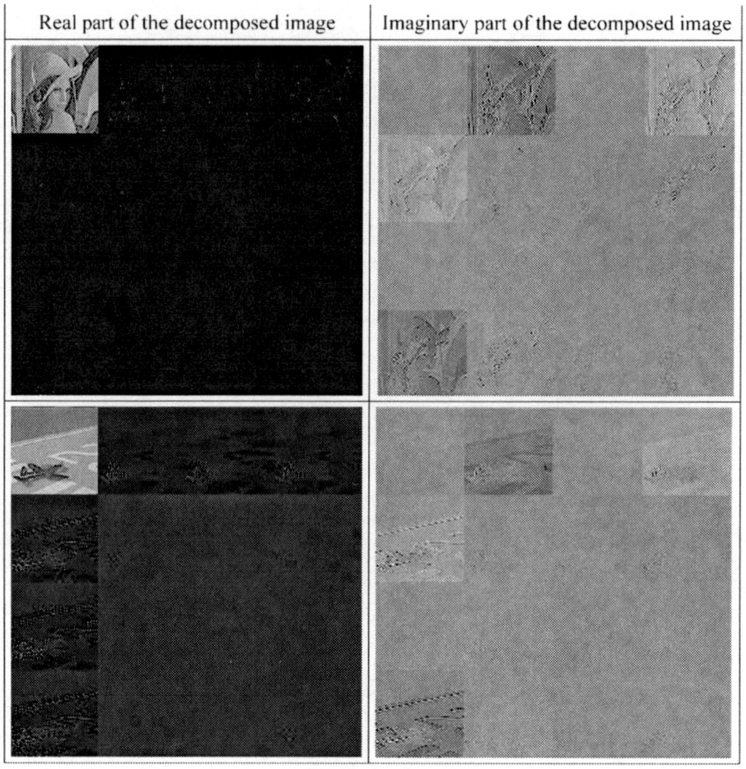

Figure 7. Reconstruction of the test images with L=4

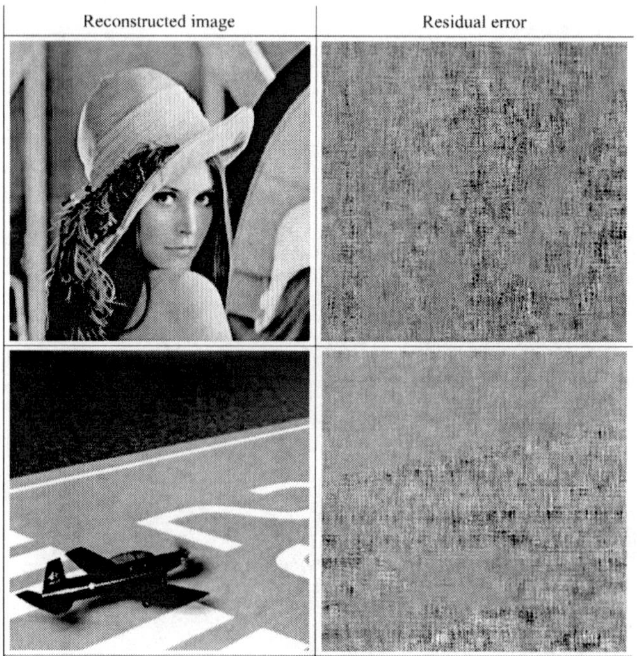

Figure 8. Gabor decomposition of the test images with L=8

Figure 9. Reconstruction of the test images with L=8

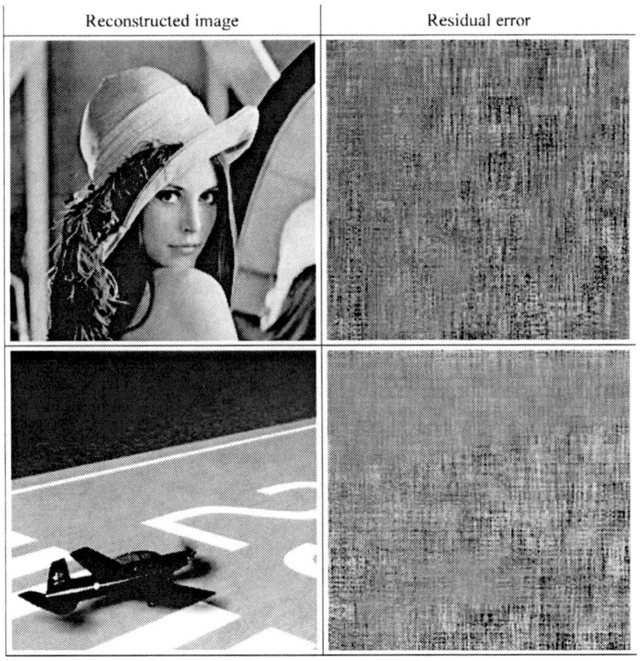

CONCLUSION

There are some benefits in using complex-valued wavelets. It has been shown that since visual mechanisms are indeed effectively bandlimited and localized in space, Gabor functions are suitable for their representation. Despite all good features of the Gabor functions, its main inconvenience is the fact that they are not orthogonal. Therefore, no straightforward method is available to extract the coefficients. In this article, an analytical method for still image decomposition and reconstruction using single-sided Gabor functions has been presented. Experimental results have shown that one can achieve perfect reconstruction with PSNR of more than 300 dB.

ACKNOWLEDGMENT

We acknowledge financial support from the Telecommunication Research Laboratories (*TRLabs*) and Mecca Media Group, Edmonton, Canada, and the National Science and Engineering Research Council (NSERC) of Canada.

REFERENCES

Bastiaans, M. J. (1981). A sampling theorem for the complex spectrogram, and Gabor's expansion of a signal in Gaussian elementary signals. *Optical Engineering (Redondo Beach, Calif.)*, *20*(2), 594–598.

Billings, A. R., & Scolaro, A. (1976). The Gabor compression-expansion system using non-Gaussian windows and its application to television coding. *IEEE Transactions on Information Theory*, 22(2), 174–190. doi:10.1109/TIT.1976.1055535

Bochum, R., & Wiskott, L. (1999). Segmentation from motion: combining Gabor and Mallat wavelets to overcome aperture and correspondence problem. *Pattern Recognition*, *32*(10), 1751–1766. doi:10.1016/S0031-3203(98)00179-4

Cowan, J. D. (1973). Some remarks on channel bandwidth for visual contrast detection. *Neurosciences Research Program Bulletin*, *15*, 1255–1267.

Daubechies, I. (1992). Ten Lectures on Wavelets. CBMS-NSF Regional Conference Series. *Applied Mathematics.*

Daugman, J. G. (1988). Complete Discrete 2-D Gabor transform by neural networks for image analysis and compression. *IEEE Transactions on Acoustics, Speech, and Signal Processing*, *36*(7), 1169–1179. doi:10.1109/29.1644

du Buf, J. M. H., & Heitkamper, P. (1991). Texture features based on Gabor phase. *Signal Processing*, *23*(3), 227–244. doi:10.1016/0165-1684(91)90002-Z

Duc, B., Fischer, S., & Bigüm, J. (1999). Face authentication with Gabor information on deformable graphs. *IEEE Transactions on Image Processing*, *8*(4), 504–516. doi:10.1109/83.753738

Gabor, D. (1946). Theory of communication. [London.]. *J. of Industrial Electrical Engineering*, *93*(3), 429–457.

Gabor, D. (1947). New possibilities in speech transmission. [London.]. *J. of Industrial Electrical Engineering*, *94*(3), 369.

Grossmann, A., & Morlet, J. (1984). Decomposition of Hardy Functions into Square Integrable Wavelets of Constant Shape. *SIAM Journal on Mathematical Analysis*, *15*(4), 723–736. doi:10.1137/0515056

Jones, J. P., & Palmer, L. A. (1987). The two dimensional spatial structure of simple receptive fields in cat striate cortex. *Journal of Neurophysiology*, *58*(6), 1187–1211. PubMed

Kulikowski, J. J., Marcelja, S., & Bishop, P. O. (1982). Theory of spatial position and spacial frequency relations in the receptive field of simple cells in the visual cortex. *Biological Cybernetics*, *43*(3), 187–198. doi:10.1007/BF00319978

Magarey, J., & Kingsbury, N. (1998). Motion estimation using a complex-valued wavelet transform. *IEEE Transactions on Signal Processing*, *46*(4), 1069–1084. doi:10.1109/78.668557

Mallat, S. G. (1999). *A wavelet tour of signal processing.* Academic Press.

Marcelja, S. (1980). Mathematical description of the responses of simple cortical cells. *Journal of the Optical Society of America*, *70*(11), 1297–1300. doi:10.1364/JOSA.70.001297

Pollen, D. A., & Ronner, S. F. (1985). Visual cortical neurons as localized spatial frequency filter. *IEEE Transactions on Systems, Man, and Cybernetics*, *15*(3), 91–101.

Porat, M., & Zeevi, Y. Y. (1989). Localized texture processing in vision: analysis and synthesis in the Gaborian space. *IEEE Transactions on Bio-Medical Engineering*, *36*(1), 115–129. doi:10.1109/10.16457

Qiu, S. (1997). Gabor-type matrix algebra and fast computation of dual and tight Gabor wavelets. *Optical Engineering (Redondo Beach, Calif.)*, *36*(1), 276–282. doi:10.1117/1.601171

Rioul, O., & Vetterli, M. (1991). Wavelets and signal processing. *IEEE Signal Processing Magazine*, *8*(4), 15–38. doi:10.1109/79.91217

Watson, A. B., Barlow, H. B., & Robson, J. G. (1983). What does eye see best. *Nature*, *302*, 419–422. doi:10.1038/302419a0

Wundrich, I. J., von der Malsburg, C., & Würtz, P. (2002). Image Reconstruction from Gabor Magnitudes. *Biologically Motivated Computer Vision* (pp. 117–126).

This work was previously published in International Journal of Cognitive Informantics and Natural Intelligence, Volume 3, Issue 4, edited by Yingxu Wang, pp. 19-33, copyright 2009 by IGI Publishing (an imprint of IGI Global)

Chapter 20
Adaptive Integrated Control for Omnidirectional Mobile Manipulators Based on Neural-Network

Xiang-min Tan
Chinese Academy of Sciences, P.R. China

Dongbin Zhao
Chinese Academy of Sciences, P.R. China

Jianqiang Yi
Chinese Academy of Sciences, P.R. China

Dong Xu
Sevenstar Electronics Co. Ltd., P.R. China

ABSTRACT

An omnidirectional mobile manipulator, due to its large-scale mobility and dexterous manipulability, has attracted lots of attention in the last decades. However, modeling and control of such systems are very challenging because of their complicated mechanism. In this paper, an unified dynamic model is developed by Lagrange Formalism. In terms of the proposed model, an adaptive integrated tracking controller, based on the computed torque control (CTC) method and the radial basis function neural-network (RBFNN), is presented subsequently. Although CTC is an effective motion control strategy for mobile manipulators, it requires precise models. To handle the unmodeled dynamics and the external disturbance, a RBFNN, serving as a compensator, is adopted. This proposed controller combines the advantages of CTC and RBFNN. Simulation results show the correctness of the proposed model and the effectiveness of the control approach.

DOI: 10.4018/978-1-60960-553-7.ch020

INTRODUCTION

A mobile manipulator, which generally consists of a mobile platform and a robot arm, provides a new direction in robot researches and applications due to its large-scale mobility and dexterous manipulability. The mobility of the mobile platform substantially increases the size of workspace, and it enables the end effector of the manipulator to reach a relatively better position to operate dexterously. The manipulability of the robot arm, which is mounted on the mobile platform, greatly improves the functionality of the mobile manipulator. Because of these distinct advantages, mobile manipulators have been applied more and more extensively.

A number of related works have been developed in this field in the last decades. In literatures (Campion and Bastin, 1996; Betourne and Campion, 1996; SicilianO, Wit, &Bastin, 1996), modeling and control of omnidirectional mobile robots were analyzed in details. The dynamic model and kinematic model of an omnidirectional mobile robot with three castor wheels were presented in (Chung, et al., 2003; Yi and Kim, 2001). Tan and Xi (2001) proposed a unified dynamic model for a mobile manipulator consisting of a Nomadic XR4000 and a Puma 560 robot arm. Holmberg and Khatib (2000), Khatib, et al. (1996), Khatib (1987) developed a holonomic mobile robot and presented a dynamic control method for a parallel redundant system. On the whole, existing works can be approximately divided into two groups: the holonomic mechanical system and the nonholonomic mechanical system. The omnidirectional mobile manipulator is a typical example of holonomic mechanical system. Compared to nonholonomic mechanical systems such as differential-driven mobile robots, the holonomic system has several advantages. Firstly, avoidance of nonholonomic constraints made the holonomic system more dexterous and easy to be controlled. Secondly, it can completely use the null space motions to improve the workspace and overall dynamic endpoint properties, and the redundant degrees of freedom can be used to accomplish accessory tasks.

However, while the omnidirectional mobile manipulator provides a lot of advantages, we suffer a lot from its complicated mechanical structure. Firstly, it is very difficult to derive the model of an omnidirectional mobile manipulator, especially the dynamic model. Secondly, more redundant degrees of freedom also implies more motors need to be controlled, therefore, path planning and motion control of such a system are very challenging. Moreover, many intelligent control methods are unable to be applied online due to the computational complexity of the model.

Existing control strategies of the omnidirectional mobile manipulator can be approximately divided into two groups: One is separate control of the mobile platform and the robot arm, i.e., the mobile manipulator is regarded as two subsystems. In this way, Liu and Lewis presented a decentralized robust controller in (Liu and Lewis, 1990). Chung and Velinsky (1998) also derived the dynamic model of the holonomic mobile platform and the manipulators separately. The other is control the mobile manipulator as a whole, that is, we consider the mobile platform as a multiple-DOFs joint. For example, Yamamoto and Yun (1994) studied a two-linked planar mobile manipulator subject to nonholonomic constraints. Considering the characteristics of our experimental setup, we adopt the latter control strategy in this paper.

As for these methods, CTC is an effective motion control strategy. However, requirements of precise model are very rigor in practical applications. Furthermore, this kind of controllers becomes unstable when the unmodeled dynamics or the external disturbance is significant. To handle these difficulties, Song and Yi (2005) adopted a fuzzy approach to compensate the uncertainties. In literatures (Kwan, et al., 1998, Wai, 2003), generalized fuzzy neural network, wavelet neural network

Figure 1. The omnidirectional manipulator and its 3-D model

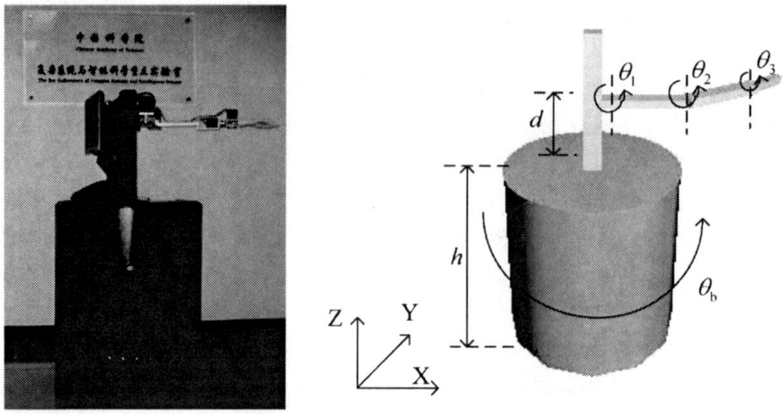

were introduced in their controller respectively. In this paper, a robust controller, based on CTC and RBFNN, is presented. As pointed in (Berni, et al.,2003; Er and Gao, 2003; Li, 2006;Hou and Tan, 2004; Fang, Zhao, &Li, 2005), artificial neural networks have been adopted extensively for their ability to perform nonlinear mappings. It has been used widely in modeling and control, artificial intelligence, cognitive informatics (Wang, Y., 2007; Wang, Y., 2006). Because neural networks provide a fast method of autonomously learning the relation between a set of output states and a set of input states, we introduce a RBFNN into the controller to approximate the unstructured or structured uncertainties of the proposed model in this paper.

In the following section, we apply Lagrangian Formalism to deduce the unified model of an omnidirectional mobile manipulator. In addition, we also introduce the structure of the mobile manipulator we have developed. Section III is devoted to controller design based on CTC and RBFNN. Section IV includes simulation results to validate feasibility and efficiency of the proposed method. Some conclusions and remarks are finally included in Section V.

SYSTEM DESCRIPTIONS

A mobile manipulator generally consists of a mobile platform and a robot arm. Figure 1 shows the omnidirectional mobile manipulator and its 3-D model in our lab. In this experimental setup, we adopt an omnidirectional mobile platform with six motors, i.e., three rolling motors and three steering motors. Therefore, it has all three DOFs (degrees of freedom) $\begin{bmatrix} x & y & \theta_b \end{bmatrix}^T$ for moving on the plane, and isn't subject to nonholonomic constraint. The robot arm, mounted on the omnidirectional platform, is similar to a SCARA manipulator. It has five DOFs $\begin{bmatrix} d & \theta_1 & \theta_2 & \theta_3 & \theta_4 \end{bmatrix}^T$. They are the height of shoulder, the joint angles of shoulder, elbow, wrist and hand.

Figure 2 shows the top view of the omnidirectional manipulator. In order to deduce the unified model, the Lagrange Formalism is applied. For the sake of simplicity, we only consider six DOFs as shown in Figure 1 and Figure 2 in this unified dynamic model, denoted by $\begin{bmatrix} x & y & \theta_b & \theta_1 & \theta_2 & \theta_3 \end{bmatrix}^T$. In Figure 2, we also define the relative coordinates.

- World frame $\sum Oxy$: inertial frame.

Figure 2. Top view of the omnidirectional mobile manipulator

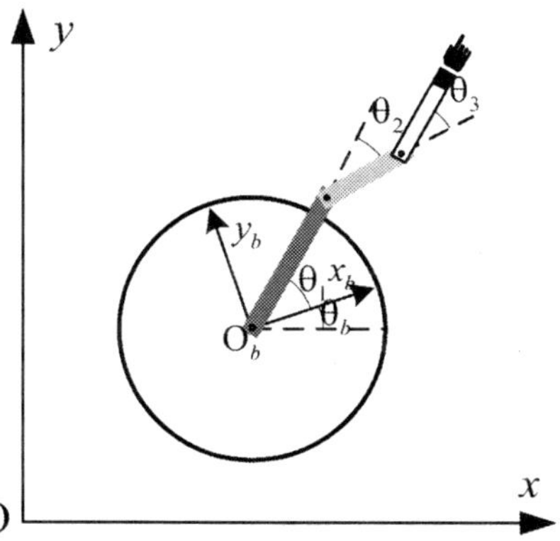

- Moving frame $\sum O_b x_b y_b$: the frame attached on the mobile platform.
- The associated coordinates of the manipulator are established by D-H method.

To facilitate modeling, it is assumed that the mobile manipulator has the following characteristics.

- The mobile platform is uniform, and its barycenter is the center of the platform, which is denoted as O_b .
- The joints between links are rigid and massless.
- All links are uniform.

According to Lagrange Formalism, the unified dynamic model can be deduced by

$$\tau_i = \frac{d}{d_t}\frac{\partial L}{\partial \dot{q}_i} - \frac{\partial L}{\partial q_i} \qquad i = 1,2,3,4,5,6 \qquad (1)$$

where τ_i represents forces or torques produced by driven motors, L denotes the energy of the system, and q_i is the joint coordinates. We can calculate the energy of the system as follows

$$\begin{aligned} L &= K - P \\ &= E_p + E_1 + E_2 + E_3 \end{aligned} \qquad (2)$$

where E_p, E_1, E_2, E_3 represent the energy of the mobile platform, link 1,link 2 and link 3, and P is the potential energy.

By substituting (2) into (1), we can get the unified dynamic model subsequently. In the absence of friction and other external disturbance, the dynamic model can be written as

$$\mathbf{M}(\mathbf{q})\ddot{\mathbf{q}} + \mathbf{C}(\mathbf{q},\dot{\mathbf{q}})\dot{\mathbf{q}} + \mathbf{G}(\mathbf{q}) = \ddot{\boldsymbol{A}} \qquad (3)$$

where $\mathbf{q} = \begin{bmatrix} x & y & \theta_b & \theta_1 & \theta_2 & \theta_3 \end{bmatrix}^T \in R^n$ denotes the generalized position, $\dot{\mathbf{q}} \in R^n$ denotes the generalized velocity, $\ddot{\mathbf{q}}$ is the generalized acceleration, $\mathbf{M}(\mathbf{q}) \in R^{n \times n}$ is the inertia matrix, $\mathbf{C}(\mathbf{q},\dot{\mathbf{q}}) \in R^{n \times n}$ is the centripetal and Coriolis matrix, and $\mathbf{G}(\mathbf{q}) \in R^n$ is the gravitational vector. $\ddot{\boldsymbol{A}} \in R^n$ is the input torques vector. Especially, n=6 and $\mathbf{G}(\mathbf{q}) = 0$ in this model.

Define m_i, l_i, J_i as the mass, the length and the Inertial of moment of link i $(i = 1,2,3)$, m_p, J_p as the mass and inertia of the mobile platform. The details of these matrixes are described as follows, where

$c_{ab} = \cos(\theta_a + \theta_b)$, $s_{ab} = \sin(\theta_a + \theta_b)$. (see Box 1)

Considering the external disturbance $\mathbf{d}$ and the unmodeled dynamics $\Delta(\mathbf{q},\dot{\mathbf{q}})$, the dynamic model of the mobile manipulator can be depicted by

$$M(q)\ddot{q} + C(q,\dot{q})\dot{q} + G(q) + \Delta(q,\dot{q}) + d = \tau \qquad (4)$$

Obviously, the above dynamic models satisfy the following three properties.

Box 1.

$$\mathbf{M}(\mathbf{q}) = \begin{bmatrix} M_{11} & M_{12} & M_{13} & M_{14} & M_{15} & M_{16} \\ M_{21} & M_{22} & M_{23} & M_{24} & M_{25} & M_{26} \\ M_{31} & M_{32} & M_{33} & M_{34} & M_{35} & M_{36} \\ M_{41} & M_{42} & M_{43} & M_{44} & M_{45} & M_{46} \\ M_{51} & M_{52} & M_{53} & M_{54} & M_{55} & M_{56} \\ M_{61} & M_{62} & M_{63} & M_{64} & M_{65} & M_{66} \end{bmatrix}$$

$$\mathbf{C}(\mathbf{q}, \dot{\mathbf{q}}) = \begin{bmatrix} 0 & 0 & C_{13} & C_{14} & C_{15} & C_{16} \\ 0 & 0 & C_{23} & C_{24} & C_{25} & C_{26} \\ 0 & 0 & C_{33} & C_{34} & C_{35} & C_{36} \\ 0 & 0 & C_{43} & C_{44} & C_{45} & C_{46} \\ 0 & 0 & C_{53} & C_{54} & C_{55} & C_{56} \\ 0 & 0 & C_{63} & C_{64} & C_{65} & 0 \end{bmatrix}$$

$$M_{11} = m_p + m_1 + m_2 + m_3$$

$$M_{12} = M_{21} = 0$$

$$M_{13} = M_{31} = -[0.5m_1l_1 \mathrm{s}_{b1} + m_2l_1s_{b1} + 0.5m_2l_2s_{b12} + m_3l_1s_{b1} + m_3l_2s_{b12} + 0.5m_3l_3s_{b123}]$$

$$M_{14} = M_{41} = -[0.5m_1l_1s_{b1} + m_2l_1s_{b1} + 0.5m_2l_2s_{b12} + m_3l_1s_{b1} + m_3l_2s_{b12} + 0.5m_3l_3s_{b123}]$$

$$M_{15} = M_{51} = -[0.5m_2l_2s_{b12} + m_3l_2s_{b12} + 0.5m_3l_3s_{b123}]$$

$$M_{16} = M_{61} = -0.5m_3l_3s_{b123}$$

$$M_{22} = m_p + m_1 + m_2 + m_3$$

$$M_{23} = M_{32} = 0.5m_1l_1c_{b1} + m_2l_1c_{b1} + 0.5m_2l_2c_{b12} + m_3l_1c_{b1} + m_3l_2c_{b12} + 0.5m_3l_3c_{b123}$$

$$M_{24} = M_{42} = 0.5m_1l_1c_{b1} + m_2l_1c_{b1} + 0.5m_2l_2c_{b12} + m_3l_1c_{b1} + m_3l_2c_{b12} + 0.5m_3l_3c_{b123}$$

$$M_{25} = M_{52} = 0.5m_2l_2c_{b12} + m_3l_2c_{b12} + 0.5m_3l_3c_{b123}$$

continued on following page

Box 1. continued

$$M_{26} = M_{62} = 0.5m_3l_3c_{b123}$$

$$M_{33} = J_p + J_1 + J_2 + J_3 + m_2l_1^2 + m_2l_1l_2c_2 + m_3l_1^2 + m_3l_2^2 + 2m_3l_1l_2c_2 + m_3l_1l_3c_{23} + m_3l_2l_3c_3$$

$$M_{34} = M_{43} = J_1 + J_2 + J_3 + m_2l_1^2 + m_2l_1l_2c_2 + m_3l_1^2 + m_3l_2^2 + 2m_3l_1l_2c_2 + m_3l_1l_3c_{23} + m_3l_2l_3c_3$$

$$M_{35} = M_{53} = J_2 + J_3 + m_3l_2^2 + 0.5m_2l_1l_2c_2 + m_3l_1l_2c_2 + 0.5m_3l_1l_3c_{23} + m_3l_2l_3c_3$$

$$M_{36} = M_{63} = J_3 + 0.5m_3l_1l_3c_{23} + 0.5m_3l_2l_3c_3$$

$$M_{44} = J_1 + J_2 + J_3 + m_2l_1^2 + m_2l_1l_2c_2 + m_3l_1^2 + m_3l_2^2 + 2m_3l_1l_2c_2 + m_3l_1l_3c_{23} + m_3l_2l_3c_3$$

$$M_{45} = M_{54} = J_2 + J_3 + 0.5m_2l_1l_2c_2 + m_3l_2^2 + m_3l_1l_2c_2 + 0.5m_3l_1l_3c_{23} + m_3l_2l_3c_3$$

$$M_{46} = M_{64} = J_3 + 0.5m_3l_1l_3c_{23} + 0.5m_3l_2l_3c_3$$

$$M_{55} = J_2 + J_3 + m_3l_2^2 + m_3l_2l_3c_3$$

$$M_{56} = M_{65} = J_3 + 0.5m_3l_2l_3c_3$$

$$M_{66} = J_3$$

$$\begin{aligned} C_{13} = &-(0.5m_1l_1c_{b1} + m_2l_1c_{b1} + 0.5m_2l_2c_{b12} + m_3l_1c_{b1} + m_3l_2c_{b12} + 0.5m_3l_3c_{b123})(\dot{\theta}_b + \dot{\theta}_1) \\ &- 0.5m_3l_3c_{b123}\dot{\theta}_3 - (0.5m_2l_2c_{b12} + m_3l_2c_{b12} + 0.5m_3l_3c_{b123})\dot{\theta}_2 \end{aligned}$$

$$\begin{aligned} C_{14} = &-(0.5m_1l_1c_{b1} + m_2l_1c_{b1} + 0.5m_2l_2c_{b12} + m_3l_1c_{b1} + m_3l_2c_{b12} + 0.5m_3l_3c_{b123})(\dot{\theta}_b + \dot{\theta}_1) \\ &- 0.5m_3l_3c_{b123}\dot{\theta}_3 - (0.5m_2l_2c_{b12} + m_3l_2c_{b12} + 0.5m_3l_3c_{b123})\dot{\theta}_2 \end{aligned}$$

$$C_{15} = -(0.5m_2l_2c_{b12} + m_3l_2c_{b12} + 0.5m_3l_3c_{b123})(\dot{\theta}_b + \dot{\theta}_1 + \dot{\theta}_2) - 0.5m_3l_3c_{b123}\dot{\theta}_3$$

$$C_{16} = -0.5m_3l_3c_{b123}(\dot{\theta}_b + \dot{\theta}_1 + \dot{\theta}_2 + \dot{\theta}_3)$$

$$\begin{aligned} C_{23} = &-(0.5m_1l_1s_{b1} + m_2l_1s_{b1} + 0.5m_2l_2s_{b12} + m_3l_1s_{b1} + m_3l_2s_{b12} + 0.5m_3l_3s_{b123})(\dot{\theta}_b + \dot{\theta}_1) \\ &- 0.5m_3l_3s_{b123}\dot{\theta}_3 - (0.5m_2l_2s_{b12} + m_3l_2s_{b12} + 0.5m_3l_3s_{b123})\dot{\theta}_2 \end{aligned}$$

$$\begin{aligned} C_{24} = &-(0.5m_1l_1s_{b1} + m_2l_1s_{b1} + 0.5m_2l_2s_{b12} + m_3l_1s_{b1} + m_3l_2s_{b12} + 0.5m_3l_3s_{b123})(\dot{\theta}_b + \dot{\theta}_1) \\ &- 0.5m_3l_3s_{b123}\dot{\theta}_3 - (0.5m_2l_2s_{b12} + m_3l_2s_{b12} + 0.5m_3l_3s_{b123})\dot{\theta}_2 \end{aligned}$$

continued on following page

Box 1. continued

$$C_{25} = -(0.5m_2 l_2 s_{b12} + m_3 l_2 s_{b12} + 0.5m_3 l_3 s_{b123})(\dot{\theta}_b + \dot{\theta}_1 + \dot{\theta}_2) - 0.5m_3 l_3 s_{b123}\dot{\theta}_3$$

$$C_{26} = -0.5m_3 l_3 s_{b123}(\dot{\theta}_b + \dot{\theta}_1 + \dot{\theta}_2 + \dot{\theta}_3)$$

$$C_{33} = -(0.5m_2 l_1 l_2 s_2 + 0.5m_3 l_1 l_3 s_{23} + m_3 l_1 l_2 s_2)\dot{\theta}_2 - (0.5m_3 l_1 l_3 s_{23} + 0.5m_3 l_2 l_3 s_3)\dot{\theta}_3$$

$$C_{34} = -(0.5m_2 l_1 l_2 s_2 + 0.5m_3 l_1 l_3 s_{23} + m_3 l_1 l_2 s_2)\dot{\theta}_2 - (0.5m_3 l_1 l_3 s_{23} + 0.5m_3 l_2 l_3 s_3)\dot{\theta}_3$$

$$C_{35} = -(0.5m_2 l_1 l_2 s_2 + 0.5m_3 l_1 l_3 s_{23} + m_3 l_1 l_2 s_2)(\dot{\theta}_b + \dot{\theta}_1 + \dot{\theta}_2) - (0.5m_3 l_1 l_3 s_{23} + 0.5m_3 l_2 l_3 s_3)\dot{\theta}_3$$

$$C_{36} = -(0.5m_3 l_1 l_3 s_{23} + 0.5m_3 l_2 l_3 s_3)(\dot{\theta}_b + \dot{\theta}_1 + \dot{\theta}_2 + \dot{\theta}_3)$$

$$C_{43} = -(0.5m_2 l_1 l_2 s_2 + 0.5m_3 l_1 l_3 s_{23} + m_3 l_1 l_2 s_2)\dot{\theta}_2 - (0.5m_3 l_1 l_3 s_{23} + 0.5m_3 l_2 l_3 s_3)\dot{\theta}_3$$

$$C_{44} = -(0.5m_2 l_1 l_2 s_2 + 0.5m_3 l_1 l_3 s_{23} + m_3 l_1 l_2 s_2)\dot{\theta}_2 - (0.5m_3 l_1 l_3 s_{23} + 0.5m_3 l_2 l_3 s_3)\dot{\theta}_3$$

$$C_{45} = -(0.5m_2 l_1 l_2 s_2 + 0.5m_3 l_1 l_3 s_{23} + m_3 l_1 l_2 s_2)(\dot{\theta}_b + \dot{\theta}_1 + \dot{\theta}_2) - (0.5m_3 l_1 l_3 s_{23} + 0.5m_3 l_2 l_3 s_3)\dot{\theta}_3$$

$$C_{46} = -(0.5m_3 l_1 l_3 s_{23} + 0.5m_3 l_2 l_3 s_3)(\dot{\theta}_b + \dot{\theta}_1 + \dot{\theta}_2 + \dot{\theta}_3)$$

$$C_{53} = (0.5m_2 l_1 l_2 s_2 + 0.5m_3 l_1 l_3 s_{23} + m_3 l_1 l_2 s_2)(\dot{\theta}_b + \dot{\theta}_1) - 0.5m_3 l_2 l_3 s_3 \dot{\theta}_3$$

$$C_{54} = (0.5m_2 l_1 l_2 s_2 + 0.5m_3 l_1 l_3 s_{23} + m_3 l_1 l_2 s_2)(\dot{\theta}_b + \dot{\theta}_1) - 0.5m_3 l_2 l_3 s_3 \dot{\theta}_3$$

$$C_{55} = -0.5m_3 l_2 l_3 s_3 \dot{\theta}_3$$

$$C_{56} = -0.5m_3 l_2 l_3 s_3 \dot{\theta}_3(\dot{\theta}_b + \dot{\theta}_1 + \dot{\theta}_2 + \dot{\theta}_3)$$

$$C_{63} = (0.5m_3 l_1 l_3 s_{23} + 0.5m_3 l_2 l_3 s_3)(\dot{\theta}_b + \dot{\theta}_1) + 0.5m_3 l_2 l_3 s_3 \dot{\theta}_1$$

$$C_{64} = (0.5m_3 l_1 l_3 s_{23} + 0.5m_3 l_2 l_3 s_3)(\dot{\theta}_b + \dot{\theta}_1) + 0.5m_3 l_2 l_3 s_3 \dot{\theta}_1$$

$$C_{65} = 0.5m_3 l_2 l_3 s_3(\dot{\theta}_b + \dot{\theta}_1 + \dot{\theta}_2)$$

$$C_{66} = 0$$

$$\mathbf{G}(\mathbf{q}) = \mathbf{0}_{6\times 1}$$

Figure 3. Block diagram of robust NN control system

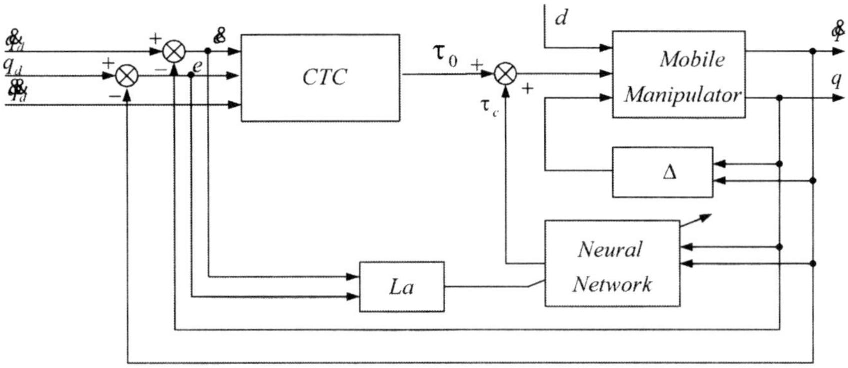

Property 1: *The inertia matrix* $\mathbf{M}(\mathbf{q})$ *is symmetric and positive definite, and satisfies*

$$0 < \lambda_m \cdot I \le \left\| M(q) \right\| \le \lambda_M \cdot I \tag{5}$$

where $\lambda_{\rm m}, \lambda_{\rm M}$ are positive scalar constants, and $\|.\|$ denotes the norm of a matrix.

Property 2: *The centripetal and Coriolis matrix* $\mathbf{C}(\mathbf{q}, \dot{\mathbf{q}})$ *is bounded as a function of* $\dot{\mathbf{q}}$, *i.e.,*

$$\left\| \mathbf{C}(\mathbf{q}, \dot{\mathbf{q}}) \right\| \le k_{\rm c} \left\| \dot{\mathbf{q}} \right\|, \ \forall \mathbf{q}, \dot{\mathbf{q}} \in \mathrm{R}^n \tag{6}$$

where $k_{\rm c}$ is a positive constant.

Property3: $\dot{\mathbf{M}}(\mathbf{q}) - 2 \cdot \mathbf{C}(\mathbf{q}, \dot{\mathbf{q}})$ *is a skew-symmetric matrix, i.e., satisfies the following relationship*

$$\mathbf{x}^T \cdot \left[\dot{\mathbf{M}}(\mathbf{q}) - 2 \cdot \mathbf{C}(\mathbf{q}, \dot{\mathbf{q}}) \right] \cdot \mathbf{x} = 0, \ \forall \mathbf{x} \in \mathrm{R}^n \tag{7}$$

CONTROLLER DESIGN AND STABILITY ANALYSIS

To control the mobile manipulator effectively, an adaptive controller, based on CTC and RBFNN, is presented in this section. Figure 3 shows the basic configuration of the proposed system, where La represents the learning algorithm; $\mathbf{d}$ denotes the external disturbance and Δ is unmodeled dynamics. It is well known that CTC is very simple and effective under giving precise models. However, requirements of precise model are too difficult to satisfy. Thus a RBFNN is adopted into the controller to compensate the uncertainties.

Problem formulation

Suppose that the desired trajectory is described by $\mathbf{q}_{\rm d}, \dot{\mathbf{q}}_{\rm d}, \ddot{\mathbf{q}}_{\rm d}$. In order to track the desired trajectory, we have to design a feedback controller. With this controller, the error between the actual trajectory $\mathbf{q}, \dot{\mathbf{q}}, \ddot{\mathbf{q}}$ and the desired trajectory satisfies

$$\lim_{t \to \infty} \left\{ \mathbf{q}(t) - \mathbf{q}_{\rm d}(t) \right\} = 0, \quad \lim_{x \to \infty} \left\{ \dot{\mathbf{q}}(t) - \dot{\mathbf{q}}_{\rm d}(t) \right\} = 0 \tag{8}$$

Without the compensating torque, According to CTC, we get

$$\tau = \mathbf{M}(\mathbf{q})(\ddot{\mathbf{q}}_{\rm d} - \mathbf{K}_{\rm V}\dot{\mathbf{e}} - \mathbf{K}_{\rm p}\mathbf{e}) + \mathbf{C}(\mathbf{q}, \dot{\mathbf{q}})\dot{\mathbf{q}} + \mathbf{G}(\mathbf{q}) \tag{9}$$

where $\mathbf{K}_{\rm P}$ and $\mathbf{K}_{\rm V}$ are proportional and derivative constant, $\mathbf{e} = \mathbf{q}(t) - \mathbf{q}_d(t)$. Substitute (9) into (4) yields

Figure 4. Three-Layer radial basis function neural network

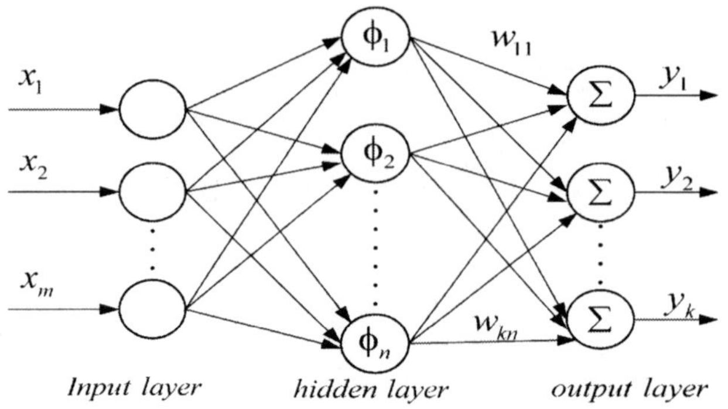

$$\ddot{\mathbf{e}} + \mathbf{K}_V\dot{\mathbf{e}} + \mathbf{K}_P\mathbf{e} = \rho \tag{10}$$

where $\rho = -\mathbf{M}^{-1}(\mathbf{q})[\Delta(\mathbf{q}, \dot{\mathbf{q}}) + \mathbf{d}]$.

It is obvious that errors will asymptotically converge to zero when ρ=0 and $\mathbf{K}_P$ and $\mathbf{K}_V$ are chosen appropriately. However, the existence of ρ influences the performance of CTC and makes the closed-loop system unsTable 1f it is significant. To handle these difficulties, A RBFNN is adopted to compensate the uncertainties. The overall control law becomes

$$\tau = \tau_0 + \tau_c \tag{11}$$

where τ_0 is the output torque of the CTC define like (9), τ_c is the compensating torque generated by RBFNN.

Radial Basis Function Neural Network

Figure 4 shows the structure of the three-layer neural network, which includes an input layer, a hidden layer, and an output layer. Assume that there are m nodes in the input layer, n nodes in the hidden layer, k nodes in the output layer. u_i^l represents the input of the node i in layer l, and o_i^l represents the output of the node i in layer l.

The relationships among these layers can be described as follows.

- **Input Layer:** signals input to the RBFNN via this layer.

$$o_i^1 = u_i^1 = x_i \qquad (i = 1, 2, 3, \cdots, m) \tag{12}$$

- **Hidden Layer:** produce nonlinearity.

$$u_j^2 = \sum_{i=1}^{m} o_i^1 \qquad (j = 1, 2, \cdots, n) \tag{13}$$

$$o_j^2 = \phi_j = \exp[-\frac{\left\| u_j^2 - c_j \right\|^2}{2\sigma_j^{\ 2}}] \qquad (j = 1, 2, \cdots, n) \tag{14}$$

$\mathbf{C} = \begin{bmatrix} c_1 & c_1 & \cdots & c_n \end{bmatrix}^T$ denotes the centre vector of Gaussian function. $\sigma = \begin{bmatrix} \sigma_1 & \sigma_2 & \cdots & \sigma_n \end{bmatrix}^T$ represents the width vector of Gaussian function.

- **Output Layer.**

$$y_h = o_h^3 = u_h^3 = \sum_{j=1}^{n} w_{hj} \cdot o_j^2 \qquad (h = 1, 2, \cdots, k) \tag{15}$$

let

$$\mathbf{W} = \begin{bmatrix} w_{11}, w_{12}, \cdots, w_{1n} \\ w_{21}, w_{22}, \cdots, w_{2n} \\ \vdots \quad \vdots \quad\quad\quad \vdots \\ w_{k1}, w_{k2}, \cdots, w_{kn} \end{bmatrix}, \Phi = \begin{bmatrix} \phi_1 \\ \phi_2 \\ \vdots \\ \phi_n \end{bmatrix}, \mathbf{Y} = \begin{bmatrix} y_1 \\ y_2 \\ \vdots \\ y_k \end{bmatrix} \tag{16}$$

In this way, the relationship of the output layer can be written as:

$$\mathbf{Y} = \mathbf{W} \cdot \Phi \tag{17}$$

where $\mathbf{W}$ is the weight matrix of the RBFNN, Φ is the excitation function vector.

RBFNN Compensating Controller

As for these reasons referred above, we adopt a RBFNN to approximate ρ. Assume that the ideal output of the neural network is

$$\rho(\mathbf{z}) = \mathbf{W}^* \cdot \Phi(\mathbf{z}) + \varepsilon(\mathbf{z}) \tag{18}$$

where $\Phi(\mathbf{z})$ is the excitation function vector of the RBFNN, $\varepsilon(\mathbf{z})$ is the reconstruction error of the RBFNN, and $\mathbf{z} = \begin{bmatrix} \mathbf{q} & \dot{\mathbf{q}} & \ddot{\mathbf{q}} \end{bmatrix}^{\mathrm{T}}$ denotes the input of the RBFNN, $\mathbf{W}^* = [w_{ij}] \in \mathrm{R}^{k \times n}$ is the optimal weight matrix satisfying

$$\mathbf{W}^* = \arg\min_{W} \left\{ \sup_{x \in D_z} \left| \hat{\rho}(\mathbf{z} \mid \hat{\mathbf{W}}) - \rho(\mathbf{z}) \right| \right\} \tag{19}$$

$\mathbf{D}_Z$ is the bounds of the input $\mathbf{z}$, which represents the limitation of position, velocity and acceleration.

$\hat{\rho}(\mathbf{z} \mid \hat{\mathbf{W}})$ is an estimation of $\rho(\mathbf{z})$, and we define it as

$$\hat{\rho}(\mathbf{z} \mid \hat{\mathbf{W}}) = \hat{\mathbf{W}} \Phi(\mathbf{z}) \tag{20}$$

Let compensative control law τ_c in (15) as

$$\tau_c = -\mathbf{M}(\mathbf{q}) \hat{\rho}(\mathbf{z} \mid \hat{\mathbf{W}}) \tag{21}$$

Combining (10), (11), (21), we write the closed-loop system as

$$\ddot{\mathbf{e}} + \mathbf{K}_{\mathrm{V}} \dot{\mathbf{e}} + \mathbf{K}_{\mathrm{P}} \mathbf{e} = \tilde{\rho}(\mathbf{z}) \tag{22}$$

where $\tilde{\rho}(\mathbf{z})$ denotes

$$\tilde{\rho}(\mathbf{z}) = \rho(\mathbf{z}) - \hat{\rho}(\mathbf{z} \mid \hat{\mathbf{W}}) = \tilde{\mathbf{W}} \Phi(\mathbf{z}) + \varepsilon(\mathbf{z}) \tag{23}$$

$\tilde{\mathbf{W}} = \mathbf{W}^* - \hat{\mathbf{W}}$ represents the error between the adjustable weight matrix and the optimal matrix.

Define the state vector as $\mathbf{x} = \begin{bmatrix} \mathbf{e}^{\mathrm{T}} & \dot{\mathbf{e}}^{\mathrm{T}} \end{bmatrix}^{\mathrm{T}}$. The state-space equation of (22) has the form as

$$\dot{\mathbf{x}} = \mathbf{A}\mathbf{x} + \mathbf{B} \tilde{\rho}(\mathbf{z}) \tag{24}$$

where $\mathbf{A} = \begin{bmatrix} \mathbf{0}_{n \times n} & \mathbf{I}_{n \times n} \\ -\mathbf{K}_{\mathrm{P}} & -\mathbf{K}_{\mathrm{V}} \end{bmatrix}$, $\mathbf{B} = \begin{bmatrix} \mathbf{0}_{n \times n} \\ \mathbf{I}_{n \times n} \end{bmatrix}$.

Then, the learning algorithm of $\mathbf{W}$ is designed as

$$\dot{\hat{\mathbf{W}}} = \alpha \Lambda^{-1} \mathbf{B}^{\mathrm{T}} \mathbf{P} \mathbf{x} \Phi^{\mathrm{T}} \tag{25}$$

where $\alpha > 0$ is the learning rate constant, $\mathbf{L} = diag(\lambda_1, \lambda_2, \cdots, \lambda_n)$ is gain matrix and $\mathbf{P}$ is positive definite solution of the following Riccati equation

$$\mathbf{A}^{\mathrm{T}} \mathbf{P} + \mathbf{P} \mathbf{A} + \mathbf{P}^{\mathrm{T}} \mathbf{B} \mathbf{B}^{\mathrm{T}} \mathbf{P} + \mathbf{Q} = 0 \tag{26}$$

where $\mathbf{Q}$ is a constant matrix with appropriate dimensions.

In order to prove the stability of the closed-loop system, the following assumptions are made:

Assumption 1: *The reconstruction error* $\varepsilon(\mathbf{z})$ *is bounded, i.e.,* $\left\|\varepsilon\right\| \leq b_{\varepsilon}$ *for* $\forall \mathbf{z} \in D_{z}$.

Assumption 2: *The norm of the optimal weight matrix is bounded, that is,* $\left\|\mathbf{W}^{*}\right\| \leq b_{w}$.

Theorem: *Consider the manipulator with the structured uncertainty involved by mobile platform and external disturbance; we apply the whole controller as (11) and the learning algorithm (25) for the compensating controller part. Based on the Riccati equation (26) and Assumption 1 and Assumption 2, we can get following result:*

The state vector $\mathbf{x}$ for the manipulator is uniformly ultimately bounded. Then, if reconstruction error of the neural network $\boldsymbol{\mu} \in L_2$, i.e., $\int_0^{\infty} \left\|\boldsymbol{\mu}(\mathrm{t})\right\|^2 \mathrm{dt} < \infty$, trajectory tracking errors of the manipulator tend to zero as time goes to infinity.

Proof: *Consider the following Lyapunov function (Tan et al., 2008)*

$$\mathbf{V} = \frac{1}{2}\mathbf{x}^{T}\mathbf{P}\mathbf{x} + \frac{1}{2\alpha}Tr(\tilde{\mathbf{W}}\Lambda\tilde{\mathbf{W}}) \tag{27}$$

By applying the properties of matrix theory, we can obtain the time derivative of $\mathbf{V}$ as

$$\begin{aligned}
\dot{V} &= \frac{1}{2}\left[\mathbf{A}\cdot\mathbf{x} + \mathbf{B}\cdot\tilde{\rho}(\mathbf{z})\right]^{\mathrm{T}}\mathbf{P}\mathbf{x} + \frac{1}{2}\mathbf{x}^{\mathrm{T}}\mathbf{P}\left[\mathbf{A}\cdot\mathbf{x} + \mathbf{B}\cdot\tilde{\rho}(\mathbf{z})\right] + \frac{1}{\alpha}Tr(\tilde{\mathbf{W}}^{\mathrm{T}}\Lambda\dot{\hat{\mathbf{W}}}) \\
&= \frac{1}{2}\mathbf{x}^{\mathrm{T}}\left(\mathbf{A}^{\mathrm{T}}\mathbf{P} + \mathbf{P}\mathbf{A}\right)\mathbf{x} + \mathbf{x}^{\mathrm{T}}\mathbf{P}\mathbf{B}\tilde{\rho}(\mathbf{z}) - \frac{1}{\alpha}Tr\left(\tilde{\mathbf{W}}^{\mathrm{T}}\Lambda\dot{\hat{\mathbf{W}}}\right) \\
&= -\frac{1}{2}\mathbf{x}^{\mathrm{T}}\left(\mathbf{P}^{\mathrm{T}}\mathbf{B}\mathbf{B}^{\mathrm{T}}\mathbf{P} + \mathbf{Q}\right)\mathbf{x} + \mathbf{x}^{\mathrm{T}}\mathbf{P}\mathbf{B}\left(\tilde{\mathbf{W}}\phi + \varepsilon\right) - \frac{1}{\alpha}Tr\left(\tilde{\mathbf{W}}^{\mathrm{T}}\Lambda\dot{\hat{\mathbf{W}}}\right) \\
&= -\frac{1}{2}\mathbf{x}^{\mathrm{T}}\left(\mathbf{P}^{\mathrm{T}}\mathbf{B}\mathbf{B}^{\mathrm{T}}\mathbf{P} + \mathbf{Q}\right)\mathbf{x} + \mathbf{x}^{\mathrm{T}}\mathbf{P}\mathbf{B}\varepsilon + \mathbf{x}^{\mathrm{T}}\mathbf{P}\mathbf{B}\tilde{\mathbf{W}}\phi - \frac{1}{\alpha}Tr\left(\tilde{\mathbf{W}}^{\mathrm{T}}\Lambda\dot{\hat{\mathbf{W}}}\right) \\
&= -\frac{1}{2}\mathbf{x}^{\mathrm{T}}\left(\mathbf{P}^{\mathrm{T}}\mathbf{B}\mathbf{B}^{\mathrm{T}}\mathbf{P} + \mathbf{Q}\right)\mathbf{x} + \mathbf{x}^{\mathrm{T}}\mathbf{P}\mathbf{B}\boldsymbol{\mu} + Tr\left(\phi^{\mathrm{T}}\tilde{\mathbf{W}}^{\mathrm{T}}\mathbf{B}^{\mathrm{T}}\mathbf{P}\mathbf{x}\right) - \frac{1}{\alpha}Tr\left(\tilde{\mathbf{W}}^{\mathrm{T}}\Lambda\dot{\hat{\mathbf{W}}}\right) \\
&= -\frac{1}{2}\mathbf{x}^{\mathrm{T}}\left(\mathbf{P}^{\mathrm{T}}\mathbf{B}\mathbf{B}^{\mathrm{T}}\mathbf{P} + \mathbf{Q}\right)\mathbf{x} + \mathbf{x}^{\mathrm{T}}\mathbf{P}\mathbf{B}\boldsymbol{\mu} + Tr\left(\tilde{\mathbf{W}}^{\mathrm{T}}\mathbf{B}^{\mathrm{T}}\mathbf{P}\mathbf{x}\phi^{\mathrm{T}}\right) - \frac{1}{\alpha}Tr\left(\tilde{\mathbf{W}}^{\mathrm{T}}\Lambda\dot{\hat{\mathbf{W}}}\right) \\
&= -\frac{1}{2}\mathbf{x}^{\mathrm{T}}\left(\mathbf{P}^{\mathrm{T}}\mathbf{B}\mathbf{B}^{\mathrm{T}}\mathbf{P} + \mathbf{Q}\right)\mathbf{x} + \mathbf{x}^{\mathrm{T}}\mathbf{P}\mathbf{B}\boldsymbol{\mu} + Tr\left[\left(\tilde{\mathbf{W}}^{\mathrm{T}}\mathbf{B}^{\mathrm{T}}\mathbf{P}\mathbf{x}\phi^{\mathrm{T}}\right) - \frac{1}{\alpha}\left(\tilde{\mathbf{W}}^{\mathrm{T}}\Lambda\dot{\hat{\mathbf{W}}}\right)\right] \\
&= -\frac{1}{2}\mathbf{x}^{\mathrm{T}}\left(\mathbf{P}^{\mathrm{T}}\mathbf{B}\mathbf{B}^{\mathrm{T}}\mathbf{P} + \mathbf{Q}\right)\mathbf{x} + \mathbf{x}^{\mathrm{T}}\mathbf{P}\mathbf{B}\boldsymbol{\mu} \\
&= -\frac{1}{2}\mathbf{x}^{\mathrm{T}}\mathbf{Q}\mathbf{x} - \frac{1}{2}\left(\mathbf{B}^{\mathrm{T}}\mathbf{P}\mathbf{x} - \varepsilon\right)^{\mathrm{T}}\left(\mathbf{B}^{\mathrm{T}}\mathbf{P}\mathbf{x} - \varepsilon\right) + \frac{1}{2}\varepsilon^{\mathrm{T}}\varepsilon \\
&\leq -\frac{1}{2}\mathbf{x}^{\mathrm{T}}\mathbf{Q}\mathbf{x} + \frac{1}{2}\varepsilon^{\mathrm{T}}\varepsilon
\end{aligned} \tag{28}$$

It is easy to obtain $\dot{V} \leq -\lambda_{\min}(Q)\left\|x\right\|^2 + \left\|\varepsilon\right\|^2$, and $V(\mathbf{x}, \tilde{\mathbf{W}})$ is negative outside the following compact set Σ_{x}:

$$\sum_{x} = \left\{ x \middle| \; 0 \leq \left\|x\right\| \leq \sqrt{\frac{1}{\lambda_{\min}(Q)}}\left\|\varepsilon\right\| \right\} \tag{29}$$

Assumption 2 ensures that $\mathbf{W}$ is bounded, i.e., $\tilde{\mathbf{W}}$ is bounded. Thus $\mathbf{x}$ is uniformly ultimately bounded. Integrating both sides of equation (28) from $t = 0$ to $t = \infty$ gives

$$\int_0^{\infty} \mathbf{x}^{\mathrm{T}}\mathbf{Q}\mathbf{x}\mathrm{dt} \leq \int_0^{\infty} \varepsilon^{\mathrm{T}}\varepsilon\mathrm{dt} + 2[\mathbf{V}(0) - \mathbf{V}(\infty)] \tag{30}$$

Then, we can easily get

$\int_0^{\infty} \left\|x\right\|^2 dt \leq a / \lambda_{\min}(Q)$, where

$a = \int_0^{\infty} \varepsilon^{\mathrm{T}}\varepsilon\mathrm{dt} + 2[\mathbf{V}(0) - \mathbf{V}(\infty)]$.

Noting that $V(t)$ is a non-increasing function of time and has low bounded, this implies $V(0) - V(\infty) < \infty$. if $\int_0^{\infty} \left\|\varepsilon(\mathrm{t})\right\|^2 \mathrm{dt} < \infty$, we know $a < \infty$ and $\mathbf{x} \in L_2$. In addition, the bound of $\mathbf{x}$ above denotes $\mathbf{x} \in L_{\infty}$. From closed-loop dynamic equation (24) and bound of $\mathbf{x}(t)$, $\tilde{\mathbf{W}}(t)$ and $\boldsymbol{\mu}(t)$, we can get $\dot{\mathbf{x}} \in L_{\infty}$. Then, $\mathbf{x} \in L_2 \cap L_{\infty}$, $\dot{\mathbf{x}} \in L_{\infty}$. Thus, $\lim_{t \to \infty} \mathbf{x}(t) = 0$ is achieved according to Barbalat's lemma.

SIMULATION RESULTS

To illustrate the validity of the proposed method, the unified model proposed above, as an example, is simulated in this section. In addition, CTC without compensating is also applied to compare with the proposed method. Simulation parameters

Table 1. Simulation parameters

	Mass(kg)		Length(m) or radius(m)		Inertial of moment($kg \cdot m^2$)	
	NV	AV	NV	AV	NV	AV
Mobile Platform	60	100	0.32	0.32	3.0720	5.1200
Link1	5	5.5	0.25	0.25	0.1042	0.1146
Link2	3	3.5	0.35	0.35	0.1225	0.1429
Link3	2	2.5	0.21	0.21	0.0294	0.0367

are shown in Table 1. In this table, NV denotes nominal value and AV represents actual value.

The initial conditions are given as follows

$$\mathbf{q}(0) = [2,\ 2, 0.5, 0.5, 0.5, 0.5]^{\mathrm{T}} \tag{31}$$

$$\dot{\mathbf{q}}(0) = [0,\ 0,\ 0,\ 0,\ 0,\ 0]^{\mathrm{T}} \tag{32}$$

Let the desired trajectory be

$$\mathbf{q}_{\mathrm{d}}(t) = \begin{bmatrix}\sin(t) & \cos(t) & 0 & \sin(t) & \sin(t) & \sin(t)\end{bmatrix}^{\mathrm{T}} \tag{33}$$

Parameters involved in the presented controller, (22) and (26) are $\mathbf{K}_{\mathrm{P}} = 19 * \mathbf{I}_6$, $\mathbf{K}_{\mathrm{V}} = 8 * \mathbf{I}_6$, $\mathbf{Q} = 50 * \mathbf{I}_{12}$, $\alpha = 1.0$. The weights of the RBFNN are initialized to zero. There are 17 nodes in the hidden layer, 6 nodes in the output layer, and the centre vector of Gaussian function is

$$\mathbf{C} = \begin{bmatrix}-4 & -3.5 & -3 & -2.5 & -2 & -1.5 & -1 & -0.5 & 0 & 0.5 & 1 & 1.5 & 2 & 2.5 & 3 & 3.5 & 4\end{bmatrix}^{\mathrm{T}} \tag{34}$$

And the width vector is

$$\sigma = \begin{bmatrix}3 & 3 & 3 & 3 & 3 & 3 & 3 & 3 & 3 & 3 & 3 & 3 & 3 & 3 & 3 & 3 & 3\end{bmatrix}^{\mathrm{T}} \tag{35}$$

In order to simulate the unmodeled dynamics caused by friction and other factors, we define the unmodeled dynamics Δ as follows

$$\Delta = \begin{bmatrix} 10sign(\dot{x})[1+\exp(-|\dot{x}|)] \\ 10sign(\dot{y})[1+\exp(-|\dot{y}|)] \\ 10sign(\dot{\theta}_b)[1+\exp(-|\dot{\theta}_b|)] \\ 0.5sign(\dot{\theta}_1)[0.5+\exp(-|\dot{\theta}_1|)] \\ 0.3sign(\dot{\theta}_2)[0.3+\exp(-|\dot{\theta}_2|)] \\ 0.2sign(\dot{\theta}_3)[0.2+\exp(-|\dot{\theta}_3|) \end{bmatrix} \tag{36}$$

The external disturbance is

$$\mathbf{d} = \begin{bmatrix} 20\sin(t) \\ 20\sin(t) \\ 20\sin(t) \\ 0.5\sin(t) \\ 0.5\sin(t) \\ 0.5\sin(t) \end{bmatrix} \text{ N or Nm} \tag{37}$$

Figure 5, Figure 6, Figure 7, Figure 8, Figure 9 and Figure 10 show the position errors of x, y, θ_b, θ_1, θ_2 and θ_3 respectively. In Figure 5-10, CTCRBFNN represents the RBFNN compensating controller we presented, and CTC denotes computed torque controller.

Figure 5. Position errors of x

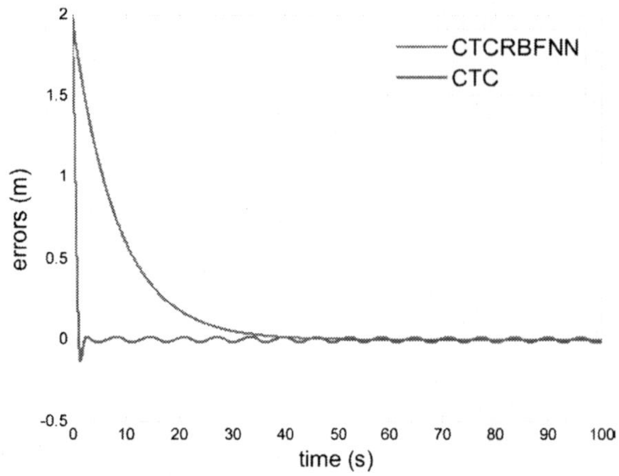

Figure 6. Position errors of y

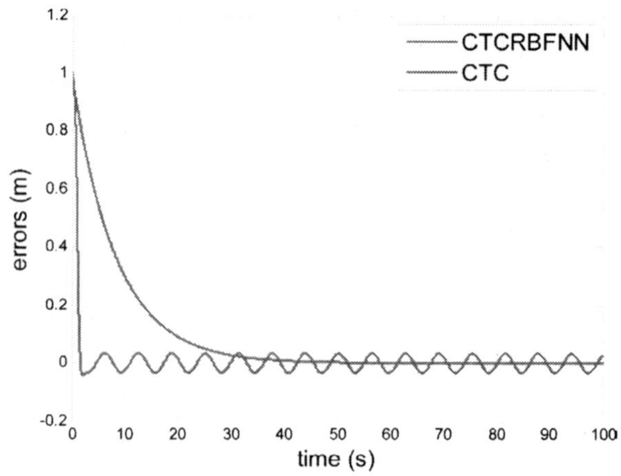

Figure 7. Position errors of θ_b

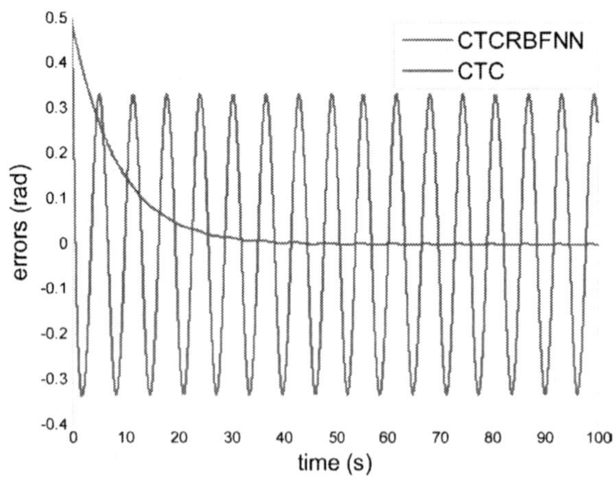

Figure 8. Position errors of θ_1

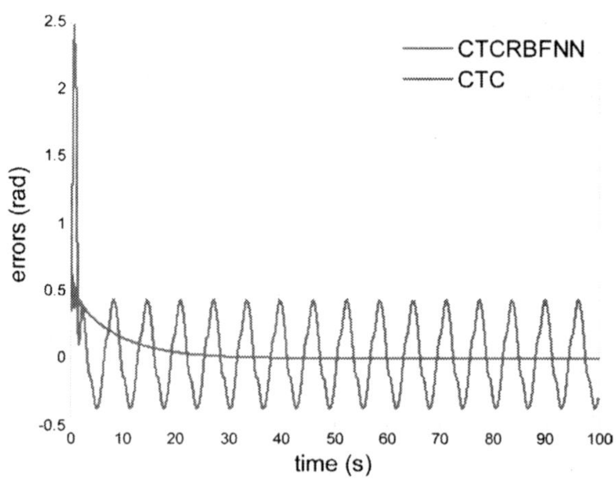

Remark 1: *From Figure 5-6, we can see CTC is basically able to tracking the given trajectory (x and y), and it converges more quickly than the CTCRBFNN. The reason is the model error and the external disturbance is relative small compared to Computed Torque (<5%). However, the transient performance is not as good as CTCRBFNN.*

Remark 2: *Figure 7 to Figure 10 show CTC is not stable when the model error or the external disturbance is significant, and CTCRBFNN is able to track the given trajectories. Define Integrate Square Error (ISE) as*

$$ISE = \int_0^{T_0} \mathbf{e}^2(t)dt \tag{38}$$

Table 2 shows the ISE of CTC and CTCRB-FNN.

As for a tracking controller, the smaller the ISE is, the better the performance is. Therefore, we can conclude that the RBFNN compensating controller we proposed greatly improves the performance.

From above, it is clear that the system quickly converges to the desired value, so the proposed

Figure 9. Position errors of θ_2

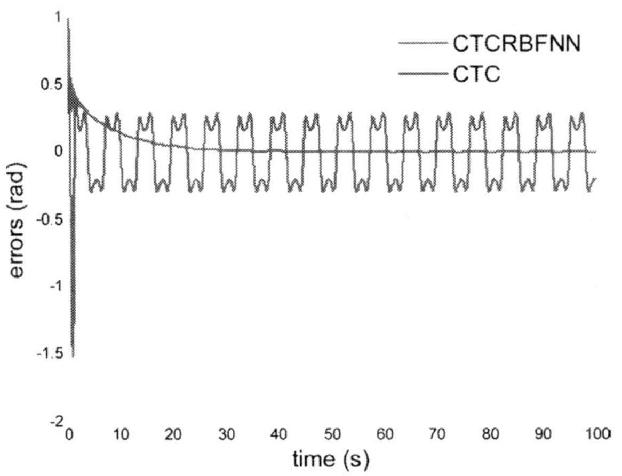

Figure 10. Position errors of θ_3

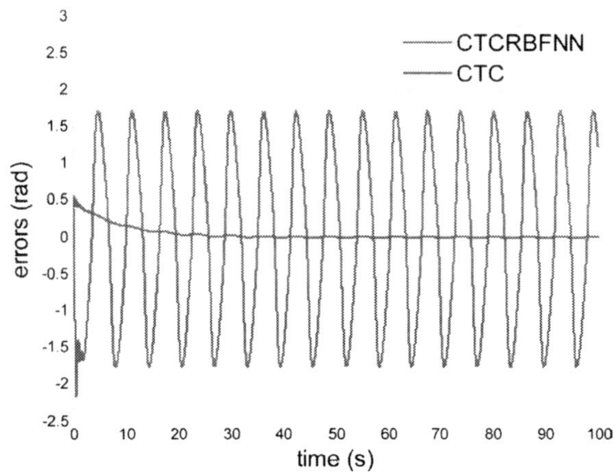

Table 2. The integrate square error (0-100s) OF CTC and CTCRBFNN

	θ_b	θ_1	θ_2	θ_3
ISE_CTC	5.6054	11.2176	6.2275	149.9676
ISE_CTCRBFNN	1.0344	1.1115	0.9736	1.0785

controller is able to achieve trajectory tracking successfully.

CONCLUSION

In this paper, we apply Lagrange Formalism to get the unified dynamic model of the omnidirectional mobile manipulator system in our lab firstly. According to this model, an adaptive integrated controller is developed. Not only it can guarantee the global stability, but also it provides a relative good transient performance. The stability analysis is proved by Lyapunov approach. From the discussion and simulation results, the following conclusions can be reached:

1. Considering the mobile platform as a joint of 3 DOFs, a novel way to get the unified dynamic model is given, and the computational complexity is greatly decreased.
2. Using a RBFNN, the proposed controller is able to track the given trajectory even if the unmodeled error and the external disturbance are significant.

Simulation results show the validity of the dynamic model and effectiveness of the developed controller.

ACKNOWLEDGMENT

This work was supported partly by NSFC Projects (No.60475030 and No.60621001), Joint Laboratory of Intelligent Sciences and Technology (No. JL0605), China.

REFERENCES

Berni, A., Ramdane-Cherif, A., Saadia, N., & Levy, N. (2003). Exploring cognitive approach through the neural network paradigm: "trajectory planning application". *Proceedings of The Second IEEE International Conference on Cognitive Informatics* (pp. 47-54).

Betourne, A., & Campion, G. (1996). Dynamic model and control design of a class of omnidirectional mobile robots. *Proceedings of the 1996 IEEE International Conference on Robotics and Automation* (pp. 2810-2815).

Campion, G., & Bastin, G. (1996). Structural properties and classification of kinematic and dynamic models of wheeled mobile robots. *IEEE Transactions on Robotics and Automation, 12*(1), 47–62. doi:10.1109/70.481750

Chung, J. H., Velinsky, S. A., & Ronald, A. H. (1998). Interaction control of a redundant mobile manipulator. *The International Journal of Robotics Research, 17*(12), 1302–1309. doi:10.1177/027836499801701203

Chung, J. H., Yi, B. J., & Kim, W. K. (2003). The dynamic modeling and analysis for an omnidirectional mobile robot with three castor wheels. *Proceedings of the 2003 IEEE International Conference on Robotics and Automation* (pp. 521-527).

Er, M. J., & Gao, Y. (2003). Robust adaptive control of robot manipulators using generalized fuzzy neural networks. *IEEE Transactions on Industrial Electronics, 50*(3), 620–628. doi:10.1109/TIE.2003.812454

Fang, R., Zhao, Y. b., & Li, W. S. (2005). A Novel Fuzzy Neural Network: The Vague Neural Network. *Proceedings of the Third IEEE International Conference on Cognitive Informatics* (pp. 94-99).

Holmberg, R., & Khatib, O. (2000). Development and control of a holonomic mobile robot for mobile manipulation tasks. *The International Journal of Robotics Research, 19*(11), 1066–1074. doi:10.1177/02783640022067977

Hou, Z. G., & Tan, M. (2004). Real-Time Optimization and Computation for Interconnected Nonlinear Systems Using Neural Networks. *Proceedings of the Third IEEE International Conference on Cognitive Informatics* (pp. 208-213).

Khatib, O. (1987). A unified approach to motion and force control of robot manipulators: the operational Space formulation. *IEEE Journal on Robotics and Automation, 3*(1), 43–53. doi:10.1109/JRA.1987.1087068

Khatib, O., Yokoi, K., Chang, K., Ruspini, D., Holmberg, R., & Casal, A. (1996). Coordination and decentralized cooperation of multiple mobile manipulators. *International Journal of Robotic System, 13*(11), 755–764. doi:10.1002/(SICI)1097-4563(199611)13:11<755::AID-ROB6>3.0.CO;2-U

Kwan, C., Lewis, F. L., & Dawson, D. M. (1998). Robust neural-network control of rigid-link electrically driven robots. *IEEE Transactions on Neural Networks, 9*(4), 581–589. doi:10.1109/72.701172

Li, Y. (2006). Robust neural networks compensating motion control of reconfigurable manipulator in geometric form. *IEEE International Conference on Mechatronics and Automation* (pp. 306-311).

Liu, K., & Lewis, F. L. (1990). Decentralized continuous robust controller for mobile robots. *Proceedings of IEEE International Conference on Robotics and Automation* (pp. 1822-1827).

Sicilian, O. B., Wit, C. C., & Bastin, G. (1996). *Theory of robot control.* Springer-Verlag.

Song, Z. S., Yi, J. Q., Zhao, D. B., & Li, X. C. (2005). A computed torque controller for uncertain robotic manipulator systems: fuzzy approach. *Fuzzy Sets and Systems*, *154*(2), 208–226. doi:10.1016/j.fss.2005.03.007

Tan, J. D., & Xi, N. (2001). Unified model approach for planning and control of mobile manipulators. *IEEE International Conference on Robotics and Automation* (pp. 3145-3152).

Tan, X. M., Zhao, D. B., Yi, J. Q., & Xu, D. (2008). Adaptive hybrid control for omnidirectional mobile manipulator based on neural network. *American Control Conference* (pp. 5174 -5179).

Wai, R. J. (2003). Robust control for nonlinear motor-mechanism coupling system using wavelet neural network. *IEEE Transactions on Systems, Man, and Cybernetics*, *33*(3).

Wang, Y. (2007). The Theoretical Framework of Cognitive Informatics. [IJCiNi]. *International Journal of Cognitive Informatics and Natural Intelligence*, *1*(1), 1–27.

Wang, Y. (2007). On Laws of Work Organization in Human Cooperation. [IJCINI]. *International Journal of Cognitive Informatics and Natural Intelligence*, *1*(2), 1–15.

Wang, Y., & Kinsner, W. (2006). Recent Advances in Cognitive Informatics. *IEEE Transactions on Systems, Man and Cybernetics. Part C, Applications and Reviews*, *36*(2), 121–123. doi:10.1109/TSMCC.2006.871120

Wang, Y., Wang, Y., Patel, S., & Patel, D. (2006). A Layered Reference Model of the Brain. *IEEE Transactions on Systems, Man and Cybernetics. Part C, Applications and Reviews*, *36*(2), 124–133. doi:10.1109/TSMCC.2006.871126

Yamamoto, Y., & Yun, X. P. (1994). Coordinating locomotion and manipulation of manipulator. *IEEE Transactions on Automatic Control*, *39*(6), 1326–1332. doi:10.1109/9.293207

Yi, B. J., & Kim, W. K. (2001). The dynamics for redundantly actuated omnidirectional mobile robots. *IEEE International Conference on Robotics and Automation* (pp. 2485-2492).

This work was previously published in International Journal of Cognitive Informantics and Natural Intelligence, Volume 3, Issue 4, edited by Yingxu Wang, pp. 34-53, copyright 2009 by IGI Publishing (an imprint of IGI Global)

Chapter 21
Knowledge Adquisition in a Cooperative and Competitive Framework[1]

Alberto de la Encina
Universidad Complutense de Madrid, Spain

Mercedes Hidalgo-Herrero
Universidad Complutense de Madrid, Spain

Natalia López
Universidad Complutense de Madrid, Spain

ABSTRACT

In this chapter, we modelize an interchange commerce system basen on the economic concept of utility function. A cognitive agent controls the interchanges of the clients in her market. When interchanges are not possible any more, the agent becomes a client of a higher market, giving place to a hierarchical market system. Now, she behaves according to what she has learned from her clients. Apart from physical resources, intangible goods such as knowledge are also interchanged. This cooperative and competitive structure is formalized via process algebra.

KNOWLEDGE ADQUISITION IN A COOPERATIVE AND COMPETITIVE FRAMEWORK

There are several literature references which deal with interchange of tangible goods (see e.g. López, Núñez, Rodríguez, & Rubio, 2002). However, when the nature of interchangeable goods is more complex, such as knowledge, some considerations must be taken into account. In case of information "interchange", it is compulsory to consider that to supply it does not imply that the initial owner losses it. This is radically different from the treatment of physical resources, since when, for instance, a person exchange a horse for a car, she has not got the horse any more. This difference, which seems simple at first sight, entails great and radical changes when dealing with interchange systems where the changeable goods may include knowledge.

When tackling interchange systems, a crucial point is the process that leads users to make their

DOI: 10.4018/978-1-60960-553-7.ch021

decisions, that is, it is important to know why they exchange a good for another one. As defined in (Wang, & Ruhe, 2007), "decision making is a process that chooses a preferred option or a course of actions from among a set of alternatives on the basis of given criteria or strategies". Obviously, decision making is also a complex issue itself, requiring different types of techniques for the different aspects it requires. In this paper we will concentrate on developing a formal framework to describe the exchange of goods (either material or intangibles) among entities, assuming that each entity can describe its own preferences.

In particular, our approach will be based on using an agent-based system. The reason is that these systems have already proved their usefulness to deal with cognitive environments (see e.g. (Yang, Lin, & Lin, 2006; Vinh, 2009; Uchiya, Maemura, Hara, Sugawara, & Kinoshita, 2009)).

In our work, the concept of utility function is very useful. A utility function returns a real number for each possible basket of goods: The bigger this number is, the happier the owner is with this basket. Intuitively, agents should act by considering the corresponding utility function (see e.g. (Rasmusson & Janson, 1999; Eymann, 2001; Dastani, Jacobs, Jonker, & Treur, 2001; Lang, Torre, & Weydert, 2002; McGeachie & Doyle, 2002; Keppens & Shen, 2002; López, Núñez, Rodríguez, & Rubio, 2002)). Besides, a formal definition of the preferences provides the entity with some negotiation capacity when interacting with other entities (Kraus, 1997; Sandholm, 1998; Lomuscio,

Wooldridge, & Jennings, 2001). Let us remark that, in most cases, utility functions take a very simple form. For instance, they may indicate that an entity E is willing to exchange the item a by the items b and c. Our framework consists of a set of agents performing exchanges of goods. Let us remark that it is not necessary to reduce all the transactions to *money*. In fact, most cognitive transactions are not based on money. Thus, an exchange is made if the involved parties are *happy* with their new goods, where the goods can be either tangibles or intangibles.

Note that, as transactions do not require money, the framework allows a richer structure of exchanges. First, money could be considered as another good, so we do not lose anything. Second, suppose a very simple circular situation where for each $1 \leq i \leq r$, agent A_i owns the good a_i and desires the good $a_{(i \bmod r)+1}$ (see Figure 1). This multi-agent transaction can be easily performed

Figure 1. Exchange of items in the presence of circular dependencies

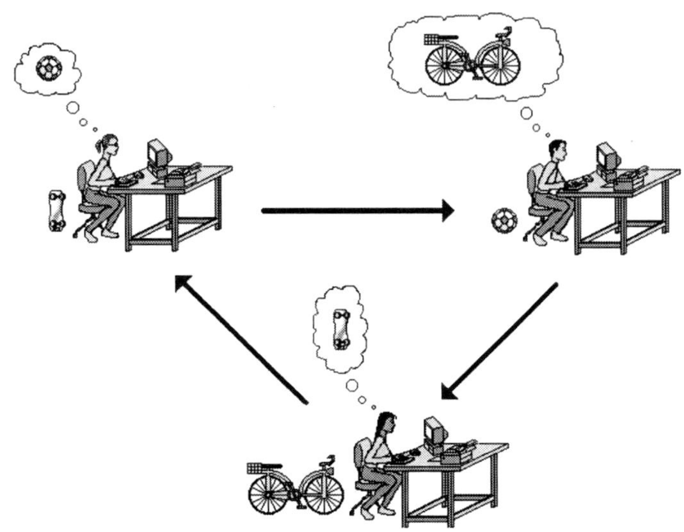

within this framework. On the contrary, it would not be so easy to perform it if these items must be first *converted* into money. In fact, in case items are to be converted into money, the agent who desires the most expensive item would be unable to obtain it. So, the whole exchange will be deadlocked, even though all the agents would get happier performing it. Actually, it could be thought that agents would be able to exchange the items provided that the price of all the items is the same, but in that case we are not really using money: If all the items have the same price, any item can be used as currency unit, and what we obtain is a barter environment where money is not needed. Moreover, in case the goods to be interchanged were intangible, the reduction to money would be quite complex.

The formalization of our system requires the notions of utility function, fair exchange, and equilibrium, borrowed from Microeconomics (see (Mas-Colell, Whinston, & Green, 1995) for a very formal and rigorous presentation of microeconomic theory). Let us suppose a system with k agents where n different tangible products and l intangible assets can be exchanged. Each agent has as information a tuple (The notation Att will be explained in the next section. It represents the possible attributes of an intangible good) $((\overline{x},\overline{\alpha}),u)$, with the pair $(\overline{x},\overline{\alpha}) \in (\mathrm{IR}\times\mathrm{Att})^n\times\mathrm{Att}^l$ and $u:(\mathrm{IR}\times\mathrm{Att})^n\times\mathrm{Att}^l \rightarrow \mathrm{IR}_+$. The first two components of the tuple denote the amount of tangible and intangible resources, respectively, that the entity owns of each good. The third component is the utility function indicating the preferences of the entity with respect to the different goods. That is, $u((\overline{x},\overline{\alpha})) < u((\overline{y},\overline{\beta}))$ denotes that the basket $(\overline{y},\overline{\beta})$ is preferred to the basket $(\overline{x},\overline{\alpha})$.

For the sake of simplicity, in the following example we assume that we are only dealing with material goods. In that case, u(2,3) < u(3,1) means that it is preferred to own 3 units of the first product and 1 unit of the second one than to own 2 and 3 units, respectively, of the corresponding goods. A subset of agents will be willing to exchange resources if none of them decreases its utility and at least one of them improves. These exchanges are called fair. Formally, let us consider $A=\{i_1,\ldots,i_m\} \subseteq \{1,\ldots,m\}$ and the tuples $((\overline{x}_i,\overline{\alpha}_i),u_i)$ for any i∈A. Let us suppose that after the exchange we have that the information associated with the agents belonging to A is given by the tuples $((\overline{y}_i,\overline{\beta}_i),u_i)$. The exchange is fair if for any i∈A we have $u_i((\overline{x}_i,\overline{\alpha}_i)) \leq u_i((\overline{y}_i,\overline{\beta}_i))$ and there exists j∈A such that $u_j((\overline{x}_j,\overline{\alpha}_j)) < u_j((\overline{y}_j,\overline{\beta}_j))$. Let us remark that a necessary condition for an exchange is that no products are created/destroyed, that is, $\sum_{i\in A}\overline{x}_i = \sum_{i\in A}\overline{y}_i$. However, this is not the case when exchanging information between entities (the producer of the info does not forget it). Eventually, the system will reach a situation where no more exchanges can be performed. In other words, it is not possible to improve the situation of one agent without deteriorating another one. Such a situation is called equilibrium (also called Pareto optimum). To determine the equilibria of a system, techniques inherited from game theory can be used (see e.g. (Tennenholtz, 2002; Parsons & Wooldridge, 2002; Stirling, Goodrich, & Packard, 2002)).

Agents are grouped hierarchically according to their localities. First, agents are combined into local markets. Once this market is saturated, that is, when no more exchanges can be performed, an agent representing the interests of all the agents in the market is created. The new agents will be grouped again into markets. This situation is repeated until a global market is created.

This hierarchical structure presents at least two advantages. First, shipping costs are diminished because agents will exchange resources as close to the location of the entity as possible. Second, by creating new (representative) agents once a market is saturated and by combining them into higher order markets, we keep a small number of agents belonging to a certain market. This is very relevant if we take into account that a big number of agents would make very difficult to find the goods that they are looking for. The reason is that the number of messages that agents send to communicate with each other dramatically increases with the number of agents in the market. Finally, let us note that if an agent does not find the good that it is looking for in a local market, there will be a new agent looking for the same good (and taking into account the preferences of the original agent) in a wider market located in a higher level.

Our formalism introduces four characteristics that do not usually appear in other exchange models. First, we consider that users may have to pay a fee depending on the goods that they exchange. Second, shipping costs will be also collected. In order to compute shipping costs we have to take into account not only the goods that the entity receives, but also the distance between the sender and the receiver. Third, we consider the possibility of dealing with public goods. Hence, we can deal with goods that can increase the utility of many users when they share them. Fourth, we also allow to deal with unmaterial goods that can be owned by several users. For instance, a resource of one agent can be its knowledge about a topic. Then, this agent A can perform an exchange with other agent B where A transmits its knowledge to B. However, A does not forget its knowledge. Note that the utility function of A can specify that the utility of its knowledge is bigger when nobody else knows it. Thus, exchanges should only be performed if no agent worsens and, at least, one agent improves.

Let us remark that the characteristics considered in our system make the problem more complicated. For instance, the set of final distributions of goods will be reduced because some fair exchanges will not be performed due to the additional costs. Moreover, we have to adapt the notion of Pareto optimum to our framework. In microeconomics terms, the problem is that we partially lose the notion of contract curve because the induced generalized Edgeworth box shrinks after an exchange. Specifically, money is taken out from the system due to the extra costs. Let us illustrate these issues with a simple example.

Example 1 Let us consider a system with two users and two physical products. In addition, we consider money as the third good. Let us suppose that the initial distributions are (0,3,5) and (2,1,10), respectively, while the corresponding utility functions are defined as $u_1(x_1,x_2,x_3)=30*x_1 + 10*x_2 + x_3$ and $u_2(x_1,x_2,x_3) = 10*x_1 + 10*x_2 + x_3$, respectively. Intuitively, the first user is indifferent between one unit of the first good and three units of the second good. If the first user gives two units of the second good in exchange for one unit of the first good, both users improve. That is, $u_1(0,3,5) < u_1(1,1,5)$ and $u_2(2,1,10) < u_2(1,3,10)$. However, this exchange could be disallowed if we consider transaction and shipping costs. In this case, we would have to decide whether we have both $u_1(0,3,5) \leq u_1(1,1,5-t(1,0)-c_{1,2}(1,0))$ and $u_2(2,1,10) \leq u_2(1,3,10-t(0,2)-c_{1,2}(0,2))$, where t is a function computing transaction costs and $c_{1,2}$ is a function computing shipping costs according to the distance between the users. Besides, in order to have a fair exchange, one of the previous inequalities must be strict. If the exchange is performed, the system will increase its amount of money by t(1,0) + t(0,2) units. Thus, the total amount of money owned by the users is reduced in $t(1,0)+t(0,2)+c_{1,2}(1,0)+c_{1,2}(0,2)$ units. ⊡

The rest of the paper is structured as follows. In the next Section we introduce some auxiliary notation, whereas the bulk of the paper is presented afterwards. First, we give an informal description of the behavior of our exchange systems. Next, we present a formalization of all the necessary concepts to specify our systems. Finally, in the last Section we present our conclusions.

BASIC DEFINITIONS

In this section we introduce some concepts that we will use during the rest of this paper. Specifically, we present the notions of baskets of resources and the notions of utility function, and we explain how operational rules for a process algebra are defined. First we present the notation used for representing the different sorts of resources. In a cognitive system we need to distinguish between *material* (cars, food) and *intangible* (knowledge) resources. On the other hand, any resource can be *owned* by only one entity, or *shared* by more than one entity, or *wanted* by the entity that does not own it, or *none* of the other three cases. The main difference between material and intangible resources is that in the case of intangible assets, once you have them they can be shared or not, but you never lose them. However, in case of material resources, if you exchange all the items that you own of a certain resource, you will have no more.

Definition 2. We consider the set of non negative real numbers $IR_+=\{x \in IR | x \geq 0\}$ to represent the amount of a tangible resources. We define the set of attributes of a resource, denoted by Att, as Att={owned,shared,wanted,none}. We will usually denote tangible resources as *vectors* in $(IR\times Att)^n$ (for $n \geq 2$) by $\bar{x}$, $\bar{y}$ ... Given $\bar{x} \in (IR\times Att)^n$, (x_i,a_i) denotes its i-th component. We define some operations over $(IR\times Att)^n$. Let $\bar{x}=((x_1,a_1),\ldots,(x_n,a_n))$ and $\bar{y}=((y_1,b_1),\ldots,(y_n,b_n)) \in (IR\times Att)^n$ be two tuple of material resources. We define $\bar{x}+_m\bar{y}$ as $\bar{z}=((z_1,c_1),\ldots,(z_n,c_n))$ where $\forall 1 \leq i \leq n$

$$(z_i,c_i)=\begin{cases}(x_i-y_i,shared) & if\ a_i\in\{owned,shared\},\ b_i=wanted,x_i>y_i\\ (0,none) & if\ a_i\in\{owned,shared\},\ b_i=wanted,x_i>y_i\\ (x_i+y_i,shared) & if\ a_i=shared,\ b_i=shared\\ (x_i,a_i) & if\ b_i=none\\ (y_i,b_i) & if\ a_i=none\end{cases}$$

We write $\bar{x} \leq \bar{y}$, if for any $1 \leq i \leq n$, $x_i \leq y_i$.

Considering intangible resources, only attributes are considered. Thus, we usually denote intangible resources as elements in Att^n (for $n \geq 2$) by $\bar{\alpha}$, $\bar{\beta}$, ... Given $\bar{\alpha} \in Att^n$, α_i represents its i-th component. We define the addition over this set. Let $\bar{\alpha}$, $\bar{\beta} \in Att^n$ be two tuple of intangible resources. We define $\bar{\alpha}+_i\bar{\beta}=\bar{\gamma}$ such that

$$\gamma_i=\begin{cases}shared\ if\ \alpha_i\in\{owned,shared\},\beta_i=wanted\\ shared\ if\ \alpha_i=shared,\beta_i=shared\\ \alpha_i\ if\ \beta_i=none\\ \beta_i\ if\ \alpha_i=none\end{cases}$$

Let $(\bar{x},\bar{\alpha}),(\bar{y},\bar{\beta}) \in (IR\times Att)^n\times Att^m$ be two basket of resources, we define $(\bar{x},\bar{\alpha})+(\bar{y},\bar{\beta})$ as the pair of tuple ($\bar{x}+_m\bar{y}$, $\bar{\alpha}+_i\bar{\beta}$).

Finally, we will usually denote matrices in $A^{n\times m}$ (for $n,m \geq 2$, and a set A) by calligraphic letters $\varepsilon, \varepsilon_1, \ldots$

The relevant characteristics of the entities of our system are their *baskets of resources* (indicating the items that they own) and their *utility functions* (indicating preference among different baskets of resources).

Definition 3. Let us suppose $m > 0$ different kinds of material resources and $l > 0$ different kinds of intangible resources. *Baskets of resources* are defined as the elements $(\bar{x},\bar{\alpha})$ of the set $BR = (IR\times Att)^m \times Att^l$. A *utility function* is a function of the function u: $BR \rightarrow IR$. ⊡

In microeconomic theory there are some restrictions that are usually imposed on utility functions (mainly strict monotonicity, convexity, and continuity). Intuitively, given a utility function u we have that $u((\bar{x},\bar{\alpha})) < u((\bar{y},\bar{\beta}))$ means that the basket $(\bar{y},\bar{\beta})$ is preferred to the basket $(\bar{x},\bar{\alpha})$.

Process Algebra

Process Algebras (see (Bergstra, Ponse, & Smolka, 2001) for a good overview on the topic) are formal languages used for the specification and verification of distributed and concurrent systems. As we pointed out in the introduction of this paper, we will use such a language to formalize our systems. The syntax of these languages is given as an EBNF expression. In order to assign meaning to syntactic terms, an operational semantics is usually defined. Operational rules will be defined as usual deduction rules. That is, a rule

$$\frac{Premise_1 \wedge Premise_2 \wedge \ldots \wedge Premise_n}{Conclusion}$$

indicates that if all of the premises hold, then the conclusion can be deduced. Premises indicate individual behavior of components of a system, while conclusions indicate how the system behaves according to individual performances. Let us remark that when a rule has not got any premise, the conclusion trivially holds.

Operational semantics is probably the simplest and more intuitive way to give semantics to any process language. In this part of the description of the language, operational behaviors will be defined by means of transitions $P \xrightarrow{\omega} P'$ that each process can execute. These are obtained in a structured way by applying a set of inference rules (Plotkin, 1981). The intuitive meaning of a transition as $P \xrightarrow{\omega} P'$ is that the process P may perform the action ω and, after this action is performed, then it behaves as P'.

From the sets of transitions of a process we can obtain their computations, which inform us about the behavior of the process in a very natural way:

$$\frac{P_1 \xrightarrow{l_1} P_2, P_2 \xrightarrow{l_2} P_3, \ldots, P_{n-1} \xrightarrow{l_n} P_n}{P_1 \xrightarrow{l_1} P_2 \xrightarrow{l_2} P_3 \xrightarrow{l_3} \ldots \xrightarrow{l_n} P_n}$$

In order to define a process algebra there exists two main decisions that have to be taken:

- The mechanism to model the choice among a set of available actions. Usually,process algebraic languages consider either a (unique) CCS-like choice operator(Milner, 1989) or a pair of choice operators as in CSP (Hoare, 1985). Let us remark that in the majority of the semantic frameworks we have that other operators, such as parallel and hiding, can be derived from the choice operator (some notable exceptions are the π-calculus and true concurrency semantics). Thus, the choice of the choice operator is usually more relevant than other design decisions.
- The semantics to assign meaning to processes. In this case, we can consider testing semantics, bisimulation semantics, trace semantics, etc.

In Figure 2 we show the semantic rules for the most typical operators used defining processes algebra. In the following, we will describe briefly some of these rules. We suppose a fixed set of visible actions Act (a, a',... to range over Act). We assume the existence of a special action $\tau \notin$ Act, which represents internal behaviour. We denote by Act_τ the set $\text{Act} \cup \{\tau\}$ (α, α',... to range over Act_τ). Finally, Id_P represents the set of process variables.

In the definition of processes, we will usually omit trailing occurrences of STOP, that denotes the process that cannot execute any action. Considering CCS-like languages, the external and internal choice operators are used to describe the choice among different actions. They are respectively denoted by

$$\sum a_i ; P_i \qquad \sum \tau ; P_i$$

where $a_i \in$ Act. The inference rules describing the behavior of these process relations are (CHO1), and (CHO2).

Figure 2. Operational semantics of the process algebra

$$\text{(CHO1)}\frac{}{\sum a_i;P_i \xrightarrow{a_i} P_i} \qquad \text{(CHO2)}\frac{}{\sum \tau;P_i \xrightarrow{\tau} P_i}$$

$$\text{(SEQ1)}\frac{P \xrightarrow{\alpha} P'}{P;Q \xrightarrow{\alpha} P';Q} \qquad \text{(SEQ2)}\frac{P \xrightarrow{\surd} P'}{P;Q \xrightarrow{\tau} Q} \qquad \text{(SEQ3)}\frac{}{\text{exit} \xrightarrow{\surd} \text{STOP}}$$

$$\text{(REC)}\frac{P[X/X:=P] \xrightarrow{\alpha} P'}{X:=P \xrightarrow{\alpha} P'}$$

$$\text{(PAR1)}\frac{P \xrightarrow{\alpha} P'}{P|||Q \xrightarrow{\alpha} P'|||Q} \qquad \text{(PAR2)}\frac{Q \xrightarrow{\alpha} Q'}{P|||Q \xrightarrow{\alpha} P|||Q'} \qquad \text{(PAR3)}\frac{P \xrightarrow{a} P',\ Q \xrightarrow{b} Q',\ a*b \neq \tau}{P|||Q \xrightarrow{a*b} P'|||Q'}$$

$$\text{(CON1)}\frac{P \xrightarrow{\alpha} P'}{P||Q \xrightarrow{\alpha} P'||Q} \qquad \text{(CON2)}\frac{Q \xrightarrow{\alpha} Q'}{P||Q \xrightarrow{\alpha} P||Q'}$$

$$\text{(CON3)}\frac{P \xrightarrow{a} P',\ Q \xrightarrow{\overline{a}} Q'}{P||Q \xrightarrow{\tau} P'||Q'} \qquad \text{(CON4)}\frac{P \xrightarrow{\overline{a}} P',\ Q \xrightarrow{a} Q'}{P||Q \xrightarrow{\tau} P'||Q'}$$

The process $\sum a_i; P_i$ will perform one of the actions a_i and after that it behaves as P_i. The term $\sum \tau; P_i$ represents the internal choice among the processes P_i. Once the choice is made, by performing an internal action τ, the process behaves as the chosen process.

Sequence is an operator in which two processes are consequently performed. It is denoted by:

$P ; Q$

The rules (SEQ1) and (SEQ2) describe the behavior of these processes. Intuitively, *P* is initially performed. Once *P* finishes, *Q* starts its performance. If the process *P* can perform an action then *P;Q* will perform it. If the process *P* finishes then the process *P;Q* will behave as *Q*. A process will finish its execution if it performs the action √ (see rule (SEQ2)).

Recursion is an operator in which the definition of a process may contain a call to itself. The most typical notation for recursion is

X:= P

where $X \in Id_P$, that is, a process identifier. The rule (REC) applies to external and internal actions. Let us also remark that *P[X/X:= P]* represents the substitution of all the free occurrences of *X* in *P* by *X:= P*.

The parallel and concurrence operators are denoted, respectively, by:

$P \;|||\; Q \qquad P \;||\; Q$

Parallel is an operator between two processes, in which the two processes are executed simultaneously, synchronized by a common system clock. The parallel process relation is designed to model behaviors of a multi-processor single-clock system. See rules (PAR1) and (PAR2) for describing the possibility of both processes to perform an action separately. We suppose that there is an operation * on the set of actions Act such that (Act,*) is a monoid and τ is its identity element. Thus, by rule (PAR3), if we have the parallel composition of *P* and *Q*, *P* may perform a, *Q* may perform b, and a * b ≠ τ then they will evolve together.

Concurrence is a process relation in which two processes are simultaneously and asynchronously executed, according to separate system clocks. The rules (CON1) and (CON2) indicate that if one of the processes of the composition can perform an action then the composition will asynchronously perform it. However, if one of the processes of the composition can perform an action and the other can perform the complementary action then there is a communication and the process relation will perform it (see rules (CON3) and (CON4)).

Although the concurrent operator is the usual one in CCS, in this chapter we will use the parallel one instead of it. Since markets must be executed in parallel and there is no communication among them, a multiprocessor system is needed. This behavior is modeled by the parallel operator.

FORMALIZING THE EXCHANGE ARQUITECTURE

In this section we present the core of our work. First, we explain intuitively the basic exchange system's algorithm. Next, we introduce the formal framework for the definition of the exchange systems.

Exchange System's Algorithm

The exchange agents need two data: Their *basket of resources* and their *utility function*. The utility function relates the preference that the entity has for the owned goods with respect to the desired goods. Besides, once an agent has reached a (possibly multilateral) deal, it must be notified, the deal will be effectively performed, transaction fees will be added and shipping costs will be computed according to both the amount of received items and to the distance between the involved entities.

From now on we concentrate on the behavior of the different entities. The behavior of an exchange system works according to the following algorithm (see its data flow diagram in Figure 3):

1. Each agent generates the barters that it would be willing to perform (according to the corresponding basket of resources and utility functions).
2. Agents exchange material and intangible goods inside their local market. A multilateral exchange will be made if (at least) one of the involved agents improves its utility and none of them decreases its utility. This is repeated until no more exchanges are possible. In this case, we say that the local market is *saturated.*
3. Once a market is saturated, their agents are combined to create a new agent. The new agent will have as basket of resources the addition of the corresponding to each agent. Its utility function will encode the utilities of the combined agents. Let us remark that this new agent behaves as a representative of the combined agents. *First order* agents will

Figure 3. Exchange system's algorithm

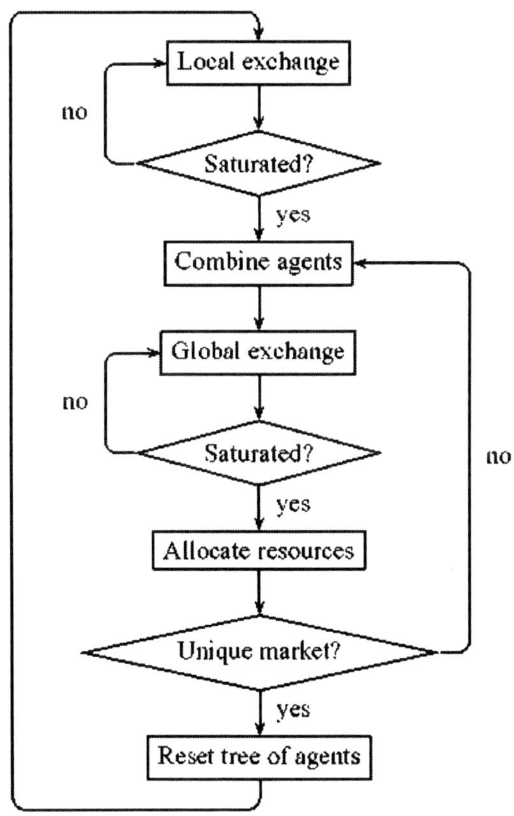

be combined again into markets, according to *proximity* reasons.

4. Higher order agents trade between them until their market is saturated.
5. Once a (higher order) market is saturated, the agents start to allocate the resources in a top-down way through the tree of markets until the resources arrive to the leaves of the tree (i.e. the *original* agents). Then, they create a new agent (as indicated in step 3).
6. Once their markets are saturated, new markets are created by combining agents until there exists a unique market. Once this market is saturated, and the resources are conveniently allocated, the whole tree of agents is reset, and we start again at the first step.

The previous algorithm ensures some good properties:

- Exchanges are made between agents located as near as possible. That is, we try to minimize possible shipping costs.
- Partial equilibria are reached in each market. That is, once a market is saturated we may assure that one (of the possible) Pareto optimum distribution of resources has been found. In other words, agents belonging to a saturated market cannot improve their utility within that market without decreasing the corresponding to another agent.
- Once the last (unique) market is saturated we may assure that one (of the possible) global equilibrium has been reached. That is, no more exchanges can be performed (according to the current utility functions and available resources).

Finally, let us comment on the advantages of using partial equilibria versus global equilibria. If we tried to reach a global equilibria, we should perform exchanges until no more exchanges are possible. Once all this (possibly) enormous amount of exchanges has been performed then the resources can be sent to their new owners. This would strongly delay some *trivial* transactions between not very distant agents (and their corresponding entities).

Formal Framework

In this section we provide a formal syntax and semantics for the definition of exchange systems. Even though we use a process algebraic notation (mainly when defining the operational rules) we do not need most of the usual operators for this kind of languages (choice, restriction, etc). In fact, our constructions remind a parallel operator as the one presented in the previous section.

Definition 4. A *market system* is given by the following EBNF:

$MS ::= ms(M)$

$M ::= A \,|\, \mathrm{unsat}((M,\ldots,M),sh,pr) \,|\, \sigma(M)$

$A ::= (S,u,(\bar{x},\bar{\alpha}),sh,pr)$

$S ::= [\,] \,|\, [A,\ldots,A]$

We will explain intuitively each term given above. First, in order to avoid ambiguity of the grammar we annotate market systems with the nonterminal symbol ms. M describes a market in three possible situations. In the first case, M=A, that is, M is the tupla $(S,u,(\bar{x},\bar{\alpha}),sh,pr)$. In this case, M represents a *saturated* market, that is, a market where no more exchanges can be performed among its agents. Regarding the first argument of M, there are two possible situations. Either S is an empty list or not. In the first case, M represents an *original* agent, that is, a direct representative of an entity (note that a single agent is trivially saturated). In the second case, if $S=[A_1,\ldots,A_n]$, then M represents an agent associated with the (possible higher order) agents $A_1,\ldots,A_n$ belonging to a saturated market. u denotes the utility function of M, $\bar{x}$ represents the basket of material resources owned by M, and $\bar{\alpha}$ the basket of intangible resources. We consider that there are m different goods[2], that is $\bar{x} \in (IR\times Att)^m$, and that the amount of *money* is placed in the last component of the tuple. We consider that there are l different intangible assets that the cognitive system is working on, that is, $\bar{\alpha} \in Att^l$, and some of them are only known by M and others are shared by more than one entity. In case of an exchange where intangible assets are involved, as the agent is not able to transmit the information, a meeting between entities is needed. Besides, sh is the *shipping function* indicating the shipping cost of each possible transaction in this market. Regarding intangible resources, the shipping cost will be computed as the amount of money needed to join the entities for sharing the information. In turn, pr is the profit collected by the market due to transaction costs. We will assume the existence of another function which will be common to all of the markets, the *transaction function*, denoted by tr. The function tr computes the transaction costs for each of the agents involved in an exchange by taking into account the goods that this agent receives. Let us remark that while the shipping costs will depend on the market in which the transaction is performed, the transaction costs will not. By doing so we can formally specify that shipping costs increase with the distance between entities.

The second possible situation of a market M is to be of the form $M=unsat((M_1,\ldots,M_n), sh,pr)$. It represents an *unsaturated* market consisting of the markets $M_1,\ldots,M_n$, the shipping function sh and the profit value pr. Let us remark that in this case some of the submarkets may be saturated. Once all the markets of the system are saturated, the whole system is turned again into unsaturated. The term $\sigma(M)$ will represent that such operation must be performed on M, and it will represent the third and last possible situation of a market.

Next, we present an example showing how an exchange system may be constructed. In this example we will also (informally) introduce operational transitions of the language.

Example 5. Let us consider a total of six agents $A_i=([],u_i,(\bar{x}_i,\bar{\alpha}_i),sh_0,0)$, for $1\leq i\leq 6$, where sh_0 denotes a dummy shipping function. We suppose that these agents are grouped into three different markets which are initially unsaturated, so we make the following definitions:

$$M_1 = unsat((A_1,A_2), sh_1, 0)$$

$$M_2 = unsat((A_3,A_4), sh_2, 0)$$

$$M_3 = unsat((A_5,A_6), sh_3, 0)$$

Suppose that the first two markets are linked, and the resulting market is also linked with the remaining M_3 We should add the following definitions:

$$M_4 = unsat((M_1,M_2), sh_4, 0)$$

$$M_5 = unsat((M_4,M_3), sh_5, 0)$$

Finally, the global market is defined as $M = ms(M_5)$.

Following the philosophy explained in the previous section, transactions will be made within a market only between saturated submarkets. So, only M_1, M_2, and M_3 are allowed to perform transactions (note that original agents are trivially saturated).

We will denote exchange of resources by $\rightsquigarrow$. Suppose that after some exchanges, M_1 becomes saturated. That is, there exists a sequence of exchanges $M_1 \rightsquigarrow M_1^1 \rightsquigarrow M_1^2 \ldots \rightsquigarrow M_1^n = M_1'$ such that $M_1' \not\rightsquigarrow$, that is, no more exchanges can be performed. In this case, the market grouping the first two agents should be labeled as saturated. So, the agents A_1 and A_2 effectively perform all the achieved transactions becoming A_1' and A_2', respectively. Then, the first market will be turned into ([A_1',A_2'],f(u_1,u_2),($\overline{x_1 +_m x_2 +_m \cos ts_1}$, $\overline{\alpha}_1 +_i \overline{\alpha}_2$),$sh_1$,$pr_1$), where f is a function combining utility functions (such a function will be formally defined), pr_1 is the amount of money that the system has obtained due to the fees applied to the exchange of goods, and $\overline{\cos ts_1}$ denotes[3] the transaction and shipping costs associated with the exchanged goods (also intangible resources) in the market M_1. In parallel, M_2 will have a similar behavior.

Once both M_1 and M_2 are saturated, the transactions between them will be allowed. Note that these transactions (inside the market M_4) will be performed according to the new utility functions, f(u_1,u_2) and f(u_3,u_4) respectively, and to the new baskets of goods from M_1 ($\overline{x_1 +_m x_2 +_m \cos ts_1}$, $\overline{\alpha}_1 +_i \overline{\alpha}_2$) and from M_2 ($\overline{x_3 +_m x_4 +_m \cos ts_2}$, $\overline{\alpha}_3 +_i \overline{\alpha}_4$), respectively.

The process will iterate until M_5 gets saturated, then all transactions are performed and M_5 becoming M_5'. Finally, we will have a market as σ(M_5'). Then, the global market is structurally *reset* to start again. ⊡

In order to simplify forthcoming operational rules we introduce the following notation to deal with utility functions. Utility functions associated with original agents (that is, A=([],u,$(\overline{x},\overline{\alpha})$,sh,0), where sh is useless) will behave as explained before, we have that $u((\overline{z},\overline{\gamma})) < u((\overline{y},\overline{\beta}))$ means that the basket $(\overline{y},\overline{\beta})$ is preferred to the basket $(\overline{z},\overline{\gamma})$. That is, $u((\overline{z},\overline{\gamma}))$ indicates the relative preference shown by A towards the basket of resources $(\overline{z},\overline{\gamma})$. Nevertheless, if A is the agent A=([A_1,…,A_n],u,$(\overline{x},\overline{\alpha})$,sh,pr) associated with the agents A_1,…,A_n then we will consider that in addition to its usual meaning, the utility function also keeps track of how a basket of resources is distributed among the (possible higher order) agents A_1,…,A_n. That is, $u((\overline{z},\overline{\gamma}))$=(r, $(\overline{z}_1,\overline{\gamma}_1)$,…, $(\overline{z}_n,\overline{\gamma}_n)$), where r still represents the utility, while $\sum_m \overline{z}_i = \overline{z}$ and $\overline{z}_i$ denotes the portion of the basket $\overline{z}$ assigned to A_i, and $\sum_m \overline{\gamma}_i = \overline{\gamma}$ and $\overline{\gamma}_i$ denotes the portion of the basket $\overline{\gamma}$ assigned to A_i. Overloading the notation, if we simply write $u((\overline{z},\overline{\gamma}))$ we are referring to the first component of the tuple, while $u((\overline{z},\overline{\gamma}))$.i denotes the (i + 1)-th component of the tuple, that is, $(\overline{z}_i,\overline{\gamma}_i)$.

In the following definitions of this section we will present the rules of the operational semantics of the exchange system. As we are working in a hierarchical system, the operational semantics is presented following this idea. On the first two rules the operational semantics of the agents will be given, that is, the exchanges that the agents can perform. The three following rules will present the operational behaviour for unsaturated markets, that is, the evolution of an unsaturated market when some exchanges can be made. After that, the following three rules will represent the evolution of an unsaturated market into a saturated one. Finally, the last five rules will present the reset of the global market.

In the next definition we present the *anchor case* of our operational semantics. In order to perform complex exchanges, agents should first indicate the barters they are willing to accept.

Definition 6. Let A=(S,u,$(\overline{x},\overline{\alpha})$,sh,pr) be a saturated market. The *exchanges* the agent A would perform are given by the following operational rules:

$$\frac{u\left(\left(\overline{x},\overline{\alpha}\right)+\left(\overline{y},\overline{\beta}\right)\right)\geq u\left(\left(\overline{x},\overline{\alpha}\right)\right)\wedge\left(\overline{x},\overline{y}\right)\geq\overline{0}}{\left(S,u,\left(\overline{x},\overline{\alpha}\right),sh,pr\right)\xrightarrow{(\overline{y},\overline{\beta})}\left(S,u,\left(\overline{x},\overline{\alpha}\right)+\left(\overline{y},\overline{\beta}\right),sh,pr\right)}$$

$$\frac{u\left(\left(\overline{x},\overline{\alpha}\right)+\left(\overline{y},\overline{\beta}\right)\right)>u\left(\left(\overline{x},\overline{\alpha}\right)\right)\wedge\left(\overline{x}+\overline{y}\right)\geq\overline{0}}{\left(S,u,\left(\overline{x},\overline{\alpha}\right),sh,pr\right)\overset{(\overline{y},\overline{\beta})}{\longmapsto}\left(S,u,\left(\overline{x},\overline{\alpha}\right)+\left(\overline{y},\overline{\beta}\right),sh,pr\right)}$$

where $(\overline{y},\overline{\beta}) \in$ BR. ⊡

Let us remark that $(\overline{y},\overline{\beta})$ will contain the barters offered by the agent. For example, considering no intangible exchanges, that is, $\overline{\beta}$ =

$\overline{none}$, if $\bar{y}$ = ((1,shared),(1,wanted),(0,none),(3,wanted)) fulfills the premise, then the agent would accept a barter where it is offered one unit of the first good in exchange of one unit of the second good and three units of money. Regarding the rules, the first premise simply indicates that the agent would not decrease (resp. would increase) its utility. The second premise indicates that the agent does not run into *red numbers*, that is, an agent cannot offer a quantity of an item if it does not own enough. Thus, a transition as $\longrightarrow$ denotes that the market does not worsen, meanwhile a transition $\longmapsto$ denotes that the market does improve. Finally, let us comment that even though transaction and shipping costs do not explicitly appear in the previous rules, they are implicitly reflected in the last component of the corresponding tuples $(\bar{y}, \bar{\beta})$. We will later formalize how these costs are assigned to the *owner* of the system and to the shipping entities.

Next we will show how the offers made by the agents are combined in an unsaturated market.

Definition 7. Let M be the unsaturated market given by M=unsat((M_1,...,M_n),sh, pr). Let I={s_1,...,s_r}⊆{1,...,n} be a set of indexes denoting the saturated markets belonging to M (that is, for any i∈I, we have M_i=(S_i,u_i,$(\bar{x}_i, \bar{\alpha}_i)$,$sh_i$,$pr_i$)). We say that the matrix $\varepsilon \in (BR)_{n\times n}$ is a *valid exchange matrix for* M *under the cost tuple* $\bar{c}$ (see footnote 2), denoted by valid(M,E,$\bar{c}$), if for any $1 \le i \le n$ we have $\sum_j \varepsilon_{ij} \preceq (\bar{x}_i - \bar{c}_i, \bar{\alpha}_i)$, $\varepsilon_{ii}=(\bar{0}, \overline{none})$, and $\forall 1 \le k \le n$ such that $k \notin I$, $\varepsilon_{ki} = (\bar{0}, \overline{none})$ and $\varepsilon_{ik} = (\bar{0}, \overline{none})$. ☐

First, let us remark that the notion of *valid* matrix is considered only in the context of unsaturated markets: If a market is saturated then no more exchanges can be performed. Second, only saturated markets belonging to an unsaturated market may perform exchanges among them. This restriction is imposed in order to give priority to transactions performed by *closer* agents belonging to unsaturated submarkets. Regarding the definition of *valid* matrix, let us note that matrixes ε have as components baskets of resources (that is, elements belonging to BR). ε_{ij} represents the basket of resources that the market M_i would give to M_j. In the tuple $\bar{c}$, c_i denotes the transaction and shipping cost that agent M_i will have to afford. So, the condition $\sum_j \varepsilon_{ij} \preceq (\bar{x}_i - \bar{c}_i, \bar{\alpha}_i)$ indicates that the total amount of resources given by market M_i must be less than or equal to the basket of resources owned by that market minus the money paid by the transaction. The second clause $\varepsilon_{ii}=(\bar{0}, \overline{none})$ for any market M_i is considered because internal exchanges are not considered in this matrix. Finally, the last rule represents that exchanges are only considered between saturated markets, and even more an exchange does not need to include all of the saturated markets. For example, if only r′ markets participate in an exchange, then the rows and columns corresponding to the remaining r−r′ saturated markets will be filled with $(\bar{0}, \overline{none})$, as they are for the unsaturated markets.

Once the notion of valid exchange matrix is given, we introduce the rules defining the exchange of resources in unsaturated markets. Intuitively, if we have a valid exchange matrix, where (at least) one of the involved agents improves and no one worsens, then the corresponding exchange will be performed.

Definition 8. Let M be the unsaturated market given by M=unsat((M_1,...,M_n),sh, pr). Let I={s_1,...,s_r}⊆{1,...,n} be a set of indexes denoting the saturated markets belonging to M (that is, for any i∈I we have M_i=(S_i,u_i,$(\bar{x}_i, \bar{\alpha}_i)$, sh_i,pr_i)). The operational transitions denoting exchanges of resources that M may perform are given by the rule shown in Figure 4.

We say that M is a *local equilibrium*, denoted by M $\not\rightsquigarrow$, if there do not exist M′ and ε such that M $\overset{\varepsilon}{\rightsquigarrow}$ M′. ☐

The operational rule shown in Figure 4 is applied under the same conditions appearing in the definition of a valid exchange matrix: It is applied to unsaturated markets and the exchange is made among a subset of the saturated submarkets. The premises indicate that (at least) an unsaturated market will improve after the exchange and that no one deteriorates. Let us remind that, in general, a market may generate both $M_i \xrightarrow{(\overline{y},\overline{\beta})} M_i'$ and $M_i \overset{(\overline{y},\overline{\beta})}{\longmapsto} M_i'$. So, the previous rule also considers situations where more than a market improves (we only require that at least one improves). Besides, let us remark that $M_i \xrightarrow{(\overline{0,none})} M_i'$ always holds. So, a market not involved in the current exchange does not disallow the exchange. The costs required to have a valid exchange matrix will be computed both from the transaction and shipping costs.

Regarding the conclusion of the rule, submarkets belonging to M are modified according to both the corresponding exchange matrix and the costs of the exchange, while unsaturated submarkets do not change. Let us remark that the costs of each exchange will be paid by the receiver. Besides, only the transaction costs will be added to the cumulated profit of the market. The following result follows from the previous definition. It indicates that exchanges allowed by the previous rule are fair.

Proposition 9. Let M be the unsaturated market given by M=unsat(($M_1,\ldots,M_n$),sh, pr). Let $I=\{s_1,\ldots,s_r\}\subseteq\{1,\ldots,n\}$ be a set of indexes denoting the saturated markets belonging to M (that is, for any i∈I we have M_i=(S_i,u_i,$(\overline{x}_i,\overline{\alpha}_i)$, sh_i,pr_i)). Let us suppose that it is possible to perform the transition unsat(($M_1,\ldots,M_n$),sh,pr) $\overset{\varepsilon}{\rightsquigarrow}$ unsat(($M_1',\ldots,M_n'$),sh,pr'), where for all i∈I we have M_i'=(S_i,u_i,$(\overline{x}_i',\overline{\alpha}_i')$,$sh_i$,$pr_i$)). Then for any i∈I we have $u_i((\overline{x}_i,\overline{\alpha}_i))\leq u_i((\overline{x}_i',\overline{\alpha}_i'))$ and there exists j∈I such that $u_j((\overline{x}_i,\overline{\alpha}_i))\leq u_j((\overline{x}_i',\overline{\alpha}_i'))$. ⊡

We need to consider two more exchanging rules to record this transformation of a market with a valid exchange matrix in any other context (see Box 1).

The first rule indicates that if an unsaturated submarket produces an exchange, then the market must take that situation into account. The second one reflects modifications in the environment of the constructor ms.

If a market reaches an equilibrium then we need to modify the attribute of the market, replacing a term of the form unsat(($M_1,\ldots,M_n$),sh,pr) by a term as (S,u,$(\overline{x},\overline{\alpha})$,sh,pr'). Once a market is saturated, the money collected in the different submarkets as transaction costs will be transferred to it. In addition, material resources are recursively moved from the corresponding agents to the leaves of the tree (indicating entities). Besides, all the meetings needed for sharing the intangible assets have to be done. Let us remark that a necessary condition for a market to be saturated is that all of its submarkets are also saturated. The

Figure 4. Operational rule for the exchange of resources in an unsaturated market

$$\frac{\exists k \in I : M_k \overset{(\overline{y_k},\overline{\beta_k})}{\longmapsto} M_k' \wedge \forall i \in I, M_i \overset{(\overline{y_i},\overline{\beta_i})}{\longrightarrow} M_i' \wedge \mathtt{valid}(M,\mathcal{E},(cost_1,\ldots,cost_n))}{M \overset{\mathcal{E}}{\rightsquigarrow} \mathtt{unsat}((M_1',\ldots,M_n'),sh,pr+\sum_i tr(\sum_j \mathcal{E}_{ji}))} \left[M_i' = \begin{cases} M_i & i \notin I \\ (S_i,u_i,(\overline{x_i},\overline{\alpha_i})+(\overline{y_i},\overline{\beta_i}),sh_i,pr_i) & \text{otherwise} \end{cases} \right]$$

where $\mathcal{E} \in (\mathtt{BR})^{n\times n}$

$$(\overline{y_i},\overline{\beta}) = \sum_j \mathcal{E}_{ji} + \sum_j \mathcal{E}_{ij} + (((0,\text{none}),\ldots,(0,\text{none}),(cost_i,\mathtt{wanted})),\overline{\text{none}})$$

$$cost_i = tr(\sum_j \mathcal{E}_{ji}) + sh(\sum_j \mathcal{E}_{ji})$$

Box 1.

$$\frac{M_k \overset{\varepsilon}{\rightsquigarrow} M'_k}{unsat\left(\left(M_1,\ldots,M_k,\ldots,M_n\right),sh,pr\right) \overset{\varepsilon}{\rightsquigarrow} unsat\left(\left(M_1,\ldots,M'_k,\ldots,M_n\right),sh,pr\right)}$$

$$\frac{M \overset{\varepsilon}{\rightsquigarrow} M'}{ms\left(M\right) \overset{\varepsilon}{\rightsquigarrow} ms\left(M'\right)}$$

following rule uses two auxiliary notions that will be formally presented in the forthcoming Definition 11.

Definition 10. Let M be the unsaturated market given by M=unsat(($M_1,\ldots,M_n$),sh, pr), where we have that M_i=(S_i,u_i, $(\bar{x}_i,\bar{\alpha}_i)$,sh_i,pr_i) for any 1≤i≤n. The following rule modifies the market from unsaturated to saturated:

$$\frac{M \not\rightsquigarrow}{M \rightsquigarrow \left(\left[M'_1,\ldots,M'_n\right],u,\sum\left(\bar{x}_i,\bar{\alpha}_i\right),sh,pr+\sum_i pr_i\right)}$$

where u=CreateUtility($u_1,\ldots,u_n$, $(\bar{x}_1,\bar{\alpha}_1)$,…, $(\bar{x}_n,\bar{\alpha}_n)$) and for any 1≤i≤n we have that M_i'=(S_i',u_i, $(\bar{x}_i,\bar{\alpha}_i)$,sh_i,0)) and S_i'= Deliver(S_i,u_i, $(\bar{x}_i,\bar{\alpha}_i)$). □

Let us remark that in this rule we do not label the transition. These transitions play a role similar to internal transitions in classical process algebras.

We need to add two more rules, as in the previous case, to record this transformation in the context of different constructors (see Box 2).

Next we present the pending functions. Intuitively, the function Deliver(S,u, $(\bar{x},\bar{\alpha})$) distributes the basket of goods $(\bar{x},\bar{\alpha})$ among the original agents appearing in the leaves of the tree S. This distribution considers both the utility functions of the agents and the quantities of goods contributed by each of the agents. In addition to that, the function CreateUtility ($u_1,\ldots,u_n$, $(\bar{x}_1,\bar{\alpha}_1)$,…, $(\bar{x}_n,\bar{\alpha}_n)$) computes a combined utility function from the ones provided as arguments, so that it is possible to negotiate for maximizing the overall profit of the represented agents. Let us remind that, in this section, utility functions associated with higher order agents do not only reveal preference, but also take into account how resources will be distributed among agents. Thus, if we are considering an agent representing n agents, a new utility function returning a tuple with n+1 com-

Box 2.

$$\frac{M_k \rightsquigarrow M'_k}{unsat\left(\left(M_1,\ldots,M_k,\ldots,M_n\right),sh,pr\right) \rightsquigarrow unsat\left(\left(M_1,\ldots,M'_k,\ldots,M_n\right),sh,pr\right)}$$

$$\frac{M \rightsquigarrow M'}{ms\left(M\right) \rightsquigarrow ms\left(M'\right)}$$

Box 3.

$$Deliver\left(S,u,\left(\overline{x},\overline{\alpha}\right)\right)=\begin{cases}\left[\;\right] & if\ S=\left[\;\right]\\ \left[M_1',\ldots,M_n'\right] & if\ S=\left[M_1,\ldots,M_n\right]\end{cases}$$ where for any $1 \le i \le n$ we have:

$M_i=(S_i,u_i,(\overline{x}_i,\overline{\alpha}_i),sh_i,pr_i)$,

$M_i'=(S_i',u_i,(\overline{x}_i,\overline{\alpha}_i),sh_i,pr_i)$, and

$S_i'=$ Deliver$(S_i,u_i,u(\overline{x}_i,\overline{\alpha}_i).i)$.

ponents will be created. The first component (the value of the utility function) will return the worst utility (0) if any of the represented agents worsens. In this way, it is guaranteed that the market does not make any exchange which deteriorates any of its clients. Otherwise, the value of the utility will be the addition of individual utilities in the distribution of resources which maximizes this value. Actually, this optimal distribution will be used to obtain the other n components of the tuple.

Definition 11. Let A=(S,u,$(\overline{x},\overline{\alpha})$,sh,pr) be an agent. The *allocation* of the basket of resources $(\overline{x},\overline{\alpha})$ among the agents belonging to S with respect to the utility function u, denoted by Deliver(S,u,$(\overline{x},\overline{\alpha})$), is recursively defined as in Box 3.

Given n tuples $(u_i,(\overline{x}_i,\overline{\alpha}_i))$ the *utility function* is defined from the utility functions $u_1,\ldots,u_n$ with respect to the *initial* baskets of resources $(\overline{x}_1,\overline{\alpha}_1)$ $,\ldots,(\overline{x}_n,\overline{\alpha}_n)$, and we denote it by using CreateUtility($u_1,\ldots,u_n$, $(\overline{x}_1,\overline{\alpha}_1),\ldots,(\overline{x}_n,\overline{\alpha}_n)$), as u_{market}, where

$$u_{market}((\overline{x},\overline{\alpha}))=\max\{(r,(\overline{x}_1,\overline{\alpha}_1),\ldots,(\overline{x}_n,\overline{\alpha}_n))\}$$

such that we have

$$r=\sum_{1\le i\le n}u_i((\overline{x}_i',\overline{\alpha}_i'))\wedge\sum_{1\le i\le n}((\overline{x}_i',\overline{\alpha}_i'))=(\overline{x},\overline{\alpha})\wedge$$
$$\wedge_{1\le i\le n}u((\overline{x}_i',\overline{\alpha}_i'))\ge u((\overline{x}_i,\overline{\alpha}_i))$$

maximizing over the first argument (representing the *utility*), and assumming that max(∅)=(0,(($\overline{0}$, *none*),…, ($\overline{0}$, *none*), *none*)). ⊡

In order to define how a market evolves, we need to be able to compose sequences of transitions, as shown in the definition below.

Definition 12. We say that a market M *evolves into* a market M′, and we write M ⇝* M′, if there exist markets $M_1,\ldots,M_{n-1}$ such that

$$M\overset{a_1}{\rightsquigarrow}M_1\overset{a_2}{\rightsquigarrow}M_2\overset{a_3}{\rightsquigarrow}\ldots M_{n-1}\overset{a_n}{\rightsquigarrow}M'$$

where for any $1 \le i \le n$ we have that a_i is an empty label or an exchange matrix. ⊡

Finally, we provide a mechanism to reset a global market. If the root of the tree becomes a saturated market, then the whole tree of markets is created again. This is done by considering the five rules in Figure 5. The first one defines how it is launched the process to turn all the markets back to unsaturated mode, provided that the global market has become saturated. In this case, the transition is labelled by the global amount of money collected by the whole market system as transaction fees. The other rules define how to recursively reset the tree from the root to the leaves (original entity).

As mentioned above, the addition of transaction and shipping costs does not allow to properly speak about Pareto optimums. First, as the following result states, if we set these costs to zero then we obtain that the last saturated market represents one of the possible Pareto optimums for the whole set of original agents (regardless their locality in the market structure).

Theorem 13. Let M, M′ be markets and A be an agent such that

$$M \rightsquigarrow^{*} M' \hookrightarrow ms(\sigma(A))$$

If transition and shipping costs are set to zero then we have that the distribution of resources provided by σ(A) represents a Pareto optimum with respect to the original agents belonging to M.

Proof Sketch: If the final situation were not a Pareto equilibrium then there would exist at least one more fair exchange. Nevertheless, according to the operational semantics, this exchange would have been performed in the lowest market which embraces all the agents involved in this hypothetical exchange, which makes a contradiction. ⊡

Regarding the general situation of our system where the transaction and shipping costs are greater than zero, the classical Pareto equilibrium concept does not apply. For instance, a desirable exchange in a free or low transaction fee environment could not be a desirable exchange in an expensive one. Then, the equilibriums are different because of the effect of the fees. This framework cannot be defined in terms of one more agent representing the effect of the market fees, because this agent would have the special ability to enable and disable some exchanges. Therefore, we need a more general concept of equilibrium.

Definition 14. A (t,s)-Pareto equilibrium is a distribution of resources in which no more fair exchanges are possible according to the transaction costs function t and the shipping costs function s. ⊡

The following result relates (t,s)-Pareto equilibria and sequences of transitions. The proof is similar to the one for Theorem 13.

Theorem 15. Let M, M′ be markets and A be an agent such that

$$M \rightsquigarrow^{*} M' \hookrightarrow ms(\sigma(A))$$

Then σ(A) represents a (t,s)-Pareto equilibrium with respect to the original agents belonging to M. ⊡

LEARNING THEORIES

In previous sections we have described how agents can negotiate among themselves following the principles of the utility functions they have been provided with. However, we have not analyzed the other important aspect: How the

Figure 5. Reseting a global market

$$\frac{}{ms((S,u,(\overline{x},\overline{\alpha}),sh,pr)) \Rightarrow_{pr} ms(\sigma((S,u,(\overline{x},\overline{\alpha}),sh,0)))} \qquad \frac{}{\sigma(([\,],u,(\overline{x},\overline{\alpha}),sh,pr)) \hookrightarrow ([\,],u,(\overline{x},\overline{\alpha}),sh,pr)}$$

$$\frac{S = [A_1,\ldots,A_n]}{\sigma((S,u,(\overline{x},\overline{\alpha}),sh,pr)) \hookrightarrow \mathtt{unsat}((\sigma(A_1),\ldots,\sigma(A_n)),sh,pr)}$$

$$\frac{M_k \hookrightarrow M'_k}{\mathtt{unsat}((M_1,\ldots,M_k,\ldots,M_n),sh,pr) \hookrightarrow \mathtt{unsat}((M_1,\ldots,M'_k,\ldots,M_n),sh,pr)} \qquad \frac{M \hookrightarrow M'}{ms(M) \hookrightarrow ms(M')}$$

utility functions are obtained? In this sense, we have to emphasize that it is not only important that each agent knows the utility function associated with her client, but it is also very useful to know any information about the utility functions of the rest of the agents, because the agent could get advantange out of this information in future negotiations. In particular, after the exchanges, agents redefine their respective utility functions. In some way, they learn how to behave in future negotiations from the experience obtained in the last ones. Therefore, an agent can be considered as an entity whose knowledge increases along time. Our two main aims are, on the one hand, to observe the external behavior of each agent, and, on the other hand, to understand how agents learn. In order to achieve both aims some preliminary notations about learning theories are needed.

The human race has been constantly worried about how learning is achieved. In the Ancient Greece, Plato and Aristotle developed two different learning theories. The a priori platonic approach states that every knowledge is innate, that is, it is already in our mind and the only task to "learn" consists in remembering. On the other hand, Aristotle completes the deductive Plato's method with induction (See Figure 6), that is, he considers that knowledge also comes from the analysis of experience. However, Aristotle does not consider that experience is the only source of knowledge. By contrast, he argued that things are considered universals via of reason.

Throughout history, both approaches has been the basis of new learning theories. In the Middle Ages, Thomas of Aquin considered that the perception was the basis of knowledge, but logic is the mental process to understand nature properly.

In the Renaissance, Galileo Galilei follows similar principles to the ones of Aristotle. Galileo explains that the way for building knowledge is based on analysis (or induction) and synthesis (or deduction). First of all, we analyze facts and, via an inductive trip, we gestate principles that could be generalized. Then, we synthesize, that is, facts are rebuilt or reorganized by using the general rules. Besides, consequences derived from these general rules are proven via experience. This Positivism has given place to observe abilities and attitudes by means of measuring and quantification.

In the Modern Era, Kant distinguished two types of propositions: analytic and synthetic. "Red flowers are red" is an example of analytic proposition, since the predicate is contained in the subject. These propositions are naturally true. By contrast, truthfulness or falseness of "The flower is red" is not derived from the sentence itself; it is an example of synthetic statement. Therefore, in order to build knowledge, Kant considers

Figure 6. Induction and deduction

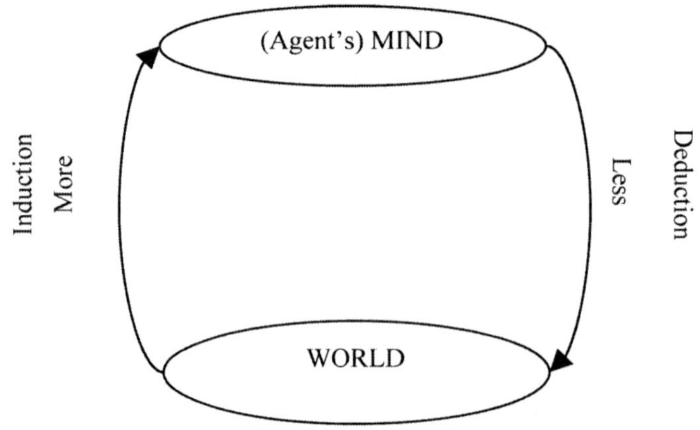

two elements: concepts (comparable with analytic statements) and facts (similar to synthetic propositions). We only learn what is understood. Reason provides a priori structures that are used to understand experience.

In the 20th century, Piaget develops his learning theory based on the principles of Kant. Piaget states that acquisition of new knowledge is based on actions and previous schemata. A schema is a mental representation of some physical or mental action that can be performed on an object, event, or phenomenon. Schemata are enhanced via the process of adaptation. It consists in two phases: assimilation and accommodation. In order to solve a problem or understand something, the fist step is to try to assimilate it with the previous schemata or internal cognitive structures. However, when this assimilation is not possible, schemata must be modified in order to make them consistent with external data or problem. Schemata are for Piaget the same as a priori structures for Kant.

Apart from Piaget, in the 20th century some others learning theories have been developed:

- **Behaviorism**: the aim of this theory is to use experimental methods to observe the behavior of the subject. Consequently, only observable facts can be studied by this theory. Since mental processes cannot be observed directly, they cannot be considered. Therefore, the studies of this theory are limited to relate stimulus and responses. With respect to learning, the premise above leads to consider that learning happens just when a correct response is given after the presentation of a stimulus, and the trainer can detect whether the subject has learnt or not by observing his or her behavior over a period of time. The main tool to motivate the learning is the use of reinforcements of learned behaviors.
- **Cognitivism**: it considers that the study of probabilistic relations among stimulus and responses is not enough to understand how we learn. Its basic principle is to consider the learning as a change of the knowledge state. The changes occur when more knowledge is acquired, and this acquisition takes place via processes of codification and structuring developed by the learner. These two tasks also guide the trainers to designate learning situations. Consequently, the learning depends on the work of the trainer, the situations that he or she designes, but also on the way the learner processes the information.
- **Constructivism**: this model states that learning takes place when the subject experiments and interacts with the environment. She learns via the action (Piaget, 1973). Therefore, knowledge is embedded in the meaningful tasks that the trainer poses to the learner. The latter assembles his or her knowledge by composing and modifying it when he or she has to solve the problem at hand.

Once the different learning theories have been described, we need to study how they can be applied in our particular case. As we have commented, we need to be able to learn two aspects. On the one hand, each agent is interested in knowing anything about the utility functions of the rest of the agents, and on the other hand, each agent need to manage properly the utility functions of the client. Although they are very similar aspects, the learning techniques needed to learn in both cases are different:

Firstly, we are interested in learning any information about the utility functions of the rest of the agents, because the agent could get advantange out of this information in future negotiations. For doing this, the only option is to analyze all the exchanges that the rest of the agents have accepted/rejected. So that, starting from this partial information we can try to rebuild to the original utility function of each agent. As the agents only can be observed as a black box, the learning theory used in this case is behaviorism. Analyzing the external behaviors we will try to infer the rules these behaviors are governed by. In this sense, we will reuse the work described in (Hidalgo-Herrero, Rodríguez, & Rubio, 2005). In that paper, we presented an experiment designed to test whether an automatic system can learn a set of rules in a similar way as human beings learn the same set of rules. Experience obtained from previous actions are used by the automatic system and human beings to act in later situations. In the present chapter, it has been developed a formal framework to help describing negotiation architectures. The actions performed by agents may be influenced by previous experience, similarly as what happens when human beings learn and act consequently. Agents in our system can be analyzed by observing their behavior following experimental techniques, that is, from a behavioral point of view. The considered conducts are the utility functions, and the analysis is based on the comparison of both functions, the one before negotiating and the one after doing it.

A very simple example, if an external agent accepts to exchange an apple for a pear, the information obtained will be used to infer the following rule of the utility function of this agent: $u(x+1,y-1)>u(x,y)$ In this case for simplicity, we consider that the basket of resources is restricted to a pair of physical products. The first component represents the number of apples while the second one represents the number of pears4. It is important to notice that this inference is not necessarily true, because the utility function could depend on the actual number of apples and pears, but it is an acceptable generalization. As more information is obtained about the exchange of apples and pears, our case study algorithm described in (Hidalgo-Herrero, Rodríguez, & Rubio, 2005) can be used to improve our supposed definition of the utility function.

Secondly, each agent has to be abled to manage properly the utility of her client. In this case, to infer the own function the most adequate learning theory is the constructivism. In this case we can analyze the client easily, due to the client is not interested in hiding information to his/her agent. In this sense, the client can initially provide a detailed description of his/her internal motivation. After that, the agent will refine it proposing different possible exchanges to the client. Let us note that in this case we can take advantage of the work described in (Hidalgo-Herrero, Rodríguez, & Rubio, 2005), but also of the work described in (Encina, Hidalgo-Herrero, Rabanal, Rubio, & Rodríguez, 2006). In that paper, an environment framework is described to analyzed the behavior of the different actors in a constructivism way. These reflections and actions could lead to modify the internal structures or pre-programmed way of behaving.

CONCLUSION

In this article, agents have been endowed formally with negotiation competencies. We have provided a formal environment to allow entities to interchange goods, which could be either physical products or inmaterial goods. We assume that the preferences of entities are described by means of utility functions. Then, we provide an exchanges framework allowing the agents to interchange their goods. The system guarantees that no agent worsens. Moreover, we can guarantee Pareto optimality. That is, the final distribution of goods among entities is optimal in the sense that there does not exist an additional transaction that could improve the utility of one entity without reducing the utility of another entity. Let us remark that using a process algebra has facilitated the definition of hierarchical transactions.

Humankind has long desired to build a machine whose behavior is similar to a human being. In (Hidalgo-Herrero, Rodríguez, & Rubio, 2005) we presented an experiment whose aim was to test if an automatic system could learn a set of rules similarly as humans learn the same set of rules. Both of them used experience obtained from previous actions to act in later situations. In the present article, a formal framework has been defined to help describing negotiation architectures. Agents have been endowed formally with negotiation competencies. The provided formal environment allows entities to interchange goods, which could be either physical products or inmaterial goods. We assume that the preferences of entities are described by means of utility functions, and agents negotiate among themselves following the principles of their respective functions. After exchanging goods, each one modifies its utility function. That is, the actions performed by agents may be influenced by previous experience. Consequently, knowledge of agents can be regarded as increasing along time, i.e. it can be considered that agents learn from experience how to behave in future negotiations. Humans act in a similar way: previous experiences can determine their subsequent actions.

The markets are formally defined by means of an exchanges framework allowing the agents to interchange their goods. The system guarantees that no agent worsens. Moreover, we can guarantee Pareto optimality. That is, the final distribution of goods among entities is optimal in the sense that there does not exist an additional transaction that could improve the utility of one entity without reducing the utility of another entity. Let us remark that the use of a process algebra has facilitated the definition of hierarchical transactions.

Our two main cognitive interests are, on the one hand, to observe each agent from an external point of view, and, on the other hand, to understand the way agents learn. In order to achieve both aims, learning theories can be used. The analysis of agents in our system can be carried out from two points of view. On the one hand, they can be observed by means of experimental techniques, that is, from a behavioral perspective. The considered conducts are the utility functions, and the analysis is based on the comparison of both functions, the one before the negotiation and the one after doing it. On the other hand, our other research interest is to investigate how agents change their way of acting, which is reflected in the changes they carry out internally in their utility functions. From a cognitive point of view, agents could analyze their internal processes and decisions to extract the advantages and drawbacks of their actions. These reflections and actions could lead to modify the internal structures or pre-programmed way of behaving.

ACKNOWLEDGMENT

The authors would like to thank Manuel Núñez, Ismael Rodríguez and Fernando Rubio for valuable comments on a previous version of the paper.

REFERENCES

Bergstra, J., Ponse, A., & Smolka, S. (Eds.). (2001). *Handbook of process algebra*. North Holland.

Dastani, M., Jacobs, N., Jonker, C., & Treur, J. (2001). Modelling user preferences and mediating agents in electronic commerce. In *Agent mediated electronic commerce, lnai 1991* (pp. 163–193). Springer. doi:10.1007/3-540-44682-6_10

de la Encina, A., Hidalgo-Herrero, M., Rabanal, P., Rubio, F., & Rodríguez, I. (2006). Testing entities in a parallel cognitive language. In *Fifth IEEE International Conference on Cognitive Informatics,* 2006 (pp. 344-355). IEEE-CS Press.

Eymann, T. (2001). Markets without makers - a framework for decentralized economic coordination in multiagent systems. In *Welcom 2001, lncs 2232* (pp. 63–74). Springer.

Hidalgo-Herrero, M., Rodríguez, I., & Rubio, F. (2005). Testing learning strategies. In *Forth IEEE International Conference on Cognitive Informatics* (pp. 212-221). IEEE-CS Press.

Hoare, C. A. R. (1985). *Communicating Sequential Processes*. Prentice-Hall.

Keppens, J., & Shen, Q. (2002). A calculus of partially ordered preferences for compositional modelling and configuration. In *AAAI Workshop on Preferences in AI and CP: Symbolic Approaches* (pp. 39–46). AAAI Press.

Kraus, S. (1997). Negotiation and cooperation in multi-agent systems. *Artificial Intelligence*, *94*(1-2), 79–98. doi:10.1016/S0004-3702(97)00025-8

Lang, J., Torre, L. v., & Weydert, E. (2002). Utilitarian desires. *Autonomous Agents and Multi-Agent Systems*, *5*(3), 329–363. doi:10.1023/A:1015508524218

Lomuscio, A., Wooldridge, M., & Jennings, N. (2001). A classification scheme for negotiation in electronic commerce. In *Agent mediated electronic commerce, lnai 1991* (pp. 19–33). Springer. doi:10.1007/3-540-44682-6_2

López, N., Núñez, M., Rodríguez, I., & Rubio, F. (2002). A formal framework for e-barter based on microeconomic theory and process algebras. In *I2CS 2002, lncs 2346*. Springer.

Mas-Colell, A., Whinston, M., & Green, J. (1995). *Microeconomic theory*. Oxford University Press.

McGeachie, M., & Doyle, J. (2002). Utility functions for ceteris paribus preferences. In *AAAI Workshop on Preferences in AI and CP: Symbolic Approaches* (pp. 33–38). AAAI Press.

Milner, R. (1989). *Communication and concurrency*. Prentice Hall.

Parsons, S., & Wooldridge, M. (2002). Game theory and decision theory in multi-agent systems. *Autonomous Agents and Multi-Agent Systems*, *5*(3), 243–254. doi:10.1023/A:1015575522401

Piaget, J. (1973). *Introduction à l'Épistemologie Genetique*. Paris: PUF.

Plotkin, G. D. (1981). A structural approach to operational semantics. Technical Report DAIMI FN-19, Computer Science Department, Aarhus University, 1981.

Rasmusson, L., & Janson, S. (1999). Agents, self-interest and electronic markets. *The Knowledge Engineering Review*, *14*(2), 143–150. doi:10.1017/S026988899914205X

Sandholm, T. (1998). Agents in electronic commerce: Component technologies for automated negotiation and coalition formation. In *CIA'98, lncs 1435* (pp. 113–134). Springer.

Stirling, W., Goodrich, M., & Packard, D. (2002). Satisficing equilibria: A non-classical theory of games and decisions. *Autonomous Agents and Multi-Agent Systems*, *5*(3), 305–328. doi:10.1023/A:1015556407380

Tennenholtz, M. (2002). Game theory and artificial intelligence. In *Foundations and applications of multiagent systems* (pp. 49–58). Springer. doi:10.1007/3-540-45634-1_4

Uchiya, T., Maemura, T., Hara, H., Sugawara, K., & Kinoshita, T. (2009). Interactive Design Method of Agent System for Symbiotic Computing. *International Journal of Cognitive Informatics and Natural Intelligence*, *3*(1), 57–74. doi:10.4018/jcini.2009010104

Vinh, P. C. (2009). Categorical Approaches to Models and Behaviors of Autonomic Agent Systems. *International Journal of Cognitive Informatics and Natural Intelligence*, *3*(1), 17–33. doi:10.4018/jcini.2009010102

Wang, Y., & Ruhe, G. (2007). The Cognitive Process of Decision Making. *International Journal of Cognitive Informatics and Natural Intelligence*, *1*(2), 73–85. doi:10.4018/jcini.2007040105

Yang, C., Lin, H., & Lin, F. O. (2006). Designing Multiagent-Based Education Systems for Navigation Training. In *IEEE International Conference on Cognitive Informatics, ICCI'06, 495-501.*

ENDNOTES

[1] Research partially supported by the Spanish MCYT projects TIN2009-14312-C02-01, TIN2009-14599-C03-01, and the Comunidad de Madrid program S2009/TIC-1465.

[2] We are assuming that all the items are *goods*. Nevertheless, agents could also trade *bads*. For example, an entity would be willing to give an apple pie if he *receives* minus s brown leaves in his garden. However, bads are usually not considered in microeconomic theory, as they can be easily turned into goods: Instead of considering the amount of leaves, one may consider the absence of them.

[3] Taking into account that the last material resource in any basket of resources is money, from now on we will use the following notation: $\overline{\cos ts_1} = ((0,none),(0,none),\ldots,(costs,wanted))$

Chapter 22
Noise Cancellation in ECG Signals with an Unbiased Adaptive Filter

Yunfeng Wu
Xiamen University, China

Rangaraj M. Rangayyan
University of Calgary, Canada

ABSTRACT

The electrocardiographic (ECG) signal is a transthoracic manifestation of the electrical activity of the heart and is widely used in clinical applications. This chapter describes an unbiased linear adaptive filter (ULAF) to attenuate high-frequency random noise present in ECG signals. The ULAF does not contain a bias in its summation unit and the filter coefficients are normalized. During the adaptation process, the normalized coefficients are updated with the steepest-descent algorithm to achieve efficient filtering of noisy ECG signals. A total of 16 ECG signals were tested in the adaptive filtering experiments with the ULAF, the least-mean-square (LMS), and the recursive-least-squares (RLS) adaptive filters. The filtering performance was quantified in terms of the root-mean-squared error (RMSE), normalized correlation coefficient (NCC), and filtered noise entropy (FNE). A template derived from each ECG signal was used as the reference to compute the measures of filtering performance. The results indicated that the ULAF was able to provide noise-free ECG signals with an average RMSE of 0.0287, which was lower than the second-best RMSE obtained with the LMS filter. With respect to waveform fidelity, the ULAF provided the highest average NCC (0.9964) among the three filters studied. In addition, the ULAF effectively removed more noise, measured by FNE, in comparison with the LMS and RLS filters in most of the ECG signals tested. The issues of adaptive filter setting for noise reduction in ECG signals are discussed at the end of this chapter.

DOI: 10.4018/978-1-60960-553-7.ch022

INTRODUCTION

The electrocardiographic (ECG) signal is the electrical manifestation of the contractile activity of the heart. Computerized ECG analysis is widely used as a reliable technique for the diagnosis of cardiovascular diseases, and the ECG signal is the most commonly used biomedical signal in clinical practice (Rangayyan, 2002; Tompkins, 1993). However, a surface recording of the ECG signal (with the frequency range of 0.05-250 Hz), obtained by placing electrodes on the subject's chest, is inevitably contaminated by several different types of artifacts. The dominant artifacts in an ambulatory ECG recording include (Rangayyan, 2002; Wu & Rangayyan, 2009):

- **Baseline wander:** Drift of the baseline is a type of low-frequency (lower than 0.5 Hz) artifact and usually caused by respiration or movement of the subject.
- **Physiological artifacts:** This type of artifact is mainly induced by muscular contractions. Electrode-motion artifact has a wide frequency range (from 1 to 5,000 Hz) and is generally considered to be the most troublesome, because it can mimic the appearance of ectopic beats and cannot be removed easily by simple filters.
- **Random noise:** Random noise could be the result of the thermal effect in the instrumentation amplifiers, the recording system, and pickup of ambient electromagnetic signals by the cables used (Rangayyan, 2002). Random noise usually appears with high frequency; its frequency range depends on the specific source. In real-time clinical monitoring systems used during surgery, electrosurgical noise is a significant obstacle to be overcome.
- **External interference:** Examples of environmental interference are those caused by 50 or 60 Hz power-supply lines, radiation from lights, and radio-frequency emissions from nearby medical devices.

The removal of artifacts is crucial for ECG monitoring, and is an essential procedure prior to further diagnostic analysis in many clinical applications, e.g., classification of ectopic beats (Afonso, Tompkins, Nguyen, & Luo, 1999; Hu, Palreddy, & Tompkins, 1997), detection of QRS complexes (Meyer, Gavela, & Harris, 2006; Hu, Tompkins, Urrusti, & Afonso, 1993), analysis of asymptomatic arrhythmia (Thakor & Zhu, 1991), extraction of the fetal ECG signal from the maternal abdominal ECG (Kanjilal, Palit, & Saha, 1997; Khamene & Negahdaripour, 2000; Zarzoso & Nandi, 2001), classification of myocardial ischemia (Silipo & Marchesi, 1998), diagnosis of atrial fibrillation (Yang, Devine, & Macfarlane, 1994), ECG-based sleep apnea detection (Mita, 2007), and ECG signal data compression (Zigel, Cohen, & Katz, 2000; Hamilton, Thomson, & Sandham, 1995). The extraction of high-resolution ECG signal from noise-contaminated recordings is an important part of the artifact removal procedure (Clifford, Azuaje, & McSharry, 2006). The goal of noise reduction in the ECG signal is to separate the valid signal components from the undesired noise, so as to present an ECG that facilitates easy and accurate interpretation (Afonso, Tompkins, Nguyen, Michler, & Luo, 1996).

Widrow *et al.* (Widrow et al., 1975) reported that an adaptive filter has the ability to adjust automatically the tap-weights to produce the desired impulse response, according to the time-varying characteristics of the input signal. Recent literature indicates that a number of adaptive filtering methods have been applied in several different clinical applications. Thakor and Zhu (Thakor & Zhu, 1991) designed a type of adaptive recurrent filter to detect normal QRS complexes in ambulatory ECG recordings, and then applied it for the analysis of arrhythmia. Xue *et al.* (Xue, Hu, & Tompkins, 1992) developed adaptive whitening and matched filters based on artificial neural networks for the detection of QRS complexes. Hamilton (Hamilton, 1996) compared the effectiveness of power-line interference removal by adaptive and nonadaptive notch filters. Chen *et al.* (Chen, Chen, & Chan, 2006) incorporated wavelet denoising and moving-average filtering methods to implement noise reduction and real-time QRS detection. Mneimneh *et al.* (Mneimneh, Yaz, Johnson, & Povinelli, 2006) used an adaptive Kalman filter that could effectively enhance the ECG signal from original recordings corrupted by baseline drift. Sayadi and Shamsollahi (Sayadi & Shamsollahi, 2008) modified the Kalman filter by adding more equations to present the governing equations of the model parameters, in order to implement simultaneous denoising and compression of the ECG. Sameni *et al.* (Sameni, Shamsollahi, Jutten, & Clifford, 2007) established a framework to update the nonlinear Bayesian model on a beat-to-beat basis to filter noisy ECG signals. Wu *et al.* (Wu, Rangayyan, Zhou, & Ng, 2009; Wu & Rangayyan, 2009) proposed two types of noise reduction methods for ECG signals. These two ECG noise reduction systems both considered the unbiased linear adaptive filter (ULAF), but one was adapted with the reference as the noise (Wu, Rangayyan, Zhou, & Ng, 2009) and the other was designed to estimate the signal component (Wu & Rangayyan, 2009). Recently, the empirical mode decomposition (EMD) method was also utilized for ECG enhancement. Chang (Chang, 2010) proposed a noise filtering algorithm based on ensemble EMD to remove power-line interference, baseline wander, and electromyographic (EMG) interference signals.

This chapter presents methods for the filtering of high-frequency random noise in ECG signals based on the ULAF with normalized coefficients (Wu & Rangayyan, 2009), and also provides a discussion on the major differences between the signal estimation approach and the noise reference method.

ECG FILTERING PROCEDURE

The ECG signal processing procedure contains the following steps: removal of baseline wander and power-line interference, detection of QRS complexes, establishment of signal templates, adaptive filtering, signal reconstruction, and performance evaluation. The flowchart of the complete ECG filtering procedure is illustrated in Figure 1. Details of the aforementioned steps are presented in the following subsections.

Baseline Wander Removal with a Derivative-Based High-pass Filter

The aim of the first step is to eliminate low-frequency baseline wander and place the output signal on the isoelectric line of the ECG record-

Figure 1. Flowchart of the adaptive ECG filtering procedure

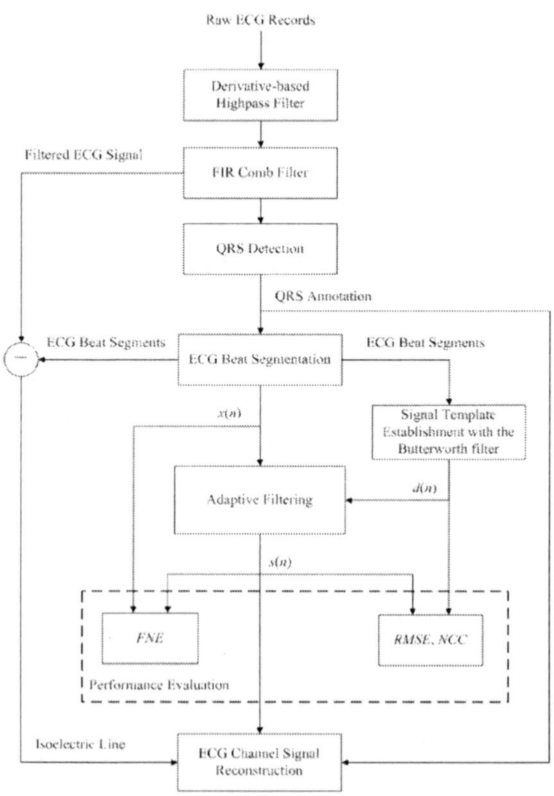

Box 1.

$$H_{\text{FIR}}(z) = 0.6312(1 - 1.8596z^{-1} + z^{-2})(1 - 0.8516z^{-1} + z^{-2})(1 + 0.6180z^{-1} + z^{-2})(1 + 1.7526z^{-1} + z^{-2})$$
$$= 0.6312 - 0.2150z^{-1} + 0.1512z^{-2} - 0.1288z^{-3} + 0.1228z^{-4} - 0.1288z^{-5} + 0.1512z^{-6}$$
$$-0.2150z^{-7} + 0.6312z^{-8}. \quad (2)$$

ing. It is implemented using an infinite-impulse-response (IIR) filter with the transfer function (Rangayyan, 2002):

$$H_{\text{IIR}}(z) = f_s \left[\frac{1 - z^{-1}}{1 - 0.995z^{-1}} \right], \quad (1)$$

where f_s denotes the sampling frequency. This derivative-based filter does not cause substantial distortion of the QRS complexes in the ECG signals.

Power-Line Interference Cancellation with a Comb Filter

Power-supply lines usually introduce periodic artifacts at 50 or 60 Hz and its harmonics in ECG recordings. To attenuate such periodic interference, we may apply a finite-impulse-response (FIR) comb filter with the transfer function (Rangayyan, 2002): (see Box 1)

This filter has zeros at 60, 180, 300, and 420 Hz, with the sampling rate at 1,000 Hz. The gain was set to 0.6312 so that the filter has unit gain at DC. For the magnitude and phase responses of such an FIR comb filter, readers are referred to the book of Rangayyan (Rangayyan, 2002).

QRS Detection

The QRS complex provides pivotal information for the analysis of ECG signals (Kohler, Hennig, & Orglmeister, 2002), and is frequently used as a real-time trigger in multichannel physiological signal processing (Pan & Tompkins, 1985). After each QRS complex has been identified in a given ECG signal, the heart rate may be calculated, the ST segment may be examined for evidence of ischemia or infarction, and the ECG waveform may be classified as normal or abnormal. For the detection of QRS complexes, we may apply the method of Murthy and Rangaraj (Murthy & Rangaraj, 1979), which contains the squared first-derivative operator, a moving-average filter, a threshold operator, and a simple peak-searching procedure. The annotation of the detected location of QRS complexes is used in the steps for the segmentation of the ECG beat (defined as a portion of the ECG signal from the beginning of a P wave to the end of the following T wave) and the reconstruction of the ECG channel signal (see Figure 1).

ECG Beat Segmentation and Signal Template Establishment

It is worth noting that the ECG signal is usually weak between the end of a T wave and the start of the next P wave, and that the signal samples between successive cardiac cycles do not provide much information in most ECG signals. Therefore, we consider the P, QRS, and T components of each cardiac cycle to compose a heart beat template in the present study. The heart beat template used as the reference for the adaptive filters is variable from one cardiac cycle to another, in consideration of the variability of the ECG signal and the accompanying artifacts due to respiration or cardiovascular abnormalities such as arrhythmia.

After the QRS detection step, the boundaries of the ECG beat template are determined with reference to the averaged RR interval of each ECG signal. The relationships between the averaged RR interval and the durations of the P-QRS and QRS-T segments were empirically derived, in the least-squares sense, as (Wu & Rangayyan, 2009):

$$\text{P-QRS} = 290 - 0.12\text{RR}, \tag{3}$$

$$\text{QRS-T} = 230 + 0.12\text{RR}, \tag{4}$$

where the P-QRS and QRS-T durations as well as the RR interval are in ms.

Adaptive Filtering

The adaptive filtering step can be performed by several different adaptive filters, such as the least-mean-square (LMS) filter, the recursive-least-squares (RLS) filter, or the ULAF. In the present work, we have implemented all of these three filters for ECG filtering and comparative evaluation of their performance. The LMS and RLS filters have been used in signal processing applications for decades, due to their simplicity in hardware implementation, stability, and robustness (Haykin, 2002).

A schematic representation of the ULAF is shown in Figure 2 (Wu & Rangayyan, 2009). The ULAF is a transversal FIR adaptive filter, whose output $s(n)$ at the time instant n can be expressed as the convolution of the input signal with the filter coefficients $w_m(n)$ as

$$s(n) = \sum_{m=1}^{M} w_m(n)x(n-m+1), \tag{5}$$

where M denotes the order of the filter, and $x(n-m+1)$ represents the input signal with a lag of $m-1$ ($1 \leq m \leq M$) samples. Compared with the LMS filter, the ULAF does not contain a constant bias, and its filter coefficients are normalized in order to provide unit gain at the direct current (DC) level, i.e.,

$$\sum_{m=1}^{M} w_m(n) = 1. \tag{6}$$

The instantaneous error $e(n)$ is defined as the difference between the filter output and the signal template at the time instant n, i.e.,

$$e(n) = d(n) - s(n) = d(n) - \sum_{m=1}^{M} w_m(n)x(n-m+1), \tag{7}$$

Figure 2. Schematic representation of the unbiased linear adaptive filter

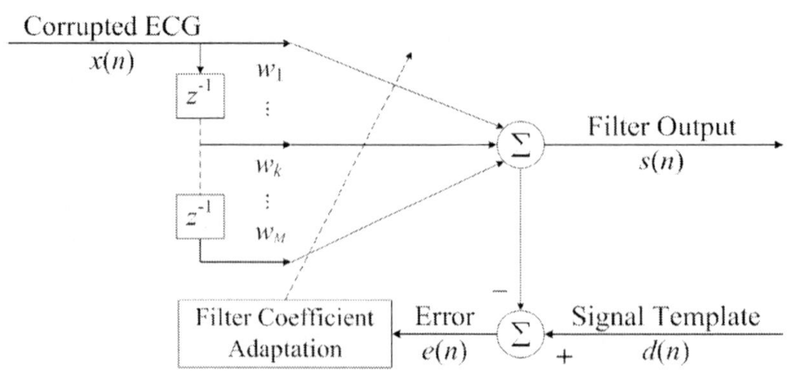

where the signal template $d(n)$ is an estimate of the ECG signal component. The filter coefficients of the ULAF can be optimized by using the steepest-descent algorithm (Haykin, 2002), which is considered to be a deterministic search method. The convergence of the squared instantaneous error follows a distinct path in the multi-dimensional filter-coefficient space provided by the corresponding negative gradient with respect to the filter coefficients, i.e.,

$$\begin{aligned} -\nabla_{w_m(n)} e^2(n) &= -\frac{\partial}{\partial w_m(n)}\left[d(n) - \sum_{m=1}^{M} w_m(n)x(n-m+1)\right]^2 \\ &= 2x(n-m+1)\left[d(n) - \sum_{k=1}^{M} w_k(n)x(n-k+1)\right]. \end{aligned} \tag{8}$$

By substituting Equations (6) and (8) into the steepest-descent adaptation algorithm (Haykin, 2002), we may update the estimated ULAF coefficients as

$$\begin{aligned} w_m(n+1) &= w_m(n) + \mu\left[-\nabla_{w_m(n)} e^2(n)\right] \\ &= w_m(n) + 2\mu x(n-m+1)\left[d(n) - \sum_{k=1}^{M} w_k(n)x(n-k+1)\right] \\ &= w_m(n) + 2\mu x(n-m+1)\left[\sum_{k=1}^{M} w_k(n)d(n) - \sum_{k=1}^{M} w_k(n)x(n-k+1)\right] \\ &= w_m(n) + 2\mu x(n-m+1)\sum_{k=1}^{M} w_k(n)\left[d(n) - x(n-k+1)\right], \end{aligned} \tag{9}$$

where $\mu(\mu>0)$ is the learning rate that indicates the search magnitude in the direction of the negative gradient. We may define $\varepsilon(n-k+1)$ to represent the difference between the current signal template sample and the filter input with a lag of k-1 samples, i.e.,

$$\varepsilon(n-k+1) = d(n) - x(n-k+1), \quad 1 \le k \le M. \tag{10}$$

Then, the ULAF adaptation algorithm of Equation (9) can be revised as

$$w_m(n+1) = w_m(n) + 2\mu x(n-m+1)\sum_{k=1}^{M} w_k(n)\varepsilon(n-k+1). \tag{11}$$

Because $\varepsilon(n-k+1)$ can be directly derived from the ULAF input and the given signal template, the ULAF coefficient adaptation process can skip the estimation of the instantaneous error $e(n)$, which enables efficient convergence of the ULAF algorithm.

According to the requirement of Equation (6), in each iteration of the ULAF coefficient adaptation process, the estimated value of the filter coefficients should be normalized in order to ensure the amplitude gain of unity as:

$$\hat{w}_m(n+1) = \frac{w_m(n+1)}{\sum_{k=1}^{M} w_k(n+1)}, \tag{12}$$

where $\hat{w}_m(n+1)$ represents the estimated coefficient value for the time instant n+1. If the instantaneous value of $w_m(n+1)$ is close to zero, i.e., $w_m(n+1)\to 0$, then the absolute value of the estimated filter coefficient, $|\hat{w}_m(n+1)|$, becomes

$$\lim_{w_m(n+1)\to 0}\left|\hat{w}_m(n+1)\right| = \frac{\left|w_m(n+1)\right|}{\left|w_m(n+1) + \sum_{k=1,k\neq m}^{K} w_k(n+1)\right|} = 0. \tag{13}$$

If $w_m(n+1)$ is estimated to be with a sufficiently large absolute value, expressed as $w_m(n+1)\to\infty$, then Equation (12) yields

$$\lim_{w_m(n+1)\to \infty}\left|\hat{w}_m(n+1)\right| = \frac{1}{\left|1 + \sum_{k=1,k\neq m}^{K} \frac{w_k(n+1)}{w_m(n+1)}\right|} = 1. \tag{14}$$

Considering the convergence of the ULAF algorithm for nonstationary ECG signal inputs, we can deduce from Equation (11) that the ULAF coefficients' convergence behavior is influenced by the value assigned to the learning rate parameter and the statistical characteristics of the M-element input signal vector

$$\mathbf{x}(n) = [x(n), x(n-1), \cdots, x(n-M+1)]^T .$$

According to Widrow's independence theory (Widrow, McCool, Larimore, & Johnson, 1976), the ULAF coefficient adaptation algorithm is convergent in the mean-squared sense provided that the learning rate $\boldsymbol{\mu}$ satisfies the condition (Haykin, 2002):

$$0 < \mu < \frac{2}{\lambda_{\max}}, \quad (15)$$

where λ_{max} is the largest eigenvalue of the correlation matrix Cx of the input signal, i.e.,

$$\mathbf{C}_{\mathbf{x}} = E\left[\mathbf{x}(n)\mathbf{x}^T(n)\right]. \quad (16)$$

For the inherently nonstationary ECG signal, the Wiener solution of the ULAF coefficients is difficult to obtain, because no prior knowledge of $\lambda_{\max}$ is available. To overcome this difficulty, the trace of C_x may be taken as a conservative estimate for $\lambda_{\max}$ (Haykin, 2002). Because the correlation matrix C_x is a square matrix, its trace is the sum of the diagonal elements, each of which equals the mean-squared value of the corresponding input signal sample. The condition for convergence of the ULAF coefficient adaptation algorithm can then be formulated as

$$0 < \mu < \frac{2}{\frac{1}{M}\sum_{m=1}^{M} x^2(n-m+1)}. \quad (17)$$

ECG Channel Signal Reconstruction

Cardiologists and physicians would prefer to view and interpret the entire ECG signal in a channel rather than a series of cardiac-cycle-to-cycle segments. For this reason, it is necessary to reconstruct the channel outputs of the filters, together with the ECG template signal in a channel. In order to reconstruct the channel signal, the isoelectric line (the DC level) was set to be the mean value of the difference between the output of the FIR comb filter (ECG signal without baseline drift or power-line interference) and the ECG signal components (ECG beat segments). Each smoothed ECG beat template was placed upon the isoelectric line at the corresponding position, where the related QRS complex was detected in the original signal, to form a template channel of the same duration as the original signal. For the channel output of a filter, the procedure of reconstruction is similar, with the only difference being that the filtered signal is utilized instead of the ECG beat template signal.

Performance Evaluation

The evaluation criteria of root-mean-squared error (RMSE), normalized correlation coefficient (NCC), and filtered noise entropy (FNE) (Wu & Rangayyan, 2009), were used for a quantitative study of the performance of the adaptive ECG filters.

The RMSE verifies the average magnitude of the noise remaining in the output of each adaptive filter, defined as:

$$\text{RMSE} = \frac{1}{N_B}\sum\nolimits_{N_B}\sqrt{\frac{1}{N_S}\sum_{n=1}^{N_S}\left[d(n)-s(n)\right]^2}, \quad (18)$$

where $s(n)$ is the filter output, $d(n)$ represents the ECG beat template, N_B denotes the number of cardiac cycles (or beats) analyzed, and N_S represents the number of samples included in each ECG beat template.

The NCC is the most popular measure of association in time-series prediction (Marques de Sa, 2003). The NCC may be used to characterize the similarity between the filtered signals and the corresponding templates, with the definition

$$\mathrm{NCC} = \frac{1}{N_B} \sum_{N_B} \left[\frac{\sum_{n=1}^{N_S} s(n) d(n)}{\sqrt{\sum_{n=1}^{N_S} s^2(n) \sum_{n=1}^{N_S} d^2(n)}} \right]. \tag{19}$$

Shannon's entropy (Shannon, 1948) of the first-order difference residual signal (original ECG signal within a cardiac cycle subtracted from an average ECG beat template) was used by Hamilton and Tompkins (Hamilton & Tompkins, 1991) for ECG data compression. Shannon's entropy could be used to characterize the nature of the noise $u(n)$ removed by the adaptive filter, obtained as

$$u(n) = x(n) - s(n), \tag{20}$$

where $x(n)$ represents the input signal. The probability density function of the noise removed can be estimated by calculating the frequencies of occurrence P_{bin} for various bins (20 bins, for example). Then, FNE is defined as

$$\mathrm{FNE} = - \sum_{\text{for all bins}} p_{\text{bin}} \log_2 p_{\text{bin}}. \tag{21}$$

EXPERIMENTAL DATA AND FILTER PARAMETER SETTINGS

The ECG signals tested in this work were obtained from the data set of Lehner and Rangayyan (Lehner & Rangayyan, 1987). The data set contains 16 limb lead II ECG signals recorded from 11 subjects (seven females and four males, aged from 4 to 29 years, including two normal subjects and nine patients with cardiovascular abnormalities). The ECG signals were simultaneously recorded with the phonocardiogram and carotid pulse signals using the three-channel HP1514B system (Hewlett-Packard, Palo Alto, CA, USA) (Lehner & Rangayyan, 1987). The ECG signals were digitized at 1 kHz with 12-bit resolution. The patients had different types of cardiovascular abnormalities, such as mitral insufficiency, ventricular septal defect, atrial septal defect, aortic stenosis, ventricular hypertrophy, and right-bundle-branch block. Some of the ECG records were severely contaminated by baseline drift, power-line interference at 60 Hz as well as its harmonics, and random noise; some of the artifacts were caused by difficulties in electrode placement on the pediatric patients.

Based on the ECG beat segmentation step, the durations from the start of the P wave until the QRS complex (P-QRS duration) and from the QRS complex until the end of the T wave (QRS-T duration) extracted from the 16 ECG recordings are listed in Table 1. The linear relationships between the averaged RR interval and the durations of the P-QRS and QRS-T segments are shown in Figures 3 and 4, respectively. The signal templates were constructed based on the ECG beat segments. Before application as the template input to the adaptive filters, the ECG beat template for each ECG signal was smoothed using a 12th-order Butterworth low-pass filter (-3 dB cutoff at 75 Hz) with unit gain at DC.

The ULAF as well as the LMS and RLS adaptive filters were implemented in the adaptive filtering procedure. The signal inputs to each adaptive filter were the time series of ECG beat segments in one heart beat followed by another, which were in accordance with the ECG beat durations tabulated in Table 1. The reference input for the filters was the ECG beat template signal for each cardiac cycle. By following the adaptation algorithm for each filter, the optimal parameters (listed in Table 2) were obtained according to the minimal RMSE criterion (Haykin, 2002). The number of decimal digits for each filter pa-

Table 1. Parameters of the ECG signal templates

ECG ID	Number of cardiac beats analyzed	Averaged RR interval (ms)	P-QRS duration (ms)	QRS-T duration (ms)
1	17	811	195	325
2	22	854	195	325
3	21	895	195	325
4	19	778	200	320
5	24	783	198	322
6	22	790	196	325
7	21	819	185	335
8	21	766	185	335
9	21	795	185	335
10	17	741	195	325
11	33	509	220	300
12	26	850	180	340
13	24	758	195	325
14	25	724	185	335
15	18	960	165	355
16	17	992	160	360

Figure 3. The linear relationship between the average RR interval and the duration of the P-QRS segments. The triangles indicate the data for the 16 ECG signals.

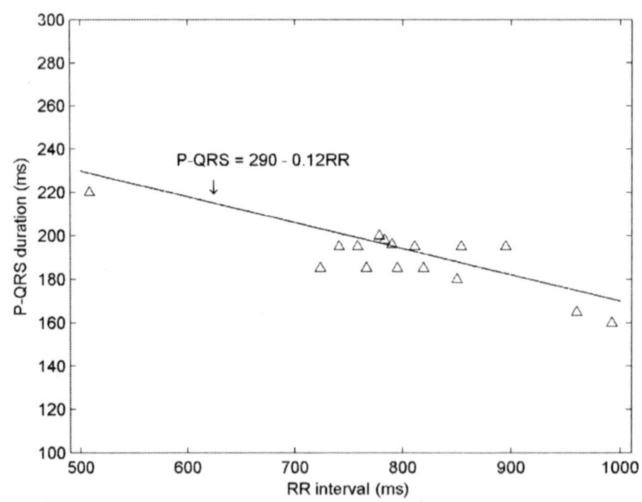

rameter was determined by the sensitivity of the corresponding filter output. It is worth noting that the ULAF provided a larger range to tune the filter parameters, because the active range of the learning rate was much wider than that of the LMS or the RLS filter.

Figure 4. The linear relationship between the average RR interval and the duration of the QRS-T segments. The triangles indicate the data for the 16 ECG signals.

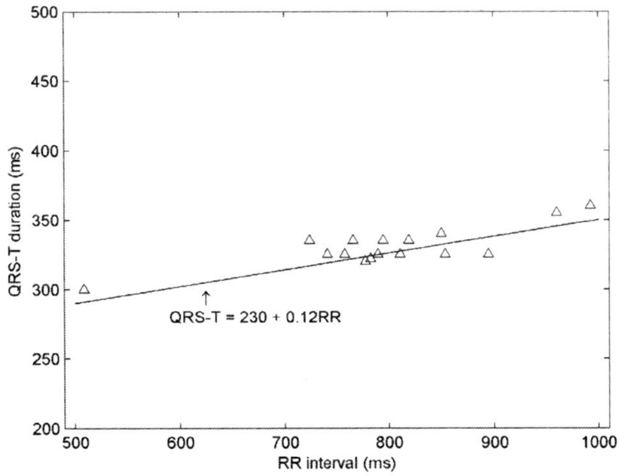

Table 2. Optimal parameters of the adaptive filters

ECG ID	Step size (the LMS filter)	Forgetting factor (the RLS filter)	Learning rate (the ULAF)
1	0.05	0.8	0.00005
2	0.04	0.9	0.001
3	0.04	0.7	0.01
4	0.02	1.0	0.0009
5	0.03	1.0	0.0005
6	0.001	0.9	0.001
7	0.014	0.8	0.001
8	0.03	0.6	0.0008
9	0.03	1.0	0.001
10	0.005	0.8	0.0005
11	0.001	1.0	0.0001
12	0.02	0.9	0.1
13	0.05	0.7	0.01
14	0.04	0.5	0.01
15	0.001	0.8	0.0005
16	0.002	1.0	0.0005

RESULTS

Figure 5(a) shows a part of the original ECG signal, with the six heart beats from 3.5 to 9.2 s, recorded from the patient ID 3. The baseline drift is visible in the time domain, and fundamental as well as the harmonic components of the power-line interference present in the ECG can be observed

Figure 5. Illustration of the effect of the combination of the derivative-based and comb filters in the time domain. (a) Original ECG signal (patient ID 3). (b) Output of the combination of the derivative-based and comb filters. AU: arbitrary units.

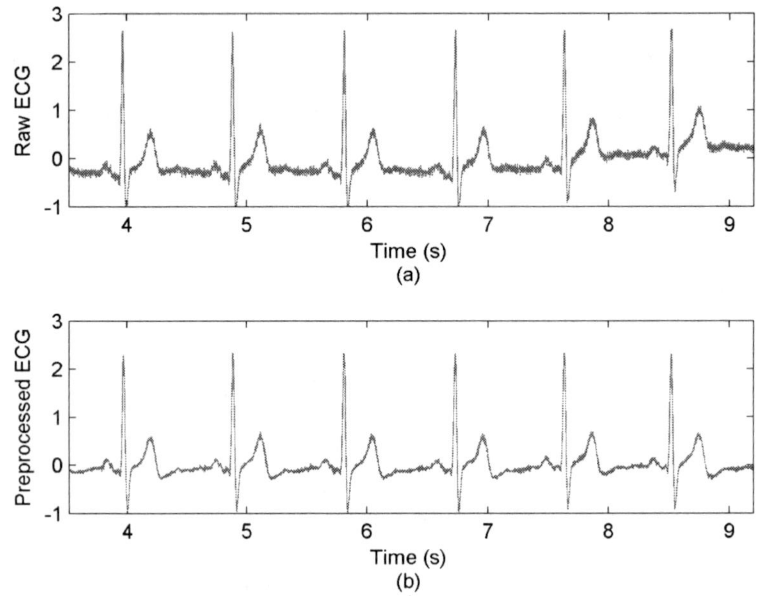

Figure 6. Effect of the combination of the derivative-based and comb filters in the frequency domain. (a) Power spectrum of the original ECG signal (patient ID 3). (b) Power spectrum of the output of the combination of the derivative-based and comb filters.

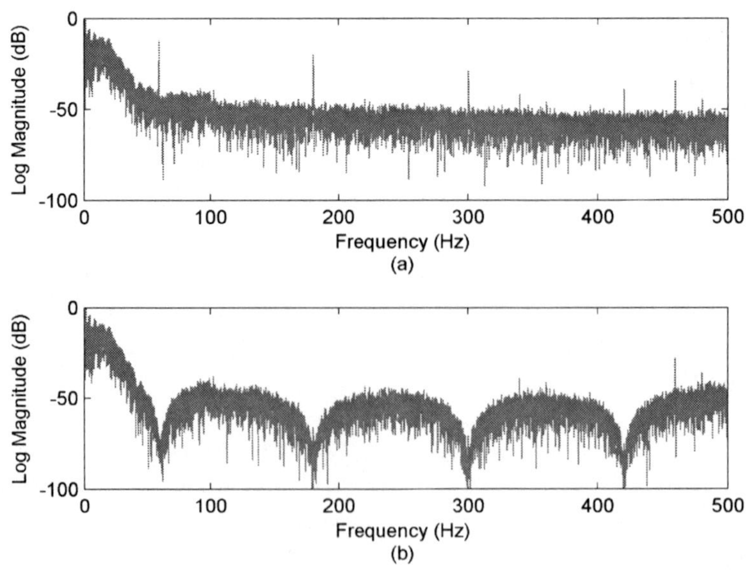

in Figure 6(a). With the effect of the combination of the derivative-based and comb filters, it can be observed that the baseline drift and power-line interference have been effectively eliminated, as depicted in Figure 5(b) and Figure 6(b).

Figure 7 illustrates the filtering results for the first 3.5 s of an ECG signal (patient ID 3) with the different adaptive filters. It can observed in Figure 7(b) that the baseline drift and power-line interference have been removed from the raw ECG recording in Figure 7(a). Figure 7(c) shows the template ECG signal reconstructed with the ECG beat segments placed upon the isoelectric line in the channel. The noise-free signal reconstructed by placing the ECG beat segment outputs of the LMS filter, the RLS filter, and the ULAF upon the isoelectric line are shown in Figure 7(d)-(f), respectively.

The evaluation of the results of the three adaptive filters is provided in Table 3. It is worth noting that the ULAF consistently outperformed the LMS filter with lower mean and standard devia-

Figure 7. Illustration of the filtering steps with the first 3.5 s of an ECG signal (patient ID 3). (a) Original signal. (b) Output of the combination of the first-order derivative-based and comb filters (with the detected locations of QRS complexes marked). (c) Reconstructed template signal in the channel. (d) Reconstructed output of the LMS filter. (e) Reconstructed output of the RLS filter. (f) Reconstructed output of the ULAF. The abscissa is marked in seconds; the ordinate is not calibrated.

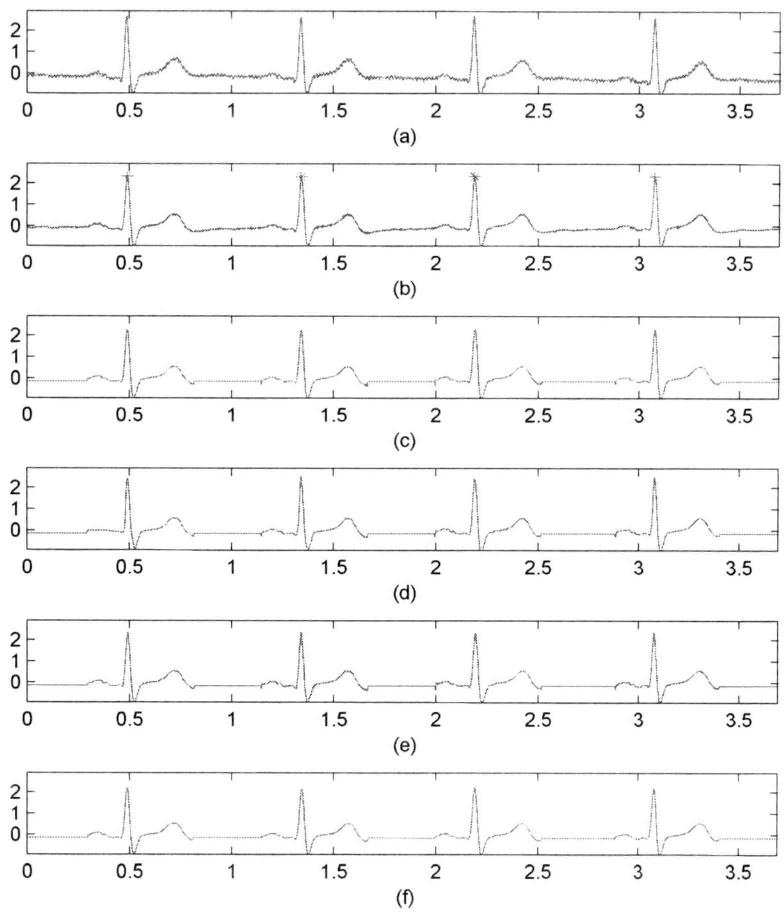

Table 3. Performance evaluation of the adaptive filters for ECG signals

ECG ID	RMSE			NCC			FNE		
	LMS	RLS	ULAF	LMS	RLS	ULAF	LMS	RLS	ULAF
1	0.0379	0.0317	0.0315	0.9951	0.9956	0.9966	2.5407	2.5924	2.5943
2	0.0391	0.0376	0.0340	0.9933	0.9932	0.9942	2.6958	2.6809	2.7177
3	0.0317	0.0365	0.0310	0.9943	0.9908	0.9941	2.3934	2.3835	2.4240
4	0.0322	0.0275	0.0282	0.9951	0.9960	0.9962	2.3102	2.2962	2.2338
5	0.0319	0.0303	0.0257	0.9952	0.9952	0.9966	2.0290	2.0461	2.0311
6	0.0353	0.0267	0.0225	0.9952	0.9965	0.9973	2.0888	2.0851	2.1013
7	0.0309	0.0478	0.0231	0.9969	0.9926	0.9982	2.3242	2.3559	2.3497
8	0.0316	0.0455	0.0274	0.9970	0.9938	0.9975	2.5296	2.5241	2.5111
9	0.0317	0.0333	0.0207	0.9965	0.9964	0.9985	2.3031	2.3420	2.3478
10	0.0446	0.0469	0.0320	0.9890	0.9874	0.9941	2.8065	2.8076	2.8627
11	0.0502	0.0388	0.0398	0.9953	0.9969	0.9969	2.5408	2.5747	2.5978
12	0.0339	0.0429	0.0306	0.9967	0.9944	0.9973	2.4622	2.4786	2.4218
13	0.0349	0.0439	0.0252	0.9912	0.9879	0.9955	2.1742	2.1531	2.1760
14	0.0336	0.0363	0.0250	0.9955	0.9942	0.9961	2.0926	2.0902	2.1210
15	0.0450	0.0389	0.0307	0.9936	0.9958	0.9965	2.6482	2.6483	2.6655
16	0.0392	0.0401	0.0316	0.9947	0.9935	0.9961	2.7671	2.7339	2.8174
Mean	0.0365	0.0378	0.0287	0.9947	0.9938	0.9964	N/A[a]	N/A	N/A
Standard deviation	0.0058	0.0066	0.0049	0.0021	0.0029	0.0013	N/A	N/A	N/A

[a] N/A: not applicable.

tion values of RMSE. Compared with the RLS filter, the ULAF produced a larger RMSE value only for two of the 16 signals. The ULAF provided the highest average NCC for all 16 ECG signals among the three filters studied, so it can be confirmed that the ULAF was able to preserve the waveform of the ECG signals with the highest fidelity. Concerning the noise removed, as indicated by FNE, the ULAF outperformed the LMS and RLS filters for 13 and 11 signals, respectively. Box plots of the RMSE and NCC measures are shown in Figure 8. It can be observed that the ULAF provided higher degrees of prediction accuracy and fidelity than the LMS and RLS filters. Similar statistical analysis of FNE is not applicable in Table 3, because the causes of random noise are multifarious and not comparable from one ECG recording to another.

DISCUSSION

FIR or IIR Comb Filter?

In contrast to IIR filters, which have internal feedback and may continue to respond indefinitely (usually decaying), the impulse response of an FIR filter settles to zero in finite time. Most physicians and medical engineers tend to use FIR comb filters for the removal of power-line interference, because FIR filters are inherently stable and require no feedback. However, there are some types of IIR filters worthy of consideration. In the work of Wu *et al.* (Wu, Rangayyan, Zhou, & Ng, 2009), an IIR comb filter with 4-decimal-digit word length was used for the cancellation of power-line interference in ECG signals. The

Figure 8. Box plots of the RMSE (a) and NCC (b) for the three adaptive filters. The "+" signs indicate outliers exceeding the range of the corresponding box by more than 1.5 times its inter-quartile range.

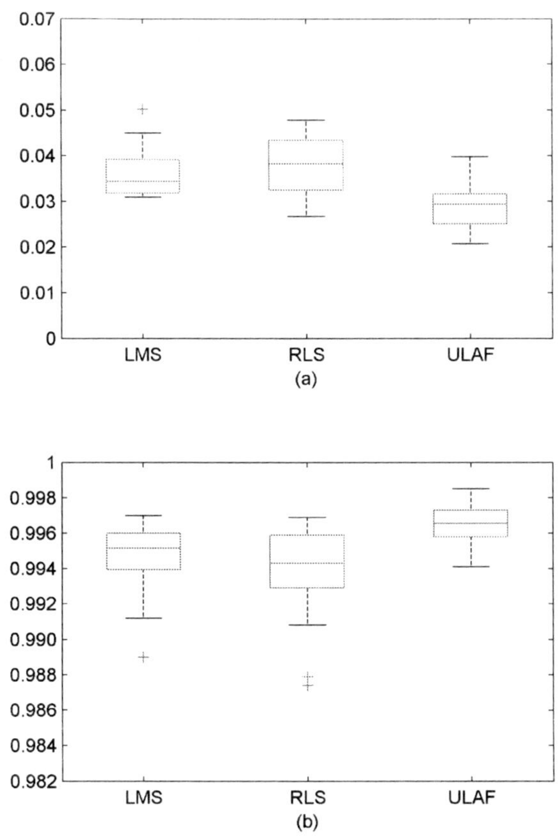

transfer function of such an IIR comb filter can be expressed as

$$H(z) = \frac{0.9502(1 - z^{-6})}{1 - 0.9004z^{-6}}. \tag{22}$$

Figure 9 shows the frequency response of the 6-th order IIR comb filter with a quality factor equal to 30, for designed for ECG signals sampled at 360 Hz. It can be observed that the rejection bands are located around 60, 120, and 180 Hz. The zeros and poles of the IIR comb filter are

Figure 9. The frequency response of the 6th-order IIR comb filter with a quality factor equal to 30, and the input ECG signal sampled at 360 Hz. (a) Magnitude response (dB). (b) Phase response (radians).

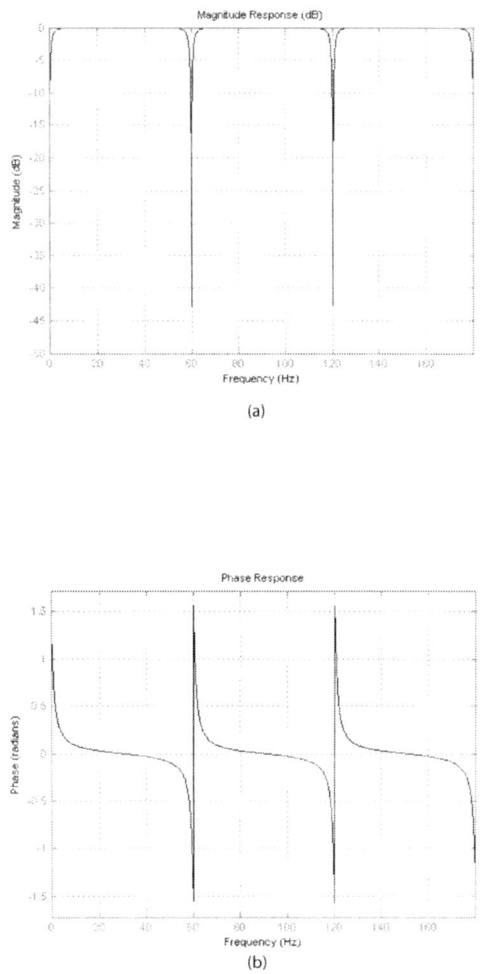

displayed in Figure 10. Because the order of the IIR comb filter is low, the regions of the zeros and poles appear to be close, but they do not overlap one another.

The filtering effect of the 6-th order IIR comb filter is shown in Figure 11. From Figure 11(b) we may observe that the power-line interference at 60, 120, and 180 Hz has been eliminated. For the input ECG signal sampled at 360 Hz, the quality factor of the IIR comb filter was set to be

Figure 10. Zeros and poles of the 6th-order IIR comb filter, with the input ECG signal sampled at 360 Hz

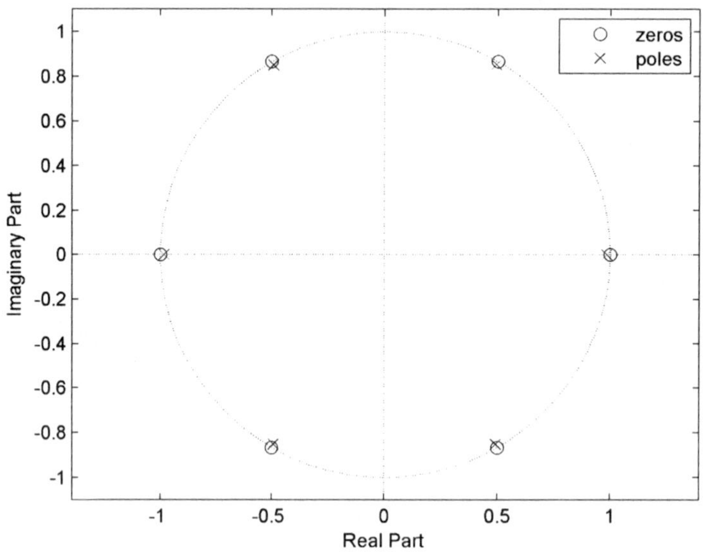

Figure 11. Power-line interference removal effect of the 6th-order IIR comb filter in the frequency domain. (a) Power spectrum of an example ECG record after baseline filtering with the sampling rate being 360 Hz. (b) Power spectrum of the output of the IIR comb filter.

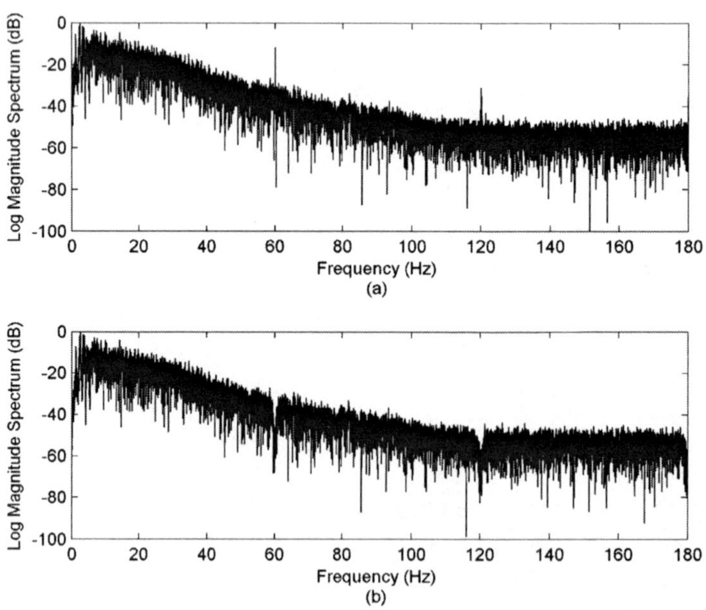

30, which rejects a range of frequencies that is narrow in comparison with the center frequencies at 60 Hz and its harmonics. Compared with the filtering results depicted in Figure 6, the IIR comb filter has performed much better than the FIR comb filter, because the narrow frequency rejection bandwidth of the IIR comb filter does not result in a significant distortion of the ECG waveform.

Fixed or Variable Signal Templates?

One strategy to generate the reference input for adaptive filtering is to create an unaltered signal template that could be obtained by averaging several initial cardiac beat segments. Such a scheme is simple to implement, but it could cause some limitations. First, the unaltered signal template being repeated from one cardiac cycle to another would emphasize the stationary characteristics of the reference input, which might cause the adaptive filter to behave somewhat similarly to a fixed filter. Second, a fixed signal template would be inappropriate if the ECG signal includes ectopic beats, e.g., premature ventricular contractions.

The use of a variable or adaptive signal template is one of a few feasible solutions: the signal template is updated for each cardiac cycle, as implemented in the present work. One possible strategy is to utilize the preceding beat's filtered signal as the template for the upcoming heart beat. However, a normal cardiac beat template will not be suitable for ectopic beats. To overcome this limitation, we applied a Butterworth low-pass filter to the current beat being processed in order to generate a smoothed signal template, with the cost of a delay of one heart beat in the filtered signal.

Signal Template or Noise Reference for Adaptive Filters?

Although a variable signal template could increase the prediction accuracy of the adaptive filter, this scheme only deals with the current beat, rather than effectively acquiring statistical knowledge from the entire available history of the signal. One alternative solution is to use noise as the reference input, and to convert the filter to be an adaptive noise canceller instead of a signal predictor. A basic requirement of such a filter is that the primary noise be not correlated with the signal of interest. Regardless of the changes in the ECG signal, the primary noise is commonly assumed to be a random variable with zero mean. Thus, the use of a noise reference input is more flexible than the approach of using a signal template. The primary noise estimated from the previous cardiac beat can be utilized to filter the upcoming heart beat, regardless of the beat being a normal or an ectopic one. An adaptive noise cancellation system (Wu, Rangayyan, Zhou, & Ng, 2009) can obviate the need for the procedures of QRS detection and ECG beat segmentation, because the reference input would be based on the primary noise, not the ECG signal. In the work of Wu *et al.* (Wu, Rangayyan, Zhou, & Ng, 2009), an unbiased and normalized adaptive noise reduction (UNANR) system was employed for the task of high-frequency random noise elimination. For the primary input to the system being from relatively noise-free (signal-to-noise ratio, SNR: 20 dB) to noisy (SNR: 5 dB), the UNANR system achieved SNR improvements increasing from 10.22 to 25.56 dB, the results of which were much superior to those obtained with the LMS filter.

CONCLUSION

The results obtained with the 16 ECG signals tested indicate that the ULAF method can provide a lower average RMSE with respect to a heart beat template derived from each ECG signal with band-pass filtering, maintain high fidelity of waveform, and attenuate random noise more effectively than the popular LMS and RLS adaptive filters. The learning rate parameter of the ULAF is on a larger scale of variation, so that the

ULAF is able to estimate better the ECG signal components. The strategy of a variable ECG signal template has been successfully implemented in the present work, although this approach requires a high level of computational complexity for QRS detection and ECG beat template establishment. The advantages of adaptive filters for ECG analysis are widely known; however, the set up of appropriate parameters in the filter design is still difficult for physicians and engineers with a lack of signal processing background. In addition, several algorithms require a detailed study of ECG features, e.g., segmentation of P-QRS-T waves (Wu & Rangayyan, 2009), windowing of QRS complexes (Sameni, Shamsollahi, Jutten, & Clifford, 2007), or filter-band reconstruction (Afonso, Tompkins, Nguyen, Michler, & Luo, 1996). These methods consume a significant amount of time for modeling, and are not flexible for application from one patient or condition to another. For further study, we believe that computational intelligence methodologies, especially the state-of-the-art tools based on artificial neural networks and evolutionary computation, have high potential in ECG signal processing and related biomedical applications.

ACKNOWLEDGMENT

This work was supported in part by the Fundamental Research Funds for the Central Universities of China under Grant No. 2010121061 and by the "University Professor" funds awarded by the University of Calgary.

REFERENCES

Afonso, V. X., Tompkins, W. J., Nguyen, T. Q., & Luo, S. (1999). ECG beat detection using filter banks. *IEEE Transactions on Bio-Medical Engineering*, *46*(2), 192–202. doi:10.1109/10.740882

Afonso, V. X., Tompkins, W. J., Nguyen, T. Q., Michler, K., & Luo, S. (1996). Comparing stress ECG enhancement algorithms. *IEEE Engineering in Medicine and Biology Magazine*, *15*(3), 37–44. doi:10.1109/51.499756

Chang, K. M. (2010). Arrhythmia ECG noise reduction by ensemble empirical mode decomposition. *Sensors (Basel, Switzerland)*, *10*(6), 6063–6080. .doi:10.3390/s100606063

Chen, S. W., Chen, H. C., & Chan, H. L. (2006). A real-time QRS detection method based on moving averaging incorporating with wavelet denoising. *Computer Methods and Programs in Biomedicine*, *82*(3), 187–195. doi:10.1016/j.cmpb.2005.11.012

Clifford, G. D., Azuaje, F., & McSharry, P. (2006). *Advanced Methods and Tools for ECG Data Analysis*. Norwood, MA: Artech House.

Hamilton, D. J., Thomson, D. C., & Sandham, W. A. (1995). ANN compression of morphologically similar ECG complexes. *Medical & Biological Engineering & Computing*, *33*(6), 841–843. doi:10.1007/BF02523019

Hamilton, P. S. (1996). A comparison of adaptive and nonadaptive filters for reduction of power line interference in the ECG. *IEEE Transactions on Bio-Medical Engineering*, *43*(1), 105–109. doi:10.1109/10.477707

Hamilton, P. S., & Tompkins, W. J. (1991). Compression of the ambulatory ECG by average beat subtraction and residual differencing. *IEEE Transactions on Bio-Medical Engineering*, *38*(3), 253–259. doi:10.1109/10.133206

Haykin, S. (2002). *Adaptive filter theory* (4th ed.). Englewood Cliffs, NJ: Prentice Hall PTR.

Hu, Y. H., Palreddy, S., & Tompkins, W. J. (1997). A patient-adaptable ECG beat classifier using a mixture of experts approach. *IEEE Transactions on Bio-Medical Engineering*, *44*(9), 891–900. doi:10.1109/10.623058

Hu, Y. H., Tompkins, W. J., Urrusti, J. L., & Afonso, V. X. (1993). Applications of artificial neural networks for ECG signal detection and classification. *Journal of Electrocardiology*, *26*(supplement), 66–73.

Kanjilal, P. P., Palit, S., & Saha, G. (1997). Fetal ECG extraction from single-channel maternal ECG using singular value decomposition. *IEEE Transactions on Bio-Medical Engineering*, *44*(1), 51–59. doi:10.1109/10.553712

Khamene, A., & Negahdaripour, S. (2000). A new method for the extraction of fetal ECG from the composite abdominal signal. *IEEE Transactions on Bio-Medical Engineering*, *47*(4), 507–516. doi:10.1109/10.828150

Kohler, B. U., Hennig, C., & Orglmeister, R. (2002). The principles of software QRS detection. *IEEE Engineering in Medicine and Biology Magazine*, *21*(1), 42–57. doi:10.1109/51.993193

Lehner, R. J., & Rangayyan, R. M. (1987). A three-channel microcomputer system for segmentation and characterization of the phonocardiogram. *IEEE Transactions on Bio-Medical Engineering*, *34*(6), 485–489. doi:10.1109/TBME.1987.326060

Marques de Sa, J. P. (2003). *Applied Statistics Using SPSS, STATISTICA, and MATLAB*. Berlin, Germany: Springer-Verlag.

Meyer, C., Gavela, J. F., & Harris, M. (2006). Combining algorithms in automatic detection of QRS complexes in ECG signals. *IEEE Transactions on Information Technology in Biomedicine*, *10*(3), 468–475. doi:10.1109/TITB.2006.875662

Mita, M. (2007). Algorithm for the classification of multi-modulating signals on the electrocardiogram. *Medical & Biological Engineering & Computing*, *45*(3), 241–250. doi:10.1007/s11517-006-0130-5

Mneimneh, M. A., Yaz, E. E., Johnson, M. T., & Povinelli, R. J. (2006). An adaptive Kalman filter for removing baseline wandering in ECG signals. *Proceedings of the 2006 Computers in Cardiology Conference (CINC '06)* (pp. 253-256). Valencia, Spain.

Murthy, I. S. N., & Rangaraj, M. R. (1979). New concepts for PVC detection. *IEEE Transactions on Bio-Medical Engineering*, *26*(7), 409–416. doi:10.1109/TBME.1979.326420

Pan, J., & Tompkins, W. J. (1985). A real-time QRS detection algorithm. *IEEE Transactions on Bio-Medical Engineering*, *32*(3), 230–236. doi:10.1109/TBME.1985.325532

Rangayyan, R. M. (2002). *Biomedical Signal Analysis: A Case-Study Approach*. New York, NY: IEEE and Wiley.

Sameni, R., Shamsollahi, M. B., Jutten, C., & Clifford, G. D. (2007). A nonlinear Bayesian filtering framework for ECG denoising. *IEEE Transactions on Bio-Medical Engineering*, *54*(12), 2172–2185. doi:10.1109/TBME.2007.897817

Sayadi, O., & Shamsollahi, M. B. (2008). ECG denoising and compression using a modified extended Kalman filter structure. *IEEE Transactions on Bio-Medical Engineering*, *55*(9), 2240–2248. doi:10.1109/TBME.2008.921150

Shannon, C. E. (1948). A mathematical theory of communication. *The Bell System Technical Journal*, *27*, 379–423, 623–656.

Silipo, R., & Marchesi, C. (1998). Artificial neural networks for automatic ECG analysis. *IEEE Transactions on Signal Processing*, *46*(5), 1417–1425. doi:10.1109/78.668803

Thakor, N. V., & Zhu, Y. S. (1991). Applications of adaptive filtering to ECG analysis: noise cancellation and arrhythmia detection. *IEEE Transactions on Bio-Medical Engineering*, *38*(8), 785–794. doi:10.1109/10.83591

Tompkins, W. J. (1993). Biomedical [*Language Examples and Laboratory Experiments for the IBM PC*. Englewood Cliffs, NJ: Prentice Hall PTR.]. *Digital Signal Processing*, C.

Widrow, B., Glover, J. R., McCool, J. M., Kaunitz, J., Williams, C. S., Hearn, R. H., & Zeidler, J. R. (1975). Adaptive noise cancelling: principles and applications. *Proceedings of the IEEE*, *63*(12), 1692–1716. doi:10.1109/PROC.1975.10036

Widrow, B., McCool, J. M., Larimore, M. G., & Johnson, C. R. Jr. (1976). Stationary and non-stationary learning characteristics of the LMS adaptive filter. *Proceedings of the IEEE*, *64*(8), 1151–1162. doi:10.1109/PROC.1976.10286

Wu, Y. F., & Rangayyan, R. M. (2009). An unbiased linear adaptive filter with normalized coefficients for the removal of noise in electrocardiographic signals. *International Journal of Cognitive Informatics and Natural Intelligence*, *3*(4), 73–90. doi:10.4018/jcini.2009062305

Wu, Y. F., Rangayyan, R. M., Zhou, Y. C., & Ng, S. C. (2009). Filtering electrocardiographic signals using an unbiased and normalized adaptive noise reduction system. *Medical Engineering & Physics*, *31*(1), 17–26. doi:10.1016/j.medengphy.2008.03.004

Xue, Q., Hu, Y. H., & Tompkins, W. J. (1992). Neural-network-based adaptive matched filtering for QRS detection. *IEEE Transactions on Bio-Medical Engineering*, *39*(4), 317–329. doi:10.1109/10.126604

Yang, T. F., Devine, B., & Macfarlane, P. W. (1994). Artificial neural networks for the diagnosis of atrial fibrillation. *Medical & Biological Engineering & Computing*, *32*(6), 615–619. doi:10.1007/BF02524235

Zarzoso, V., & Nandi, A. K. (2001). Noninvasive fetal electrocardiogram extraction: blind separation versus adaptive noise cancellation. *IEEE Transactions on Bio-Medical Engineering*, *48*(1), 12–18. doi:10.1109/10.900244

Zigel, Y., Cohen, A., & Katz, A. (2000). ECG signal compression using analysis by synthesis coding. *IEEE Transactions on Bio-Medical Engineering*, *47*(10), 1308–1316. doi:10.1109/10.871403

Compilation of References

Aarts, E. (2004). Ambient intelligence: a multimedia perspective. *IEEE MultiMedia*, *11*(1), 12–19. doi:10.1109/MMUL.2004.1261101

Abar, S., Konno, S., & Kinoshita, T. (2008). Autonomous network monitoring system based on agent-mediated network information. *The International Journal of Computer Science and Network Security*, *8*(2), 326–333.

Adamek, J., Herrlich, H., & Strecker, G. (2009). *Abstract and Concrete Categories*. Dover Publications.

Afonso, V. X., Tompkins, W. J., Nguyen, T. Q., & Luo, S. (1999). ECG beat detection using filter banks. *IEEE Transactions on Bio-Medical Engineering*, *46*(2), 192–202. doi:10.1109/10.740882

Afonso, V. X., Tompkins, W. J., Nguyen, T. Q., Michler, K., & Luo, S. (1996). Comparing stress ECG enhancement algorithms. *IEEE Engineering in Medicine and Biology Magazine*, *15*(3), 37–44. doi:10.1109/51.499756

Agosti, M., & Smeaton, A. (1996). *Information retrieval and hypertext*. New York: Kluwer.

Alda, S., Cramers, A. B., Bilek, J., & Hartmann, D. (2004), Support of Collaborative Structural Design Processes through the Integration of Peer-to-Peer and Multi-agent Architectures, *in Proceedings of the 10th International Conference on Computing in Civil and Building Engineering (ICCCDE-X)*, Weimar, Germany.

Alto, H., Rangayyan, R. M., & Desautels, J. E. L. (2005). Content-based retrieval and analysis of mammographic masses. *Journal of Electronic Imaging*, *14*(2), 1–17. doi:10.1117/1.1902996

Amir, Y., & Wool, A. (1998). Optimal availability quorum systems: Theory and Practice. *Information Processing Letters*, *65*(5), 223–228. doi:10.1016/S0020-0190(98)00017-9

Anderson, J. A., & Rosenfeld, E. (Eds.). (1988). *Neurocomputing: Foundations of Research*, Cambridge.

André, T. C. S. S., & Rangayyan, R. M. (2006). Classification of tumors and masses in mammograms using neural networks with shape and texture features. *Journal of Electronic Imaging*, *15*(1), 1–10. doi:10.1117/1.2178271

Aschoff, J. (1984). Circadian Timing . *Annals of the New York Academy of Sciences*, *423*, 442–468. doi:10.1111/j.1749-6632.1984.tb23452.x

Asperti, A., & Longo, G. (1991). *Categories, Types and Structures*. M.I.T. Press.

Axelrod, R. (1977). *The Complexity of Cooperation: Agent-Based Models of Competition and Collaboration*. Princeton, NJ: Princeton Univ. Press.

Ayars, J., Bulterman, D., Cohen, A., Day, K., Hodge, E., Hoschka, P., et al. (2005). *Synchronized Multimedia Integration Language (SMIL 2.0) - [Second Edition]* (W3C Recommendation).

Azevedo, R., & Lajoie, S. P. (1998). The cognitive basis for the design of a mammography interpretation tutor. *International Journal of Artificial Intelligence in Education*, *9*, 32–44.

Barr, A., & Feigenbaum, E. A. (Eds.). (1981). *The Handbook of Artificial Intelligence* (*Vol. 1*). Stanford and Los Altos, CA: HeurisTech Press and Kaufmann.

Bartle, R. A. (2005). In Media, C. R. (Ed.), *Why People Play, Massively Multiplayer Game Development 2, Thor Alexander* (pp. 3–18). Hingham, MA: Virtual Worlds.

Bartle, R. A. (1996), Hearts, Clubs, Diamonds, Spades: Players Who Suit MUDs, from http://mud.co.uk/richard/hcds.htm

Bastiaans, M. J. (1981). A sampling theorem for the complex spectrogram, and Gabor's expansion of a signal in Gaussian elementary signals. *Optical Engineering (Redondo Beach, Calif.)*, *20*(2), 594–598.

Battista, S., Casalino, F., & Lande, C. (1999). MPEG-4: A Multimedia Standard for the Third Millennium, Part 1. *IEEE MultiMedia*, *6*(4), 74–83. doi:10.1109/93.809236

Battista, S., Casalino, F., & Lande, C. (2000). MPEG-4: A Multimedia Standard for the Third Millennium, Part 2. *IEEE MultiMedia*, *7*(1), 76–84. doi:10.1109/93.839314

Belalem, G., & Slimani, Y. (2007). A hybrid approach to replica management in data grids [IJWGS]. *International Journal Web and Grid Services*, *3*(1), 2–18. doi:10.1504/IJWGS.2007.012634

Belalem G., Benotmane Z. & Benhallou K. (2009). Self Adjustable Negotiation Mechanism for Convergence and Conflict Resolution of Replicas in Data Grids, *International Journal of Cognitive Informatics and Natural Intelligence, (IJCINI*). 3(1), 95-110

Belalem, G. (2008). Economic Model for Consistency Management of Replicas in Data Grids with OptorSim Simulator, *Networks for Grid Applications, Second International Conference - (GridNets 2008),* (pp. 121-129), Beijing, China, October 8-10, 2008, Lecture Notes of the Institute for Computer Sciences, Social Informatics and Telecommunications Engineering, Vol. 2, Springer.

Belalem, G., Tayeb, F. Z., & Zaoui, W. (2010). Approaches to Improve the Resources Management in the Simulator CloudSim, *First International Conference Information Computing and Applications - (ICICA'2010),* (pp. 189-196), Tangshan, China, October 15-18, Lecture Notes in Computer Science, Vol. 6377, Springer.

Bellifemine, F., Poggi, A., & Rimassa, G. (1999). JADE – a FIPA-compliant agent framework. In *Proceedings of Practical Application of Intelligent Agents and Multi Agents* (PAAM '99), (pp.97-108).

Bender, E. A. (2000). *Mathematical Methods in Artificial Intelligence*. Los Alamitos, CA: IEEE CS Press.

Benjamins, V. R., Centreras, J., Corcho, O., & Gomez-Perez, A. (2002). Six Challenges for the Semantic Web. *ISWC2002*.

Berger, J. (1990). *Statistical Decision Theory – Foundations, Concepts, and Methods*. Springer-Verlag.

Bergman, G. M. (1998). *An Invitation to General Algebra and Universal Constructions. 15 the Crescent, Berkeley CA 94708*. US: Henry Helson.

Bergstra, J., Ponse, A., & Smolka, S. (Eds.). (2001). *Handbook of process algebra*. North Holland.

Berners-Lee, T., & Fischetti, M. (1999). *Weaving the Web: The Original Design and Ultimate Destiny of the World Wide Web by Its Inventor*. Harper San Francisco.

Berners-Lee, T., Hawke, S., & Connolly, D. (2004). *Semantic Web Tutorial Using N3*. Turorial.

Berners-Lee, T. (n.d.). *Semantic Web Road Map*. World Wide Web consortium, http://www.w3.org/DesignIssues/Semantic.html.

Berners-Lee, T., Hendler, J., & Lassila, O. (2001). *The Semantic Web*. Scientific American.

Berni, A., Ramdane-Cherif, A., Saadia, N., & Levy, N. (2003). Exploring cognitive approach through the neural network paradigm: "trajectory planning application". *Proceedings of The Second IEEE International Conference on Cognitive Informatics* (pp. 47-54).

Betourne, A., & Campion, G. (1996). Dynamic model and control design of a class of omnidirectional mobile robots. *Proceedings of the 1996 IEEE International Conference on Robotics and Automation* (pp. 2810-2815).

Beveridge, W. I. (1957). *The Art of Scientific Investigation*. UK: Random House Trade Paperbacks.

Billings, A. R., & Scolaro, A. (1976). The Gabor compression-expansion system using non-Gaussian windows and its application to television coding. *IEEE Transactions on Information Theory*, *22*(2), 174–190. doi:10.1109/TIT.1976.1055535

Bjork, S., Hansson, J., & Ljungstrand, P. (2001), Pirates! - Using the Physical World as a Game Board, Proceedings of INTERACT IFIP TC.13 Conference on Human-Computer Interaction, 2001.

Björk, S., Holopainen, J., Ljungstrand, P., & Åkesson, K.-P. (2002), Designing Ubiquitous Computing Games - A Report from a Workshop Exploring Ubiquitous Computing Entertainment, Personal and Ubiquitous Computing, January '02, Volume 6, Issue 5-6, pp. 443-458.

Bochum, R., & Wiskott, L. (1999). Segmentation from motion: combining Gabor and Mallat wavelets to overcome aperture and correspondence problem. *Pattern Recognition*, *32*(10), 1751–1766. doi:10.1016/S0031-3203(98)00179-4

Bond, A. H., & Gasser, L. (1988). *Readings in Distributed Artificial Intelligence*. San Mateo, CA: Morgan Kaufmann.

Bronson, R., & Naadimuthu, G. (1997). *Schaum's Outline of Theory and Problems of Operations Research* (2nd ed.). NY: McGraw-Hill.

Brooks, R. A. (1970). New Approaches to Robotics . *American Elsevier, NY*, *5*, 3–23.

Broomhead, D. S., & Lowe, D. (1988). Multi-variable functional interpolation and adaptive networks. *Complex Systems*, *2*(3), 269–303.

Brumitt, B., Meyers, J., & Krumm, A. Kern and Shafer, S. (2000), EasyLiving: Technologies for Intelligent Environments, Proceedings of the International Conference on Handheld and Ubiquitous Computing, Springer, 2000, pp.12-29.

Bu, D. B., Bai, S., & Li, G. J. (2002). Principle of Granularity in Clustering and Classification. [in Chinese]. *Chinese Journal of Computers*, *25*(8), 810–816.

Burges, C. J. C. (1998). A Tutorial on Support Vector Machines for Pattern Recognition. *Data Mining and Knowledge*, *2*(2), 121–167. doi:10.1023/A:1009715923555

Buschmann, F., Meunier, R., Rohnert, H., Sommerlad, P., & Stal, M. (1996). *Pattern-Oriented Software Architecture, Volume 1: A System of Patterns*: John Wiley & Sons, Inc.

Butera, W. (2007, 9-11 July). Text Display and Graphics Control on a Paintable Computer. In G. Serugendo, J. Flatin, & M. Jelasity (Eds.), *Proceedings of 1st international conference on self-adaptive and self-organizing systems (saso'07)* (pp.45–54). Boston, Massachusetts, USA: IEEE Computer Society Press.

Buyya, R., & Vazhkudai, S. (2001). Compute power market: Towards a market-oriented Grid, CCGRID'01, *First International Symposium on Cluster Computing and the Grid*, (pp. 574-581), Brisbane, Australia.

Calisti, M., Meer, S., & Strassner, J. (Eds.). (2008). *Advanced Autonomic Networking and Communication.* Springer-Verlag. doi:10.1007/978-3-7643-8569-9

Calvary, G., Coutaz, J., & Nigay, L. (1997). From single-user architectural design to PAC*: a generic software architecture model for CSCW. *CHI'97 Conference*, 242-249.

Campion, G., & Bastin, G. (1996). Structural properties and classification of kinematic and dynamic models of wheeled mobile robots. *IEEE Transactions on Robotics and Automation*, *12*(1), 47–62. doi:10.1109/70.481750

Carneiro, G., & Vasconcelos, N. (2005). A Database Centric View of Semantic Image Annotation and Retrieval. *Proceedings of ACM Conference on Research and Development in Information Retrieval.*

Carroll, L., & Chorpenning, C. B. (1958). *Alice in Wonderland*. Dramatic Publishing Co., Woodstock.

Cavazza, M., Charles, F., & Mead, S. J. (2002), Character-based Interactive Storytelling. In IEEE Intelligent Systems, special issue on AI in Interactive Entertainment, pp. 17-24.

Chaib-Draa, B. Moulin, R. Mandiau, and P. Millot. (1992). Trends in Distributed Artificial Intelligence . *Artificial Intelligence Review*, *6*, 35–66. doi:10.1007/BF00155579

Chang, K. M. (2010). Arrhythmia ECG noise reduction by ensemble empirical mode decomposition. *Sensors (Basel, Switzerland)*, *10*(6), 6063–6080. .doi:10.3390/s100606063

Chang, R.-S., & Chang, J.-S. (2006). Adaptable Replica Consistency Service for Data Grids. *Third International Conference on Information Technology: New Generations (ITNG'06)*, pp. 646-651, Las Vegas, Nevada, USA.

Chen, S. W., Chen, H. C., & Chan, H. L. (2006). A real-time QRS detection method based on moving averaging incorporating with wavelet denoising. *Computer Methods and Programs in Biomedicine*, *82*(3), 187–195. doi:10.1016/j.cmpb.2005.11.012

Cheng, X. M. (2003). The Method Analysis of Formant Parameters Picked-up in Sensibility Speech Communication. *Journal of Huzhou Teachers College*, *25*(6), 76–80.

Cheok, A. D., et al. (2003), Human Pacman: A mobile entertainment system with ubiquitous computing and tangible interaction over a wide outdoor area, Proceedings of the 17th Annual Human Computer Interaction Conference, England, Sept. 2003, Springer-Verlag LNCS press.

Chiew, V., & Wang, Y. (2003). From cognitive psychology to cognitive informatics. In *Second IEEE International Conference on Cognitive Informatics, ICCI'03*, London, UK, (pp. 114-120).

Chorafas, D. N. (1998). *Agent Technology Handbook.* NY: McGraw-Hill.

Chung, J. H., Velinsky, S. A., & Ronald, A. H. (1998). Interaction control of a redundant mobile manipulator. *The International Journal of Robotics Research*, *17*(12), 1302–1309. doi:10.1177/027836499801701203

Chung, J. H., Yi, B. J., & Kim, W. K. (2003). The dynamic modeling and analysis for an omnidirectional mobile robot with three castor wheels. *Proceedings of the 2003 IEEE International Conference on Robotics and Automation* (pp. 521-527).

Clifford, G. D., Azuaje, F., & McSharry, P. (2006). *Advanced Methods and Tools for ECG Data Analysis*. Norwood, MA: Artech House.

Coaen, S., Waad, L. M., & Enns, J. T. (1993). *Sensation and Perception* (4th ed.). Fort Worth, USA: Harcourt Brace College Publishers.

Coaen, S., Ward, L. M., & Enns, J. T. (1994). *Sensation and Perception* (4th ed.). NY: Harcourt Brace College Pub.

Codognet, P. (1998), "Artificial Nature and Natural Artifice". Presented at the Art & Technology Conference, Tate Gallery, Liverpool, and at ARCO'02, Madrid, panel discussion on "Art and New Media".

Costa, P. T. Jr, & McCrae, R. R. (1992). *The NEO-PI-R: Professional manual*. Odessa, FL: Psychological Assessment Resources.

Costa, P.C.G., Laskey, K.B., & Laskey, K.J. (2005). PR-OWL: A Bayesian Ontology Language for the Semantic Web. *URSW'05.*

Coutaz, J. (1987). PAC, an Implemention Model for Dialog Design. *Interact*, *87*, 431–436.

Coutaz, J., Lachenal, C., Berard, F., & Barralon, N. (2002). Quand les Surfaces Deviennent Interactives, Les Cahiers du Numérique . *Lavoisier*, *3*(4), 101–126.

Cover, T. M. (1965, June). Geometrical and statistical properties of systems of linear inequalities with applications in pattern recognition. *IEEE Transactions on Electronic Computers*, *EC-14*, 326–334. doi:10.1109/PGEC.1965.264137

Cowan, J. D. (1973). Some remarks on channel bandwidth for visual contrast detection. *Neurosciences Research Program Bulletin*, *15*, 1255–1267.

Cowie, R., Douglas-Cowie, E., Tsapatsoulis, N., Votsis, G., Kollias, S., & Fellenz, W. (2001). Emotion recognition in human-computer interaction. *IEEE Signal Processing Magazine*, *18*(1), 32–80. doi:10.1109/79.911197

Csikszentmihalyi, M. (1996). *Creativity: Flow and the Psychology of Discovery and Invention.* New York: HarperCollins.

Cubaud, P., Dupire, J., & Topol, A. (2005), Digitization and 3D Modeling of Movable Books, ACM-IEEE Joint Conference on Digital Libraries, Denver, USA, June, 2005.

Cutnell, J. C., & Johnson, K. W. (1998). *Physics* (4th ed.). NY: John Wiley & Sons.

Dastani, M., Jacobs, N., Jonker, C., & Treur, J. (2001). Modelling user preferences and mediating agents in electronic commerce . In *Agent mediated electronic commerce, lnai 1991* (pp. 163–193). Springer. doi:10.1007/3-540-44682-6_10

Daubechies, I. (1992). Ten Lectures on Wavelets. CBMS-NSF Regional Conference Series. *Applied Mathematics.*

Daugman, J. G. (1988). Complete Discrete 2-D Gabor transform by neural networks for image analysis and compression. *IEEE Transactions on Acoustics, Speech, and Signal Processing, 36*(7), 1169–1179. doi:10.1109/29.1644

Davies, J., Fensel, D., & Harmelen, F. V. (2003). *Towards the Semantic Web -- Ontology-Driven Knowledge Management*. Wilq.

De, S. A., Loach, W. O., & Matson, E. (2008, February). A Capabilities-based Model for Adaptive Organizations. *Autonomous Agents and Multi-Agent Systems, 16*(1), 13–56. doi:10.1007/s10458-007-9019-4

de la Encina, A., Hidalgo-Herrero, M., Rabanal, P., Rodríguez, I., & Rubio, F. (2008). Testing the behaviour of entities in a cognitive language. *International Journal of Cognitive Informatics and Natural Intelligence, 2*(1), 29–43. doi:10.4018/jcini.2008010103

de la Encina, A., Hidalgo-Herrero, M., Rabanal, P., Rubio, F., & Rodríguez, I. (2006). Testing entities in a parallel cognitive language. In *Fifth IEEE International Conference on Cognitive Informatics,* 2006 (pp. 344-355). IEEE-CS Press.

Dellaert, F., Polzin, T., & Waibel, A. (1996). Recognizing Emotion in Speech. In . *Proceedings of the ICSLP, 96*, 1970–1973.

DeLoach, S. A., Wood, M., & Sparkman, C. H. (2001). Multiagent systems engineering. *International Journal of Software Engineering and Knowledge Engineering, 11*(3), 231–258. doi:10.1142/S0218194001000542

Demazeau, Y., & Costa, A. C. R. (1996). Populations and Organizations in Open Multi-Agent Systems, *in Symposium on Parallel and Distributed Artificial Intelligence (PDAI'96)*, Hyderabad, India.

Denko, M. K., Yang, L. T., & Zhang, Y. (Eds.). (2009). *Autonomic Computing and Networking* (1st ed.). Springer USA. (452 pages)

Ding, Z., & Peng, Y. (2004). A Probabilistic Extension to Ontology Language OWL. *Proceedings of the 37th Hawaii International Conference on System Sciences.*

do Espírito Santo, R., de Deus Lopes, R., & Rangayyan, R. M. (2005). Classification of mammographic masses using radial basis functions and simulated annealing with shape, acutance, and texture features. In *Proc. 3rd IASTED International Conference on Biomedical Engineering*, Innsbruck, Austria, (pp. 164-167).

du Buf, J. M. H., & Heitkamper, P. (1991). Texture features based on Gabor phase. *Signal Processing, 23*(3), 227–244. doi:10.1016/0165-1684(91)90002-Z

Dubos, R. J. (1950). *Louis Pasteur: Freelance of Science*. Boston: Little, Brown & Co.

Duc, B., Fischer, S., & Bigüm, J. (1999). Face authentication with Gabor information on deformable graphs. *IEEE Transactions on Image Processing, 8*(4), 504–516. PubMeddoi:10.1109/83.753738

Edwards, W. K. (2000). *Core JINI*. Prentice Hall PTR.

Einstein, A. (1995). *Relativity: The Special and the General Theory*. Reprint, Three Rivers Press.

Einstein, A., & Besso, M. (1972). *Correspondence, 1903–1955, Translated by P. Speziali from French*. Paris: Hermann.

Einstein, A. (1905), On the Electrodynamics of Moving Bodies, *Annalen der Physik,* 17(891), June, (English translation in 1922).

Einstein, A. (1916), The Foundation of the General Theory of Relativity, *Annalen der Physik,* 49.

El-Nasr, M. S., & Vasilakos, T. (2006). DigitalBeing: An Ambient Intelligent Dance Space. *Fuzzy Systems, 2006 IEEE International Conference on*, 907-914.

Er, M. J., & Gao, Y. (2003). Robust adaptive control of robot manipulators using generalized fuzzy neural networks. *IEEE Transactions on Industrial Electronics, 50*(3), 620–628. doi:10.1109/TIE.2003.812454

Eymann, T. (2001). Markets without makers - a framework for decentralized economic coordination in multiagent systems. In *Welcom 2001, lncs 2232* (pp. 63–74). Springer.

Fang, R., Zhao, Y. b., & Li, W. S. (2005). A Novel Fuzzy Neural Network: The Vague Neural Network. *Proceedings of the Third IEEE International Conference on Cognitive Informatics* (pp. 94-99).

Feijs, L. M. G., & Hu, J. (2004). Component-wise Mapping of Media-needs to a Distributed Presentation Environment. *The 28th Annual International Computer Software and Applications Conference (COMPSAC 2004)*, 250-257.

Flax, L. (2007). Cognitive modelling applied to aspects of schizophrenia and autonomic computing. *International Journal of Cognitive Informatics and Natural Intelligence*, *1*(2), 58–72. doi:10.4018/jcini.2007040104

Foner, L. (1993), What ís an Agent, Anyway? A Sociological Case Study, *Agents Memo 93-01*, MIT Media Lab, Cambridge, MA.

Foster, I., & Kesselmann, C. (Eds.). (2004). *The Grid 2: Blueprint for a new computing infrastructure*. Elsevier Series in Grid Computing. Morgan Kaufmann Publishers.

Fougeres, A.-J. (2004). Agents to cooperate in distributed design. *IEEE International Conference on Systems, Man and Cybernetics, 3*, 2629-2634.

Fujita, S., Hara, H., Sugawara, K., Kinoshita, T., & Shiratori, N. (1998). Agent-based design model of adaptive distributed system. *Applied Intelligence*, *9*(1), 57–70. doi:10.1023/A:1008299131268

Gabor, D. (1946). Theory of communication. [London.]. *J. of Industrial Electrical Engineering*, *93*(3), 429–457.

Gabor, D. (1947). New possibilities in speech transmission. [London.]. *J. of Industrial Electrical Engineering*, *94*(3), 369.

Gagne, R. (1985). *The conditions of learning* (4th ed.). New York: Holt, Rinehart and Winston.

Gallais, Henry, Saphores, Rapine, Guillebon, Roubinet, et al (2007), NSRC gamedoc, available on request at: contact@enjmin.fr

Gamma, E., Helm, R., Johnson, R., & Vlissides, J. (1995). *Design Patterns - Elements of Reusable Object-oriented Software*. Addison-Wesley.

Garlan, D., Siewiorek, A. and Steenkiste, P. (2002), Project Aura: Toward Distraction-Free Pervasive Computing, IEEE Pervasive Computing.

Genesereth, M. R., & Ketchpel, S. P. (1994). Software Agents . *Communications of the ACM*, *37*(7), 48–53. doi:10.1145/176789.176794

Genvo, S. (2006), Le game design de jeux video, approche communicationnelle et interculturelle, PhD thesis, University of Metz, October 2006. Available at: http://www.omnsh.org/article.php3?id_article=97

Ghafoor, A., & Rehman, M. ur, Khan, Z. Abbas, Ali, A., Ahmad, H. Farooq and Suguri, H. (2004). SAGE: next generation multi-agent system. In *Proceedings of Parallel and Distributed Processing Techniques and Applications*, (pp.139-145).

Giarrantans, J., & Riley, G. (1989). *Expert Systems: Principles and Programming*. Boston: PWS-KENT Pub. Co.

Glickstein, M. (1988). The Discovery of the Visual Cortex. *Scientific American*, *259*, 118–127. doi:10.1038/scientificamerican0988-118

Goel, S., Sharda, H., & Taniar, D. (2005). Replica synchronisation in grid databases. [IJWGS]. *International Journal Web and Grid Services*, *1*(1), 87–112. doi:10.1504/IJWGS.2005.007551

Goldstein, E. B. (1999). *Sensation and Perception*, 5th ed. NY: Brooks/Cole Publishing Co., ITP.

Google. http://www.google.com/press/pressrel/6billion.html.

Google Achieves Search Milestone with Immediate Access To More Than 6 Billion Items.

Gray, P. (1994). *Psychology* (2nd ed.). New York: Worth Publishers, Inc.

Gray, J., Helland, P., Neil, P. O., & Shasha, D. (1996), The dangers of replication and a solution. *In ACM SIGMOD International Conference on Management of Data*, (pp. 173-182), Montreal, Quebec, Canada, 4-5 June 1996. ACM Press.

Griffith, D., & Greitzer, F. (2007). Neo-Symbiosis: The Next Stage in the Evolution of Human Information Interaction. *International Journal of Cognitive Informatics and Natural Intelligence*, *1*(1), 39–52. doi:10.4018/jcini.2007010103

Grossmann, A., & Morlet, J. (1984). Decomposition of Hardy Functions into Square Integrable Wavelets of Constant Shape. *SIAM Journal on Mathematical Analysis*, *15*(4), 723–736. doi:10.1137/0515056

Grünvogel, S. M., Vega, L., & Natkin, S. (2004), A new Methodology for Spatiotemporal Game Design, Proc of the Fifth Game-On International Conference on Computer Games: Artificial Intelligence, Design and Education CGAIDE'2004, pp. 109-113.

Guardiola, E. (2000). *Ecrire pour le jeu: Techniques scénaristiques du jeu informatique et vidéo*. Ed. Dixit.

Guiford, J. P. (1967). *The Nature of Human Intelligence*. NY: McGraw-Hill.

Gustavo, M., & Talmud, I. (2006). The Quality of Online and Offline Relationships, the role of multiplexity and duration . *The Information Society*, 2006.

Haddad, C., & Slimani, Y. (2007). Economic model for replicated database placement in Grid. *In Proceedings of* Seventh *IEEE International Symposium on Cluster Computing and the Grid (CCGrid'07)*, (pp. 283-292), Rio de Janeiro, Brazil.

Hadjiiski, L., Sahiner, B., Chan, H.-P., Petrick, N., & Helvie, M. (1999). Classification of malignant and benign masses based on hybrid ART2LDA approach. *IEEE Transactions on Medical Imaging*, *18*(12), 1178–1193. doi:10.1109/42.819327

Hall, M. A. (1999). *Correlation based Feature Selection for Machine Learning.* Doctoral dissertation, Department of Computer Science, The University of Waikato, Hamilton, New Zealand.

Hamilton, D. J., Thomson, D. C., & Sandham, W. A. (1995). ANN compression of morphologically similar ECG complexes. *Medical & Biological Engineering & Computing*, *33*(6), 841–843. doi:10.1007/BF02523019

Hamilton, P. S. (1996). A comparison of adaptive and nonadaptive filters for reduction of power line interference in the ECG. *IEEE Transactions on Bio-Medical Engineering*, *43*(1), 105–109. doi:10.1109/10.477707

Hamilton, P. S., & Tompkins, W. J. (1991). Compression of the ambulatory ECG by average beat subtraction and residual differencing. *IEEE Transactions on Bio-Medical Engineering*, *38*(3), 253–259. doi:10.1109/10.133206

Hara, H., Sugawara, K., Kinoshita, T., & Uchiya, T. (2002). Flexible distributed agent system and its application. In *Proceedings of the Fifth Joint conference of Knowledge-based Software Engineering*, (pp.72-77), IOS Press.

Hauser, L. (1997). Searle's chinese box: Debunking the chinese room argument. *Minds and Machines*, *7*, 199–226. doi:10.1023/A:1008255830248

Hayes-Roth, B. (1995). An Architecture for Adaptive Intelligent Systems . *Artificial Intelligence*, *72*(1-2), 329–365. doi:10.1016/0004-3702(94)00004-K

Haygood, R., & Bourne, R. (1965). Attribute- and rule-learning aspects of conceptual behavior. *Psychological Review*, *72*(3), 175–195. doi:10.1037/h0021802

Haykin, S. (1999). *Neural Networks: A Comprehensive Foundation*. Upper Saddle River, NJ: Prentice Hall.

Haykin, S. (2002). *Adaptive filter theory* (4th ed.). Englewood Cliffs, NJ: Prentice Hall PTR.

Heckel, P. (1991). *The Elements of Friendly Software Design*.

Hewitt, C., & Inman, J. (1991). *DAI Betwixt and Between: From Intelligent Agents to Open Systems Science*. IEEE Trans. on System, Man, and Cybernetics, Nov/Dec.

Hewitt, C., Bishop, R., & Steiger, R. (1973), A Universal Modular Actor Formalism for Artificial Intelligence, *Proc. 3rd Int. Joint Conf. on Artificial Intelligence*, Stanford, CA, Aug.

Hidalgo-Herrero, M., Rodríguez, I., & Rubio, F. (2005). Testing learning strategies. In *Forth IEEE International Conference on Cognitive Informatics* (pp. 212-221). IEEE-CS Press.

Hoare, C. A. R. (1985). *Communicating Sequential Processes.* Prentice-Hall.

Holland, J. H. (1992). Genetic Algorithms . *Scientific American*, *267*, 66–72. doi:10.1038/scientificamerican0792-66

Holland, J. H., Holyoak, K. J., Nisbett, R. E., & Thagard, P. R. (1986). *Induction: Processes of Inference, Learning, and Discovery.* Cambridge, MA: MIT Press/ Bradford Books.

Holmberg, R., & Khatib, O. (2000). Development and control of a holonomic mobile robot for mobile manipulation tasks. *The International Journal of Robotics Research*, *19*(11), 1066–1074. doi:10.1177/02783640022067977

Hou, Z. G., & Tan, M. (2004). Real-Time Optimization and Computation for Interconnected Nonlinear Systems Using Neural Networks. *Proceedings of the Third IEEE International Conference on Cognitive Informatics* (pp. 208-213).

Hu, J. (2006). *Design of a Distributed Architecture for Enriching Media Experience in Home Theaters*. Technische Universiteit Eindhoven.

Hu, Y. H., Palreddy, S., & Tompkins, W. J. (1997). A patient-adaptable ECG beat classifier using a mixture of experts approach. *IEEE Transactions on Bio-Medical Engineering*, *44*(9), 891–900. doi:10.1109/10.623058

Hu, Y. H., Tompkins, W. J., Urrusti, J. L., & Afonso, V. X. (1993). Applications of artificial neural networks for ECG signal detection and classification. *Journal of Electrocardiology*, *26*(supplement), 66–73.

Hu, J. (2003). StoryML: Enabling Distributed Interfaces for Interactive Media. *The Twelfth International World Wide Web Conference*.

Hu, J., & Bartneck, C. (2005). Culture Matters - A Study on Presence in an Interactive Movie. *PRESENCE 2005, The 8th Annual International Workshop on Presence*, 153-159.

Hu, J., & Feijs, L. M. G. (2003). An Adaptive Architecture for Presenting Interactive Media onto Distributed Interfaces. *The 21st IASTED International Conference on Applied Informatics (AI 2003)*, 899-904.

Hu, J., Janse, M. D., & Kong, H. (2005). User Evaluation on a Distributed Interactive Movie. *HCI International 2005, 3 - Human-Computer Interfaces: Concepts, New Ideas, Better Usability, and Applications*, 735.731-710.

Hubel, D., & Wiesel, T. N. (1959). Receptive Fields of Single Neurons in the Cat's Visual Cortex. *The Journal of Physiology*, *148*, 574–591.

Hubel, D., & Wiesel, T. N. (1979). Brain Mechanisms of Vision. *Scientific American*, *82*, 84–97.

Huhns, M., & Singh, M. (Eds.). (1997). *Readings in Agents*. San Francisco: Kaufmann.

Hyvonen, E., Saarela, S., & Viljanen, K. (2003). Intelligent Image Retrieval and Browsing Using Semantic Web Techniques – A Case Study. *The International SEPIA Conference*

IBM. (2001). Autonomic Computing Manifesto. Retrieved from http://www.research. ibm.com/ autonomic/.

IBM. (2006), *Autonomous Computing White Paper: An Architectural Blueprint for Autonomous Computing*, 4th ed., June, 1-37.

Illmann, T., Weber, M., Martens, A., & Seitz, A. (2000). A Pattern-Oriented Design of a Web-Based and Case Oriented Multimedia Training System in Medicine. *The 4th World Conference on Integrated Design and Process Technology*.

Imai, S., Kitagata, G., Konno, S., Suganuma, T., & Kinoshita, T. (2004). Developing a knowledge-based videoconference system for non-expert users. *Journal of Distance Education Technologies*, *2*(2), 13–26. doi:10.4018/jdet.2004040102

Jain, A. K., & Chandrasekaran, B. (1983). Dimensionality and sample size considerations. In P.R. Krishnaiah & L.N. Kanal (Eds.), *Pattern Recognition Practice, 2*(39), 835-855.

Janse, M. D., van der Stok, P., & Hu, J. (2005). Distributing Multimedia Elements to Multiple Networked Devices. *User Experience Design for Pervasive Computing, Pervasive 2005*.

Jazar, R. N. (2007). *Theory of Applied Robotics: Kinematics, Dynamics, and Control*. Berlin: Springer.

Jennings, N. R. (2000). On Agent-Based Software Engineering . *Artificial Intelligence*, *17*(2), 277–296. doi:10.1016/S0004-3702(99)00107-1

Jensen, R., & Shen, Q. (2007). Rough set based feature selection: A review . In *Rough Computing*. Theories, Technologies and Applications. doi:10.4018/9781599045528.ch003

Jensen, F.V., Lauritzen, S.L., & Olesen, K.G. (1990). *Bayesian Updating in Causal Probabilistic Network by Local Computation.*

Jeon, H., Petrie, C., & Cutkosky, M. (2000). JATLite: a java agent infrastructure with message routing. *IEEE Internet Computing*, *4*(2), 87–96. doi:10.1109/4236.832951

Jin, X., & Liu, J. (2004, April). From Individual Based Modeling to Autonomy Oriented Computation. In Nickles, M., Rovatsos, M., & Weiss, G. (Eds.), *Agents and computational autonomy: Potential, risks, and solutions* (*Vol. 2969*, pp. 151–169). Springer Berlin. doi:10.1007/978-3-540-25928-2_13

Jones, J. P., & Palmer, L. A. (1987). The two dimensional spatial structure of simple receptive fields in cat striate cortex. *Journal of Neurophysiology*, *58*(6), 1187–1211. PubMed

Kanizsa, G. (1979). *Organization in Vision: Essays on Gestalt Perception*. NY: Praeger.

Kanjilal, P. P., Palit, S., & Saha, G. (1997). Fetal ECG extraction from single-channel maternal ECG using singular value decomposition. *IEEE Transactions on Bio-Medical Engineering*, *44*(1), 51–59. doi:10.1109/10.553712

Kant, S., & Mamas, E. (2005). Statistical Reasoning – A Foundation for Semantic Web Reasoning. *URSW'05*.

Kephart, J., & Chess, D. (2003). The Vision of Autonomic Computing, IEEE. *Computer*, *26*(1), 41–50. doi:10.1109/MC.2003.1160055

Keppens, J., & Shen, Q. (2002). A calculus of partially ordered preferences for compositional modelling and configuration. In *AAAI Workshop on Preferences in AI and CP: Symbolic Approaches* (pp. 39–46). AAAI Press.

Kermarrec, A.-M., Rowstron, A., Shapiro, M., & Druschel, P. (2001). The IceCube approach to the reconciliation of divergent replicas. *PODC '01: Proceedings of the twentieth annual ACM symposium on Principles of distributed computing*, (pp. 210-218), Newport, Rhode Island, USA.

Khamene, A., & Negahdaripour, S. (2000). A new method for the extraction of fetal ECG from the composite abdominal signal. *IEEE Transactions on Bio-Medical Engineering*, *47*(4), 507–516. doi:10.1109/10.828150

Khamis, A., Rodriguez, F. J., & Salichs, M. A. (2003). Remote Interaction with Mobile Robots. *Autonomous Robots*, *15*(3). doi:10.1023/A:1026268504593

Khamis, A., Rivero, D. M., Rodriguez, F., & Salichs, M. (2003). Pattern-based Architecture for Building Mobile Robotics Remote Laboratories. *IEEE International Conference on Robotics and Automation (ICRA'03), 3*, 3284-3289.

Khatib, O. (1987). A unified approach to motion and force control of robot manipulators: the operational Space formulation. *IEEE Journal on Robotics and Automation*, *3*(1), 43–53. doi:10.1109/JRA.1987.1087068

Khatib, O., Yokoi, K., Chang, K., Ruspini, D., Holmberg, R., & Casal, A. (1996). Coordination and decentralized cooperation of multiple mobile manipulators. *International Journal of Robotic System*, *13*(11), 755–764. doi:10.1002/(SICI)1097-4563(199611)13:11<755::AID-ROB6>3.0.CO;2-U

Khoo, E. T., & Cheok, A. D. (2006). Age Invaders: Inter-generational Mixed Reality Family Game . *The International Journal of Virtual Reality*, *5*(2), 45–50.

Kim, H. Kinoshita, T., Lim, Y. and Kim, T. (2010). A bankruptcy problem approach to load-shedding in multiagent-based microgrid operation. Sensors Vol.10, No.10, (pp.8888-8898), MDPI Publishing.

Kinoshita, T., & Sugawara, K. (1998). ADIPS framework for flexible distributed systems. In *Proceedings of Pacific Rim International Workshop on Multi-Agents* (PRIMA'98 in PRICAI'98), (pp.161-175).

Kinsner, W. (2007, January). Towards Cognitive Machines: Multiscale Measures and Analysis. [IJCINI]. *International Journal of Cognitive Informatics and Natural Intelligence*, *1*(1), 28–38. doi:10.4018/jcini.2007010102

Kinsner, W. (2007). Is entropy suitable to characterize data and signals for cognitive informatics? *International Journal of Cognitive Informatics and Natural Intelligence*, *1*(2), 34–57. doi:10.4018/jcini.2007040103

Kirkpatrick, S., Gelatt, C. D., & Vecchi, M. P. (1983). Optimization by simulated annealing. *Science*, *220*(4598), 671–680. doi:10.1126/science.220.4598.671

Kistler, J. J., & Satyanarayanan, M. (1992). Disconnected operation in the coda file. *ACM Transactions on Computer Systems*, *10*(1), 3–25. doi:10.1145/146941.146942

Kitagata, G., Matsushima, Y., Hasegawa, D., Kinoshita, T., & Shiratori, N. (2005). An agent-based middleware for communication service on ad hoc network. In *Proceedings of the 19th International Conference on Advanced Information Networking and Application*, (pp.363-367).

Kitagata, G., Sekiba, J., Suganuma, T., Kinoshita, T., & Shiratori, N. (2000). Agent-based flow control mechanism for flexible asynchronous messaging system FAMES. In *Proceedings of the 14th Int. Conf. on Information Networking (ICOIN-14)*, (pp.2B-2.1-8).

Kleene, S.C. (1956), Representation of Events by Nerve Nets, in C.E. Shannon and J. McCarthy eds., *Automata Studies*, Princeton Univ. Press, 3-42.

Ko, S., Gupta, I., & Jo, Y. (2007, 9-11 July). Novel Mathematics-Inspired Algorithms for Self-Adaptive Peer-to- Peer Computing. In G. Serugendo, J. Flatin, & M. Jelasity (Eds.), *Proceedings of 1st international conference on self-adaptive and self-organizing systems (saso'07)* (pp. 3–12). Boston, Massachusetts, USA: IEEE Computer Society Press.

Kohler, B. U., Hennig, C., & Orglmeister, R. (2002). The principles of software QRS detection. *IEEE Engineering in Medicine and Biology Magazine*, *21*(1), 42–57. doi:10.1109/51.993193

Konno, S., Iwaya, Y., Abe, T., & Kinoshita, T. (2004). Design of network management support system based on active information resource. In *Proceedings of the 18th International Conference on Advanced Information Networking and Application*, (pp.102-106).

Krasner, G. E., & Pope, S. T. (1988). A cookbook for using the model-view controller user interface paradigm in Smalltalk-80. *Journal of Object Oriented Program*, *1*(3), 26–49.

Kraus, S. (1997). Negotiation and cooperation in multi-agent systems. *Artificial Intelligence*, *94*(1-2), 79–98. doi:10.1016/S0004-3702(97)00025-8

Kuenning, G. H., Bagrodia, R., Gay, R. G., Popek, G. J., Reiher, P. L., & Wang, A.-I. (1998). Measuring the Quality of Service of Optimistic Replication, *ECOOP '98: Workshops on Object-Oriented Technology*, pp. 319-320, Brussels, Belgium.

Kulikowski, J. J., Marcelja, S., & Bishop, P. O. (1982). Theory of spatial position and spacial frequency relations in the receptive field of simple cells in the visual cortex. *Biological Cybernetics*, *43*(3), 187–198. PubMed-doi:10.1007/BF00319978

Kupinski, M. A., & Anastasio, M. A. (1999). Multiobjective genetic optimization of diagnostic classifiers with implication for generating receiver operating characteristic curves. *IEEE Transactions on Medical Imaging*, *18*(8), 675–685. doi:10.1109/42.796281

Kurzweil, R. (1990). *The Age of Intelligent Machines*. Cambridge, MA: MIT Press.

Kwan, C., Lewis, F. L., & Dawson, D. M. (1998). Robust neural-network control of rigid-link electrically driven robots. *IEEE Transactions on Neural Networks*, *9*(4), 581–589. doi:10.1109/72.701172

Kwon, O. W., Chan, K., Hao, J., & Lee, T. W. (2003). Emotion Recognition by Speech Signals. *EuroSpeech*, *2003*, 125–128.

Lang, J., Torre, L. v., & Weydert, E. (2002). Utilitarian desires. *Autonomous Agents and Multi-Agent Systems*, *5*(3), 329–363. doi:10.1023/A:1015508524218

Lau, C. G. Y. (Ed.). (1991). *Neural Networks: Theoretical Foundation and Analyses*. Piscataway, NJ: IEEE Press.

Lawvere, F., & Schanuel, S. (1997). *Conceptual Mathematics: A First Introduction to Categories* (1st ed.). Cambridge University Press.

Le Prado, C., & Natkin, S. (2007), "Listen Lisboa: scripting languages for interactive musical installations". Sound and Music Computing Conference, SMC'07, Lefkada Greece.

Leahey, T. H. (1997). *A History of Psychology: Main Currents in Psychological Thought* (4th ed.). Upper Saddle River, NJ: Prentice- Hall Inc.

Lehner, R. J., & Rangayyan, R. M. (1987). A three-channel microcomputer system for segmentation and characterization of the phonocardiogram. *IEEE Transactions on Bio-Medical Engineering*, *34*(6), 485–489. doi:10.1109/TBME.1987.326060

Levine, M. (1998). Categorical Algebra . In Benkart, G., Ratiu, T., Masur, H., & Renardy, M. (Eds.), *Mixed motives* (*Vol. 57*, pp. 373–499). USA: American Mathematical Society.

Li, Y., & Lan, Z. (2005). *A survey of load balancing in grid computing. High Performance Computing and Algorithms, Lecture Notes in Computer Science* (*Vol. 3314*, pp. 280–285). LNCS.

Li, J., & Wang, J. (2003). Automatic Linguistic Indexing of Pictures by a Statistical Modeling Approach. *IEEE Trans. on Pattern Analysis and Machine Intelligence*.

Li, J., & Wang, J. (2004). Studying digital imagery of ancient paintings by mixtures of stochastic models. *IEEE Transactions on Image Processing*.

Li, X., Uchiya, T., Konno, S., & Kinoshita, T. (2008). Proposal for agent platform dynamic interoperation facilitating mechanism. In *Proceedings of the 21th Int. Conf. Industrial, Engineering and Other Applications of Applied Intelligent Systems* (IEA/AIE 2008), LNAI5027, (pp.825-834).

Li, Y. (2006). Robust neural networks compensating motion control of reconfigurable manipulator in geometric form. *IEEE International Conference on Mechatronics and Automation* (pp. 306-311).

Lin, T. Y., Yao, Y. Y., & Zadeh, L. A. (2002). *Data mining, Rough Sets and Granular Computing*. Heidelberg: Physica-Verlag.

Lin,T.Y.(1997). Granular computing, announcement of the BISC Special Interest Group on Granular Computing.

Little, T. D. C., & Ghafoor, A. (1990). Synchronization and Storage Models for Multimedia Objects. *IEEE Journal on Selected Areas in Communications*, *8*(3), 413–427. doi:10.1109/49.53017

Liu, R. J., & Huang, X. W. (2005). The granular theorem of quotient space in image segmentation. [in Chinese]. *Chinese Journal of Computers*, *28*(10), 37–40.

Liu, R. J., Huang, X. W., & Meng, J. (2005). Texture Image Segmentation Based on Quotient Space Granularity Synthesis. *Asian Journal of Information Technology*, *4*(3), 61–67.

Liu, K., & Lewis, F. L. (1990). Decentralized continuous robust controller for mobile robots. *Proceedings of IEEE International Conference on Robotics and Automation* (pp. 1822-1827).

Liu, R.J., Huang, X.W., Meng, J., & Zhong, X.R. (2004). Texture image segmentation based on quotient space. *Computer applications, 14*(7), 37-40(in Chinese).

Lomuscio, A., Wooldridge, M., & Jennings, N. (2001). A classification scheme for negotiation in electronic commerce . In *Agent mediated electronic commerce, lnai 1991* (pp. 19–33). Springer. doi:10.1007/3-540-44682-6_2

Long, F., Zhang, H., & Feng, D. D. (2003). *Fundamentals of Content-based image retrieval. Multimedia Information Retrieval and Management- Technological Fundamentals and Applications*. Springer.

López, N., Núnez, M., & Pelayo, F. L. (2007). A formal specification of the memorization process. *International Journal of Cognitive Informatics and Natural Intelligence*, *1*(4), 47–60. doi:10.4018/jcini.2007100104

López, N., Núñez, M., Rodríguez, I., & Rubio, F. (2002). A formal framework for e-barter based on microeconomic theory and process algebras . In *I2CS 2002, lncs 2346*. Springer.

López, N., Rodríguez, I., & Rubio, F. (2003). Defining meta-adaptable living agents. *Second IEEE International Conference on Cognitive Informatics* (pp. 161–170). IEEE-CS Press.

Maes, P. (Ed.). (1991). *Designing Autonomous Agents: Theory and Practice from Biology to Engineering and Back*. London: The MIT press.

Magarey, J., & Kingsbury, N. (1998). Motion estimation using a complex-valued wavelet transform. *IEEE Transactions on Signal Processing*, *46*(4), 1069–1084. doi:10.1109/78.668557

Magerko, B., & Laird, J. E. (2003), Building an Interactive Drama Architecture, Proc of First International Conference on Technologies for Interactive Digital Storytelling and Entertainment, TIDSE'03. Darmstadt, Germany, pp. 226-237.

Mallat, S. G. (1999). *A wavelet tour of signal processing*. Academic Press.

Mao, J. J., Wu, T., Zheng, T. T., & Zhang, L. (2005). Algorithm of hierarchical competitive covering networks based on quotient space. [in Chinese]. *Microcomputer Development*, *14*(4), 37–39.

Mao, J. J., Zhang, L., & Xu, Y. S. (2004a). Fuzzy clustering analysis based on quotient space and information granularity. [in Chinese]. *Operations Research and Management Science*, *13*(4), 25–29.

Mao, J. J., Zheng, T. T., & Zhang, L. (2004b). Biological sequence alignments based on quotient space. [in Chinese]. *Computer Engineering And Applications, 34*(14), 15–17.

Marcelja, S. (1980). Mathematical description of the responses of simple cortical cells. *Journal of the Optical Society of America, 70*(11), 1297–1300. PubMed-doi:10.1364/JOSA.70.001297

Marques de Sa, J. P. (2003). *Applied Statistics Using SPSS, STATISTICA, and MATLAB*. Berlin, Germany: Springer-Verlag.

Mas-Colell, A., Whinston, M., & Green, J. (1995). *Microeconomic theory*. Oxford University Press.

Mateas, M., & Stern, A. (2003), Integrating Plot, Character and Natural Language Processing in the Interactive Drama Facade, Proceedings of the TIDSE'03, Darmstadt, Germany, Fraunhofer IRB Verlag.

Matlin, M. W. (1998). *Cognition* (4th ed.). Orlando, FL: Harcourt Brace College Publishers.

McBride, B. (2004). *RDF Primer* (W3C Recommendation).

McBride, B. (n.d.). *Jena: Implementing the RDF Model and Syntax Specification,* http://www.hpl.hp.com/personal/bwm/papers/20001221-paper.

McCarthy, J. (1963). *Situations, Actions, and Causal Laws, Memo 2*. Stanford, CA: Stanford University Artificial Intelligence Project.

McCarthy, J., Minsky, M. L., Rochester, N., & Shannon, C. E. (1955), *Proposal for the 1956 Dartmouth Summer Research Project on Artificial Intelligence*, Dartmouth College, Hanover, NH, USA, http://www.formal.stanford.edu/jmc/history/dartmouth/*dartmouth.html.*

McCulloch, W. S. (1965). *Embodiments of Mind*. Cambridge, MA: MIT Press.

McCulloch, W. S., & Pitts, W. (1943). A Logical Calculus of the Ideas Immanent in Nervous Activity . *The Bulletin of Mathematical Biophysics, 5*, 115–137. doi:10.1007/BF02478259

McGeachie, M., & Doyle, J. (2002). Utility functions for ceteris paribus preferences. In *AAAI Workshop on Preferences in AI and CP: Symbolic Approaches* (pp. 33–38). AAAI Press.

McGuinness, D. L., & Harmelen, F. V. (2004). OWL Web Ontology Language Overview. *W3C,* http://www.w3.org/TR/owl-features/.

Mednich, S. A., & Mednich, M. T. (1967). *Examiner's Manual, Remote Associates Test*. Boston: Houghton Mifflin.

Metz, C. (1986, September). ROC methodology in radiologic imaging. *Investigative Radiology, 21*, 720–733. doi:10.1097/00004424-198609000-00009

Meyer, C., Gavela, J. F., & Harris, M. (2006). Combining algorithms in automatic detection of QRS complexes in ECG signals. *IEEE Transactions on Information Technology in Biomedicine, 10*(3), 468–475. doi:10.1109/TITB.2006.875662

Meystel, A. M., & Albus, J. S. (2002). *Intelligent Systems, Architecture, Design, and Control*. John Wiley & Sons.

Michael Jeronimo, J. W. (2003). *UPnP Design by Example: A Software Developer's Guide to Universal Plug and Play*. Intel Press.

Microsoft Bayesian Network Toolkit. http://research.microsoft.com/adapt/MSBNx/.

Milner, R. (1989). *Communication and concurrency*. Prentice Hall.

Minkowski, H. (1908), *Space and Time*, Address, 80th Assembly of German Natural Scientists and Physicians, Cologne, Sept.

Mita, M. (2007). Algorithm for the classification of multi-modulating signals on the electrocardiogram. *Medical & Biological Engineering & Computing, 45*(3), 241–250. doi:10.1007/s11517-006-0130-5

Mneimneh, M. A., Yaz, E. E., Johnson, M. T., & Povinelli, R. J. (2006). An adaptive Kalman filter for removing baseline wandering in ECG signals. *Proceedings of the 2006 Computers in Cardiology Conference (CINC'06)* (pp. 253-256). Valencia, Spain.

Moody, J., & Darken, C. J. (1989). Fast learning in networks of locally-tuned processing units. *Neural Computation, 1*(2), 281–294. doi:10.1162/neco.1989.1.2.281

Moore, E. F. (1962). Machine models of self reproduction. *American Mathematical Society . Proceedings of Symposia in Applied Mathematics, 14*, 17–33.

Mudigonda, N. R., Rangayyan, R. M., & Desautels, J. E. L. (2000). Gradient and texture analysis for the classification of mammographic masses. *IEEE Transactions on Medical Imaging, 19*(10), 1032–1043. doi:10.1109/42.887618

Mudigonda, N. R., Rangayyan, R. M., & Desautels, J. E. L. (2001). Detection of breast masses in mammograms by density slicing and texture flow-field analysis. *IEEE Transactions on Medical Imaging, 20*(12), 1215–1227. doi:10.1109/42.974917

Murch, R. (2004). *Autonomic Computing*. London: Person Education.

Murthy, I. S. N., & Rangaraj, M. R. (1979). New concepts for PVC detection. *IEEE Transactions on Bio-Medical Engineering, 26*(7), 409–416. doi:10.1109/TBME.1979.326420

Mylopoulos, J., Kolp, M., & Giorgini, P. (2002). Agent oriented software development. In *Proceedings of the 2nd Hellenic Conference on Artificial Intelligence* (SETN-02).

Nandi, R. J., Nandi, A. K., Rangayyan, R. M., & Scutt, D. (2006). Classification of breast masses in mammograms using genetic programming and feature selection. *Medical & Biological Engineering & Computing, 44*(8), 683–694. doi:10.1007/s11517-006-0077-6

Natkin, S. (2006), Video Games & Interactive Media, A glimpse at new Digital Entertainment, AK Peters Ed, Wesley MA, USA, March, 2006.

Natkin, S., & Yan, C. (2005), Analysis of Correspondences between Real and Virtual Worlds in General Public Applications, Proc of Computer Graphics, Imaging and Visualization (CGIV05), Beijing, July 25-28, 2005, IEEE, 2005, pp. 223-231.

Natkin, S., & Yan, C. (2006), User Model in Multiplayer Mixed Reality Entertainment Applications, Proc of International Conference on Advances in Computer Entertainment Technology ACE'06, California, USA, June, 2006, ACM SIGCHI press.

Newell, A. (1990). *Unified Theories of Cognition*. Cambridge, MA: Harvard University Press.

Newell, A., & Simon, H. A. (1972). *Human Problem Solving*. NJ: Prentice-Hall Englewood Cliffs.

Niemelä, E., Kalaoja, J., & Lago, P. (2005). Toward an Architectural Knowledge Base for Wireless Service Engineering. *IEEE Transactions on Software Engineering, 31*(5), 361–379. doi:10.1109/TSE.2005.60

Niemelä, E., & Marjeta, J. (1998). Dynamic Configuration of Distributed Software Components. *ECOOP '98: Workshop ion on Object-Oriented Technology*, 149-150.

Nilsson, N. J. (1998). *Artificial Intelligence: A New Synthesis*. San Mateo, CA: Morgan Kaufmann.

Nwana, H. S., Ndumu, D. T., Lee, L. C., & Collins, J. C. (1999). ZEUS: a toolkit for building distributed multi-agent systems. *Applied Artificial Intelligence Journal, 13*(1), 129–186. doi:10.1080/088395199117513

Olston, C., & Widom, J. (2005). Efficient Monitoring and Querying of Distributed, Dynamic Data via approximate Replication . *IEEE Data Eng. Bull, 28*(1), 11–18.

Orlowska, E. (1997). *Incomplete Information: Rough Set Analysis*. Springer Verlag.

Oudeyer, P. Y. (2003). The Production and Recognition of Emotions in Speech: features and algorithms. *International Journal of Human-Computer Studies, 59*(1), 157–183. doi:10.1016/S1071-5819(02)00141-6

Pacheco, O. (2004, April). Autonomy in an Organizational Context . In Nickles, M., Rovatsos, M., & Weiss, G. (Eds.), *Agents and computational autonomy: Potential, risks, and solutions* (*Vol. 2969*, pp. 195–208). Springer Berlin. doi:10.1007/978-3-540-25928-2_16

Pacitti, E., & Ozsu, M. T. (2003). Replica Consitency for Lazy Multi-Master Configurations in a Cluster of Autonomous Databases [Lyon, France.]. *DBA, 03*, 318–327.

Pacitti, E., Minet, P., & Simon, E. (1999). Fast Algorithms for Maintaining Replica Consistency in Lazy Master Replicated Databases, *Int. Conf. on Very Large Databases*, Edinburgh, UK.

Pan, J., & Tompkins, W. J. (1985). A real-time QRS detection algorithm. *IEEE Transactions on Bio-Medical Engineering*, *32*(3), 230–236. doi:10.1109/TBME.1985.325532

Parashar, M., & Hariri, S. (Eds.). (2006). *Autonomic Computing: Concepts, Infrastructure and Applications* (1st ed.). CRC Press.

Parsons, S., & Wooldridge, M. (2002). Game theory and decision theory in multi-agent systems. *Autonomous Agents and Multi-Agent Systems*, *5*(3), 243–254. doi:10.1023/A:1015575522401

Pawlak, Z. (1984). Rough Classification. *International Journal of Man-Machine Studies*, *20*(5), 469–483. doi:10.1016/S0020-7373(84)80022-X

Pawlak, Z. (1991). *Rough Sets: Theoretical Aspects of Reasoning about Data*. Dordrecht: Kluwer Academic Publishing.

Payne, D. G., & Wenger, M. J. (1998). *Cognitive Psychology*. Boston: Houghton Mifflin Co.

Pearl, J. (1988). *Probabilistic Reasoning in Intelligent Systems: Networks of Plausible Inference*, (pp. 1-20).

Peinado, F., & Gervás, P. (2004), Transferring Game Mastering Laws to Interactive Digital Storytelling, Proceedings of the 2nd International Conference on Technologies for Interactive Digital Storytelling and Entertainment (TIDSE'04), 24-26 June, Darmstadt, Germany, LNCS, 3105, 2004, pp. 48-54.

Peter, J., & Skowron, A. (2002). A rough set approach to knowledge discovery. *International Journal of Intelligent Systems*, *17*, 109–112. doi:10.1002/int.10010

Petersen, K., Spreitzer, M., Terry, D., & Theimer, M. (1996). Bayou: replicated database services for world-wide applications, *EW 7: Proceedings of the 7th workshop on ACM SIGOPS European workshop*, (pp. 275-280), Connemara, Ireland.

Piaget, J. (1973). *Introduction à l'Épistemologie Genetique*. Paris: PUF.

Pierce, J. H., & Jane, M. H. (2003), The Five Factor Model: An Introduction to the Five-Factor Model of Personality, Center for Applied Cognitive Studies (CentACS), Charlotte, North Carolina, 2003. Available: http://www.childrenofmillennium.org/eugenics/pages/articles/bigfive.htm

Pinel, J. P. J. (1997). *Biopsychology* (3rd ed.). Needham Heights, MA: Allyn and Bacon.

Plotkin, G. D. (1981). A structural approach to operational semantics. Technical Report DAIMI FN-19, Computer Science Department, Aarhus University, 1981.

Poggio, T., & Girosi, F. (1990). Regularization algorithms for learning that are equivalent to multilayer networks. *Science*, *247*(4945), 978–982. doi:10.1126/science.247.4945.978

Pollen, D. A., & Ronner, S. F. (1985). Visual cortical neurons as localized spatial frequency filter. *IEEE Transactions on Systems, Man, and Cybernetics*, *15*(3), 91–101.

Polzin, T. S., & Waibel, A. (2000). Emotion-Sensitive Human Computer Interfaces. In *Proceedings of the ISCA Workshop on Speech and Emotion,* (pp. 201-206).

Poole, D., Mackworth, A., & Goebel, R. (1997). *Computational Intelligence: A Logical Approach. Oxford.* Oxford, UK: Oxford University Press.

Porat, M., & Zeevi, Y. Y. (1989). Localized texture processing in vision: analysis and synthesis in the Gaborian space. *IEEE Transactions on Bio-Medical Engineering*, *36*(1), 115–129. PubMeddoi:10.1109/10.16457

Powell, M. J. D. (1985). Radial basis functions for multivariable interpolations: A review. *IMA Conference on Algorithms for the Approximations of Functions and Data.* RMCS, Shrivenham, UK, (pp. 143-167).

Press, H. P., Teukolsky, S. A., Vetterling, W. T., & Falnnerry, B. P. (1992). *Numerical Recipes in C: The art of Scientific Computing*. UK: Cambridge University Press.

Qiu, S. (1997). Gabor-type matrix algebra and fast computation of dual and tight Gabor wavelets. *Optical Engineering (Redondo Beach, Calif.)*, *36*(1), 276–282. doi:10.1117/1.601171

Quinlan, J. R. (1993). *C4.5: Programs for Machine Learning. The Morgan Kaufmann Series in Machine Learning.* San Mateo, CA: Morgan Kaufmann Publishers.

Rajlich, V., & Xu, S. (2007). Constructivist learning during software development. *International Journal of Cognitive Informatics and Natural Intelligence, 1*(3), 78–101. doi:10.4018/jcini.2007070106

Ranganathan, K., & Foster, I. (2001). *Identifying Dynamic Replication Strategies for a High-Performance Data Grid* (pp. 75–86). In GRID.

Rangayyan, R. M., Mudigonda, N. R., & Desautels, J. E. L. (2000). Boundary modeling and shape analysis methods for classification of mammographic masses. *Medical & Biological Engineering & Computing, 38*, 487–495. doi:10.1007/BF02345742

Rangayyan, R. M. (2002). *Biomedical Signal Analysis: A Case-Study Approach.* New York, NY: IEEE and Wiley.

Rasmusson, L., & Janson, S. (1999). Agents, self-interest and electronic markets. *The Knowledge Engineering Review, 14*(2), 143–150. doi:10.1017/S026988899914205X

Razak, A. A., Komiya, R., & Abidin, M. I. Z. (2005). Comparison Between Fuzzy and NN Method for Speech Emotion Recognition. *Proceedings of the ICITA'05*, Sydney, (pp. 297-302).

Reed, S. (1972). Pattern Recognition and Categorization. *Cognitive Psychology, 3*, 383–407. doi:10.1016/0010-0285(72)90014-X

Reed, S., Ernst, G., & Banerji, R. (1974). The Role of Analogy in Transfer between Similar Problem States. *Cognitive Psychology, 6*, 436–450. doi:10.1016/0010-0285(74)90020-6

Reisberg, D. (2001), Cognition, second edition, Exploring the science of the mind, W.W. Norton & Company, Inc.

Renals, S. (1989). Radial basis functions network for speech pattern classification. *Electronics Letters, 25*(7), 437–439. doi:10.1049/el:19890300

Renquist, N. R., Andrew, H., & Andrew, L. (2001). Jack – summary of an agent infrastructure. In *Proceedings* of *5th International Conference on Autonomous Agents.* Reticular Systems. AgentBuilder – An integrated toolkit for constructing intelligence software agents, available at http://www.agentbuilder.com/

Ribeiro-Neto, B., Silva, I., & Muntz, R. (2000). *Bayesian Network Models for Information Retrieval. Soft Computing in Information Retrieval: Techniques and Applications* (pp. 259–291). Springer.

RIEMANN (Research on Intelligent Media Annotation), Pennsylvania State University, http://wang.ist.psu.edu/IMAGE/.

Rioul, O., & Vetterli, M. (1991). Wavelets and signal processing. *IEEE Signal Processing Magazine, 8*(4), 15–38. doi:10.1109/79.91217

ROCKIT 0.9 B – Beta Version: www.radiology.uchicago.edu/krl/KRL_ROC/software_index.htm.

Rodrigues, L., & Raynal, M. (2003). Atomic broadcast in asynchronous crash-recovery distributed systems and its use in quorum-based replication. *IEEE Transactions on Knowledge and Data Engineering, 15*(5), 1206–1217. doi:10.1109/TKDE.2003.1232273

Rusak, B., & Zucker, I. (1979). Neural Regulation of Circadian Rhythms. *Physiological Reviews, 59*, 449–526.

Russell, S. J., & Norvig, P. (1995). *Artificial Intelligence: A Modern Approach.* Englewood Cliffs, NJ: Prentice-Hall.

Russell, B. (1901), Is Position in Time and Space Absolute or Relative? *Mind*, July, London.

Rutledge, L. (2001). SMIL 2.0: XML for Web Multimedia. *IEEE Internet Computing, 5*(5), 78–84. doi:10.1109/4236.957898

Rutten, J. (2001). Elements of Stream Calculus (An Extensive Exercise in Coinduction). [Elsevier Science Publishers Ltd.]. *Electronic Notes in Theoretical Computer Science*, 45.

Sahiner, B. S., Chan, H.-P., Petrick, N., Helvie, M. A., & Goodsitt, M. M. (1998). Computerized characterization of masses on mammograms: The rubber band straightening transform and texture analysis. *Medical Physics, 25*(4), 516–526. doi:10.1118/1.598228

Sahiner, B. S., Chan, H.-P., Petrick, N., Helvie, M. A., & Hadjiiski, L. M. (2001). Improvement of mammographic mass characterization using spiculation measures and morphological features. *Medical Physics*, *28*(7), 1455–1465. doi:10.1118/1.1381548

Saito, Y., & Shapiro, M. (2005). Optimistic replication. *ACM Computing Surveys*, *37*(1), 42–81. doi:10.1145/1057977.1057980

Sameni, R., Shamsollahi, M. B., Jutten, C., & Clifford, G. D. (2007). A nonlinear Bayesian filtering framework for ECG denoising. *IEEE Transactions on Bio-Medical Engineering*, *54*(12), 2172–2185. doi:10.1109/TBME.2007.897817

Sandholm, T. (1998). Agents in electronic commerce: Component technologies for automated negotiation and coalition formation. In *CIA'98, lncs 1435* (pp. 113–134). Springer.

Sayadi, O., & Shamsollahi, M. B. (2008). ECG denoising and compression using a modified extended Kalman filter structure. *IEEE Transactions on Bio-Medical Engineering*, *55*(9), 2240–2248. doi:10.1109/TBME.2008.921150

Saygin, A., Cicekli, I., & Akman, V. (2000). Turing test: 50 years later. *MANDMS: Minds and Machines*, *10*, 463–518. doi:10.1023/A:1011288000451

Searle, J. (1980). Minds, brains and programs. *The Behavioral and Brain Sciences*, *3*, 417–424. doi:10.1017/S0140525X00005756

Searle, J. (1990). Is the brain's mind a computer program? *Scientific American*, *3*(262), 26–31. doi:10.1038/scientificamerican0190-26

Shafer, G.R., & Shenoy, P.P. (1990). *Probability Propagation Annals of Mathematics and Artificial Intelligence.*

Shannon, C. E. (Ed.). (1956). *Automata Studies*. Princeton: Princeton University Press.

Shannon, C. E. (1948). A mathematical theory of communication. *The Bell System Technical Journal*, *27*, 379–423, 623–656.

Shiratori, N., et al. (2005). Symbiotic Computing Project, http://symbiotic.agent-town.com/

Sicilian, O. B., Wit, C. C., & Bastin, G. (1996). *Theory of robot control.* Springer-Verlag.

Silipo, R., & Marchesi, C. (1998). Artificial neural networks for automatic ECG analysis. *IEEE Transactions on Signal Processing*, *46*(5), 1417–1425. doi:10.1109/78.668803

Smith, R. G. (1980). The contract net protocol: high-level communication and control in a distributed problem solver. *IEEE Transactions on Computers*, *29*(12), 1104–1113. doi:10.1109/TC.1980.1675516

Smith, R. E. (1993). *Psychology*. St. Paul, MN: West Publishing Co.

Smith, S. M. (1995). Fixation, Incubation, and Insight in Memory and Creative Thinking. In S.M. Smith, T.B. Ward, & R.A. Finke (Eds.), *The Creative Cognition Approach*. Cambridge, MA: MIT Press.

Song, Z. S., Yi, J. Q., Zhao, D. B., & Li, X. C. (2005). A computed torque controller for uncertain robotic manipulator systems: fuzzy approach. *Fuzzy Sets and Systems*, *154*(2), 208–226. doi:10.1016/j.fss.2005.03.007

Sternberg, R. J. (1997). The Concept of Intelligence and the its Role in Lifelong Learning and Success . *The American Psychologist*, *52*(10), 1030–1037. doi:10.1037/0003-066X.52.10.1030

Sternberg, R. J., & Lubart, T. I. (1995). *Defying the Crowd: Cultivating Creativity in a Culture of Conformity*. NY: Free Press.

Stirling, W., Goodrich, M., & Packard, D. (2002). Satisficing equilibria: A non-classical theory of games and decisions. *Autonomous Agents and Multi-Agent Systems*, *5*(3), 305–328. doi:10.1023/A:1015556407380

Suganuma, T., Imai, S., Kinoshita, T., Sugawara, K., & Shiratori, N. (2003). A flexible videoconference system based on multiagent framework. *IEEE Trans. on Systems, Man, and Cybernetics – Part A . Systems and Humans*, *33*(5), 633–641.

Suganuma, T., Uchiya, T., Konno, S., Kitagata, G., Hara, H., Fujita, S., et al. (2006). Bridging the E-Gaps: towards post-ubiquitous computing. In *Proceedings of the 20th International Conference on Advanced Information Networking and Applications* (AINA'06), FINA 2006 Symposium, Vol.2, (pp.780-784).

Sugawara, K., et al. (2007). A concept of symbiotic computing and its application to telework. In *Proceedings of the IEEE 2007 International Conference on Cognitive Informatics* (ICCI'07), (pp.302-311).

Sun, X., & Qian, W. (2002). System-oriented optimization of CAD for mass detection in digital mammography. In D.P. Chakraborty & E.A. Krupinski (Eds.), *Proc. SPIE Medical Imaging 2002: Image Perception, Observer Performance, and Technology Assessment, 4686*, 273-278. Bellingham, WA: SPIE.

Szilas, N., Rety, J. H., & Marty, O. (2003), Authoring highly generative Interactive Drama, Proc of International Conference of Virtual Storytelling (ICVS), Toulouse (France), November, 2003.

Takahashi, H., Izumi, S., Suganuma, T., Kinoshita, T., & Shiratori, N. (2009). Multi-agent system for user-oriented healthcare support. [IJIS]. *The International Journal of Informatic Society*, *1*(3), 32–41.

Takahashi, A., Suganuma, T., Abe, T. and Kinoshita, T. (2006). Dynamic construction scheme of multimedia processing system based on multiagent framework. *The International Journal of Wireless and Mobile Computing*, Vol.2, No.1.

Tan, J. D., & Xi, N. (2001). Unified model approach for planning and control of mobile manipulators. *IEEE International Conference on Robotics and Automation* (pp. 3145-3152).

Tan, X. M., Zhao, D. B., Yi, J. Q., & Xu, D. (2008). Adaptive hybrid control for omnidirectional mobile manipulator based on neural network. *American Control Conference* (pp. 5174 -5179).

Tennenholtz, M. (2002). Game theory and artificial intelligence . In *Foundations and applications of multiagent systems* (pp. 49–58). Springer. doi:10.1007/3-540-45634-1_4

Terdiman, D. (n.d.). 'Tagging' gives Web a human meaning. *CNET News.com*. http://news.com.com/Tagging+gives+Web+a+human+meaning/2009-1025_3- 5944502.html.

Thakor, N. V., & Zhu, Y. S. (1991). Applications of adaptive filtering to ECG analysis: noise cancellation and arrhythmia detection. *IEEE Transactions on Bio-Medical Engineering*, *38*(8), 785–794. doi:10.1109/10.83591

Thomas, B., Close, B., Donoghue, J., Squires, J., De Bondi, P., Morris, M., & Piekarski, W. (2000), ARQuake: An Outdoor/Indoor Augmented Reality First Person Application, Proc of 4th International Symposium on Wearable Computers, Atlanta, GA, Oct 2000, pp 139-146.

Tompkins, W. J. (1993). Biomedical [*Language Examples and Laboratory Experiments for the IBM PC*. Englewood Cliffs, NJ: Prentice Hall PTR.]. *Digital Signal Processing*, C.

Topaloglu, U., & Bayrak, C. (2008, February). Secure Mobile Agent Execution in Virtual Environment. *Autonomous Agents and Multi-Agent Systems*, *16*(1), 1–12. doi:10.1007/s10458-007-9018-5

Tovee, M. J. (1996). *An Introduction to the Visual System*. Cambridge, UK: Cambridge, University Press.

Tucker, A. B. Jr., (Ed.). (1997). *The Computer Science and Engineering Handbook*. FL: CRC Press.

Turing, A. M. (1950). Computing Machinery and Intelligence . *Mind*, *59*, 433–460. doi:10.1093/mind/LIX.236.433

Tversky, B., N. Franklin, H.A. Taylor, D.J. Bryant (1999), Spatial mental models from descriptions, *Journal of the American Society for Information Science*, Jan., 45(9), *pp. 656 – 668*.

Uchiya, T., Maemura, T., Hara, H., Sugawara, K., & Kinoshita, T. (2009). Interactive Design Method of Agent System for Symbiotic Computing. *International Journal of Cognitive Informatics and Natural Intelligence*, *3*(1), 57–74. doi:10.4018/jcini.2009010104

Uchiya, T., Maemura, T., Li, X., & Kinoshita, T. (2007). Design and Implementation of Interactive design environment of Agent System. In *Proceedings of the 20th Int. Conf. Industrial, Engineering and Other Applications of Applied Intelligent Systems* (IEA/AIE2007), LNAI4570, AAAI/ACM, (pp.1088-1097).

Uchiya, T., Suganuma, T., Kinoshita, T., & Shiratori, N. (2002). An architecture of active agent repository for dynamic networking. In *Proceedings of the First International Joint Conference on Autonomous Agents and Multiagent Systems* (AAMAS2002), (pp.1266-1267).

Uchiya, T., Takeda, A., Suganuma, T., Kinoshita, T., & Shiratori, N. (2003). A method for realizing user-oriented service with repository-based agent framework. In *Proceedings of the 1st Int. Forum on Information and Computer Technology* (IFICT2003), IPSJ, (pp.119-124).

Veltkamp, R., & Tanase, M. (2000). Content-based Image Retrieval Systems: A Survey. Technical Report UU-CS-2000-34, Utrecht University.

Ververidis, D. C. Kotropou, & Pitas, I. (2004). Automatic Emotional Speech Classification. *Proceedings of the ICASSP2004*, Canada, (pp. 593-596).

Vidot, N., Cart, M., Ferrié, J., & Suleiman, M. (2000). Copies convergence in a distributed real-time collaborative environment, *CSCW '00: Proceedings of the 2000 ACM conference on Computer supported cooperative work*, (pp. 171-180), Philadelphia, Pennsylvania, USA.

Vinh, P. (2009b). *Dynamic Reconfigurability in Reconfigurable Computing Systems: Formal Aspects of Computing* (1st ed.). Saarbrucken, Germany: VDM Verlag Dr. Muller.

Vinh, P. (2009d, January). Categorical Approaches to Models and Behaviors of Autonomic Agent Systems. [IJCiNi]. *International Journal of Cognitive Informatics and Natural Intelligence*, *3*(1), 17–33. doi:10.4018/jcini.2009010102

Vinh, P., & Bowen, J. (2008, June). Formalization of Data Flow Computing and a Coinductive Approach to Verifying Flowware Synthesis. *LNCS Transactions on Computational Science*, *4750*(1), 1–36. doi:10.1007/978-3-540-79299-4_1

Vinh, P. C. (2009). Categorical Approaches to Models and Behaviors of Autonomic Agent Systems. *International Journal of Cognitive Informatics and Natural Intelligence*, *3*(1), 17–33. doi:10.4018/jcini.2009010102

Vinh, P. (2009a, May). Formal Aspects of Self-* in Autonomic Networked Computing Systems . In Denko, M. K., Yang, L. T., & Zhang, Y. (Eds.), *Autonomic Computing and Networking* (1st ed., pp. 381–410). Springer, USA. doi:10.1007/978-0-387-89828-5_16

Vinh, P. (2007). Homomorphism between AOMRC and Hoare Model of Deterministic Reconfiguration Processes in Reconfigurable Computing Systems. *Scientific Annals of Computer Science* (XVII), 113-145.

Vinh, P. (2009c, May). Formalizing Parallel Programming in Large Scale Distributed Networks: From Tasks Parallel and Data Parallel to Applied Categorical Structures. In F. Xhafa (Ed.), *Parallel Programming, Models and Applications in Grid and P2P Systems* (1st ed., Vol. 17, pp. 24–53). IOS Press.

Vinh, P. (2010, May). Aspect-Oriented Self-configuring P2P Networking in Mobile Environments: A Formal Specification and Verification. In P. Alencar and D. Cowan (Ed.), *Handbook of Research on Mobile Software Engineering: Design, Implementation and Emergent Applications* (1st ed.). IGI Global.

Vinh, P., & Bowen, J. (2007, 6–8 June). A Formal Approach to Aspect-Oriented Modular Reconfigurable Computing. In *Proceedings of 1st ieee & ifip international symposium on theoretical aspects of software engineering (tase)* (pp. 369–378). Shanghai, China: IEEE Computer Society Press.

von Neumann, J. (1958). *The Computer and the Brain, Yale Univ*. New Haven: Press.

von Neumann, J., & Burks, A. W. (1966). Theory of Self-Reproducing Automata, Univ. of Illinois Press . *Urbana (Caracas, Venezuela)*, IL.

von Neumann, J. (1946), The Principles of Large-Scale Computing Machines, reprinted in *Annals of History of Computers*, 3(3), 263-273.

von Neumann, J. (1963), *General and Logical Theory of Automata,* A.H. Taub ed., Collected Works, Vol. 5, Pergamon, 288-328.

Wai, R. J. (2003). Robust control for nonlinear motor-mechanism coupling system using wavelet neural network. *IEEE Transactions on Systems, Man, and Cybernetics*, *33*(3).

Wald, A. (1950). *Statistical Decision Functions*. John Wiley & Sons.

Wallas, G. (1926). *The Art of Thought.* New York: Harcourt-Brace.

Walter, F.E., S. B., & Schweitzer, F. (2008, February). A Model of a Trust-based Recommendation System on a Social Network. *Autonomous Agents and Multi-Agent Systems*, *16*(1), 57–74. doi:10.1007/s10458-007-9021-x

Wang, Y. (2007a). *Software Engineering Foundations: A Software Science Perspective* (*Vol. II*). Aurebach Publications, NY, USA: CRC Book Series in Software Engineering.

Wang, Y. (2007a, January). The Theoretical Framework of Cognitive Informatics. [IJCiNi]. *International Journal of Cognitive Informatics and Natural Intelligence*, *1*(1), 1–27. doi:10.4018/jcini.2007010101

Wang, Y. (2007b, July–September). Toward Theoretical Foundations of Autonomic Computing. [IJCiNi]. *International Journal of Cognitive Informatics and Natural Intelligence*, *1*(3), 1–16. doi:10.4018/jcini.2007070101

Wang, Y., & Kinsner, W. (2006, March). Recent Advances in Cognitive Informatics. *IEEE Transactions on Systems, Man and Cybernetics. Part C, Applications and Reviews*, *36*(2), 121–123. doi:10.1109/TSMCC.2006.871120

Wang, Y., & Kinsner, W. (2006). Recent advances in cognitive informatics. *IEEE Transactions on Systems, Man and Cybernetics. Part C, Applications and Reviews*, *36*(2), 121–123. doi:10.1109/TSMCC.2006.871120

Wang, Y. (2002b). The Real-Time Process Algebra (RTPA), *Annals of Software Engineering: An International Journal . Baltzer Science Publishers, Oxford*, *14*(Oct), 235–274.

Wang, Y. (2007c). On The Cognitive Processes of Perception with Emotions, Motivations, and Attitudes, *International Journal of Cognitive Informatics and Natural Intelligence . IGI Publishing, USA*, *1*(4), 1–13.

Wang, Y. (2008b). RTPA: A Denotational Mathematics for Manipulating Intelligent and Computational Behaviors, *International Journal of Cognitive Informatics and Natural Intelligence . IGI Publishing, USA*, *2*(2), 44–62.

Wang, Y. (2008d). On the Big-R Notation for Describing Iterative and Recursive Behaviors, *International Journal of Cognitive Informatics and Natural Intelligence . IGI Publishing, USA*, *2*(1), 17–28.

Wang, Y., & Wang, Y. (2006). Cognitive Informatics Models of the Brain [C]. *IEEE Transactions on Systems, Man, and Cybernetics*, *36*(2), 203–207. doi:10.1109/TSMCC.2006.871151

Wang, Y., Wang, Y., Patel, S., & Patel, D. (2006). A Layered Reference Model of the Brain (LRMB) [C]. *IEEE Transactions on Systems, Man, and Cybernetics*, *36*(2), 124–133. doi:10.1109/TSMCC.2006.871126

Wang, Y. (2007a). *Software Engineering Foundations: A Software Science Perspective. CRC Book Series in Software Engineering* (*Vol. II*). NY, USA: Auerbach Publications.

Wang, Y. (2008a). On Contemporary Denotational Mathematics for Computational Intelligence [Springer.]. *Transactions of Computational Science*, *2*, 6–29. doi:10.1007/978-3-540-87563-5_2

Wang, Y. (2008b). On System Algebra: A Denotational Mathematical Structure for Abstract System modeling. [IGI Publishing, USA.]. *International Journal of Cognitive Informatics and Natural Intelligence*, *2*(2), 20–42. doi:10.4018/jcini.2008040102

Wang, Y. (2009a). On Abstract Intelligence: Toward a Unified Theory of Natural, Artificial, Machinable, and Computational Intelligence. [IGI, USA, Jan.]. *International Journal of Software Science and Computational Intelligence*, *1*(1), 1–17. doi:10.4018/jssci.2009010101

Wang, Y. (2009b). On Visual Semantic Algebra (VSA): A Denotational Mathematical Structure for Modeling and Manipulating Visual Objects and Patterns. *International Journal of Software Science and Computational Intelligence*, *1*(4), 1–18. doi:10.4018/jssci.2009062501

Wang, Y., Kinsner, W., & Zhang, D. (2009). Contemporary Cybernetics and its Facets of Cognitive Informatics and Computational Intelligence. [B]. *IEEE Transactions on Systems, Man, and Cybernetics*, *39*(2), 1–11.

Wang, Y., & Wang, Y. (2006). Cognitive Informatics Models of the Brain. [C]. *IEEE Transactions on Systems, Man, and Cybernetics*, *36*(2), 203–207. doi:10.1109/TSMCC.2006.871151

Wang, G., Liu, Q., Yao, Y. Y., & Skowron, A. (2003). *Rough sets, Fuzzy sets, Data mining, and Granular Computing*. Berlin: Springer. doi:10.1007/3-540-39205-X

Wang, G. Y., Zheng, Z., & Zhang, Y. (2002). RIDAS-A Rough Set Based Intelligent Data Analysis System. *Proceedings of, ICMLC2002*, 646–649.

Wang, Y. (2007). On the Cognitive Processes of Human Perception with Emotions, Motivations, and Attitudes. *International Journal of Cognitive Informatics and Natural Intelligence*, *1*(4), 1–13. doi:10.4018/jcini.2007100101

Wang, Y. J., & Guan, L. (2005). Recognition human emotion from audiovisual information. *Proceedings of the ICASSP*, *05*, 1125–1128.

Wang, Y. (2007). The Theoretical Framework of Cognitive Informatics. [IJCiNi]. *International Journal of Cognitive Informatics and Natural Intelligence*, *1*(1), 1–27.

Wang, Y. (2007). On Laws of Work Organization in Human Cooperation. [IJCINI]. *International Journal of Cognitive Informatics and Natural Intelligence*, *1*(2), 1–15.

Wang, Y., & Kinsner, W. (2006). Recent Advances in Cognitive Informatics. *IEEE Transactions on Systems, Man and Cybernetics. Part C, Applications and Reviews*, *36*(2), 121–123. doi:10.1109/TSMCC.2006.871120

Wang, Y., Wang, Y., Patel, S., & Patel, D. (2006). A Layered Reference Model of the Brain. *IEEE Transactions on Systems, Man and Cybernetics. Part C, Applications and Reviews*, *36*(2), 124–133. doi:10.1109/TSMCC.2006.871126

Wang, Y., & Ruhe, G. (2007). The Cognitive Process of Decision Making. *International Journal of Cognitive Informatics and Natural Intelligence*, *1*(2), 73–85. doi:10.4018/jcini.2007040105

Wang, L. W. (2004). *Applications of quotient space and the constructive learning method in the communication countermeasure reconnaissance* [Ph. D dissertation]. Anhui University, Hefei, China(in Chinese).

Wang, L. W., Zhang, L., & Zhang, M. (2003). A method of pattern classification based on RS and NCA. In *Proceedings of International Conference on Machine Learning and Cybernetics* (pp. 3090-3094), Xi'an, China.

Wang, Y. (2002). Keynote: On Cognitive Informatics. *Proc. 1st IEEE International Conference on Cognitive Informatics* (ICCI'02), Calgary, Canada, IEEE CS Press, August, (pp. 34-42).

Wang, Y. (2002). On cognitive informatics. *1st IEEE International Conference on Cognitive Informatics* (pp. 34–42). IEEE.

Wang, Y. (2002a), Keynote: On Cognitive Informatics, *Proc. 1st IEEE International Conference on Cognitive Informatics* (ICCI'02), Calgary, Canada, IEEE CS Press, August, 34-42.

Wang, Y. (2002b), The Real-Time Process Algebra (RTPA), *Annals of Software Engineering: An International Journal*, 14, USA, 235-274.

Wang, Y. (2003), On Cognitive Informatics, *Brain and Mind: A Transdisciplinary Journal of Neuroscience and Neurophilosophy*, Kluwer Academic Publishers, August, 4(3), pp. 151-167.

Wang, Y. (2003). On Cognitive Informatics. *Brain and Mind: A Transdisciplinary Journal of Neuroscience and Neurophilosophy, 4*(3), 151-167.

Wang, Y. (2003a), Cognitive Informatics: A New Transdisciplinary Research Field, *Brain and Mind: A Transdisciplinary Journal of Neuroscience and Neurophilosophy,* 4(2), 115-127.

Wang, Y. (2003b), Keynote: Cognitive Informatics Models of Software Agent Systems, Proc. 1st International Conference on Agent-Based Technologies and Systems (ATS'03), Univ. of Calgary Press, Calgary, Canada, August, 25.

Wang, Y. (2003b). On Cognitive Informatics. *Brain and Mind: A Transdisciplinary Journal of Neuroscience and Neurophilosophy, 4*(2), 151-167.

Wang, Y. (2004), Keynote: On Autonomic Computing and Cognitive Processes, *Proc. 3rd IEEE International Conference on Cognitive Informatics* (ICCI'04), Victoria, Canada, IEEE CS Press, August, 3-4.

Wang, Y. (2005). On cognitive properties of human factors in engineering. In *Proceedings of the IEEE 2005 International Conference on Cognitive Informatics* (ICCI'05), (pp.174-182). IEEE CS Press.

Wang, Y. (2006). Cognitive Informatics - Towards the Future Generation Computers that Think and Feel, Keynote, *Proc. 5th IEEE International Conference on Cognitive Informatics* (ICCI'06), Beijing, China, IEEE CS Press, July, pp. 3-7.

Wang, Y. (2007a), *Software Engineering Foundations: A Software Science Perspective*, CRC Book Series in Software Engineering, Vol. II, Auerbach Publications, NY., USA, July.

Wang, Y. (2007a). *Software Engineering Foundations: A Software Science Perspective*. CRC Book Series in Software Engineering (Vol. II). New York: Auerbach Publications.

Wang, Y. (2007b), Keynote: Cognitive Informatics Foundations of Nature and Machine Intelligence, *Proc. 6th International Conference on Cognitive Informatics* (ICCI'07), IEEE CS Press, Lake Tahoe, CA., Aug., 3-12.

Wang, Y. (2007b). The Theoretical Framework of Cognitive Informatics. *International Journal of Cognitive Informatics and Natural Intelligence*, *1*(1), 1–27.

Wang, Y. (2007d), Exploring Machine Cognition Mechanisms for Autonomic Computing, *International Journal on Cognitive Informatics and Natural Intelligence*, March, 1(2), i - v.

Wang, Y. (2007e), The Cognitive Processes of Formal Inferences, *International Journal of Cognitive Informatics and Natural Intelligence*, IGI, USA, Dec., 1(4), 75-86.

Wang, Y. (2008a), Keynote: On Abstract Intelligence and Its Denotational Mathematics Foundations, *Proc. 7th IEEE International Conference on Cognitive Informatics* (ICCI'08), Stanford University, CA., USA, IEEE CS Press, August, 5-15.

Wang, Y. (2008a), On Contemporary Denotational Mathematics for Computational Intelligence, *Transactions on Computational Science, 2*, Springer, Sept., pp. 6-29.

Wang, Y. (2008a). On Contemporary Denotational Mathematics for Computational Intelligence. *Transactions of Computational Science*, *2*, 6–29. doi:10.1007/978-3-540-87563-5_2

Wang, Y. (2008b), Toward a Generic Mathematical Model of Abstract Game Theories, *Transactions of Computational Science, 2,* Springer, June, 205-223.

Wang, Y. (2008b). Keynote: Abstract Intelligence and Its Denotational Foundations. *Proceedings 7th International Conference on Cognitive Informatics (ICCI'08).* CA, USA: Stanford University.

Wang, Y. (2008c), A Cognitive Informatics Theory for Visual Information Processing, *Proc. 7th International Conference on Cognitive Informatics* (ICCI'08), IEEE CS Press, Stanford University, CA., Aug., pp.317-323.

Wang, Y. (2008c), On Contemporary Denotational Mathematics for Computational Intelligence, *Transactions of Computational Science, 2,* Springer, June, 6-29.

Wang, Y. (2008d), On Concept Algebra: A Denotational Mathematical Structure for Knowledge and Software Modeling, *International Journal of Cognitive Informatics and Natural Intelligence,* IGI, USA, April, 2(2), 1-19.

Wang, Y. (2008d). On System Algebra: A Denotational Mathematical Structure for Abstract System modeling. *International Journal of Cognitive Informatics and Natural Intelligence*, *2*(2), 20–42.

Wang, Y. (2008f), RTPA: A Denotational Mathematics for Manipulating Intelligent and Computational Behaviors, *International Journal of Cognitive Informatics and Natural Intelligence,* IGI, USA, April, 2(2), 44-62.

Wang, Y. (2009). On Abstract Intelligence: Toward a Unified Theory of Natural, Artificial, Machinable, and Computational Intelligence, *International Journal of Software Science and Computational Intelligence,* IGI, USA, Jan., 1(1), 1-18.

Wang, Y. (2009b). On Visual Semantic Algebra (VSA): A Denotational Mathematical Structure for Modeling and Manipulating Visual Objects and Patterns. *International Journal of Software Science and Computational Intelligence*, *1*(4).

Wang, Y. and G. Ruhe (2007), The Cognitive Process of Decision Making, *International Journal of Cognitive Informatics and Natural Intelligence*, IGI, USA, March, 1(2), 73-85.

Wang, Y., & Chiew, V. (2009). On the Cognitive Process of Human Problem Solving. *Cognitive Systems Research: An International Journal, 9*(4). UK: Elsevier.

Wang, Y., & Wang, Y. (2008), The Cognitive Processes of Consciousness and Attention, *Proc. 7th International Conference on* Cognitive Informatics (ICCI'08), IEEE CS Press, Stanford University, CA., Aug, 30-39.

Wang, Y., & Wang, Y. (2002). Cognitive models of the brain. *First IEEE International Conference on Cognitive Informatics* (pp. 259–269). IEEE-CS Press.

Wang, Y., & Wang, Y. (2006). Cognitive Informatics Models of the Brain. [C]. *IEEE Transactions on Systems, Man, and Cybernetics*, *36*(2), 203–207. doi:10.1109/TSMCC.2006.871151

Wang, Y., Kinsner, W., & Zhang, D. (2009a). Contemporary Cybernetics and its Faces of Cognitive Informatics and Computational Intelligence. *IEEE Trans. on System, Man, and Cybernetics (B)*, *39*(4), 823–833. doi:10.1109/TSMCB.2009.2013721

Wang, Y., Kinsner, W., Anderson, J. A., Zhang, D., Yao, Y., Sheu, P., et al. (2009b). A Doctrine of Cognitive Informatics. *Fundamenta Informaticae*, *90*(3), 203–228.

Wang, Y., Wang, Y., Patel, S., & Patel, D. (2006). A Layered Reference Model of the Brain (LRMB). *IEEE Transactions on Systems, Man and Cybernetics. Part C, Applications and Reviews*, *36*(2), 124–133. doi:10.1109/TSMCC.2006.871126

Watson, A. B., Barlow, H. B., & Robson, J. G. (1983). What does eye see best. *Nature*, *302*, 419–422. PubMeddoi:10.1038/302419a0

Wellman, M. P. (1999), Multiagent Systems, in R.A. Wilson and C.K. Frank eds., *The MIT Encyclopedia of the Cognitive Sciences*, MIT Press, MA.

Westen, D. (1999). *Psychology: Mind, Brain, and Culture* (2nd ed.). NY: John Wiley & Sons, Inc.

Wever, R. A. (1979). *The Circadian System of Man: Results of Experiments Under Temporal Isolation*. NY: Springer.

Widrow, B., Glover, J. R., McCool, J. M., Kaunitz, J., Williams, C. S., Hearn, R. H., & Zeidler, J. R. (1975). Adaptive noise cancelling: principles and applications. *Proceedings of the IEEE*, *63*(12), 1692–1716. doi:10.1109/PROC.1975.10036

Widrow, B., McCool, J. M., Larimore, M. G., & Johnson, C. R. Jr. (1976). Stationary and nonstationary learning characteristics of the LMS adaptive filter. *Proceedings of the IEEE*, *64*(8), 1151–1162. doi:10.1109/PROC.1976.10286

Widrow, B., & Lehr, M. A. (1990), 30 Years of Adaptive Neural Networks: Perception, Madeline, and Backpropagation, *Proc. of the IEEE*, Sept., 78(9), 1415-1442.

Wilson, R. A., & Keil, F. C. (Eds.). (1999). *The MIT Encyclopedia of the Cognitive Sciences*. Cambridge, Mass: The MIT Press

Witkowski, M., & Stathis, K. (2004, April). A Dialectic Architecture for Computational Autonomy . In Nickles, M., Rovatsos, M., & Weiss, G. (Eds.), *Agents and computational autonomy: Potential, risks, and solutions* (*Vol. 2969*, pp. 261–273). Springer Berlin. doi:10.1007/978-3-540-25928-2_21

Wittig, T. (Ed.). (1992). *ARCHON: An Architecture for Multi-Agent Systems*. London: Ellis Horwood.

Wolf, T. D., & Holvoet, T. (2006) *Autonomic Computing: Concepts, Infrastructure and Applications* (1st ed.). A Taxonomy for Self-* Properties in Decentralized Autonomic Computing (pp. 101–120). CRC Press.

Wolfgang Münch and Kiyoshi Furukawa, (2000), Bubbles, Prototype at Schloss Wahn, Theaterwissenschaftliche Sammlung, Universität Köln, July 2000.

Woods, K., & Bowyer, K. W. (1997). Generating ROC curves for artificial neural networks. *IEEE Transactions on Medical Imaging*, *16*(3), 329–337. doi:10.1109/42.585767

Wooldridge, M. (2002). *An Introduction to Multiagent Systems*. John Wiley & Sons.

Wooldridge, M., & Jennings, N. (1995). Intelligent Agents: Theory and Practice . *The Knowledge Engineering Review*, *10*(2), 115–152. doi:10.1017/S0269888900008122

Wu, T., Zhang, L., & Zhang, Y. (2005). Kernel Covering Algorithm for Machine Learning. [in Chinese]. *Chinese Journal of Computers*, *28*(8), 1295–1300.

Wu, Y. F., & Rangayyan, R. M. (2009). An unbiased linear adaptive filter with normalized coefficients for the removal of noise in electrocardiographic signals. *International Journal of Cognitive Informatics and Natural Intelligence*, *3*(4), 73–90. doi:10.4018/jcini.2009062305

Wu, Y. F., Rangayyan, R. M., Zhou, Y. C., & Ng, S. C. (2009). Filtering electrocardiographic signals using an unbiased and normalized adaptive noise reduction system. *Medical Engineering & Physics*, *31*(1), 17–26. doi:10.1016/j.medengphy.2008.03.004

Wu, M. (2000). *The research on design of the classifier for large scale pattern recognition* [Ph. D dissertation]. Tsinghua University, Beijing, China (in Chinese).

Wu, M., Zhang, B., & Zhang, L. (2000). A neural network based classifier for handwritten Chinese character recognition. In *Proceedings of the 15th International Conference on Pattern Recognition* (pp. 561-568), Barcelona.

Wundrich, I. J., von der Malsburg, C., & Würtz, P. (2002). Image Reconstruction from Gabor Magnitudes. *Biologically Motivated Computer Vision* (pp. 117–126).

Xu, J., Li, B., & Li, D. (2002). Placement problems for transparent data replication proxy services. *IEEE Journal on Selected Areas in Communications*, *7*, 1383–1398.

Xu, F., & Zhang, L. (2005). An analysis of uneven granules clustering based on quotient space. [in Chinese]. *Computer Engineering*, *31*(3), 26–28.

Xue, Q., Hu, Y. H., & Tompkins, W. J. (1992). Neural-network-based adaptive matched filtering for QRS detection. *IEEE Transactions on Bio-Medical Engineering*, *39*(4), 317–329. doi:10.1109/10.126604

Yagoubi, B., & Slimani, Y. (2007). Task Load Balancing Strategy for Grid Computing . *Journal of Computer Science*, *3*(3), 186–194. doi:10.3844/jcssp.2007.186.194

Yamamoto, Y., & Yun, X. P. (1994). Coordinating locomotion and manipulation of manipulator. *IEEE Transactions on Automatic Control*, *39*(6), 1326–1332. doi:10.1109/9.293207

Yan, C. (2007), Jeux Vidéo Multijoueurs Ubiquitaires: principes de conception et architecture d'exécution, PhD dissertation, CNAM, Paris, December 2007.

Yang, T. F., Devine, B., & Macfarlane, P. W. (1994). Artificial neural networks for the diagnosis of atrial fibrillation. *Medical & Biological Engineering & Computing*, *32*(6), 615–619. doi:10.1007/BF02524235

Yang, B., & Liu, J. (2007, 9-11 July). An Autonomy Oriented Computing (AOC) Approach to Distributed Network Community Mining. In G. Serugendo, J. Flatin, & M. Jelasity (Eds.), *Proceedings of 1st international conference on self-adaptive and self-organizing systems (saso'07)* (pp. 151–160). Boston, Massachusetts, USA: IEEE Computer Society Press.

Yang, C., Lin, H., & Lin, F. O. (2006). Designing Multiagent-Based Education Systems for Navigation Training. In *IEEE International Conference on Cognitive Informatics, ICCI'06, 495-501.*

Yao, Y. Y., & Zhong, N. (1999). Potential applications of granular computing in knowledge discovery and data mining . In *Proceedings of World Multi-conference on Systems* (pp. 573–580). Cybernetics and Informatics.

Yao, Y. Y., Zhao, Y., & Wang, J. (2006). On reduct construction algorithms, Rough Sets and Knowledge Technology. *Proceedings of RSKT*, *2006*, 297–304.

Yao, Y. Y. (2000). Granular computing: basic issues and possible solutions. In *Proceedings of the 5th Joint Conference on Information Sciences* (pp. 186-189), Atlantic City, New Jersey, USA.

Yao, Y. Y. (2005). Perspectives of granular computing. In *Proceedings of IEEE International Conference on Granular Computing* (pp. 85-90), Beijing, China.

Yi, B. J., & Kim, W. K. (2001). The dynamics for redundantly actuated omnidirectional mobile robots. *IEEE International Conference on Robotics and Automation* (pp. 2485-2492).

Yu, H., Wang, G. Y., Yang, D. C., & Wu, Z. F. (2002). Knowledge Reduction Algorithms Based on Rough Set and Conditional Information Entropy. *Proceedings of the Society for Photo-Instrumentation Engineers*, *4730*, 422–431.

Yu, H., & Vahdat, A. (2001). The Costs and Limits of Availability for Replicated Services, *In SOSP '01: Proceedings of the Eighteenth ACM Symposium on Operating Systems Principles*, pp. 29-42, New York.

Zadeh, L. A. (1997). Towards a theory of fuzzy information granulation and its centrality in human reasoning and fuzzy logic. *Fuzzy Sets and Systems*, *19*(1), 111–127. doi:10.1016/S0165-0114(97)00077-8

Zadeh, L. A. (1998). Some reflections on soft computing, granular computing and their roles in the conception, design and utilization of information/intelligent systems. *Soft Computing*, *2*(1), 23–25. doi:10.1007/s005000050030

Zadeh, L. A. (1965). Fuzzy Sets and Systems. In J. Fox (Ed.), *Systems Theory* (pp. 29-37). Brooklyn, NY: Polytechnic Press.

Zadeh, L. A. (1973). Outline of a New Approach to Analysis of Complex Systems. *IEEE Transactions on Systems, Man, and Cybernetics*, *1*(1), 28–44.

Zambonelli, F., Jennings, N., & Wooldridge, M. (2003). Developing multiagent systems: the gaia methodology. *ACM Transactions on Software Engineering and Methodology*, *12*(3), 317–370. doi:10.1145/958961.958963

Zarzoso, V., & Nandi, A. K. (2001). Noninvasive fetal electrocardiogram extraction: blind separation versus adaptive noise cancellation. *IEEE Transactions on Bio-Medical Engineering*, *48*(1), 12–18. doi:10.1109/10.900244

Zhang, C. J., Li, Y., & Zhang, L. (2004a). Realizing the high-precision fuzzy control based on the theory of quotient space methods of granular computing. [in Chinese]. *Computer Engineering And Applications*, *40*(11), 37–39.

Zhang, L., & Zhang, B. (1989a). Quotient space model (I) of qualitative reasoning. [in Chinese]. *Anqing Normal College Journal*, *7*(1), 1–8.

Zhang, L., & Zhang, B. (1989b). Mathematic model of quotient space of problem description. [in Chinese]. *Chizhou College Journal*, *8*(1), 15–20.

Zhang, L., & Zhang, B. (1990a). Quotient space model (II) of qualitative reasoning. [in Chinese]. *Anqing Normal College Journal*, *8*(1), 15–20.

Zhang, L., & Zhang, B. (1990b). Computational complexity of problem solving of quotient space model. [in Chinese]. *Anqing Normal College Journal*, *8*(2), 1–7.

Zhang, L., & Zhang, B. (1992). *Theory of Problem Solving and Its Applications. North-Holland.* Elsevier Science Publishers.

Zhang, L., & Zhang, B. (1999). A geometrical representation of Mc-Culloch-Pitts neural model and its applications. *IEEE Transactions on Neural Networks*, *10*(4), 925–929. doi:10.1109/72.774263

Zhang, L., & Zhang, B. (2003). Theory of fuzzy quotient space (methods of fuzzy granular computing). [in Chinese]. *Journal of Software*, *14*(4), 770–776.

Zhang, L., & Zhang, B. (2005a). A quotient space approximation model of multiresolution signal analysis. *Journal of Computer Science & Technology*, *20*(1), 92–108.

Zhang, L., & Zhang, B. (2005b). Fuzzy reasoning model under quotient space structure. *Information Sciences*, *173*(4), 353–364. doi:10.1016/j.ins.2005.03.005

Zhang, L., & Zhang, B. (2005c). The structure analysis of fuzzy sets. *International Journal of Approximate Reasoning*, *40*(1-2), 92–108. doi:10.1016/j.ijar.2004.11.003

Zhang, Y. P. (2002). A repeated cover algorithm of achieving characteristic rule. [in Chinese]. *Journal of Anhui University*, *26*(2), 9–13.

Zhang, Y. P., Zhang, L., & Wu, T. (2003). A constructive self-adjusting and probabilistic decision-making classifier. [in Chinese]. *Microcomputer Development*, *13*(7), 85–87.

Zhang, Y. P., Zhang, L., & Wu, T. (2004b). The representation of different granular worlds: a quotient space. [in Chinese]. *Chinese Journal of Computers*, *27*(3), 328–333.

Zhang, Y. P., Zhang, L., & Wu, T. (2005). A multiside increase by degrees algorithm at machine learning. [in Chinese]. *ACTA Electronica Sinica*, *33*(2), 327–331.

Zhang, Y. P., Zhang, L., & Xia, Y. (2004c). To compare the theory of quotient space with rough set. [in Chinese]. *Microcomputer Development*, *14*(10), 21–24.

Zhang, L., & Zhang, B. (2004). The quotient space theory of problem solving. *Fundamenta Informaticae, 59*(2,3), 278-298.

Zhao, W., & Kearney, D. (2003). Deriving Architectures of Web-Based Applications. *Lecture Notes in Computer Science*, *2642*, 301–312. doi:10.1007/3-540-36901-5_31

Zhao, L., & Zhang, L. (2006). Research in Quotient Space Theory Based on Structure. In *Proceedings of 5th IEEE International Conference on Cognitive Informatics,* (pp. 309-313). Beijing, China. IEEE CS Press.

Zhoun, W., Wang, L., & Jia, W. (2004). An analysis of update ordering in distributed replication systems. *Future Generation Computer Systems*, *20*(4), 565–590. doi:10.1016/S0167-739X(03)00174-2

Zigel, Y., Cohen, A., & Katz, A. (2000). ECG signal compression using analysis by synthesis coding. *IEEE Transactions on Bio-Medical Engineering*, *47*(10), 1308–1316. doi:10.1109/10.871403

Zoethout, K., W. J., & Molleman, E. (2008, February). Task Dynamics in Self-organising Task Groups: Expertise, Motivational, and Performance Di erences of Specialists and Generalists. *Autonomous Agents and Multi-Agent Systems*, *16*(1), 75–94. doi:10.1007/s10458-007-9022-9

About the Contributors

Yingxu Wang is professor of cognitive informatics and software engineering, Director of International Center for Cognitive Informatics (ICfCI), and Director of Theoretical and Empirical Software Engineering Research Center (TESERC) at the University of Calgary. He is a Fellow of WIF, a P.Eng of Canada, a Senior Member of IEEE and ACM, and a member of ISO/IEC JTC1 and the Canadian Advisory Committee (CAC) for ISO. He received a PhD in Software Engineering from The Nottingham Trent University, UK, in 1997, and a BSc in Electrical Engineering from Shanghai Tiedao University in 1983. He has industrial experience since 1972 and has been a full professor since 1994. He was a visiting professor in the Computing Laboratory at Oxford University in 1995, Dept. of Computer Science at Stanford University in 2008, and the Berkeley Initiative in Soft Computing (BISC) Lab at University of California, Berkeley in 2008, respectively. He is the founder and steering committee chair of the annual IEEE International Conference on Cognitive Informatics (ICCI). He is founding Editor-in-Chief of International Journal of Cognitive Informatics and Natural Intelligence (IJCINI), founding Editor-in-Chief of International Journal of Software Science and Computational Intelligence (IJSSCI), Associate Editor of IEEE Trans on System, Man, and Cybernetics (A), and Editor-in-Chief of CRC Book Series in Software Engineering. He is the initiator of a number of cutting-edge research fields and/or subject areas such as cognitive informatics, abstract intelligence, denotational mathematics, cognitive computing, theoretical software engineering, coordinative work organization theory, cognitive complexity of software, and built-in tests. He has published over 105 peer reviewed journal papers, 193 peer reviewed conference papers, and 12 books in cognitive informatics, software engineering, and computational intelligence. He is the recipient of dozens international awards on academic leadership, outstanding contribution, research achievement, best paper, and teaching in the last 30 years. He can be reached at: yingxu@ucalgary.ca.

* * *

Ghalem Belalem graduated from university of Oran, Algeria, where he received PhD degree in computer science in 2007. He is now a research fellow of management of replicas in data grid. His current research interests are: distributed systems, grid computing and data grid, placement of replicas and consistency in large scale systems and mobile environment.

Khadidja Benhallou graduate from University of Oran (Es Senia)- Algeria, where she received Master degree in computer science in June 2006 from Faculty of Sciences, University of Oran, Algeria. His research interests are consistency management and multi-agents systems.

Zineb Benotmane graduated from University of Oran (Es Senia)- Algeria, where she received Master degree in computer science in June, 2006 from Faculty of Sciences, University of Oran, Algeria. His research interests are replication strategies, collaborative environments.

Phan Cong-Vinh has a Ph.D. in Computer Science from London South Bank University (LSBU) in UK, a B.S. in Mathematics and an M.S. in Computer Science from Vietnam National University (VNU) at Ho Chi Minh City, a B.A in English from Hanoi University of Foreign Languages Studies in Vietnam. He finished his Ph.D. dissertation with the title "Formal Aspects of Dynamic Reconfigurability in Reconfigurable Computing Systems" supervised by Prof. Jonathan P. Bowen at LSBU where he was affiliated with Centre for Applied Formal Methods (CAFM), Institute for Computing Research (ICR). From 1983 to 2000, he was a lecturer in Mathematics and Computer Science at VNU, Posts and Telecommunications Institute of Technology (PTIT) and several other Universities in Vietnam before he joined research with Dr. Tomasz Janowski at International Institute for Software Technology (IIST) in Macao SAR, China as a Fellow in 2000. From 2001 to present he does research together with Prof. Jonathan P. Bowen as a research scholar and then visiting researcher at CAFM. He is also an IEEE member. His research interests center on all aspects of Formal Methods, Self-* Computing and Applied Categorical Structures in Computer Science.

Seif El-Nasr is an Assistant Professor in the School of Interactive Arts and Technology at the Simon Fraser University. She earned her PhD from the Northwestern University. Her research work focuses on designing tools and techniques that enhance the engagement of interactive environments used for education and entertainment. She received several grants and awards, including 2nd Best Paper Award at the International Conference of Virtual Storytelling, Best Student Paper Award at the Autonomous Agents conference 1999. She is on the editorial board of several journals, including Journal of Game Development, Int. J. Intelligent Games and Simulation and ACM Computers in Entertainment.

Alberto de la Encina is a Lecturer in the Computer Systems and Computation Department, Universidad Complutense de Madrid (Spain). He obtained his MS degree in Mathematics in 1999 and his PhD in the same subject in 2008. Dr. de la Encina has published more than 20 papers in international refereed conferences and journals. His research interests cover functional programming, learning strategies, debugging techniques, and formal methods.

Rafaeldo Espírito-Santo is an Adjunct Professor of the Department of Computer Science at the Universidade Paulista (UNIP) and Professor of the Departamento de Exatas of the Universidade Nove de Juliio (UNINOVE) in São Paulo, Brazil. He is also a Researcher of the Laboratório de Sistemas Integráveis da Escola Politécnica da Universidade de São Paulo and a Researcher of the Instituto Israelita de Ensino e Pesquisa Albert Einstein. He received the Bachelor of Engineering degree in Electronics and Process Control in 1989 from the Universidade Federal da Bahia (UFBA), Salvador, Bahia, Brazil, and the Ph.D. degree in Electrical Engineering from the Universidade de São Paulo (USP) in 2004.

Lisa Fan is an associate professor at Department of Computer Science, Faculty of Science, University of Regina. She received her Ph.D. from the University of London, U.K. Dr. Fan's main research areas include Web Intelligence, Intelligent Learning Systems, Web-based Adaptive Learning, Cognitive Modeling. She is also interested in intelligent systems applications in engineering (intelligent manufacturing system and intelligent transportation system).

Loe Feijs studied Electrical Engineering at TU/e and has a Ph.D. in Computer Science. He is a full professor in the Designed Intelligence group, the department of Industrial Design of Eindhoven University of Technology. His research interests include semantics, artificial languages, ambient intelligence and embedded systems. He is the author of several books on formal specification and design.

Hideki Hara is an associate professor of the Department of Information and Network Science, Chiba Institute of Technology, Chiba, Japan. He received his Ph.D. degree from Chiba Institute of Technology in 1999. His research interests include Computer Network and Software Agent. Dr. Hara is a member of IEICE and IPSJ.

Fumio Hattori received the B.E. and the M.E. degree from Waseda University, Tokyo Japan, in 1973 and 1975 respectively. He also received the doctor degree in informatics from Kyoto University in 2000. In 1975, he joined Nippon Telegraph and Telephone public corporation (currently NTT Corporation), Tokyo, Japan. At NTT Laboratories, he engaged in the research and development of database systems, knowledge engineering, expert systems, agent-based communication, and so on. From 1996 to 1998, he was the Director of the Computer Science Laboratory at NTT Communication Science Laboratories. From 1998 to 2004, he had been with NTT Software Corporation, where he serves the Director of the Advanced Systems Development Center. Since 2004, he is the Professor of College of Information Science and Engineering, Ritsumeikan University, Japan. His research interest includes socialware, community computing, multi-agent system, and so on.

Mercedes Hidalgo-Herrero is a Lecturer in the Mathematics Education Department, Universidad Complutense de Madrid (Spain). She obtained her MS degree in Mathematics in 1997 and her PhD in the same subject in 2004. Dr. Hidalgo-Herrero has published more than 20 papers in international refereed conferences and journals. Her research interests cover functional programming, learning strategies, and formal methods.

Jun Hu is an assistant professor in the Department of Industrial Design at the Eindhoven University of Technology. He has a background in Mathematics, Computer Science and Human-Computer Interaction. His expertise and research interests are in interactive multimedia, software architecture and formal methods. He is a qualified system analyst and senior programmer. He worked for several companies and institutes including the Institute of Geophysics of Jiangsu Oil Exploration Corp (Nanjing, China), the information center of Shaanxi Construction Machinery Co. Ltd. (Xi'an, China), the Institute of Visualization of Northwest University (Xi'an, China) and Philips Research (Eindhoven, The Netherlands).

Tetsuo Kinoshita is a professor of Cyberscience Center and Graduate School of Information Science, Tohoku University, Japan. He received his Dr.Eng. degree in information engineering from Tohoku University in 1993. His research interests include agent engineering, knowledge engineering, knowledge-based systems and agent-based systems. He received the IPSJ Research Award, the IPSJ Best Paper Award and the IEICE Achievement Award in 1989, 1997 and 2001, respectively. Dr. Kinoshita is a member of IEEE, ACM, AAAI, IEICE, IPSJ, JSAI, and Society for Cognitive Science of Japan.

Witold Kinsner (S'72-M'73-SM'88) is Professor and Associate Head at the Department of Electrical and Computer Engineering, University of Manitoba, Winnipeg, Canada. He is also Affiliate Professor at the Institute of Industrial Mathematical Sciences, and Adjunct Scientist at the Telecommunications Research Laboratories, Winnipeg. He obtained the Ph.D. degree in Electrical Engineering from McMaster University in 1974. He was Assistant Professor in Electrical Engineering at McMaster University and McGill University. He is a co-founder of the first Microelectronics Centre in Canada, and was its Director of Research from 1979 to 1987. He also was designing configurable self-synchronizing CMOS memories at the ASIC Division, National Semiconductor, Santa Clara, California. Dr. Kinsner has been involved in research on algorithms and software/hardware computing engines for real-time multimedia, using wavelets, fractals, chaos, emergent computation, genetic algorithms, rough sets, fuzzy logic, neural networks. Applications included signal and data compression, signal enhancement, classification, segmentation, and feature extraction in various areas such as real-time speech compression for multimedia, wideband audio compression, aerial and space ortho image compression, biomedical signal classification, severe weather classification from volumetric radar data, radio and power-line transient classification, image/video enhancement, and modeling of complex processes such as dielectric discharges. He also spent many years in VLSI design (configurable high-speed CMOS memories, as well as magnetic bubble memories), and computer-aided engineering of electronic circuits (routing and placement for VLSI, and field-programmable gate arrays). He has authored and co-authored over 500 publications in the above areas. Dr. Kinsner is a senior member of the Institute of Electrical & Electronics Engineers (IEEE), a member of the Association of Computing Machinery (ACM), a member of the Mathematical and Computer Modeling Society, a member of Sigma Xi, and a life member of the Radio Amateurs of Canada.

Natalia López is an Associate Professor in the Computer Systems and Computation Department, Universidad Complutense de Madrid (Spain). She obtained her MS degree in Mathematics in 1997 and her PhD in Computer Science in 2003 from the Universidad Complutense de Madrid. Dr. López has published more than 20 papers in international refereed conferences and journals. Her topics of interest include process algebras, stochastic temporal systems, and formal testing methodologies.

Jean-Claude Latombe is Kumagai professor and former chairman of computer science in the Department of Computer Science at Stanford University. He received an Engineering Degree in Electrical Engineering from National Polytechnic Institute of Grenoble, France, in June 1969; an Engineering Degree in Computer Science from National Polytechnic Institute of Grenoble, France, in June 1970; An M.S. (Docteur-Ingenieur) in Electrical Engineering from National Polytechnic Institute of Grenoble, France, in November 1972; and a Ph.D. (Docteur d'Etat) in Computer Science from University of Grenoble, November 1977. He published numerals peer reviewed journal and conference papers, and many of them are widely cited. Prof. Latombe's research is to create autonomous agents that sense, plan, and act in real and/or virtual worlds. His work focuses on designing architectures and algorithms to represent, sense, plan, control, and render motions of physical objects. A key underlying issue is to efficiently capture the connectivity of configuration or state spaces that are both high-dimensional and geometrically complex. Specific topics include: motion planning in the presence of multiple constraints (obstacle avoidance, maintenance of equilibrium, as well as kino-dynamic, visibility, and contact constraints), assembly sequence planning, making decisions under sensing and control uncertainty, construction of 3-D geometric models of complex environments, visual tracking of rigid, articulated and deforming objects, and reasoning in multiple-agent worlds. Applications of his work include robot-assisted medical surgery, integration of design and manufacturing, graphic animation of digital actors, study of molecular motions (folding, binding). His current projects study legged robots navigating on steep terrain, sensing and manipulation of deformable objects, structure and motion of proteins, and surgical simulation.

Botang Li is a graduate student at Department of Computer Science, University of Regina. He received his bachelor degree from Guangdong, China and has more than two years' industrial experiences in Canada. Image Retrieval, Web Mining, and Social Network Analysis are his current research areas and interests.

Roseli de Deus Lopes is associate professor with the Department of Electronic Systems Engineering of the Polytechnic School at the University of São Paulo (USP), São Paulo, Brazil. Currently, she is also the Director of Estação Ciência (a Science Center) at USP. She received her Bachelor of Electrical Engineering degree in 1987, her Master's degree in Electrical Engineering in 1993, and her Ph.D. degree in Electrical Engineering in 1998, all from the Polytechnic School at USP. She works as a scientist at the Laboratory of Integrated Systems (LSI) at USP, where she investigates Interactive Electronic Media, including computer graphics, image processing, interactive devices, visualization techniques, and virtual reality. She has been involved in the research and development of high-performance graphics systems (hardware and software), visualization in engineering, and image-processing techniques applied to medical and dentistry data. Her current interests include: multi-modality image-processing techniques and advanced collaborative work/learning/entertainment environments.

Takahide Maemura is a doctoral student of Graduate School of Information Sciences, Tohoku University, Japan. He received M.E. degree from Chiba Institute of Technology in 2004.

Ryohei Nakatsu, National University of Singapore, Singapore. Ryohei Nakatsu received the B.S. (1969), M.S. (1971) and Ph.D. (1982) degrees in electronic engineering. After joining NTT in 1971, he mainly worked on speech recognition technology. 1994-2002, he joined ATR (Advanced Telecommunications Research Institute) as the president of ATR Media Integration & Communications Research Laboratories. From the spring of 2002 until 2007 he was full professor at School of Science and Technology, Kwansei Gakuin University in Sanda (Japan). At the same time he established a venture company, Nirvana Technology Inc., and became the president of the company. Since Spring 2008 he is professor at the National University of Singapore, Interactive Digital Media Institute.

Stéphane Natkin is professor in the department of Computer Science in charge of the system multimedia chair, a member of the administrative board at the Conservatoire National des Arts et Métiers (CNAM) in Paris France. He is at the head of the French High School on Games and Interactive Media (ENJMIN) (hppt://www.enjmin.fr) created by the French prime ministry in 2003, and in charge the Network System and Multimedia research group at the CEDRIC (the computer science research laboratory of the CNAM http://cedric.cnam.fr). He teaches in particular Computer Games principles and Multimedia Systems, Computer Networks and Security. He has worked in the field multimedia systems, video games and critical computer system and both from the research and the industrial point of view and is the author of numerous publications and communications in these fields. From 1998 to

2005 he was the director of the CEDRIC. He act, as a scientific advisor, for France Telecom R&D for research programs related to entertainments and games. He is the founder of a security software editor and he was also the manager of an art gallery located in the centre of Paris. He is the author of the books "Internet Security Protocols" Dunod 2001 and "and "Video Games and Interactive Media, A Glimpse at New Digital Entertainment", AKPeters, 2006. He is also the producer and one of the authors of the book "Sol LeWitt Black Gouaches".

Mario Piattini is MSc and PhD in Computer Science by the Polytechnic University of Madrid. Certified Information System Auditor by ISACA (Information System Audit and Control Association). Full Professor at the Escuela Superior de Informática of the Castilla-La Mancha University. Author of several books and papers on databases, software engineering and information systems. He leads the ALARCOS research group of the Department of Computer Science at the University of Castilla-La Mancha, in Ciudad Real, Spain. His research interests are: advanced database design, database quality, software metrics, object oriented metrics, software maintenance.

Javier Portillo is currently studying the last course of the Computer Science Degree and he is also a technician working at the Alarcos Research Group, (University of Castilla-La Mancha) Spain. His research interests include Agent Technology and Knowledge Management.

Rangaraj M. Rangayyan received the B.E. degree in electronics and communication from the University of Mysore at the People's Education Society College of Engineering, Mandya, India, in 1976, and the Ph.D. degree in electrical engineering from the Indian Institute of Science, Bengaluru, India, in 1980. He is currently a Professor in the Department of Electrical and Computer Engineering, University of Calgary, Calgary, AB, Canada. Dr. Rangayyan received the 1997 and 2001 Research Excellence Awards of the Department of Electrical and Computer Engineering, the 1997 Research Award of the Faculty of Engineering, and the IEEE Third Millennium Medal by in 2000. He was appointed as a "University Professor" at the University of Calgary in 2003. He was elected as a Fellow of the Engineering Institute of Canada in 2002, the American Institute for Medical and Biological Engineering in 2003, the International Society for Optical Engineering (SPIE) in 2003, the Society for Imaging Informatics in Medicine in 2007, the Canadian Medical and Biological Engineering Society in 2007, and the Canadian Academy of Engineering in 2009. He was awarded the Killam Resident Fellowship thrice in 1998, 2002, and 2007. His current research interests include areas of digital signal and image processing, biomedical signal analysis, biomedical image analysis, and computer-aided diagnosis.

Matthias Rauterberg received a B.S. in Psychology (1978), a B.A. in Philosophy (1981) and a B.S. in Computer Science (1983), a M.S. in Psychology (1981) and a M.S. in Computer Science (1986), and a Ph.D. in Computer Science/ Mathematics (1995). He was a senior lecturer at the Swiss Federal Institute of Technology (ETH) in Zurich, where later he was heading the Man-Machine Interaction research group (MMI). Since 1998 he is professor first at IPO, Centre for User System Interaction Research, and later at the Department of Industrial Design at the Eindhoven University of Technology (TU/e, The Netherlands). From 1999 till 2001 he was director of IPO. He is now the head of the Designed Intelligence research group.

Fazel Rezai received his BSc. and M.Sc. in Electrical Engineering and Biomedical Engineering in 1990 and 1993, respectively. He received his Ph.D. in Electrical Engineering from the University of Manitoba in Winnipeg, Canada in 1999. From 2000 to 2002, he worked in industry as a senior research scientist and research team manager. After gaining a couple of years industrial experience he joined academia at Sharif University of Technology and later the University of Manitoba as assistant professor in 2002 and 2004, respectively. Currently, he is assistant Professor and the Director of Biomedical Signal Processing Laboratory at the Department of Electrical Engineering, University of North Dakota, USA. His research interests include biomedical signal and image processing, brain computer interface, EEG signal processing, seizure detection and prediction, neurofeedback, and human performance evaluation based on EEG signals.

Ismael Rodríguez is an associate professor in the Department of Computer Systems and Computation, Universidad Complutense de Madrid (Spain). He obtained his MS degree in Computer Science in 2001 and his PhD in the same subject in 2004. Dr. Rodríguez received the Best Thesis Award of his faculty in 2004. He has published more than 50 papers in conferences and journals. His research interests cover formal methods, testing techniques, nature-based computation, and e-commerce.

Fernando Rubio is an associate professor in the Department of Computer Systems and Computation, Universidad Complutense de Madrid (Spain). He obtained his MS degree in Computer Science in 1997 and his PhD in the same subject in 2001. Dr Rubio received the National Degree Award on the subject of Computer Science from the Spanish Ministry of Education in 1997, as well as the Best Thesis Award of his faculty in 2001. His research areas cover functional programming, testing techniques, and nature-based computation.

Benjamin Salem, Eindhoven University of Technology, The Netherlands. Ben Salem received a Diploma of Architecture (1987) and a Master of Architecture (1993) before a Ph.D. in Electronics (2003). From 2001 to 2003 he was director of Polywork Ltd (UK), a research consultancy. From 2003 to 2006 he had a PostDoc position at the Department Industrial Design of the Technical University Eindhoven (The Netherlands). From 2006 to 2007 he was a research fellow at the School of Science and Engineering at the Kwansei Gakuin University, Japan. Since 2007 he is Assistant Professor at the Department of Industrial Design, Eindhoven University of Technology. His main research interests are game and game controller design, robotic and ambient intelligence.

Norio Shiratori was bone in 1946 in Miyagi Prefecture. He received his doctoral degree from Tohoku University in 1977. He is currently a Professor at RIEC. Before moving to RIEC in 1993, he was the Professor of Information Engineering at Tohoku University from 1990 to 1993. Prior to that, he served as an Associate Professor and Research Associate at RIEC. He received IEEE Fellow in 1998, IPSJ Fellow in 2000 and IEICE Fellow in 2002. He is the recipient of many awards including, IPSJ Memorial Prize Wining Paper Award in 1985, IPSJ Best Paper Award in 1996, IPSJ Contribution Award in 2007, IEICE Achievement Award in 2001, IEICE Best Paper Award, IEEE ICOIN-11 Best Paper Award in 1997, IEEE ICOIN-12 Best Paper Award in 1998, IEEE ICPADS Best Paper Award in 2000, IEEE 5-th WMSCI Best Paper Award in 2001, UIC-07 Outstanding Paper Award in 2007, Telecommunication Advancement Foundation Incorporation Award in 1991, Tohoku Bureau of Telecommunications Award in 2002, etc. He was the vice president of IPSJ in 2002, IFIP representative from Japan in 2002, an associate member of Science Council of Japan in 2007. He is working on methodology and technology for symbiosis of human and IT environment.

Juan Pablo Soto is a Computer Science Engineer by the University of Baja California (México). He is a PhD student in the Computer Science Department of the University of Castilla - La Mancha (Spain) and the topic of his thesis is based on the goal of improving Knowledge Management Systems by using Multi-Agent technology. His research interests are Knowledge Management and Multi-Agent Systems.

Takuo Suganuma is an associate professor of Research Institute of Electrical Communication of Tohoku University, Japan. He received a Dr.Eng. degree from Chiba Institute of Technology. His research interests include agent-based computing, flexible network, and symbiotic computing. He is a member of IEICE, IPSJ and IEEE.

Kenji Sugawara is a professor of the Department of Information and Network Science, Chiba Institute of Technology, Japan. He received his doctorate degree in Engineering from Tohoku University in 1983. His research interests include Agent-Oriented Computing and Web Computing. Prof. Sugawara is a member of IEEE, ACM, IEICE and IPSJ.

Xiangmin Tan is a Ph.D student in the Institute of Automation, Chinese Academy of Sciences, China. His research interests lie in the area of robotics, industrial control, fuzzy systems, and neural networks, modeling and control of hypersonic crafts.

Takahiro Uchiya is an associate professor of the Information Technology Center, Nagoya Institute of Technology, Nagoya, Japan. He received his Ph.D. degree from Tohoku University in 2004. His research interests include knowledge engineering and design methodologies of agent system. Dr.Uchiya is a member of IEICE and IPSJ.

Athanasios Vasilakos is a Professor at the Department of Computer and Telecommunications Engineering, University of Western Macedonia, Greece and a Professor at the Graduate Programme of the Department of Electrical and Computer Engineering, National Technical University of Athens (NTUA) and Department of Theatre Studies, University of Peloponnese. He is a Co-author of several books and has published over 150 articles in top international journals and conferences. He is the Editor-in-chief of several journals, including Int. J. Adaptive and Autonomous Communications Systems. He was and is on the editorial board of several international journals, including IEEE Communications Magazine and IEEE Transactions on Systems, Man and Cybernetics.

Aurora Vizcaíno is an associate professor at the Escuela Superior de Informática of the University of Castilla-La Mancha, Spain. She is MSc and has a European PhD in Computer Science by the University of Castilla-La Mancha. Her PhD work was based on the use of a Simulated Student in collaborative environments. Her research interests include Collaborative Learning, Agents, Simulated Student and Knowledge Management. She has numerous publications in important International Conferences and is part of the program committee of many of them.

Guoyin Wang is a professor of computer science, born in Chongqing, China in 1970. He worked at the University of North Texas, USA, and the University of Regina, Canada, as a visiting scholar during 1998-1999. Since 1996, he has been working at the Chongqing University of Posts and Telecommunications, where he is currently a professor and PhD supervisor, the Chairman of the Institute of Computer Science and Technology (ICST), and the Dean of the College of Computer Science and Technology. He is also a part-time professor with the Xi'an Jiaotong University, Shanghai Jiaotong University, Southwest Jiaotong University, Xidian University, and University of Electronic Science and Technology of China. Professor Wang is the Chairman of the Advisory Board of International Rough Set Society (IRSS), Chairman of the Rough Set Theory and Soft Computation Society, Chinese Association for Artificial Intelligence. He serves as a program committee member for many international conferences and editorial board member of several international journals. Professor Wang has been awarded several government medals, and was named as a national excellent teacher and a national excellent university key teacher by the Ministry of Education, China, in 2001 and 2002 respectively. In 2004, Professor Wang was elected into the Program for New Century Excellent Talents in University by the Ministry of Education of P R China. He has given many invited talks at international and national conferences, and has given many seminars in many universities in USA, Canada, Poland, and China. The institute (ICST) directed by Professor Wang was elected as one of the top ten outstanding youth organizations of Chongqing, China. Professor Wang is the author of 2 books, the editor of many proceedings of international and national conferences, and has over 100 research publications. His research interests include data mining, machine learning, rough set, neural network, soft computing, etc.

Yunfeng Wu received the B.E. degree in Information Engineering and the Ph.D. degree in Signal and Information Processing from the Beijing University of Posts and Telecommunications (BUPT), Beijing, China, in 2003 and 2008, respectively. Dr. Wu currently works as a Post-Doctoral Fellow in the Department of Electrical and Computer Engineering, Ryerson University, Toronto, ON, Canada. His research interests cover machine learning, biomedical signal and image processing, pattern recognition, and neural engineering. Dr. Wu is a member of the Institute of Electrical and Electronics Engineers (IEEE), Association for Computing Machinery (ACM), and Biomedical Engineering Society (BMES). Dr. Wu was the recipient of 2007 Visiting Scholar grant sponsored by the Croucher Foundation in Hong Kong, and also received the travel grants of ICONIP'06, ISSS-MDBS'06, and IJCNN'04; Second Place of the 2002 IEEE Industry Applications Society (IAS) Myron Zucker Student Design Contest; Meritorious award of the 2002 Interdisciplinary Contest in Modeling; and the IBM Chinese Excellent Student Scholarship in 2002.

Dong Xu received the Ph.D degree in July 2008 from Institute of Automation, Chinese Academy of Sciences, China. Now he is a research engineer at the IC Processing Equipment R&D Center of Beijing Sevenstar Electronoics Co.Ltd.. His research interests lie in the area of robotics, industrial control, fuzzy systems, and neural networks.

Chen Yan, Ph.D., Dean of Game School, of Jilin Animation Institute in China. From 2004 to 2007, she was an engineer researcher in the department of human-computer interface at France Telecom R&D in France, mainly worked in the conception and the implementation technology for multiplayer pervasive game systems. In 2007, she obtained her Ph.D. in computer science of the Computer Sciences Research

Center (CEDRIC) of Conservatoire National des Arts et Metiers (CNAM) in Paris. From 2007 to 2009, she was a lecturer responsible for teaching and research works related to game design and development in CNAM, and she also worked as game designer and project director in a French urban game company Xilabs. In 2009, she went back to China and now works for Jilin Animation Institute.

Yong Yang was born in Yunnan, China, in 1976. He is currently a vice professor at the Chongqing University of Posts and Telecommunications. His research interests include affective computing, intelligent information system, pattern recognition information security, etc.

Jianqiang Yi received the B.E. degree from the Beijing Institute of Technology, Beijing, China, in 1985, and the M.E. and Ph.D. degrees from the Kyushu Institute of Technology, Kitakyushu, Japan, in 1989 and 1992, respectively. From 1992 to 1994, he was with the Computer Software Development Company, Tokyo, Japan. From 1994 to 2001, he was a Chief Researcher at MYCOM Inc., Kyoto, Japan. Currently, he is a Full Professor in the Institute of Automation, Chinese Academy of Sciences. His research interests include theories and applications of intelligent control, intelligent robotics, underactuated system control, sliding-mode control, flight control, etc. He is an Associate Editor for the IEEE Computational Intelligence Magazine, Journal of Advanced Computational Intelligence and Intelligent Informatics, and Journal of Innovative Computing, Information and Control. He worked as a Research Fellow at CSD, Inc., Tokyo, Japan from 1992 to 1994, and a Chief Engineer at MYCOM, Inc., Kyoto, Japan from 1994 to 2001. Since 2001 he has been with the Institute of Automation, Chinese Academy of Sciences, China, where he is currently a Professor. His research interests lie in the area of intelligent control and robotics.

Dongbin Zhao received the B.S., M.S., Ph.D. degrees in Aug. 1994, Aug. 1996, and Apr. 2000 respectively, in materials processing engineering/welding robotics and automation from the State Key Laboratory of Advanced Welding Production Technology, Harbin Institute of Technology, China. He was a postdoctoral fellow in humanoid robot at the Department of Mechanical Engineering, Tsinghua University, China, from May 2000 to Jan. 2002. He is currently an associate professor at the Laboratory of Complex Systems and Intelligence Science, Institute of Automation, Chinese Academy of Sciences, China. His current research interests are in the area of computational intelligence, robotics, and intelligent transportation systems. He is an associate editor of IEEE Intelligent Transportation System Magazine.

Liquan Zhao is an associate professor of computer application at the Nanjing University of Finance and Economics. He received a PHD in artificial intelligence from Anhui University., China, in 2007. His research interests include electronic commerce, commerce intelligence, artificial neural networks, granular computing, computational intelligence and their applications. He has published over thirty papers in these fields.

Jian Zhou was born in Anhui, China, in 1981.He has been working at the Anhui University since 2007.His research interests include signal processing, machine learning, etc.

Index

Symbols

A

B

C

O

P

Q

R

S

CPSIA information can be obtained at www.ICGtesting.com
263785BV00008B/63-122/P

9 781609 605